Human–Computer Interaction Series

Editors-in-chief

Desney Tan
Microsoft Research, Redmond, WA, USA

Jean Vanderdonckt
Louvain School of Management, Université catholique de Louvain,
Louvain-La-Neuve, Belgium

The Human-Computer Interaction Series, launched in 2004, publishes books that advance the science and technology of developing systems which are effective and satisfying for people in a wide variety of contexts. Titles focus on theoretical perspectives (such as formal approaches drawn from a variety of behavioural sciences), practical approaches (such as techniques for effectively integrating user needs in system development), and social issues (such as the determinants of utility, usability and acceptability).

HCI is a multidisciplinary field and focuses on the human aspects in the development of computer technology. As technology becomes increasingly more pervasive the need to take a human-centred approach in the design and development of computer-based systems becomes ever more important.

Titles published within the Human–Computer Interaction Series are included in Thomson Reuters' Book Citation Index, The DBLP Computer Science Bibliography and The HCI Bibliography.

More information about this series at http://www.springer.com/series/6033

Mark Blythe · Andrew Monk

Editors

Funology 2

From Usability to Enjoyment

Second Edition

 Springer

Editors
Mark Blythe
Department of Design
Northumbria University
Newcastle upon Tyne
UK

Andrew Monk
Department of Psychology
University of York
York
UK

ISSN 1571-5035
Human–Computer Interaction Series
ISBN 978-3-030-09825-4   ISBN 978-3-319-68213-6   (eBook)
https://doi.org/10.1007/978-3-319-68213-6

Printed on acid-free paper

This Springer imprint is published by the registered company Springer International Publishing AG part of Springer Nature
The registered company address is: Gewerbestrasse 11, 6330 Cham, Switzerland

In memory of Kees Overbeeke and John Karat

PREFACE - HOW TO USE THE SECOND EDITION

Mark: Well here we are again, nestling cosily inside the pages of a book.

Andrew: Or momentarily assembled on the screen of a device that someone is poking at. Make a sales pitch or they'll be off to check how many "likes" their last remark got. Quick, Funology 2, what is it?

Mark: Well, I'll tell you what it isn't! It's not just the same old book with a new preface slapped on and some references to social media.

Andrew: It was quite the loss for advertising when you became an academic wasn't it Mark? We need to say what it is, not what it isn't. Fifteen years ago we were all arguing that we had to look beyond usability to enjoyment because computers had moved from the workplace to the home.

Mark: And now computers are in our pockets and pretty much everywhere else.

Andrew: I don't think anybody would argue that fun isn't important anymore would they? It's all about user experience today.

Mark: Yes, the people that wanted computing to be more engaging certainly got their wish. But in this edition it sometimes feels as if someone has asked Dr Frankenstein if his experiment was a success or not.

Andrew: What, after the creature had been on a few rampages you mean?

Mark: Yes, you know, in some respects the reanimation of dead tissue is quite the achievement but everyone concerned would probably have to admit that the results were, at best, mixed.

Andrew: So we have fourteen new chapters!

Mark: Yes, the organisation is more or less the same as last time – theory, methods and case studies. The theory chapters are grouped as "critique" because there is often a critical sensibility in the discussion of what's happening. Lots of the methods were about generating alternative approaches and creativity so they're under "ideation". In the first edition, there were mostly single case studies but the new chapters look at collections of work, almost emerging schools of design, so that section is called "directions and approaches".

Andrew: Sounds like a must-have book! And the authors of the old chapters have added commentaries reflecting on what has happened since they wrote them. Most of them anyway, what were we calling the chapters with no additional material? "Classics" wasn't it?

Mark: Yes, not too many of those though because so much has changed.

Andrew: It's incredible how different things were in 2003 when the first edition came out isn't it?

Mark: Back in the day, as our American friends would say – there was no Facebook! Imagine that, it didn't happen until 2004! YouTube didn't arrive until 2005 and there was no such thing as Twitter either, not until 2006.

Andrew: Good God can that be right? However did we manage? Hang on… yes that is right, according to Wikipedia, which did exist back then – it had only been around for two years.

Mark: Makes you feel old doesn't it?

Andrew: There was no such thing as an iPhone! That wasn't spawned until 2010 and an iPad would have sounded like something to do with incontinence until 2012 Ha!

Mark: Will Self calls 2004 the "break point year" because this was when wireless broadband made interaction "frictionless". He also says this was the beginning of the end for the codex – the paper sandwich that used to be the only way of making a book.

Andrew: No wonder publishers are so touchy.

Mark: Well they've made it very clear that they don't want any cheeky remarks about their profession, like in the first preface, so you just behave yourself this time Andrew.

Andrew: All I said was that publishers are ba -

Mark: Battling against a rapidly changing technological and economic land-scape? Yes and so they are.

Andrew: No, publishers are bas -

Mark: Bassoon players? Yes, some of them, probably, I expect that might well be true.

Andrew: What I'm trying to say is that publishers are bast -

Mark: Bastions of integrity and fair dealing. Quite right, I couldn't agree more.

Andrew: Well if they won't let me say it people can always look at the preface to the first edition.

Mark: Yes but not the one in this volume because they've made us change it, the basta -

Andrew: Batstalysts?

Mark: What?

Andrew: Bastalysts.

Mark: That's cheating! It's a made up word.

Andrew: So is Funology.

Mark: Fair point. But bastalyst doesn't mean anything.

Andrew: Bastalysts are the makers of hybrid forms. We are all bastalysts these
          days, especially publishers. They are massive bastalysts.

Mark: LOL. Nobody said LOL in 2003 did they? At least I didn't. Good God,
          what have we become? Shall we introduce the new chapters do you
          think?

Andrew: Alright but let's be quick about it. I've retired you know.

Contents

Contributors

Kristina Andersen TU/e, Eindhoven University of Technology, Eindhoven, The Netherlands

Jeffrey Bardzell Informatics, Computing, and Engineering, Indiana University, Bloomington, USA

Shaowen Bardzell Informatics, Computing, and Engineering, Indiana University, Bloomington, USA

Steve Benford The Mixed Reality Laboratory, The University of Nottingham, Nottingham, UK

Erik Blankinship Media Modifications, Cambridge, USA

Mark Blythe School of Design, Northumbria University, Newcastle, UK

Petter Bae Brandtzæg SINTEF Digital, Oslo, Norway

Jo Briggs Northumbria University, Newcastle upon Tyne, UK

Barry Brown University of Stockholm, Stockholm, Sweden

Paul Cairns Department of Computer Science, University of York, York, UK

John M. Carroll The Pennsylvania State University, State College, PA, USA

Paul Coulton Lancaster University, Lancaster, England, UK

Françoise Decortis Université de Vincennes - Paris 8, Paris, France

Pieter Desmet Technical University of Delft, Delft, The Netherlands

Alan Dix University of Birmingham, Birmingham, UK

Tom Djajadiningrat Eindhoven University of Technology, Eindhoven, Noord-Brabant, The Netherlands

Abigail Durrant Northumbria University, Newcastle upon Tyne, UK

Pilapa Esara Carroll Department of Anthropology, The College at Brockport, New York, USA

Jennica Falk NNIT, Copenhagen, Denmark

Ylva Fernaeus KTH Royal Institute of Technology, Stockholm, Sweden

Joep Frens Eindhoven University of Technology, Eindhoven, Noord-Brabant, The Netherlands

Florence Fu Northwestern University, Evanston, USA

Bill Fulton Amazon, Seattle, USA

Asbjørn Følstad SINTEF Digital, Oslo, Norway

Elizabeth M. Gerber Northwestern University, Evanston, USA

Gabriella Giannachi Centre for Intermedia, The University of Exeter, Exeter, UK

Michael Golembewski Microsoft Research, Cambridge, UK

Sukeshini Grandhi Eastern Connecticut State University, Willimantic, CT, USA

Chris Greenhalgh The Mixed Reality Laboratory, The University of Nottingham, Nottingham, UK

Marc Hassenzahl Ubiquitous Design/Experience and Interaction, University of Siegen, Siegen, Germany

Jan Heim SINTEF Digital, Oslo, Norway

Lars Erik Holmquist Northumbria University, Newcastle upon Tyne, UK

Jonathan Hook University of York, York, UK

Richard Hull Hewlett-Packard Laboratories, Bristol, UK

Caroline Hummels Eindhoven University of Technology, Eindhoven, Noord-Brabant, The Netherlands

Kristina Höök KTH Royal Institute of Technology, Stockholm, Sweden

Margot Jacobs Mia Leher and Associates, Los Angeles, USA

Quentin Jones New Jersey Institute of Technology, Newark, NJ, USA

Oskar Juhlin University of Stockholm, Stockholm, Sweden

Clare-Marie Karat IBM TJ Watson Research Center, Hawthorne, NY, USA

John Karat IBM TJ Watson Research Center, Hawthorne, NY, USA

David Kirk Northumbria University, Newcastle upon Tyne, UK

Ben Kirman University of York, York, England, UK

Shaun Lawson Northumbria University, Newcastle upon Tyne, England, UK

Joseph Lindley Lancaster University, Lancaster, England, UK

Conor Linehan University College Cork, Cork, Republic of Ireland

Sara Ljungblad University of Gothenburg, Gothenburg, Sweden

Joe Marshall The Mixed Reality Laboratory, The University of Nottingham, Nottingham, UK

Patrizia Marti University of Siena, Siena, Italy

Deborah Maxwell University of York, York, England, UK

John McCarthy School of Applied Psychology, University College Cork, Cork, Ireland

Lisa Meekison http://lisameekison.com/

William Odom School of Interactive Arts + Technology, Simon Fraser University, Surrey, BC, Canada

Kees Overbeeke Eindhoven University of Technology, Eindhoven, Noord-Brabant, The Netherlands

Dan O'Hara New College of the Humanities, London, England, UK

Randy J. Pagulayan Microsoft, Redmond, USA

Laura Polazzi Experientia, Milan, Italy

Darren J. Reed Department of Sociology, University of York, York, UK

Jo Reid Hewlett-Packard Laboratories, Bristol, UK

Antonio Rizzo University of Siena, Siena, Italy

Tom Rodden The Mixed Reality Laboratory, The University of Nottingham, Nottingham, UK

Ramon L. Romero Sony Interactive Entertainment America, San Diego, USA

Mary Beth Rosson The Pennsylvania State University, State College, PA, USA

Job Rutgers Philips Design, Eindhoven, Nederland

Phoebe Sengers Information Science and Science and Technology Studies, Cornell University, Ithaca, USA

Tobias Skog DIRECTV Latin America, Dallas, USA

Keith R. Steury Microsoft, Redmond, USA

Miriam Sturdee Lancaster University, Lancaster, England, UK

Anna Ståhl RISESICS, Kista, Sweden

Jonathan Sykes Brighton, UK

Paul Tennent The Mixed Reality Laboratory, The University of Nottingham, Nottingham, UK

Vanessa Thomas Lancaster University, Lancaster, England, UK

Paul Thursfield Philips Design, Eindhoven, Nederland

Aadjan Van Der Helm Delft University of Technology, Delft, The Netherlands

John Vines Northumbria University, Newcastle upon Tyne, England, UK

Ron Wakkary School of Interactive Arts + Technology, Simon Fraser University, Surrey, BC, Canada; Industrial Design, Eindhoven University of Technology, Eindhoven, Netherlands

Brendan Walker The Mixed Reality Laboratory, The University of Nottingham, Nottingham, UK

Stephan Wensveen Eindhoven University of Technology, Eindhoven, Noord-Brabant, The Netherlands

Richard Wiseman University of Hertfordshire, Hatfield, UK

Peter Wright Open Lab, Newcastle University, Newcastle upon Tyne, UK

Doug Zytko New Jersey Institute of Technology, Newark, NJ, USA

Part I
Funology 2

Chapter 1
Funology 2: Critique, Ideation and Directions

Mark Blythe and Andrew Monk

The word academic is sometimes used to signify an occupation, a person who works in a university and studies a particular field, as in—"do you see that stoop-backed fellow over there? He's an academic I bet, wouldn't last long in the wild would he?" But it is also means "pointless" or "unnecessary" as in "well it's all academic in any case because I've already hit the send button." Of course the two senses of the word often collide in figures like *The Onion's* Professor at the "Center for Figuring Out Really Obvious Things" whose research links teen sex to drugs and alcohol (www. TheOnion.com). The idea of studying fun and enjoyment can seem like a project devised in that centre at *The Onion*. Fun, after all, is something that any five year old understands perfectly well. A group of children confronted with the limb of a tree that's blown down outside their school need no instruction in how to have fun with it. They run straight over and clamber onto the branches, bouncing up and down, hiding in the leaves, and playing with the oldest and most popular toys in the world– sticks, which become drumsticks, guns, or lightsabres. One child claims he has killed another one but the accuracy of the imaginary shot is disputed. The kids are playing as if someone might stop them and one shouts "This is the new playground!" But of course the grown ups clear it away the next day. What is left of the tree is cut down because another limb might have fallen, it was a little bit dangerous but that was part of the fun. We know fun when we see it, we are all, as Brown and Juhin insist, already experts, though maybe we lose some of our virtuosity as we get older. Fun is often serendipitous, unplanned and spontaneous. So how do we design for fun, enjoyment or, for that matter, any other kind of experience?

M. Blythe (✉)
School of Design, Northumbria University, Newcastle, UK
e-mail: mark.blythe@northumbria.ac.uk

A. Monk
Department of Psychology, University of York, York, UK
e-mail: andrew.monk@york.ac.uk

© Springer International Publishing AG, part of Springer Nature 2018
M. Blythe and A. Monk (eds.), *Funology 2*,
Human–Computer Interaction Series,
https://doi.org/10.1007/978-3-319-68213-6_1

The first edition of this book described the shift in the study of human–computer interaction away from usability (ease of use, ease of learning, time on task, accuracy and efficiency) towards fun, enjoyment and broader conceptions of user experience. It turned out that usability had been easier to measure and design for than amorphous experiences like enjoyment. At the time of publication psychology had little to say about fun. Czikszentmihalyi had studied activities like rock climbing, chess and programming and discovered that they all shared periods of intense absorption he described as "flow". But aside from this psychologists had not paid much attention to enjoyment. This changed in 2002 when Martin Seligman became the President of the American Psychological Association and used the position to argue that psychologists had been studying psychosis, neuroses, depression and other forms of anguish for long enough, it was now time to focus attention on people who were thriving and happy to see what could be learned from them (Seligman 2011). Seligman called this approach "positive psychology" and a range of empirical studies have been undertaken over the last decade which have deepened our understanding of well being. Seligman summarises this with the acronym PERMA which stands for: positive emotion, engagement, relationships, meaning and achievement. Researchers in HCI are increasingly applying models of well being in their work (e.g. Thieme et al. 2012). There are also now many off the shelf technologies to monitor our physical and mental health and lots of apps encouraging mediation and spirituality.

Now more than ever, enjoyment is a serious business. Even our momentary distractions are the basis of multi billion dollar industries. Facebook has been described as an "attention merchant" providing us with fleeting gratification in return for permission to sell our data to advertisers (Wu 2017). We like Facebook "likes" so much it is fuelling a "moral panic" (Cohen 1972) that social media is addictive. Yet anxiety is real enough to make some people quit social media altogether including the Facebook engineer who invented the "like" button (Lewis 2017). Slavoj Zizek has long argued that enjoyment is a political factor. He illustrates this notion with *Pokemon Go* where users look at the world through their phones and see imaginary creatures layered on top of the camera view. For Zizek this is a perfect illustration of the way ideology functions: it adds an extra layer of meaning onto the world and shapes the way that we look at it. He claims there is a pirate version of *Pokemon Go* on sale in Germany that pictures not Pokemon monsters but immigrants and terrorists lurking around every corner (Zizek 2017). The immigrant Pokemon story may be apocryphal, there are versions of a game like this attributed to the UK independence party as well as Germany. Whether it is true or not is not the point: it could be, and if it were there are no doubt people who would find it enjoyable.

Reading a piece of investigative journalism that challenges our preconceptions is probably not going to be as enjoyable as reading some fake news which confirms our preconceptions (Lanchester 2017). A recent neuroimaging study found that the brain when presented with evidence that challenges our beliefs responds in the same way it does when presented with a threat (Kaplan et al. 2016). There is then a politics of fun but *Pokemon Go*, as an activity not a metaphor, is not easy to explain. Although academics like Steve Benford have studied augmented reality games for many years the precise nature of what success would look like in this field was not easy to predict.

For Linehan and Kirman (2017) the main appeal of *Pokemon Go* was nostalgia for the nineteen eighties and the speed with which the craze died off suggests that they might be right. So we need not only theory but different kinds of theory: psychology, sociology, anthropology, politics and philosophy are all relevant.

We have put the fourteen new chapters of this book into three parts. There are five chapters of theory, under the heading "Critique", these illustrate not only the diverse disciplines that now inform HCI but also how frequently these theories directly contradict one another. There follow four chapters on methods and techniques, under the heading "Ideation" and finally five chapters that are best described as design case studies under the heading "Approaches and directions".

Theory and Concepts: Critique

The original version of Marc Hassenzahl's "The Thing and I" is the most downloaded and cited chapter of the first edition of this book. Marc revised the chapter so thoroughly we decided to place it here with the entirely new work (Chap. 2). He begins with a personal reflection on the turn of the millennium when he was working as a usability engineer at Siemens and the idea of studying fun was "hilarious". He recalls an early "computers and fun" workshop organised by Andrew where Bill Gaver presented design concepts like a "dream communicator" and a "worry stone". This was radically different to traditional (and indeed much current) HCI. In this "remix" of his 2003 paper Marc summarises the pragmatic and hedonic qualities of an interaction, distinguishes between "be" goals and "do" goals and considers design in terms of *what*, *how* and *why* questions. The "what" questions are concerned with form (what is it, what does it do?) the "how" questions with interaction (how does it function, how do I work it?) and the "why" questions consider goals and motivation (why are we doing this? Does it really matter?). It is often argued that it is not possible to design an experience rather we design *for* an experience. But Marc takes issue with this using a plan for a party as an example: if there is enough music, beer and wine then the guests should have a good time as the host intends. Some of the authors in this volume who take a more phenomenological approach would disagree. This may be because they don't get invited to the right sort of parties or it may be that they look at parties in a different sort of way.

In *Meditations on a Toolshed* CS Lewis (1996) illustrated two distinct kinds of thought by describing himself looking at beam of light coming in through a crack in the door of a dark garden shed. Motes of dust are clearly visible in the sunbeam, and while he is staring at it everything else is in deep shadow. But when he puts his eye to the beam and looks along it this picture vanishes and he sees leaves on a tree, the sky above and the sun itself some ninety million miles away. For Lewis, those who look *at* things (in a detached way) browbeat those who preferred looking *along* them (in a subjective or experience based way). He argues that stepping outside one experience always involves stepping into another (looking at or looking along) and that neither is more valid or true than another. Paul Cairns' chapter (Chap. 3) questions how much we can ever know about engagement. He is particularly interested in immersion but not of the intense and absorbing kind Csiktzentmihalyi described as "flow". He thinks about the kind of engagement we have momentarily playing a game on a phone while commuting. What's enjoyable about that, if you're not in the kind of flow state that expert chess players might feel? He points out that many claims are made about games in terms of their potential for helping us to learn or their potentially harmful effects in, for example, making children more aggressive or violent. But the chapter shows how difficult it is to know anything much about gaming by looking at it. Experience cannot be measured through a questionnaire and immersion cannot be easily observed in a lab.

Psychology and economics have long tried to measure pleasure but for Brown and Juhin (Chap. 4) these disciplines have had little success. They are also sceptical of Marxist analyses which posit the notion of "false enjoyment" to be unravelled by academics like Zizek. They agree with the philosopher Gilbert Ryle that we do not need philosophy to understand fun: nobody is an expert and liking and enjoyment are not technical terms. Yet, they maintain, that the simple account of pleasure as a sensation in the brain is inadequate, pointing out the difference between the pleasure found in long term endeavours like learning to play an instrument and the pleasure of sipping a nice cup of tea. They agree with fifties sociologist Howard Becker that getting high is a social accomplishment, we have to learn to enjoy sensations that might otherwise be felt as alarming and scary rather than pleasurable. (A contemporary of Becker's wrote an article in response to this argument called: Howard Becker Change Your Dealer!) Brown and Juhin make a four part argument: (1) pleasure is in the world and not in the head, (2) it is a set of skills, (3) it is ordinary and routinely describable, (4) it is felt and emotional. Ending with an account of the pleasures of family life they call for empirical study through ethnography.

Some recent ethnographic studies of the impact of technology on our social life suggest that while it connects us to each other it is also, somewhat paradoxically, making us more isolated. Sherry Turkle has charted our evolving relationship with new technology for over twenty years. Although her early work celebrated "life on screen" she is now concerned that we expect more from technology and less from each other (Turkle 2012). While we are sitting in bars with friends who are actually there we stare at screens tapping out messages to Facebook friends who are not. Anxiety around social media is everywhere in the culture and forms the basis of many stand up routines and sitcoms. The comedian Aziz Ansari (2016) recently collaborated with academics (including Turkle) to write a book called *Modern Romance*. This charts the radical changes between generations in how they met their romantic partners and how long these relationships lasted: fifty percent of American marriages will end in divorce. The sociologist Zygmunt Bauman (2003) calls this state of affairs "liquid love": we are in relationships, like we are in our jobs, until further notice. These new social formations are supported by technologies like Tinder which enable loose ties that are easily broken.

Location based dating services have revolutionised the way we meet one another and form relationships but, according to Doug Zytko and colleagues (Chap. 5), they are often far from fun. We may enjoy the game like aspects of swiping yes or no on Tinder but this is not necessarily worth the existential angst we feel creating our profile or "relationshopping" by playing romantic games of "stick or twist". Zytko and colleagues reflect on the reasons why we keep using these apps despite sometimes deeply problematic user experiences. Women's inboxes are swamped with many times more messages than those of men but how we frame this as a problem depends on gender politics. Design problems and solutions co-evolve (Dorst and Cross 2001); if we frame the messaging imbalance as a problem for men then one approach is provocative messages as recommended by online coaches in (Zytko et al. 2016) study "The Coaches Said What?". If however it is framed as a problem for women then this suggests approaches like Bumble where men are not allowed to message women until women have first messaged them.

The importance of gender politics is very clearly illustrated in the chapter by Shaowen and Jeffrey Bardzell (Chap. 6). They describe the design and socio political campaign work around the *Mooncup*, a menstruation cup made of reusable silicon that lasts for five to ten years. The design involved not just careful product design research but also fund raising through a crowd funding website and political campaign work because mooncups were illegal. This is a fascinating account of how concerns were addressed by the design team and their supporters through social media and activism.

1 Methods and Techniques: Ideation

Fifteen years ago we were already describing the pace of technological development as dizzying, looking back the changes we were living through then seem quite slow and steady. Undergraduates today remember the phones they had when they were fourteen with the same kind of wry nostalgia previously associated with retirees talking about the olden days. Moore's law still holds but the idea of computing power more or less doubling every eighteen months is difficult to comprehend. Kurzweill (1999) provided a powerful illustration of the exponential changes that Moore's law implies with the fable of the Chinese emperor who wished to reward an inventor with anything he wanted. The inventor asks for one grain of rice to be doubled for every square on a chessboard. This sounded like a modest request and the emperor agreed but after doubling for just 21 squares he found there were already one million grains. By the middle of the board the grains of rice need to be described with the mathematical terms we now use to describe our data storage capacities—giga, tera, and peta. On the 64th square the number the prefix is exa (ten to the power of 18) and the grains of rice total 18,446,744,073,709,551,615. For Kurzweill we reached the middle of the board some time around 2005, since then we have been presented with world changing technologies with punctual regularity. Interaction designers cannot be trained solely by studying previous practice. Undergraduates in their third year are confronted with digital products and services that did not exist in their first year. Generating ideas around new technological possibilities is then increasingly important.

Coming up with new ideas can be difficult. The blank page can be as intimidating for a designer as it is for a writer. There are many definitions of ideation, one popularised by the company IDEO is coming up with lots of bad ideas in order to get one good one. In the book "Gamestorming" Gray et al. (2010) estimate that it takes about one hundred bad ideas to get one good one. This kind of calculation informs the "1000 ideas" workshop devised by Jon Rogers of Dundee University. The title of this workshop can be scary and Jon begins it by asking how many of the students think they will be able to come up with this many ideas in one afternoon, there are no hands in the air at this point; then he asks how many think they can do one hundred and a few hands go up; when he asks if they could do ten most (but not all) hands are in the air. The only resources needed for the workshop are sticky tape and a large pile of paper. There are no "post it" notes, but rather A4 photocopy paper torn in half. First Jon asks them to draw the Internet. They do, he walks round commenting often amused, there are lots of circles, some connected dots. He tells the students that what they have just drawn is an idea. Ideas are cheap, ideas are easy. The workshop proceeds from small beginnings. He asks for ten ideas roughly related to whatever brief they are working on. These initial ideas become jumping off points: the students swap ideas and do ten variations, then ten opposites, they imagine the design for ten different kinds of people. Quickly the pieces of paper are filled with sketches and Jon seizes on them offering comments, questions and praise. By the end of the afternoon an entire wall is papered with ideas.

When students get stuck in this kind of exercise it is helpful to use some of the ideas that Elizabeth Gerber and colleagues have culled from their studies of improvisation: *Be obvious! Accept Invitations! Fail cheerfully!* Keith Johnston's was a renowned teacher of drama and over many years he developed a variety of techniques to help actors improvise in theatre. He pointed out that one of the first problems for any actor standing on a stage in front of an audience with no script is fear. They can become paralysed as they try and think of something to say that will be funny, or clever, or original. Being obvious helps overcome this fear but more important than that—what is obvious to you is not obvious to me. When a group of people are being obvious together interesting and unpredictable things happen. Being obvious is important in improv but so too is accepting invitations. Accepting invitations means going with ideas. And failing cheerfully is just that, improvising actors don't desperately try to make something work if it is going nowhere, they try something else instead. Elizabeth Gerber has been applying these techniques to design ideation for many years. The chapter by Gerber and Fu (Chap. 7) identifies key skills like collaboration and exploration and also details improv activities to develop them. In "door in the floor" for example, players mislabel objects, pointing to a computer and claiming it is, say, a book. The chapter provides great insights into the process of ideation and also practical techniques for doing it.

One of the best definitions of writer's block is "trying to be critical and creative at the same time". As Pinker (2014) points out: the first draft of anything is nearly always very bad, it is made better through successive edits. Similarly designers can become blocked if they are trying to think only of good ideas. Bad ideas are useful not only because having them allows better ones to come but also because sometimes they are interesting. The Japanese practice of Chindogu involves the creation of "unuseless" objects. A toilet roll hat, for example, is not useless, because the wearer can reach up and get a tissue whenever they need to blow their nose, but it is not useful either because it creates other larger problems (like having to wear a toilet roll hat). In an early study of domestic technology we proposed a number of unuseless design concepts to explore the themes of an ethnography of domestic technology (Blythe and Monk 2002). Anxieties around the gendered division of labour were illustrated with an iron that drew on the aesthetics of powertools so that it might make a direct appeal to men. The tedium of ticking boxes in online shopping lists was reimagined as a shopping game along the lines of Sonic the Hedgehog bouncing on top of cereal boxes to select them. These were not serious proposals but they did illustrate real problems and possibilities. John Vines (Chap. 8) reflects on the uses of provocation as a strategy in participatory design workshops. He reports on three techniques that researchers can use to provoke participants into engaging in design. *Questionable Concepts* are illustrations of deliberately flawed designs that are given to participants for critique and discussion. *Invisible Design* is a film technique where characters discuss an artefact that the viewer cannot see. *Experience Design Theatre* also features actors discussing unseen artefacts in improvisations which are then re-scripted with participants. Vines makes the point that it is not always easy to make designs that are provocative and admits that previously this had sometimes been achieved accidently.

Accidental discovery is also a feature of the chapter by Abi Durrant and colleagues (Chap. 9) who discuss their use of comic strips in the design of digital souvenirs for theme parks. The importance of sketching has long been recognised in HCI (e.g. Buxton 2007) and Scott Mcloud gave a CHI keynote on the power of comic books as a medium. But Durrant and colleagues show the way that comics can be used to create engaging and inspiring character driven scenarios. Their cartoons are beautifully simple, they are almost but not quite stick figures: the heads are articulated and coloured (rather beautifully) and they have no noses (significantly perhaps in the context of a theme park). These stylised and deliberately non representative cartoons achieve a depth of character in the way that xkcd strips often do with minimal line and gestures. Adding a beret or a pony tail to a stick figure can sometimes suggest character and individual quirkiness more effectively than yards of persona based description. The storyboards here include multiple voices and perspectives—the researchers, the system, the characters—and the work is discussed as a dialogical process.

The development of prototyping technologies like Arduino and Raspberry Pi mean that moving from concept design to prototype development is quicker and cheaper than ever before. Aside from using maker kits researchers are also appropriating off the shelf technologies to make new designs. This has been described as the "phone in a box" strategy, where a smartphone is presented in such a way that it no longer looks like one (by for example putting it into a wooden box), its functionality is restricted and becomes a prototype to be presented to a user group. Those engaged with design fiction take it as read that this or that prototype could be made and that it might also be deployed with a particular group. But rather than doing this they first explore possible outcomes through fiction. This builds on practices in physical science where an experiment can be written up as a full paper with gaps left for the results. There have been imaginary abstracts, full papers and now a whole fictional conference. Kirman and colleagues (Chap. 10) provide a fictional programme for a conference on design fiction. This gives them an opportunity to give a snapshot of the world of design fiction which is exaggerated, but only slightly.

Along with fictional designs there has also been a massive increase in the development of actual prototypes. The first edition of Funology featured several design case studies. The chapters here seldom feature a single device, rather they are concerned with groups of designs, approaches and directions, perhaps even schools of design.

2   Design Case Studies: Approaches and Directions

Prototype studies have dominated the field of HCI but recently there has been some anxiety around Research through Design: what do we learn from making? (see for example Zimmerman and Forlizzi 2008; Gaver 2012). Wakkary and Odom (Chap. 11) make an important distinction between a prototype and a "research artefact". Unlike prototypes "research artefacts" are presented as being of interest for what they are rather than what they might become. Their chapter brings together a number of fascinating examples from their work on *slowness*. Most technology increases the speed and scale of modern life but the slow movement encourages us to take our time. *The Photobox* is an oak box which prints selections from the user's photo collection just four or five times a month; *Slow Game* is a wooden cube with moving lights that represent a game of Snake which is played one move a day; *Table Non Table* is a piece of furniture made from paper which moves short distances every six to twelve hours; the *Tilting Bowl* is a receptacle that tilts three or four times a day; *Morse Things* are ceramics that periodically communicate with one another. This body of work constitutes a reaction against industrial and commercial design that aims to increase speed and efficiency. Rather than simply critique the bewildering pace of modern life, as an essay might, these artefacts show alternative design spaces. Through these "material speculations" the chapter articulates an emerging design vocabulary around "slowness, unawareness and thingness".

The available forms for human–computer interaction are no longer dominated by keyboards and screens. Increasingly the body itself is the site of human–computer interaction: we are wearing our computers, working out with them, even dancing. Kia Höök and her colleagues have become advocates of design approaches that draw on "somaesthetics" or body consciousness. The originator of somaesthetics

Richard Shusterman uses the term "soma" rather than body, to indicate a living, feeling sentience rather than a merely physical thing, and "aesthetics" to emphasise the soma's role in perception and appreciation (Shusterman 2007). As with slowness, somaesthetic design represents a movement in opposition to the dominant cultural and industrial direction of design. Shusterman points out that everyday we are subjected to "floods of signs, images and factoids", somaesthetics suggests instead a focus on our bodily experience. Fernaeous, Höök and Ståhl (Chap. 12) describe novel design work in this space: *The Nebula Cloak* creates a soundtrack based on how the user moves; *The Breathing light* dims or brightens with the user's breath. Shusterman's body consciousness is closely related to mindfulness and the final design directly addresses meditation: a heat mat is used to draw attention to particular parts of the body in a guided meditation. These designs are discussed as alternative strategies: the magic and mystery of movement based feedback, performance and full body interaction. They also discuss the ways that discomfort and even fear can be thrilling and enjoyable and illustrate this notion with *Breathless* a work also discussed by Steve Benford and colleagues in their chapter.

Breathless is a playground swing controlled by the breath of a rider wearing a gas mask fitted with a respiration sensor. Benford and colleagues (Chap. 13) point out that any swing causes an element of discomfort, swings are not entirely under our control, we cannot stop instantly and sometimes we go so high it feels as if we'll fall off. *Breathless* intensifies discomfort through the smell, claustrophobia and wider cultural associations of the gas mask. The chapter brings together several examples of designs like this which represent "the dark side" of fun. *Oscilate* is another swing coupled with a VR headset that causes the floor to drop out of a virtual environment as the swing moves higher. *Bronco* is a bucking bronco where the rider's breathing increases the speed of the ride making it more difficult to stay on. These designs are discussed in terms of their visceral aspects as well as the way they push notions of control and intimacy. Benford and colleagues have often collaborated with *Blast Theory* to create work that is challenging in terms of both form and content. Novel and innovative technologies are deployed to addresses challenging and difficult subjects like the gulf war (*Desert Rain*), surveillance society (*Uncle Roy All Around*) and terrorism (*Ulrike and Eamon Compliant*).

Difficult subjects also interest Kirman and colleagues (Chap. 14) who take mischief and fun as starting points to design with location based data. In the *Blowtooth* game travellers at airports hide virtual illicit objects on other passengers who have their Bluetooth settings enabled on their mobile phones. *Feckr* is a social tagging system that allows the user to surreptitiously tag the people nearby with entertaining insults. *Fearsquare* shows users how much crime has occurred in the areas they have just been to. *Getlostbot* suggests new locations when it detects patterns and routines in the user's life, for example a hotel worker was advised to skip work and go to the pub while a church goer was told to go to a mosque. These challenging designs raise serious issues not only about what might happen in the future with this kind of technology but with what is happening now. Many of these designs began as jokes, with someone in a meeting saying—"it would be funny if…". The work reflects not only the politics of the designers but also their sense of humour, but this does not mean it is frivolous or trivial. A sombre tone should not be mistaken for seriousness and, as Chesterton noted, angels can fly only because they take themselves lightly.

These kinds of prototypes are clearly very different to those developed in usability studies which were designed to investigate whether this or that interface would increase speed, accuracy or productivity. When the scope of a design is very specifically defined evaluation is relatively straightforward and often quantitative—which one was faster and by how much? But this kind of work defies traditional evaluation techniques. Blythe and colleagues (Chap. 15) reflect on different strategies of evaluation from the kind of "crits" developed in Design schools to open ended conversation which–

Andrew   That'll do won't it? If anybody wants to know any more about your rejected conference paper they can go and read it can't they? Unless you want to add a paragraph about enjoying the sound of your own voice?

Mark   Well Noel Coward did say that work was more fun than fun.

Andrew   Yes but he never had a job did he? Not one that anybody ever retires from anyway. I don't know if I mentioned it but I'm retired you know.

Mark   Yes, yes, alright, alright. Off to play golf are you?

Andrew   Certainly not. I'm going to go and do something fun.

References

Ansari A (2016) Modern romance. Penguin Random House

Bauman Z (2003) Liquid love. Polity Press

Blythe M, Monk A (2002) Notes towards an ethnography of domestic technology. DIS 2002, London, pp 277–281

Buxton W (2007) Sketching user experiences: getting the design right and the right design. Morgan Kaufmann Publishers Inc., San Francisco, CA, USA

Cohen S (1972) Folk devils and moral panics. MacGibbon and Kee, London

Dorst K, Cross N (2001) Creativity in the design process: co-evolution of problem–solution. Des Stud 22(5):425–437

Gaver W (2012) What should we expect from research through design. In: CHI 2012, pp 937–946

Gray D, Brown S, Macanufo J (2010) Gamestorming: a playbook for innovators, rulebreakers, and changemakers. O' Reilly Media Inc

Kaplan JT, Gimbel SI, Harris S (2016) Neural correlates of maintaining one's political beliefs in the face of counterevidence. Nature.com. Scientific Reports 6, Article number: 39589

Kurtzweil R (1999) The age of spiritual machines: when computers exceed human intelligence. Penguin, New York, p 37

Lanchester J (2017) You are the product. London Rev Books

Lewis CS (1996) Meditations in a toolshed in god in the dock. William B Eerdmans Publishing Co

Lewis P (2017) "Our Minds Could be Hi-Jacked": the tech insiders who fear a smartphone dystopia. The Guardian. https://www.theguardian.com/technology/2017/oct/05/smartphone-addiction-silicon-valley-dystopia?CMP=share_btn_fb

Linehan C, Kirman B (2017) MC Hammer presents: the Hammer of transformative nostalgifi-cation—designing for engagement at scale. In: CHI 2017

Pinker S (2014) The sense of style: the thinking person's guide to writing in the 21st century. Penguin Books

Seligman MEP (2011) Flourish: a visionary new understanding and well being. Nichoals Brealey Publishing

Shusterman R (2007) Body consciousness: a philosophy of mindfulness and somaesthetics. Cambridge University Press, Cambridge

The Onion. Teen Sex Linked to Drugs and Alcohol Reports. Center for Figuring Out Really Obvious Things. http://www.theonion.com/article/teen-sex-linked-to-drugs-and-alcohol-reports-cente-1417

Thieme A, Balaam M, Wallace J, Coyle D, Lindley S (2012) Designing wellbeing. In: Proceedings of DIS'12. ACM, pp 789–790

Turkle S (2012) Alone together: why we expect more from technology and less from each other. Kindle Edition

Wu T (2017) The attention merchants: from the daily newspaper to social media, how our time and attention is harvested and sold. Atlantic

Zimmerman J, Forlizzi J (2008) The role of design artifacts in design theory construction. Human Computer Interaction Institute. Paper 37

Zizek S (2017) From surplus value to surplus enjoyment. https://www.youtube.com/watch?v=qB0m64u3N7M

Zytko D, Grandhi SA, Jones Q (2016) The coaches said...What?: analysis of online dating strategies recommended by dating coaches. In: Proceedings of the 19th international conference on supporting group work (GROUP '16). ACM, New York, NY, USA, 385–39

Part II
"Critique"

Chapter 2
The Thing and I (Summer of '17 Remix)

Marc Hassenzahl

Editors note, 2018 Edition
This New Chapter is based on Marc Hassenzahl's Chap. 3 in the 2003 Edition. Marc has so thoroughly revised the original chapter that we have classified this as a new chapter.

1 Introduction

More than 25 years ago, the Human-Computer Interaction (HCI) community found an interest in what was at that time considered a slightly esoteric concept: *user experience* (UX). Up to then, effective and efficient goal achievement had been HCI's prime objective and HCI experts mocked apparent follies of designers committed in the name of aesthetics. But doubts crept in. There were so many examples of more mature, consumer-oriented technologies, such as automobiles, which had to be usable and at the same time beautiful, exciting and ever new. In fact, already in 1998, *Apple* had been successful with an unusual consumer-oriented industrial design for a computer, the colorful *iMac G3 s*. The "experiential" made its appearance.

In the 1990s, experiential marketing pointed out that customers want products "that dazzle their senses, touch their hearts and stimulate their minds" (Schmitt 1999, p. 22), the experience economy (Pine and Gilmore 1999) highlighted the economic value of experiences, and the experience society (Schulze 1992) discussed a general societal shift from the material to the experiential. Suddenly, experience mattered. HCI became interested in how people "feel" while and as a consequence of engaging with technology. For some time, the mainstream of HCI defended its image as consisting of the most sober and reasonable (some would say

M. Hassenzahl (✉)
Ubiquitous Design/Experience and Interaction, University of Siegen, Siegen, Germany
e-mail: marc.hassenzahl@uni-siegen.de

© Springer International Publishing AG, part of Springer Nature 2018   17
M. Blythe and A. Monk (eds.), *Funology 2*,
Human–Computer Interaction Series,
https://doi.org/10.1007/978-3-319-68213-6_2

boring) people around. Those pointed out that goal achievement feels good as well, that satisfaction *is* an emotion and that it is all already defined in an International Standard (ISO 9241-11).

However, others felt the need to broaden their perspective. Notions such as "fun" (Carroll and Thomas 1988; Draper 1999), "pleasure" (Jordan 2000), "emotional usability" (Logan 1994), the "hedonic" (Hassenzahl et al. 2000) and the "ludic" (Gaver and Martin 2000) entered the HCI literature. What they had in common was a call for a holistic perspective on technology and an enrichment of traditional quality models with non-utilitarian concepts.

While I became more and more enthusiastic about this novel approach, the engineers and clients around me remained sceptic. What was needed was a way to describe what this enrichment (fun, pleasure, experience joy) could be, how it relates to good ol' usability, as well as a way to measure it and its presumed impact on users' acceptance. As a consequence, I formulated a model which was the focus of the original chapter "The Thing and I" (Hassenzahl 2003). The present chapter is a remix of this chapter. Beloved *Wikipedia* defines a remix as "a piece of media, which has been altered from its original state by adding, removing, and/or changing pieces of the item." Similar to a good remix, this chapter is not to replace the original. It is maybe the most fun to read both versions in one go—supposed you'll find the time.

2 A First, Process-Oriented Model of User Experience

Figure 1 shows the key elements of the initial, process-oriented model from (a) a designer perspective and (b) a user perspective.

When designing a product, designers choose and combine features (content, presentation, functionality, interaction) to convey a product character (or gestalt; Janlert and Stolterman 1997). Note that the particular character is only *intended*. There is no guarantee that users will actually perceive and appreciate the product the way designers wanted it to be perceived and appreciated. When users come in contact with a product, they take its features in and construct an *apparent* product character. Product characters can be described by attributes, such as simple, predictable, novel or interesting.

In general, the model distinguishes two broad groups of attributes. *Pragmatic quality* refers to a product's perceived ability to support the effective and efficient achievement of tasks (i.e., a subjective version of usability). *Hedonic quality* refers to a product's perceived ability to create "pleasure" through use. In the first version of the model, I identified *stimulation* (e.g., through a product's novelty), *identification* (e.g., through a product's apparent professionalism) and *evocation* (e.g., through precious memories associated with the product) as the main sources of the hedonic. To research the model, colleagues and I devised a questionnaire to measure pragmatic and hedonic quality perceptions, most notably the *AttrakDiff2*

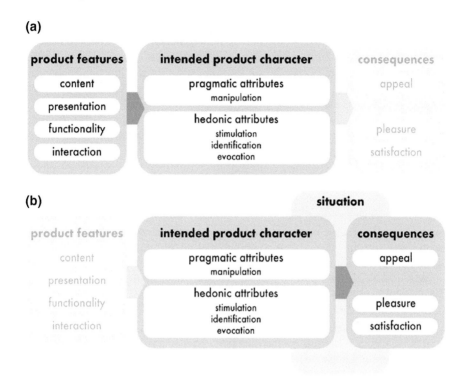

Fig. 1 Key elements of the first process-oriented model from **a** a designer perspective and **b** a user perspective (for details refer to text)

(Hassenzahl et al. 2003, www.attrakdiff.de) and an abridged version (Hassenzahl and Monk 2010; see Diefenbach et al. 2014, for an overview and alternative questionnaires).

Perceptions of character can vary. What seems novel and stimulating to one person, may appear unexciting to the other. Such variations *between* individuals can be explained by differing standards and expectations. The apparent character can also change *within* a person over time. For example, a product perceived as new and stimulating in the beginning may lose some ability to stimulate over time. Conversely, with increasing expertise, products initially perceived as hard to use may become more familiar and, thus, might be perceived as easier. There is a growing body of research, which takes a dynamic perspective on user experience (e.g., Karapanos et al. 2009) with methods such as retrospective interviewing (e.g., *iScale,* Karapanos et al. 2012) or true longitudinal measurement (Harbich and Hassenzahl 2017). The suggestion is that the quality of an interactive product is best captured by the *shape* of change in perceptions over time (see Hassenzahl 2010, p. 19 for more examples).

The apparent product character will mediate a number of consequences, such as potential judgments about the product's general goodness (e.g., "It is good/bad"), emotional consequences (e.g., pleasure, satisfaction) and behavioral consequences (e.g., increased time spent with the product). Originally, I simply wanted to show that pragmatic and hedonic quality are different and that *both* are important for predicting consequences. Overall, this was the case. For instance, in Hassenzahl (2001) pragmatic and hedonic quality perceptions of different screen types were independent of each other. While pragmatic quality was highly correlated with a measure of mental effort ($r = -0.61$), hedonic quality was not ($r = 0.01$). But both predicted appeal with an almost equal weight (pragmatic: $\beta = 0.62$, hedonic: $\beta = 0.61$). This explained a puzzling personal experience. Back then, I sat through hours of usability tests, watching participants struggling with various problems. While they were aware of a lack of usability, they often remained quite enthusiastic about the product. My professional tunnel vision left me clueless: Why do people bother with a piece of technology that obviously does not perform as best as it could? Now I began to understand that there might be other sources of appeal. Overall, usability might not be as important as I always thought.

While the notion that pragmatic and hedonic quality equally contributes to consequences appeared a good start, it seemed obvious that the relative importance of the one or the other will vary with context. Some situations may call for novelty and stimulation, others for highly efficient task achievement. As a result, emotional responses toward a product and judgments of appeal or "goodness" become susceptible to variation caused by the situation. This is an argument for separating potentials for consequences (i.e., the perceived product character) from the actual consequences (i.e., appeal, pleasure, and satisfaction).

The situation as a crucial moderator of the link between product perceptions and consequences poses a serious challenge. Designers need to have an idea of how a particular character will be experienced in a particular situation. However, situations are plenty. As a solution to this, I proposed to focus on the metamotivational state of users—their *usage mode*—rather than the situation.

I distinguished two modes: a *goal mode* and an *action mode* (Hassenzahl et al. 2002) inspired by Apter's (1989) *Reversal Theory* (see Fokkinga and Desmet 2014 for more on *Reversal Theory* in the context of design). The notion of usage modes assumes that interaction *always* consists of behavioral goals and action to fulfil these goals. However, in *goal mode,* goal fulfilment is in the fore. Individuals try to be effective and efficient. They describe themselves as "serious" and "planning". Low arousal is preferred. If arousal increases (e.g., because of a usability problem, which circumvents goal fulfilment), it is experienced as mounting anxiety (frustration). In *action mode* the action is in the fore. People determine goals "on the fly". Effectiveness and efficiency do not play an important role. Individuals describe themselves as "playful" and "spontaneous". High arousal is preferred. If arousal decreases (because of a lack of stimulation) it is experienced as increasing boredom.

While the perception of a product as primarily pragmatic or hedonic will not be influenced by the usage mode, consequences will depend on the product's momentary fit to the usage mode. While in the original chapter this remained a hypothesis, a number of later studies explored the idea empirically (Hassenzahl et al. 2002, 2008; Hassenzahl and Ullrich 2007; van Schaik and Ling 2011).

In one study, Hassenzahl and Ullrich (2007) let half of the participants use software with a number of predetermined goals (i.e., goal mode, a typical usability testing set up). The other half was free to explore and to set their own goals (i.e., action mode). In goal mode, the more mental effort people experienced, the more negative they felt and the poorer the product evaluation. Spontaneity was experienced as negative. In action mode, however, spontaneity was experienced as positive and became highly related to positive product evaluation, while mental effort became dissociated from product evaluation. What led to a positive evaluation in one mode, led to a negative in the other.

In sum, the initial model was a process model. It defined crucial constructs and their relationships and then described how users perceive and value a product. While certainly helpful at that time, from a design perspective the process is not the most interesting. It is rather the *content* that matters. The crude distinction between hedonic and pragmatic qualities was certainly a first step into that direction. However, it seemed necessary to better understand potentially different types of positive experiences and how those could be shaped through technology by design.

3   A Second, Content-Oriented Model of User Experience

3.1   Overview of the Model

The second model of user experience (Hassenzahl 2010) started from the notion of action (Fig. 2).

Fig. 2 Key elements of the second, content-oriented model (for details refer to text)

Many psychological theories of action distinguish different levels of action regulation. Carver and Scheier (1998), for example, postulate be-goals, do-goals and motor-goals. *Motor-goals* are on the sensomotoric level. They address concrete, detailed operations, such as pressing a button or reading an instruction. This level of action is heavily dependent on the physical product and the momentary context of use. It is the level of "How" and in most cases the most obvious concern of interaction design. If a designer devises a particular arrangement of graphical elements on a touchscreen to be swiped and touched by a user, the designer designed action on a motor-level, i.e., "arranged" motor-goals. Arrangements of motor goals, however, are operated to fulfill a do-goal, such as making a telephone call. This is the level of "What". From a product perspective, this level is more concerned with the functionality offered by a product than with details of the interaction with this functionality. These two levels, do-goals and motor-goals, the "What" and the "How", capture the mainstream understanding of what HCI was supposed to be about.

However, there is a third level of goals, the so-called be goals, such as to feel close to relevant others. They are the motives behind do-goals, lending meaning to action. They are the level of "Why". In HCI the "Why", that is, motives, needs, or be-goals, never received much attention. Two chapters of Kaptelinin and Nardi's (2006) book on *Activity Theory*, an in HCI popular theory of action, only begin to address basic questions of the relationship between motives (i.e., be-goals) and actions (for example, whether there is a one-to-one relation between motive and action or a many-to-one). McCarthy and Wright's (2004, p. 85) model of experience only broadly asserts that all our activities are "shot through with values, needs, desires, and goals." I argued elsewhere (Hassenzahl 2010, p. 45) that "[u]nderstanding action in terms of motives fulfilled is crucial to an experiential approach to design. Brewing the morning coffee because of a need for routine is different from brewing it because of expressing your affection for your loved one through caring. Although the action appears the same, the newly designed, especially efficient coffee maker will certainly fail to support a need for routine (it actually may take it away), whereas the same coffeemaker may give me some extra minutes to spend with my spouse in the morning (but may actually ruin the symbol of caring because the less invested the less it might be valued)." Unfortunately, Carver and Scheier's (1998) model is a process model and silent about the potential content of be-goals. The same is true for *Activity Theory* or McCarthy and Wright's (2004) model of experience. They postulate motives or needs as theoretical entities, but remain silent about the potential content of those motives.

Fortunately, there is research into the different type of needs (i.e., motives). Sheldon and colleagues (2001) reviewed a number of need theories and provided 10 candidates: autonomy/independence, competence/effectance, relatedness/belongingness, self-actualizing/meaning, security/control, money/luxury, influence/popularity, physical thriving/bodily, self-esteem/self-respect, pleasure/stimulation. They not only provided a neat list of needs, but also a questionnaire to measure the saliency of experienced need fulfillment retrospectively, with items such as "During the experience, I felt a sense of contact with people who care for me, and whom I care for" (relatedness) or "During the experience, I felt a sense of deeper purpose" (meaning).

Inspired by this work, we focused on autonomy, competence, relatedness, stimulation, popularity, security, and meaning and began to apply it to the experience of interactive products (Hassenzahl 2008; Hassenzahl et al. 2010, 2015). We asked people to remember a positive experience in which an interactive technology played a role and measured experienced affect and need fulfilment with questionnaires. In general, we found people capable of distinguishing between different types of needs. In addition, need fulfillment was correlated with affect—the more intense need fulfillment, the more positive the experience. Particular needs revealed particular patterns. For example, competence experience was not only related to positive affect, but also to negative affect. This underlines that feeling competent calls for a certain amount of challenge, which implies episodes of inconvenience and extra effort as well as being afraid of failures. In contrast, security showed the lowest correlation with positive affect, highlighting its nature as a "deficiency need". Security may be the most salient, when it becomes threatened. In addition, all studies revealed a clear link between need fulfillment and the perception of the product as hedonic, while need fulfillment was less clearly or not at all related to pragmatic quality. In Hassenzahl et al. (2015), for example, the correlation between need fulfillment and hedonic quality was significant ($r = 0.27$), but the correlation between need fulfillment and pragmatic quality was not ($r = 0.04$). All in all, needs provide a collection of discernible sources of positive experience related to hedonic product perception.

Other researchers in HCI took up the concept of needs and provided a number of further studies. For example, Partala and Kallinen (2012) extended our earlier research to negative experiences. In another study, Partala (2011) compared virtual life (in Linden Lab's *Second Life*) and real life in terms of need fulfillment. In virtual life, people experienced more autonomy, more luxury and more physical thriving (since one can shape one's virtual body in line with personal preference). In real life, there was more competence, relatedness, security and popularity. Karapanos and colleagues studied the meaning of *Facebook*, *WhatsApp* and different activity trackers through the lens of psychological needs (Karapanos et al. 2016a, b). Needs are useful to better understand the different qualities of enjoyable and meaningful experiences. Of course, the particular selection of needs provided by the second model can be debated. Partala und Kujala (2016) recently provided a study based on ten universal values (Schwartz and Bilsky 1987), such as achievement, stimulation, power, security, or benevolence. However, since needs are conceptualized as universal, as "innate psychological nutriments that are essential for ongoing psychological growth, integrity, and well-being" (Deci and Ryan 2000, p. 229), they are closely tied to our very human nature. As a consequence, need theories (Deci and Ryan 2000; Maslow 1954; Reiss 2004; Schwartz and Bilsky 1987) tend to substantially overlap. I believe that the seven categories proposed by the second model (i.e., autonomy, competence, relatedness, stimulation, popularity, security, physical thriving, and meaning) cover substantial ground.

3.2 User Experience and Wellbeing

While the first, process-oriented model justified hedonic quality implicitly by promising improved acceptance by users/consumers, the second model sets wellbeing as an explicit goal. "Experiences make us happier" was one out of three reasons, I provided to justify caring about user experience (Hassenzahl 2010). Note that Deci and Ryan's definition of a need as "essential for growth, integrity and wellbeing" already provides an explicit link to psychological wellbeing (i.e., happiness). In fact, Diener et al. (2009) offered "need and goal satisfaction theories" as one of the two major theoretical explanations for differences in levels of happiness among people (besides genetic predisposition). The concept of wellbeing used in the present model is one that understands everyday activity as an important source of happiness (Lyubomirsky et al. 2005). The right activities ("What"), done in the right way ("How") are likely to touch upon need fulfillment and will create moments of pleasure and meaning. Psychological wellbeing is understood as the consequence of experiencing these moments regularly. Since almost all human activity is not only mediated by technology, but actively construed in terms of the technology (Kaptelinin and Nardi 2006; Reckwitz 2002), design provides a powerful mechanism to shape activities, experiences and ultimately wellbeing. Not surprisingly, the *wellbeing-oriented* design of technology ("design for wellbeing", "positive design") has become broadly discussed, with a growing body of research, examples and design-oriented tools (Calvo and Peters 2014; Desmet and Pohlmeyer 2013; Hassenzahl et al. 2013).

3.3 On the Relationship Between Experience and Interaction

The hierarchy of the "Why", the "What" and the "How" featured by the model implies a relationship between levels. From a design perspective, we assume that the "Why", that is, needs (e.g., to feel stimulated, to feel related) and the envisioned experience, must be determined first to become able to choose functionality (the "What") and to determine the appropriate form and interaction (the "How") in line with the experience. In other words, without a clear picture of the experience to be created, the "material" cannot be specified.

Such a notion assumes that there are no generally valid design solution on lower levels. For example, efficiency may be an important aspect of many technology-mediated activities, but not all. The experience of closeness between couples in long-distance relationships enabled by technology may require less attention to efficiency than a time-critical, competence-oriented activity. In fact, applying an overall rule to all situations may even be harmful. The sometimes thoughtless application of efficiency-oriented practices acquired at work to private life are certainly at the heart of phenomena, such as "time poverty", and counter movements, such as practicing mindfulness and cherishing the moment.

Recently, we engaged in a line of studies to more systematically explore the relationship between the experiential level (the "Why") and lower levels (predominantly the "How") (Diefenbach et al. 2017; Lenz et al. 2013, 2017). In one study (Diefenbach et al. 2017), we designed a digital picture frame for an office workplace, which allows to keep and consume a secret picture. Keeping the small secret and revealing it in an undisturbed moment during a busy office day was supposed to create an experience of autonomy and privacy. Specifically, we compared two different ways of interacting with the picture. One was quite technical: the user could reveal and hide the picture with the press of a button. For the other, we first asked individuals to describe how interacting with a secret feels good and especially secretive. They described how they approach the secret slowly and appreciatively, that they require an instantaneous way of hiding the secret and that they feel an urge to physically touch it. We transformed this into a particular touch interaction for the picture frame. Upon putting the finger on the screen of the frame, a part of the secret picture appears just under the fingertip. Moving the finger reveals the picture further. The moment the finger leaves the screen, the exposed parts of the secret picture disappear again. Participants saw video prototypes of both interactions and were asked to imagine using the frame in their office with a picture of their choice. The results showed that participants not only believed that the experience would be more positive when using the touch interaction, but also that the experience in itself would provide a more intense feeling of autonomy and privacy. This supports the notion of a necessary fit between low-level interaction and envisioned experience.

3.4   Three Points for Discussion

Finally, I want to briefly clarify my position on three points for discussion, which constantly crept up over the years in personal communications and other authors' writings about user experience: (1) The role of instrumentality, (2) the difference between experiences *mediated through* a product and the experience *of* a product and (3) whether experiences are designable.

The role of instrumentality. In Hassenzahl (2010), I used the example of high heels to show how easy it is to misunderstand instrumentality. Since high heels belong to the category of shoes, one may readily associate them with the major purpose of shoes—to support walking. Obviously, high heels are bad for this and thus utterly useless and unusable. However, for many, the real purpose of a high heel is to look good, to be admired and desired. Consequently, whether a particular high heel "functions" or not, that is, whether it is instrumental or not, rests solely on its ability to create the experience of being admired and desired (i.e., popularity, relatedness). Comfy walking is simply not an issue here. This underlines the importance of making explicit the envisioned experiences to be created through a product, since instrumentality can only be understood in relation to these experiences.

Experiences mediated through a product versus experience of a product. In my practice, I found industries' interest in user experience mainly driven by the desire to reshape the experience *of* their product. Quite understandably, they look for a "wow" to differentiate from competitors. Personally, I don't believe products should be an experience in themselves. They should provide the functionality and interaction to *create* and *mediate* enjoyable and meaningful everyday experiences. That's how they become significant in the long run. Of course, especially early adopters thrive on the stimulation, popularity and competence reaped from trying out cool, new gear. People buy activity trackers, smart watches, robots, virtual reality goggles or apps just to check them out, to stay on the ball. One could argue that there is nothing to object to. If relishing the novel (stimulation), appearing cool (popularity) and evolving one's skills to master technology (competence) is the outcome of all this, people simply work their happiness levels. However, this strategy will eventually lead (or has already led) to an unhealthy succession of the ever new. Often, we do not long for a new smartphone because it improves (although the rhetoric around it suggests so), but simply because it is new. This is a wasteful strategy in terms of resources and bound to make people unhappy. Focusing on the experiences emerging from activities may inspire less material-intensive ways of creating these experiences and emphasizes alternative, less fleeting sources of wellbeing beyond stimulation.

Are experiences designable? Some in the field of HCI understand experiences as not designable. For example, Kaptelinin and Bannon (2012, p. 296) state that "[o]ne cannot design—or 'give shape to'—something that is outside one's direct control. Therefore, one cannot design human activities and experiences: They are personal, situated, emergent phenomena that cannot be shaped, or even completely antici-pated in advance" (p. 296). One can only design *for* an experience (e.g., provide the necessary ingredients and infrastructures), but not the experience in itself. I tend to disagree. All in all, we should understand an experience as an immaterial outcome, which can be at least envisioned in an ideal "form". This is already an act of design. If then all elements pliable are arranged in a way to create this experience, it becomes inscribed into and expressed through material arrangements. Again an act of design, even if not all elements are under full control. And while there is no guarantee that an experience emerges exactly as envisioned, it is not unlikely that due to a shared cultural understanding, experiences become meaningful for those people, they had been designed for. This is at least the basic hope of everybody throwing a birthday party: That the guests will relish the experience arranged for them. If this does not happen, there is certainly a lack of understanding between the "designer" and the "user". But this is not a novel phenomenon, and basically the reason for working empirically in experience design and HCI.

4 Conclusion

Nowadays, "good" user experience is a broadly accepted goal of design. While maybe not always completely understood and still riddled with conceptual inconsistencies, researchers and practitioners of HCI agree that experiences emerging from the interaction with technology are important, that these experiences should be worthwhile (i.e., positive), and that they are situated and dynamic, yet somehow "designable" through particular functionality, interaction and form bundled into a technology (Law et al. 2009).

In this chapter, I presented my view on user experience. I understand psychological wellbeing as the ultimate goal of technology design, not efficiency. In his *Letter to Humanity*, van Mensvoort (2017) from *Next Nature* writes: "Technology not only alters our environment, it ultimately alters us. The changes to come will allow you to be more human than ever before. What if we used technology to magnify our best human qualities and support us in our weaknesses? We could call such technology humane, for lack of a better word. Humane technology takes human needs as its starting point. It would play to our strengths rather than rendering us superfluous. It would expand our senses rather than blunting them. It would be attuned to our instincts; it would feel natural. Humane technology would not only serve individuals but, first of all, humanity as a whole." This neatly summarizes, what experience design is all about.

To make it happen requires an idea of where the positivity stems from. The set of psychological needs described above are an appropriate starting point, representing potential "Why's" of technology use. They can serve as inspiration, guides and as a way to characterize or "evaluate" experiences. Ultimately, though, design requires a more specific understanding. Need fulfillment becomes only real when situated through activities. In turn, most of these activities involve technology. Technology is, thus, not only instrumental to the activity, but also substantially shapes it. Determining what to do in which way (the "What", the "How") to preserve and even strengthen need fulfillment and wellbeing (the "Why") is the overarching goal of my flavor of experience design.

The present models rely on a number of well-studied concepts mainly borrowed from psychology, applied and adapted to HCI. In itself, basic assumptions received some empirical validation beyond what psychology already knew (e.g., that need fulfillment is linked in a certain way to product perceptions). This certainly distinguishes it from many other frameworks whose basic assumptions never became systematically tested.

Admittedly, models are always reductions. They cannot and will not replace detailed inquiries into situations, practices and experience. But they systemize, spurn debates about what belongs and what not. Maybe more important, we found especially the second model to be inspiring for design work. There are a number of case studies ranging from automotive interaction design (e.g., Eckoldt et al. 2013; Knobel et al. 2012) to kitchen appliances (Klapperich and Hassenzahl 2016) as well as from communication in distributed families (Lenz et al. 2016) to communication among teenage boys (Laschke et al. 2010). In all these cases, the model presented above served as grounding and inspiration. I sincerely hope that the ideas presented in this remix chapter will inspire your work, too—be it the design of technology-mediated experiences or attempts to further challenge, falsify, operationalize or improve models of experience.

5   Finally, on a More Personal Note

On a personal note, the change of perspective in HCI had been thrilling for me. In 1999, I was working as a usability engineer in the competence center for user interface design at *Siemens* corporate technology. The whole idea of "fun" seemed absurd in this environment. Yet, my colleague Michael Burmester headed a research project on computer games and operator interfaces. Likewise, my colleague Axel Platz worked on beautiful interfaces for *Siemens'* medical imaging systems (*syngo*), which posed challenges to my notion of usability and the tools, I used to conceptualize and measure quality. I remember travelling to York in 1999 to attend one of Andrew Monk's "Computers and Fun" workshops, feeling at home in a small crowd of HCI researchers interested in technology and positive emotions. It was a stimulating experience, already hinting at what could be gained from going "beyond usability." Briefly after this, I presented the first paper on hedonic quality (Hassenzahl et al. 2000) on the CHI conference in The Hague in 2000 in the very same session as Bill Gaver presented a sketchbook full of alternative conceptual devices, such as a dawn chorus, a (de)tour guide, an intimate view, a dream communicator, or a worry stone (Gaver and Martin 2000). (If you have no clue what all these things are, please take a look at the paper.) And of course, there was the workshop on CHI in Minneapolis leading to the first version of the Funology book. These had been powerful influences.

Acknowledgements While I chose to write this chapter from a personal perspective, I am of course indebted to many colleagues, students and friends, who inspired me. My work is certainly the result of many hours of personal conversations, reading impressive papers and being involved in concrete design work. I nevertheless like to especially thank all current and past members of my work group: Sarah Diefenbach, Judith Dörrenbächer, Kai Eckoldt, Stefanie Heidecker, Anne Karrenbrock, Holger Klapperich, Matthias Laschke, Eva Lenz, Thies Schneider, Alarith Uhde, Julika Welge und Tim Zum Hoff. Their work was and is essential.

References

Apter MJ (1989) Reversal theory: motivation, emotion and personality. Routledge, London
Calvo RA, Peters D (2014) Positive computing: technology for wellbeing and human potential [Kindle Edition]. MIT Press, Cambridge
Carroll JM, Thomas JMC (1988) Fun. ACM SIGCHI Bull 19(3):21–24

Carver CS, Scheier MF (1998) On the self-regulation of behavior. Cambridge University Press, Cambridge

Deci EL, Ryan R (2000) The "What" and "Why" of goal pursuits: human needs and the self-determination of behavior. Psychol Inq 11(4):227–268

Desmet PMA, Pohlmeyer AE (2013) Positive design: an introduction to design for subjective well-being. Int J Des 7(3):5–19

Diefenbach S, Kolb N, Hassenzahl M (2014) The "Hedonic" in human-computer interaction: history, contributions, and future research directions. In: Proceedings of the 2014 conference on designing interactive systems—DIS 14. ACM Press, NY, pp 305–314

Diefenbach S, Hassenzahl M, Eckoldt K, Hartung L, Lenz E, Laschke M (2017) Designing for well-being: a case study of keeping small secrets. J Positive Psychol 12(2):151–158

Diener E, Oishi S, Lucas R (2009) Subjective well-being: the science of happiness and life satisfaction. In: Lopez SJ, Snyder CR (eds) Oxford handbook of positive psychology, 2nd edn. Oxford University Press, Oxford, pp 187–194

Draper SW (1999) Analysing fun as a candidate software requirement. Pers Ubiquit Comput 3(3): 117–122

Eckoldt K, Hassenzahl M, Laschke M, Knobel M (2013) Alternatives: exploring the car's design space from an experience-oriented perspective. In: Proceedings of the 6th international conference on designing pleasurable products and interfaces—DPPI'13. ACM Press, NY, pp 156–164

Fokkinga S, Desmet P (2014) Reversal theory from a design perspective. J Motiv Emot Pers 2(2): 12–26

Gaver B, Martin H (2000) Alternatives: exploring information appliances through conceptual design proposals. In: Proceedings of the SIGCHI conference on human factors in computing systems—CHI 00. ACM Press, NY, pp 209–216

Harbich S, Hassenzahl M (2017) User experience in the work domain: a longitudinal field study. Interact Comput 29(3):306–324

Hassenzahl M (2001) The Effect of Perceived Hedonic Quality on Product Appealingness. Int J Hum Comput Interac 13(4):481–499

Hassenzahl M (2003) The thing and I: understanding the relationship between user and product. In: Blythe M, Overbeeke C, Monk AF, Wright PC (eds) Funology: from usability to enjoyment. Kluwer Academic Publishers, Dordrecht, pp 31–42

Hassenzahl M (2008) User experience (UX): towards an experiential perspective on product quality. In: Proceedings of the 20th international conference of the association francophone d'interaction home-machine. ACM Press, NY, pp 11–15

Hassenzahl M (2010) Experience design: technology for all the right reasons. Morgan & Claypool, USA

Hassenzahl M, Monk A (2010) The inference of perceived usability from beauty. Hum Comput Interact 25(3):235–260

Hassenzahl M, Ullrich D (2007) To do or not to do: differences in user experience and retrospective judgments depending on the presence or absence of instrumental goals. Interact Comput 19(4):429–437

Hassenzahl M, Platz A, Burmester M, Lehner K (2000) Hedonic and ergonomic quality aspects determine a software's appeal. In: Proceedings of the SIGCHI conference on human factors in computing systems—CHI 00. ACM Press, NY, pp 201–208

Hassenzahl M, Kekez R, Burmester M (2002) The importance of a software's pragmatic quality depends on usage modes. In: Proceedings of the 6th international conference on work with display units—WWDU 2002, pp 275–276

Hassenzahl M, Burmester M, Koller F (2003) AttrakDiff: Ein Fragebogen zur Messung wahrgenommener hedonischer und pragmatischer Qualität. In: Mensch & Computer 2003. Teubner, Stuttgart, pp 187–196

Hassenzahl M, Schöbel M, Trautmann T (2008) How motivational orientation influences the evaluation and choice of hedonic and pragmatic interactive products: the role of regulatory focus. Interact Comput 20(4–5):473–479

Hassenzahl M, Diefenbach S, Göritz A (2010) Needs, affect, and interactive products—facets of user experience. Interact Comput 22(5):353–362

Hassenzahl M, Eckoldt K, Diefenbach S, Laschke M, Lenz E, Kim J (2013) Designing moments of meaning and pleasure. Experience design and happiness. Int J Des 7(3):21–31

Hassenzahl M, Wiklund-Engblom A, Bengs A, Hägglund S, Diefenbach S (2015) Experience-oriented and product-oriented evaluation: psychological need fulfillment, positive affect, and product perception. Int J Hum Comput Interact 31(8):530–544

Janlert L-E, Stolterman E (1997) The character of things. Des Stud 18(3):297–314

Jordan P (2000) Designing pleasurable products. An introduction to the new human factors. Taylor & Francis, London

Kaptelinin V, Bannon LJ (2012) Interaction design beyond the product: creating technology-enhanced activity spaces. Hum Comput Interact 27(3):277–309

Kaptelinin V, Nardi BA (2006) Acting with technology. Activity theory and interaction design. MIT Press, Cambridge

Karapanos E, Zimmerman J, Forlizzi J, Martens J (2009) User experience over time: an initial framework. In: Proceedings of the SIGCHI conference on human factors in computing systems. ACM Press, NY, pp 729–738

Karapanos E, Martens J-B, Hassenzahl M (2012) Reconstructing experiences with iScale. Int J Hum Comput Stud 70(11):849–865

Karapanos E, Gouveia R, Hassenzahl M, Forlizzi J (2016a) Wellbeing in the making: peoples' experiences with wearable activity trackers. Psychol Theory Res Pract 6(4). http://doi.org/10.1186/s13612-016-0042-6

Karapanos E, Teixeira P, Gouveia R (2016b) Need fulfillment and experiences on social media: a case on Facebook and WhatsApp. Comput Hum Behav 55(Feb):888–897

Klapperich H, Hassenzahl M (2016) Hotzenplotz: reconciling automation with experience. In: Proceedings of the Nordic conference on human-computer interaction—NordiCHI 16 (Article 39). ACM Press, NY

Knobel M, Hassenzahl M, Lamara M, Sattler T, Schumann J, Eckoldt K, Butz A (2012) Clique trip: feeling related in different cars. In Proceedings of the 9th ACM conference on designing interactive systems—DIS 12. ACM Press, NY, pp 29–37

Laschke M, Hassenzahl M, Mehnert K (2010) Linked—A relatedness experience for boys. In: Proceedings of the 6th Nordic conference on human-computer interaction—NordiCHI 10. ACM Press, NY, pp 839–844

Law E, Roto V, Hassenzahl M, Vermeeren A, Korte J (2009) Understanding, scoping and defining user experience: a survey approach. In: Proceedings of the SIGCHI conference on human factors in computing systems—CHI 09. ACM Press, NY, pp 719–728

Lenz E, Diefenbach S, Hassenzahl M (2013) Exploring relationships between interaction attributes and experience. In: Proceedings of the 6th international conference on designing pleasurable products and interfaces—DPPI'13. ACM Press, pp 126–135

Lenz E, Hassenzahl M, Adamow W, Beedgen P, Kohler K, Schneider T (2016) Four stories about feeling close over a distance. In: Proceedings of the tenth international conference on tangible, embedded, and embodied interaction—TEI 16. ACM Press, NY, pp 494–499

Lenz E, Hassenzahl M, Diefenbach S (2017) Aesthetic interaction as fit between interaction attributes and experiential qualities. New Ideas Psychol 47:80–90

Logan RJ (1994) Behavioral and emotional usability: Thomson consumer electronics. In: Wiklund M (ed) Usability in practice. Academic Press, London

Lyubomirsky S, Sheldon KM, Schkade D (2005) Pursuing happiness: the architecture of sustainable change. Rev Gen Psychol 9(2):111–131

Maslow AH (1954) Motivation and personality. Harper & Row, NY

McCarthy J, Wright P (2004) Technology as experience. MIT Press, Cambridge

Partala T (2011) Psychological needs and virtual worlds: case second life. Int J Hum Comput Stud 69(12):787–800

Partala T, Kallinen A (2012) Understanding the most satisfying and unsatisfying user experiences: emotions, psychological needs, and context. Interact Comput 24(1):25–34

Partala T, Kujala S (2016) Exploring the role of ten universal values in using products and services. Interact Comput 28(3):311–331

Pine BJ, Gilmore JH (1999) The experience economy: work is theatre & every business a stage. Harvard Business School Press, Cambridge

Reckwitz A (2002) Toward a theory of social practices: a development in culturalist theorizing. Eur J Soc Theory 5(2):243–263

Reiss S (2004) Multifaceted nature of intrinsic motivation: the theory of 16 basic desires. Rev Gen Psychol 8(3):179–193

Schmitt BH (1999) Experiential marketing. Free Press, NY

Schulze G (1992) Die Erlebnisgesellschaft: Kultursoziologie der Gegenwart. Campus

Schwartz SH, Bilsky W (1987) Toward a universal psychological structure of human values. J Pers Soc Psychol 53(3):550–562

Sheldon KM, Elliot AJ, Kim Y, Kasser T (2001) What is satisfying about satisfying events? Testing 10 candidate psychological needs. J Pers Soc Psychol 80(2):325–339

van Mensvoort K (2017) Letter to humanity. Retrieved 28 Apr 2017, from https://www.nextnature.net/2017/04/letter-to-humanity/

van Schaik P, Ling J (2011) An integrated model of interaction experience for information retrieval in a web-based encyclopaedia. Interact Comput 23(1):18–32

Chapter 3
Can Games Be More Than Fun?

Paul Cairns

1 Games Are Very Engaging

Digital games are undoubtedly hugely engaging. Any statistics on how many people are playing games and for how long are staggering. For example, the UK Association for Interactive Entertainment, UKIE, reports that 20M people in the UK aged between 6 and 64 play games, that's about 42% of that age bracket. Furthermore, the age group 11–64 plays on average about 8–9 hrs of games a week (UKIE 2017). In the USA, 63% of all households have at least one frequent gamer and there is an even split of men and women.

Given playing games is a voluntary activity that requires a significant investment of time and effort, it is not surprising that people have wondered if such engagement could be directed to more productive outcomes. Serious games, here, very loosely refers to games that are intended to achieve some form of meaningful real world outcome (Ritterfeld et al. 2009). This can range from improving health (Wattanasoontorn et al. 2013), increasing donations to humanitarian issues (Steinemann et al. 2015), training people (*Virtual Battlespace 3*[1]) and of course education, where there has been a long interest in using games to improve learning (Malone 1981). McGonigal (2011) even goes so far as to say that modern approaches to learning, work and productivity are dysfunctional and games hold the key to remedying these problems.

However, though there is considerable enthusiasm amongst researchers and developers for using games to promote out of game outcomes, success in serious games has been mixed, for example Sherry (2015). While studies have shown

---

[1]https://bisimulations.com/.

P. Cairns (✉)
Department of Computer Science, University of York,
Deramore Lane, York YO10 5GH, UK
e-mail: paul.cairns@york.ac.uk

© Springer International Publishing AG, part of Springer Nature 2018
M. Blythe and A. Monk (eds.), *Funology 2*,
Human–Computer Interaction Series,
https://doi.org/10.1007/978-3-319-68213-6_3

increased engagement as a result of using games, these results do not always translate into real world success. Energy meters are ignored, exercise apps remain switched off and teachers revert to more traditional methods. There seems to be a substantial disconnect between the high volume of voluntary engagement with games and exploiting that voluntary engagement to achieve real world outcomes.

The goal then of this chapter is to help discuss some of the problems of transferring the success of digital games into other domains and the challenges of even researching this area. This chapter discusses three specific topics: flow, learning and research design. Flow is regularly seen as a key outcome of game that may have important psychological benefits for players. However, flow experience and what is called flow in players of games may be quite different things. Games also seem to have a lot of potential for use in education as a way of getting players to engage, voluntarily, with new or difficult concepts. This chapter describes how recent research undermines naive attempts at learning in games because mere exposure to concepts in games is not enough to initiate the learning process via priming players to think about those concepts.

What is needed is more research but we also discuss how players in research projects are susceptible to all sorts of influences that can be very hard to remove and which can influence research results. At the end of the day, it seems that the games are meant to be played and players can see the fun in all sorts of games. But moving games to provide more meaningful experiences presents substantial challenges that we are not yet equipped to address.

2   A Note About "Games"

Throughout this chapter, the term "games" is used predominantly to mean digital games though it may apply to other sorts of games with caution. This is primarily because digital games are the core research area of the author and it would be over-reaching the current research in digital games to apply its findings to other sorts of games such as card games, board games, sports and so on. This may be something of an artificial distinction because after all many of the original video games were versions of existing games transferred to computers, for instance, solitaire, chess and football. However, digital games have two fundamental differences from other games. First, the game itself has agency in a way that other games do not. This is most obvious in first person shooter games, like *Call of Duty* where the computer can control a literal army of opponents. But even where much of the action in a game is due to playing against other real players, say in *Starcraft*, the computer also acts independently to enforce complex rule sets such as who owns what things, what can be done with those things and whether you really did shoot the other player or not.

Secondly, digital games offer game worlds that do not rely on the limitations or possibilities of the physical world. Golf in the real world is a balancing act between the design of the course, the skills of the player and the vicissitudes of nature such as a breeze or a particularly springy bit of grass. Digital golf could be all of these things and further offer impossible contexts such as variable gravity, aliens and torpedoes as well, such as in *Wonderputt*. This is not confined to games with an obvious physical underpinning: it is also true of many casual and self-paced games. Take for example *Two Dots*. While it is possible to imagine a table top version of this game, the digital *Two Dots* works because without any set-up cost to the player, there is a board full of dots that refresh themselves during play and there are challenges ready to be tackled. Moreover, the challenges progress far further than any finite stack of challenge cards could.

Thus digital games in this chapter are those distinct games which work because the games themselves have agency and action on the game world. Furthermore, such a game world might not be otherwise realisable. As Wittgenstein famously noted,[2] there is not a crisp categorisation of games but prototypical digital games such as first person shooters, open world real-time strategy games and puzzle platformers are sufficient to characterise our focus in this chapter.

3 It's Not All About Flow

In recent times, psychology has developed a strand of work, called positive psychology, looking to understand what makes people happy (Boniwell 2012). Much of this work has sprung up as a consequence of some seminal work in the 1970s into the experience of flow, the sense of being fully engaged in a task. Csikszentmihalyi (2002) found that flow provides a sense of fulfillment and satisfaction in doing a task and that this leads to longer term happiness and fulfillment in life. Moreover, people deliberately seek out flow experiences in order to achieve these positive outcomes.

Flow is characterised as an optimal psychological experience arising from eight components (Csikszentmihalyi 2002), a formulation which has changed little since its inception and is summarised as six aspects here (Engeser and Schiepe-Tiska 2012):

1. Intense concentration on a limited stimulus
2. Merging of action and awareness
3. A loss of self-consciousness
4. A feeling of control over the situation or activity
5. Coherence in actions, feedback and progression to goals
6. Experience of the activity as intrinsically rewarding (autotelic)

Each of these aspects may occur separately but only when they occur together as a unified holistic experience do you get flow.

Looking at these attributes, there is considerable overlap with the properties of playing a game: games offer clear goals, specific actions and good feedback in a

---

[2]https://en.wikipedia.org/wiki/Family_resemblance.

very coherent way. The result is players develop a strong focus on games to the exclusion of the external world, often called immersion (Cairns et al. 2014). Indeed, Chen (2007) argue that games have evolved to generate flow and it is flow that brings people into play.

Thus, flow is held to be the basis for describing player enjoyment of digital games (Sweetser and Wyeth 2005) leading to models to help designers build flow into their games. Specific measures of flow are used to study player experiences, for example in the work of Vella et al. (2013). Flow is also a constituent of game specific measures of player experiences, for example in the game engagement questionnaire (Brockmyer et al. 2009). Moreover, if games can offer flow and flow can lead to psychological well-being then there is the possibility that merely playing games could be of positive psychological benefit (Vella and Johnson 2012).

However, even in the early characterisations of immersion (Brown and Cairns 2004), it was evident that the very intense experiences of immersion in games were fleeting and that often playing a game was immersive but without all of the conditions that might lead to flow. In some sense, flow may be the optimal experience of playing a game but many experiences of play are much more prosaic and sub-optimal (though still valued by players). A quick game of *Candy Crush* or *Monument Valley* as a break from work can still be engaging, somewhat immersive and valued even if flow is not in any sense achieved.

Furthermore, games do not always attempt to engender this optimal experience. Some games are hugely frustrating and yet equally popular with some players, typical such games being *Super Meat Boy* and *Dark Souls*. And even when frustration is not such a central feature of the game, players can find their failure enjoyable (Ravaja et al. 2008) or important in building up the sum enjoyment of the game (Petralito et al. 2017). Juul (2013) also points out that failure is an important constituent of digital games and this perhaps goes against the coherence aspect of flow where actions are meant to lead towards goals not the failure to achieve them. Even the emotions of instances of play do not match the positive experience of flow but it is only in sum and perhaps on reflection that players find a game to have been a positive experience (Triberti 2016). There is also growing evidence that players are using games to distract themselves from other concerns and so bring about well-being (Collins and Cox 2014). It is not clear if flow is necessary or even sought after in these contexts.

The focus on flow also comes with another problem which is how to measure this optimal experience in digital games. Moneta (2012) identifies the three main ways to measure flow. The earliest is the flow questionnaire (FQ) but that really is tailored to understanding the general level of flow experiences that people have rather than their flow in relation to specific experiences and in particular in response to playing a game. The experience sampling methods (ESM) improves upon the FQ by sampling people in their daily tasks and asking about their experiences. However, ESM prompts for flow experiences explicitly which may be biasing and also it is very hard to validate that, when people did report flow, it was in fact a flow experience. The third approach looks like more typical psychometrics in that questionnaires are used to collect people's experiences and identify latent concepts in the questionnaires that correspond to flow. The most well-established such questionnaire is the revised Flow State Scale (FSS-2) (Jackson and Eklund 2002) and this has been used in several studies of player experience.

Questionnaires like FSS-2 have an intuitive appeal and fit with many other measures used in player experience study such as the immersive experience questionnaire (Jennett et al. 2008). However, if flow is an optimal experience and, for the sake of argument, a questionnaire measures flow on the scale of 1–10. What does a flow score of 5 mean? Is this sub-optimal flow (and so not really flow)? Or does it mean it is not flow, in which case attributing it to a flow experience or even aggregating it across players to give a mean level of flow would be meaningless. What if the score of 5 is due to a high score on some factors of flow but very low score on other factors? Then the score is not an indicator of flow at all as flow would need all or at least several factors to be present. Alternatively, the score indicates a player moving towards flow, if not necessarily achieving it. But this interpretation is rarely seen in the player experience literature and even if it were, would a movement towards increased flow in a study actually be a useful measure if flow were never actually achieved?

Thus, despite the prevalence of questionnaire approaches to measuring flow in games, there is substantial research needed to establish that it is indeed flow that players are experiencing. It should be noted, the problems of capturing flow in activities are not unique to player experience in digital games. For instance in elite sports, where flow is believed to be an important constituent of peak performance, it is still unclear which components of flow are essential for a flow experience and what intensity each component needs to attain (Swann et al. 2012). And in the field of music, there are indications that musicians may be experiencing flow differently from athletes and also differently depending on their level of musicianship (Sinnamon et al. 2012).

If we step back somewhat from flow as central to gaming experiences, then flow takes its place alongside other constituent experiences that players have and seek when they play. The promise of games to bring about psychological well-being may still be realised but flow may not be so central to achieving it. For example, self-determination theory has shown promise in explaining the enduring attraction of games (Ryan et al. 2006) and also has the potential to bring about well-being in players (Vella and Johnson 2012). However, we are still a long way from establishing whether games could be good for people and whether flow in fact has any part to play in bringing about that good.

4   Learning in Games

Another way in which games might move beyond mere fun is to use them as educational tools. The rationale for this is, at first glance, self-evident. People clearly have to learn to play games, whether what is learned are motor skills like the rapid responses needed in a first-person shooting game like *Overwatch*, or whether it is explicit, factual knowledge such as the ordering of technologies in the "technology tree" in *Civilization*. Thus, people who play games incidentally have to learn something if they want to play. Secondly, people voluntarily play games and enjoy learning them, something which is not always true of formal education. Bringing the necessity of learning and the joy of learning together in a game seems like a natural opportunity which was identified very early on in the history of computer games (Malone 1981).

Simulators are effectively a form of game where, for example, a player can fly a simulated Boeing 747 for fun. Often desktop simulators like *Flight Simulator X* have settings and scenarios that allow players to fly planes where they would never be allowed to do so in real life, for example over the Great Pyramids of Giza. At the same time, if the player does play a realistic simulator, then in order to play they must learn the controls of the real aircraft. Learning is happening in a game context. There are clearly games, such as *World without Oil* (in this case non-digital), where simulation is intended to bring home a very real educational message (McGonigal 2011). And of course, in professional contexts, particularly the aerospace industry, simulators are used as an essential part of training. Even there, though, simulators are only ever a part (albeit substantial) of wider training programmes including traditional classroom learning and formal assessments.

The distinction between simulators and games can be blurry. The game *Eurotrucker* takes professional levels of commitment as the player must do long-distance goods haulage across a virtual model of Europe in real-time. Conversely, the strategic war game, *ARMA 2*, has such realistic battle simulation that there is a professional version, *Virtual Battle Space (VBS)*, that is used worldwide in military training.

Aside from simulators, games hold the promise to teach things that are otherwise difficult in a traditional classroom, such as persistence and moral viewpoints (Gee 2004). However, it is a different matter to bring about learning of a specific curriculum of knowledge through a game. While a player might well learn about the complexities of societies and their development through playing *Civilization*, it is very hard to direct what exactly players do learn. Indeed, it may need a radical re-thinking of what the goals of formal education are in order to effectively exploit the learning that happens in such games (Squire 2005).

Nonetheless, there is a still a persistent assumption that mere exposure to ideas in games might be sufficient to bring about learning of those ideas. This is possibly most evident in the general aggression model (GAM), where mere exposure to aggressive concepts in a game are held to lead, first, to aggressive thoughts and then, over time, to aggressive behaviours (Anderson and Bushman 2001). The specific route to aggressive behaviours from exposure to violent concepts in digital games comes through cognition. That is, exposure to violent concepts leads players to think about those concepts, which under the influence of other factors increases their propensity to be aggressive.

Learning aggressive concepts may seem a very specific form of learning but the GAM has been expanded almost wholesale to the general learning model (GLM) (Buckley and Anderson 2006). The GLM can be taken as a suitable model for a wide-range of game-based learning (Tang et al. 2009). By analogy, under the GLM, exposure to any concepts through a game leads to cognition around the concept and this leads, again alongside other factors, to learning the concept. While this seems quite natural, it needs careful consideration, particularly going back to the original model where violent games are believed to lead to aggressive behaviours. The key question is: does exposure to in-game concepts really lead to thinking about those concepts?

4.1 Learning Via Priming

To see whether games can lead to learning through mere exposure to games, we need to know what people are thinking about during or immediately after playing games. Psychologists have refined good methods for identifying what people are thinking about through the principle of priming (Sternberg 1999). In the context of learning in games, the priming we are concerned with is conceptual priming (Eysenck and Keane 2010). This occurs whenever a person thinks of a concept. Other concepts related to that concept are made easier to access, that is, they are primed to be used. More concretely, if I refer to concept of "spider" then concepts related to spider like "fly", "web", "hairy legs" and so on are primed.

Priming can be measured in various ways but one simple way is that when shown images of concepts related to a primed concept, people are faster to categorise those images. This is called the image categorisation task (Tipper and Driver 1988). It should be noted that detecting priming is tricky because the ICT is looking to see just a small difference in reaction time to images in an experimental context against a background of all the other stuff rattling around inside our brains. To give an analogy, it is like trying to detect whether there are more hippopotamuses inside one group of lorries over another group of lorries when the only measurement you can make is to weigh each lorry once for a specific instant...And the lorries are all different shapes and sizes...And we have no idea how many hippos could be in each lorry...And the hippos are dancing.

To examine priming through digital games, David Zendle, in his doctorate, conducted a series of experiments where different concepts were represented in different games and then he measured the priming of those concepts in players (Zendle 2016). Of course, games can vary in a huge number of ways including graphics, sounds, controls, challenges, gameplay and so on. Any one of these could have a subtle effect on the concepts in a player's mind. Thus, it was important to take tight experimental control and only manipulate the representation of the concepts in very specific ways.

For example, in one of Zendle's experiments, he had a maze game where players were required to find the exit within a give time limit. To ensure experimental control, there was no exit for players to find but fortunately the time limit was not too long either. To manipulate the concepts, in one version the game was "skinned" to be about a mouse running through a maze, Fig. 1 and in the other version the

Fig. 1 Mouse version of maze game, from Zendle (2016)

game was about a car driving through a city. Each game had otherwise identical gameplay and controls so it was really only the representation of the concepts in the game that was changed.

Priming was measured using the image categorisation task where there were two types of images presented, those related to animals and those related to vehicles. If priming happens in games then people exposed to the mouse version of the game should react more quickly to animal related images than vehicle related images. And conversely, those who played the car version of the game should react more quickly to the vehicle related images. The results are shown on an interaction plot in Fig. 2.

The most obvious difference is the gap between the two lines. People react more slowly to vehicle related images than animal related images but this is not remarkable. Not all concepts are equally salient and so people react less quickly to some concepts than others. What is important in these results is that the lines are not parallel. This reflects a small but significant interaction effect. Further, and more surprisingly, the direction of the interaction is the wrong way for priming: players who played the mouse version of the game reacted more slowly to animal related images than those who played the car version of the game; and players who played the car version of the game reacted more slowly to vehicle related images.

This is the opposite of priming, named negative priming (Tipper 1985), where the concepts in the game are actually reducing the ease with which people respond to related concepts. This strongly suggests that if anything, the representation of

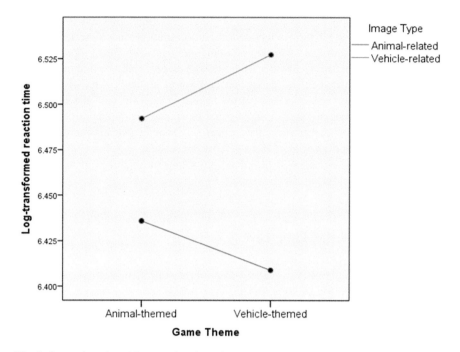

Fig. 2 Interaction plot of log reaction times for the two different types of images and the two different versions of the game, from Zendle (2016)

concepts in the game means that players are less likely to think about them. This strongly undermines the first step in the GLM/GAM of how games lead to learning.

In all fairness, there does need to be some caution. Note the scales on the sides of the graph in Fig. 2. The effects seen are small. This is typical of priming experiments because of all the other things a person might be thinking about (remember the dancing hippos). However, across a series of studies, Zendle found no evidence for positive priming of concepts through games. If anything, negative priming was more likely (Zendle 2016).

4.2   Learning What in Games?

Thus, while people clearly do learn while playing games and simulation games in particular do lead to genuine development of skills, it is not enough to put something in a game and expect players to learn about it. The basic assumptions of the GLM/GAM or any model of learning that relies on mere exposure must be called into question. If learning, and certainly the learning of aggressive behaviours, is attributable to games then there must be other factors that make the learning happen.

Interestingly, many serious games do rely on representing concepts in games as the basis for delivering their message. Indeed, this probably needs to be the first step in any serious game to have a meaningful outcome. However, mere representation seems unlikely to be sufficient for learning to take place. It may be that there needs to be other links between the game and real world situation otherwise players actually suppress the concepts represented in order to get on with the process of playing the game. For instance, it may be that representation in games needs to be tied to the actions of players in games, effectively moving players more to simulating some real world concept. However, we are a long way from having clear models of how this might work let alone proposing mechanisms to bring about effective learning.

5   Where Does the Time Go?

Clearly then, games have the potential to promote well-being in players but is it through flow? And games have the capacity for supporting learning but what exactly are the mechanisms that lead to good learning? And can we achieve these real world outcomes with diminishing the fun that makes people engage with games in the first place? Though it is certainly the possible to have isolated systems that use games well to achieve real world outcomes, without good underpinning theories and concepts of how games work, there is little chance of achieving systematic, reliable development of games that are more than just fun. Good research is needed first to identify the concepts and then the theories that explain how the popularity and engagement in games can be transferred to achieving something outside of fun.

However, there are numerous challenges to manipulating games to elucidate what makes games so engaging. The huge variety of components in games including sound, controllers, graphics, mechanics, feedback and so on all interact in complex ways to build up the experience of play. To illustrate this more concretely, let's take the issue of time perception in digital games.

It is commonly stated that, when people are playing games, they lose a sense of time. This is noted in lots of highly engaging activities and is a constituent of flow (Csikszentmihalyi 2002), but is also seen in other specific measures of player experience including immersion (Jennett et al. 2008) and engagement (Brockmyer et al. 2009). Thus, to understand player experiences, it would seem useful to isolate what influences players' perception of time while they play.

There are various ways in which people are able to perceive time, depending on whether time is being considered on the very small scales of milliseconds or on the lifetime scales of years (Hammond 2012). In the context of playing a game though, the timescales involved are in the order of minutes to hours. On these scales, two particular paradigms of time perception are relevant (Block and Zakay 1997):

- retrospective time perception: where players are asked to estimate the duration of playing after having finished playing and without knowing that they would be asked
- prospective time perception: where players are told they will be asked to estimate the duration of play ahead of playing

Both paradigms have been used extensively in psychology research and though challenging, some robust effects do seem to present themselves in how both paradigms are influenced differently. Thus, in order to examine time perception in digital games, Imran Nordin set about a series of experiments intended to manipulate players' time perception over short periods of play of 5–10 min (Nordin 2014). Despite a wide variety of manipulations across nearly a dozen experiments covering both paradigms, Nordin was unable to find any systematic influence of digital games on time perception even when there were differences in levels of immersion due to the experimental manipulations.

It seems then that, though players report losing track of time when involved in playing games, this does not naturally map to known mechanisms of time perception. Of course, it could simply be that players are actually insensitive to time when playing and that "losing track of time" is in some way figurative rather than a literal experience.

Alena Denisova therefore investigated this explicitly to see whether time could be used to manipulate players' experiences (Denisova and Cairns 2015). Players were set to play a survival/shooting game for a fixed amount of time with the timer counting down during play. However, unbeknownst to the players, in one condition of the experiment, the ticking of the timer was adjusted depending on whether players were performing well or performing badly. Better players got less time, worse players got more time. The result was that the players for whom the timer adapted were more immersed in the game than those players where time ran at a steady rate. Thus, players were sensitive to the passing of time but not one player in the experimental condition noticed the adaptation of the timer.

One explanation then of both these findings is that players are indeed sensitive to time but that our experimental measurement techniques were not sufficiently sensitive to detect them. So Denisova did a further experiment where she told some players that there was an adaptive timer and others that there was just a normal timer. And for half of each of these groups, what they were told was true and for the other half false (Denisova 2017). What she found here was that the players who were told about the adaptation experienced increased immersion *whether or not it was true*. Additionally, as with the previous experiment, players who experienced an adaptive timer were more immersed in the game on top of any effect due to what they were told. This seems to suggest that players perception of time is both reliable and can be fooled at the same time!

After all these studies, we do not know how players' sense of time is altered by playing digital games only that time is relevant to the play experience. The best explanation is that players do experience time in some way but they are not in any position to articulate this clearly as a systematic effect. It may be that there are mechanisms of time perception being used while playing (or indeed engaged in any other task) that are inaccessible to the existing experimental paradigms. What is clear that we are a long way from understanding the experience of time while playing.

Thus, there are considerable challenges in researching how games might have real world outcomes before even turning to developing ways to bring about those outcomes. A concept that is reportedly central to immersive experience of playing games, namely the perception of time, is elusive in experimentation. What then of other concepts that are also said to be crucial to the experience of games, such as challenge and fantasy (Sherry 2015)? And when players' experiences of a game are even sensitive to what they are told, whether or not it is true, then how should researchers instruct participants of studies without biasing their results? This is not an issue of sloppy methodology in games-related research but rather a deep problem about how people respond flexibly and openly to the experiences games offer.

6 Conclusions

The success of games is alluring. Where other interactive systems may have lukewarm or positively chilly receptions from users, games have huge popularity and seem to go from strength to strength. It would be marvellous to draw on the success of games to provide useful or meaningful real world experiences to players merely as a result of playing. Yet, as this chapter hopefully shows, we are really a long way from understanding what players feel and think about when they are playing. This chapter is primarily a call for more research because not only do we not know how games engender player experiences, we do not even really know what those experiences are or even how to research them. For now, the best we can say is that games are just a bit of fun.

References

Anderson CA, Bushman BJ (2001) Effects of violent video games on aggressive behavior, aggressive cognition, aggressive affect, physiological arousal, and prosocial behavior: a meta-analytic review of the scientific literature. Psychol Sci 12(5):353–359

Block RA, Zakay D (1997) Prospective and retrospective duration judgments: a meta-analytic review. Psychon Bull Rev 4(2):184–197

Boniwell I (2012) Positive psychology in a nutshell: the science of happiness, 3rd edn. Open University Press

Brockmyer JH, Fox CM, Curtiss KA, McBroom E, Burkhart KM, Pidruzny JN (2009) The development of the game engagement questionnaire: a measure of engagement in video game-playing. J Exp Soc Psychol 45(4):624–634

Brown E, Cairns P (2004) A grounded investigation of game immersion. In: CHI'04 extended abstracts on human factors in computing systems. ACM, pp 1297–1300

Buckley KE, Anderson CA (2006) A theoretical model of the effects and consequences of playing video games. In: Vorderer P, Bryant J (eds) Playing video games: motives, responses, and consequences, pp 363–378

Cairns P, Cox A, Nordin AI (2014) Immersion in digital games: review of gaming experience research. In: Handbook of digital games. Wiley Online Library, pp 337–361

Chen J (2007) Flow in games (and everything else). Commun ACM 50(4):31–34

Collins E, Cox AL (2014) Switch on to games: can digital games aid post-work recovery? Int J Hum Comput Stud 72(8):654–662

Csikszentmihalyi M (2002) Flow: the classic work on how to achieve happiness. Random House

Denisova A (2017) Adaptive technologies in digital games: the influence of perception of adaptivity on immersion. PhD thesis, University of York. URL http://etheses.whiterose.ac.uk/16463/

Denisova A, Cairns P (2015) Adaptation in digital games: the effect of challenge adjustment on player performance and experience. In: Proceedings of the 2015 annual symposium on computer-human interaction in play. ACM, pp 97–101

Engeser S, Schiepe-Tiska A (2012) Historical lines and an overview of current research on flow. In: Advances in flow research. Springer, pp 1–22

Eysenck MW, Keane MT (2010) Cognitive psychology: a student's handbook, 6th edn. Psychology Press

Gee JP (2004) What video games have to teach us about learning and literacy. Macmillan

Hammond C (2012) Time warped: unlocking the mysteries of time perception. Canongate Books

Jackson SA, Eklund RC (2002) Assessing flow in physical activity: the flow state scale–2 and dispositional flow scale–2. J Sport Exerc Psychol 24(2):133–150

Jennett C, Cox A, Cairns P, Dhoparee S, Epps A, Tijs T (2008) Measuring and defining the experience of immersion in games. Int J Hum Comput Stud 66(9):641–666

Juul J (2013) The art of failure: an essay on the pain of playing video games. MIT Press

Malone TW (1981) Toward a theory of intrinsically motivating instruction. Cognitive Sci 5 (4):333–369

McGonigal J (2011) Reality is broken: why games make us better and how they can change the world. Penguin

Moneta GB (2012) On the measurement and conceptualization of flow. In: Advances in flow research. Springer, pp 23–50

Nordin A (2014) Immersion and players' time perception in digital games. PhD thesis, University of York

Petralito S, Brühlmann F, Iten G, Mekler ED, Opwis K (2017) A good reason to die: how avatar death and high challenges enable positive experiences. In: ACM CHI 2017, pp 5087–5097

Ravaja N, Turpeinen M, Saari T, Puttonen S, Keltikangas-Järvinen L (2008) The psychophysiology of James Bond: phasic emotional responses to violent video game events. Emotion 8(1):114

Ritterfeld U, Cody M, Vorderer P (2009) Introduction. In: Serious games: mechanisms and effects. Routledge, pp 3–9

Ryan RM, Rigby CS, Przybylski A (2006) The motivational pull of video games: a self-determination theory approach. Motiv Emot 30(4):344–360

Sherry JL (2015) Formative research for stem educational games. Z Psychol

Sinnamon S, Moran A, O'Connell M (2012) Flow among musicians: measuring peak experiences of student performers. J Res Music Educ 60(1):6–25

Squire K (2005) Changing the game: what happens when video games enter the classroom. Innovate: J Online Educ 1(6)

Steinemann ST, Mekler ED, Opwis K (2015) Increasing donating behavior through a game for change: the role of interactivity and appreciation. In: Proceedings of the 2015 annual symposium on computer-human interaction in play. ACM, pp 319–329

Sternberg RJ (1999) Cognitive psychology, 2nd edn. Harcourt Brace

Swann C, Keegan RJ, Piggott D, Crust L (2012) A systematic review of the experience, occurrence, and controllability of flow states in elite sport. Psychol Sport Exerc 13(6):807–819

Sweetser P, Wyeth P (2005) Gameflow: a model for evaluating player enjoyment in games. Comput Entertainment 3(3):3

Tang S, Hanneghan M, El Rhalibi A (2009) Introduction to games-based learning. In: Games-based learning advancements for multi-sensory human computer interfaces: techniques and effective practices. IGI Global, pp 1–17

Tipper SP (1985) The negative priming effect: inhibitory priming by ignored objects. Q J Exp Psychol 37(4):571–590

Tipper SP, Driver J (1988) Negative priming between pictures and words in a selective attention task: evidence for semantic processing of ignored stimuli. Mem Cognition 16(1):64–70

Triberti S (2016) This drives me nuts! In: Villani D, Cipresso P, Gaggioli A, Riva G (eds) Psychology and mental health: concepts, methodologies, tools, and applications: concepts, methodologies, tools, and applications. IGI Global, pp 269–292

UKIE (2017) UK videogames factsheet. URL https://ukie.org.uk/research(hashtag)fact (underscore)sheet

Vella K, Johnson D (2012) Flourishing and video games. In: Proceedings of the 8th Australasian conference on interactive entertainment: playing the system. ACM, p 19

Vella K, Johnson D, Hides L (2013) Positively playful: when videogames lead to player wellbeing. In: Proceedings of the first international conference on gameful design, research, and applications. ACM, pp 99–102

Wattanasoontorn V, Boada I, García R, Sbert M (2013) Serious games for health. Entertainment Comput 4(4):231–247

Zendle D (2016) Priming and negative priming in violent video games. PhD thesis, University of York. URL http://etheses.whiterose.ac.uk/16463/

Chapter 4
What Is Pleasure?

Barry Brown and Oskar Juhlin

One of the biggest surprises about modern technology is not how productive it makes us, or how it has revolutionized the workplace, but how *enjoyable* it is. The great success of new technologies—such as social networking, computer graphics, wireless networks—are in how they create pleasure in our lives. People taking a walk in a forest can use a GPS device to track where they are, or while watching a football match use a phone to record a video clip of the game. Entire categories of leisure activities (such as sport and television) depend upon technology for their very existence.

If we are interested in technology, then, it is natural that we should turn to understanding enjoyment, as the chapters in this volume do. Yet enjoyment presents some interesting challenges to serious investigation; by its very nature for much of the time it is itself distinctly *not* serious. However, its importance forces us to be "serious about the non-serious". For this chapter we focus on a fairly basic question in this investigation: *what is enjoyment?*

Even while our focus is on technology and pleasure, it is worth taking some conceptual space to build up an understanding of what pleasure is, and the different forms it takes. Although answering this question might seem straightforward, even trivial, it presents considerable complexity and leads to a host of other questions. For example, can we trust others' reports of enjoyment? Perhaps we only have access to our own enjoyment and cannot trust others' reports? Can we measure enjoyment? Is there only one form of enjoyment, or are there many? How is enjoyment shared with others?

In this chapter we develop our own perspective of pleasure that is both empirically informed but also enables a deeper understanding of some of the many different forms that happiness takes. It lets us engage with all the different activities and practices that seem to be involved in pleasure at different times. Central to our

B. Brown (✉) · O. Juhlin
University of Stockholm, Stockholm, Sweden
e-mail: barry.brown@me.com

© Springer International Publishing AG, part of Springer Nature 2018
M. Blythe and A. Monk (eds.), *Funology 2*,
Human–Computer Interaction Series,
https://doi.org/10.1007/978-3-319-68213-6_4

argument is that pleasure should be understood as a set of skills, activities, expectations, and actions that form enjoyable experiences, rather than as an event that occurs "in the brain" or as a psychological feature of our consciousness. It is not that we are skeptical of understandings of the latter kind, but rather that they give only a partial account of the complex *practical* undertaking that enjoyment is. Doing this enables an empirical study of enjoyment that is about studying the different places where enjoyment takes place, where we find technology and enjoyment working together. In our book (Brown and Juhlin 2015) we go into more detail about this empirical work, as well as giving a wide literature review of contemporary work on pleasure and technology.

So, what *is* enjoyment? There are four parts to our argument. We will start by arguing that pleasure is *worldly*. As we have mentioned our interest is in understanding pleasure as something that is 'in the world' rather than 'in the head'. Second, we argue that pleasure is a *skill*, something both learned and embedded in culture. Many pleasures are 'acquired tastes' where we have to learn how and what is enjoyable about an activity. Third, pleasure is *ordinary and routinely describable*. We mean by this that there is no need for exotic methods; we are all experts on pleasure, since we have lived since birth in a world in which pleasure is ever present. Analyzing pleasure does not rely on philosophy or advanced methods, but rather on returning to what we already understand, see, and do. Lastly, pleasure is something *felt and emotional*. By this we mean that it is not understandable simply as the movement of objects, or as reactions and interactions. We feel pleasure as an immediate part of our lived experience. Our accounts, then, should not shy away from describing how we feel about things and doings.

These four 'foundations' of thinking about pleasure in particular motivate the empirical study of enjoyment, and understanding how technology and enjoyment work together. We conclude the chapter by outlining some of this work and how this enables an interest in the technology behind enjoyable experiences.

1 Pleasure in the Literature

Let us start though by discussing how some of the different academic disciplines have taken to examining pleasure. The two dominant approaches to understanding pleasure come from psychology and economics (Schroeder 2004). While there are differences in the research reported in each discipline, they both share much in how they conceptualize what enjoyment is. Psychology has attempted to measure pleasure and to find its correspondence with brain activity (Berridge and Kringelbach 2008). Meanwhile economics has focused exclusively on *measuring* enjoyment [and presenting curious artefacts like the 'world map of wellbeing' (Layard 2005)]. Lastly, in sociology while enjoyment has been much less of a topic, we do find much more critical perspectives on the role that pleasure plays in everyday life, developing some skepticism to even the possibility of enjoyment in contemporary society (Rehberg 2000).

In terms of theories of enjoyment, and while there is a rich variety in the accounts of enjoyment given in the literature, the "simple view" of enjoyment has come to dominate: that pleasure is a mental event that varies only in intensity and is the opposite of pain. This view has its roots in nineteenth-century thought, crystallized by Jeremy Bentham and other pragmatist philosophers (White 2008) in utilitarianism (the philosophical motivation behind much modern economic thought). Yet the "simple view" leaves many tricky problems of pleasure unexamined. It reduces pleasure to an event in the brain—a mental experience that has a concurrent physical event in the brain. Resorting to brain chemistry to find out what pleasure is, or to stimulation of different parts of the brain as an explanation for what pleasure "really is," presents the problem of failing to explain the great diversity of experiences that are pleasurable in different ways.

This said, the simple view has enabled considerable investigation of the 'organization' of enjoyment worldwide. In economics in particular this has led to debate on this general organization of happiness on a societal level, and particularly on measuring enjoyment across populations and countries (Kahnemann et al. 1998). What government policies might we choose to encourage the enjoyment of different people as measured by nationwide surveys? How have enjoyment and pleasure changed as people have become wealthier? Economists have documented paradoxes in which the decisions we make, either collectively (such as favoring economic growth over equality) or individually (commuting a long distance to work), end up with both individuals and societies in a suboptimal state (Bruni 2007).

In contrast, in sociology, we find talk of "false needs." Drawing on Marx's term "false consciousness," this engages with the ways in which we are led into desiring what we cannot have-a "euphoria in unhappiness," as Marcuse put it (Marcuse 2013). As is sometimes the case in such critical analysis, we find that the job of disentangling *false* from *real* pleasures falls to the academically trained (Vowinckel 2000). We have some skepticism about their expertise in this matter—and this at times falls into the trap of erasing any sort of possibility of pleasure replaced by a narrative of life as corporate control (as one book puts it: "the fear of enjoyment").

Yet this is not to say that the reliance of measurement in economics is without problem. Can we really reduce enjoyment to an easily measurable "mass variable,"—avoiding treating participants' accounts, with a single number all we need to understand the phenomena? While we would distance ourselves from the sort of skepticism toward pleasure shown in sociology, we also find the economic reduction unsatisfying. Pleasure is grossly observable in many of the things that are done in contemporary society—and we are all experts in some way in our own and others' pleasure. We do not need a jury to tell us that we enjoy a pint of ale or that heroin addiction can be problematic. Pleasure is everywhere, whatever its unequal societal distribution.

2 Pleasure: What Is in the Word?

This leads us to consider the importance of what term is being used. What difference is there between pleasure and enjoyment, or between well-being and happiness? These terms are related, yet they highlight different characteristics. Well-being seems more a goal than a state; enjoyment seems to be something that goes with an activity; pleasure seems to be an evaluation. But what about the other terms that we might use? Should we be careful to differentiate and disambiguate all the different terms?

Ryle (2015, 54–55) is thoughtful on how we need not spend too much time differentiating them from one another:

> Just as the wicket-keeper cannot keep wicket unless other cricketers perform their cricketing functions too, so the business of such words as 'enjoy', 'like' and 'pleasure' is ex officio meshed in with the variegated business of countless other words.... The notions of enjoying and disliking are not technical notions. Everybody uses them and there is no coterie of experts, who by dint of their special training or calling, are the ultimate authority on their use. We know as a rule quite well, though without using any special methods of research, whether we enjoyed something or not this morning, and even more generally whether we prefer cricket to football.

We can use other terms, but, as Ryle explains, this does not mean we are confused about how they fit together. As with cricket, we should understand how the different concepts work together. Happiness is about much more than the differences between terms such as 'enjoyment', 'pleasure', and 'well-being', and we shouldn't waste too much time on differentiating the different terms. What we are after is how "the game" itself works. What is involved in being happy, and in working with others on their happiness and our own? We must focus on understanding what enjoyment is in-itself before we spend our energies differentiating the various "enjoyment words."

This said, one important distinction worth making is that between enjoyment as a *disposition* and enjoyment as experienced. That is the difference between talking about how one generally enjoys football, compared to actually enjoying today's game. Bennett and Hacker put it this way:

> One must distinguish between emotional character traits (which are not feelings), emotions as episodic perturbations, and emotions as longer-standing attitudes (both of which are said to be feelings). Many emotion terms have a use as names of character traits: we speak of people as having a compassionate or loving nature, as being of a jealous or envious disposition, or as being irascible, timid or timorous by nature... When we characterize a person as being 'emotional', we do not mean that he feels love or hatred for many people, harboring numerous fears and hopes, etc. Rather, what we have in mind is that he is prone to emotional perturbation, is given to outbursts of feeling, expresses his anger, indignation, love or hate freely, and per- haps to excess, and tends to allow his emotions to affect his judgment deleteriously. (Bennett and Hacker 2003)

3   Pleasure Is Wordly

Let us dig a little deeper then, not by focusing on what word to use, but asking *where* does enjoyment happen? As we have discussed above, perhaps the most central philosophical question about enjoyment has been the question of whether it can be reduced to a mental sensation or not. If pleasure is an inward mental experience and cannot easily be shared, studying enjoyment may be difficult to imagine. If enjoyment is an internal and private object, can only I know about my own pleasure?

The most common view of pleasure in psychology is what has come to be known as the "simple view of pleasure", developed most fully in the work of the utilitarian philosophers. Under this view pleasure is a simple and indescribable feeling in momentary consciousness. Pleasure is present when we feel happy—and at its core is the experience of happiness. In this rendition, pleasure, as a feeling, is something that is triggered by the stimulation of various parts of the brain. In some versions of present-day neuroscience, a part of the brain known as the 'perigenual region of the anterior cingulate cortex' (PGAC) is seen as the site of pleasure in the brain and as the real source of pleasure for human beings (Schroeder 2004). As the argument goes, certain outside stimuli (enjoyable experiences, drugs, etc.) stimulate brain activity, which cascades in the brain until eventually there is stimulation of the PGAC. The result of all this is a feeling of pleasure.

Though there is much that is attractive about a purely "mental" picture of pleasure, there are serious problems with this account. While we would not deny that there are different brain activities that sometimes correspond with reports of pleasure, pleasure as pure sensation lacks any of the complexity that comes from how we talk about, share, describe, and learn about pleasure with others. While the brain has something to do with pleasure (just as the body does), but it is not the complete story.

Turning to Ryle again, he points out that if pleasure is a sensation then it should (in principle) be possible to separate the sensation from the enjoyment:

> If pleasure was correctly classified as a sensation, we should expect it to be possible correspondingly to describe some of these sensations, too, as pleasant, some as neutral, and others as unpleasant, and yet this palpably will not do. The last two would be contradictions.... If I have been enjoying a game, there need not have been something else in progress, additional to the game, which I also disliked or enjoyed, namely some special sensation or feeling engendered in me by the game.... The enjoyment and the seeing of a joke are not in this way two different phenomena, even though other things than jokes are enjoyed and even though some jokes are seen and not enjoyed. Though thunder-claps never do occur in the absence of lightning, we can conceive of them doing so. We cannot conceive of enjoyment occurring on its own. We could not make sense of the statement that someone had been just enjoying, any more than we could of the statement that he had been simply being interested or merely absorbed.... When I enjoy or dislike a conversation, there is not, besides the easily clockable stretches of the conversation, something else, stretches of which might be separately clocked, some continuous or intermittent introspectible

phenomenon which is the agreeableness or disagreeableness of the conversation to me.
I might indeed enjoy the first five minutes and the last ten minutes of the conversation,
detest one intermediate stage of it and not care one way or the other about another stage.
(Ryle 2015)

Ryle is trying to separate talk about "mental sensations" from the variety of different ways we talk about pleasure in our lives. If pleasure were a sensation, an impression, or a mental phenomenon, it would have to be something that goes along with the "actual phenomena of pleasure" we talk about. Yet this leads to nonsense-when we talk about pleasure, we aren't talking about events in the brain, however much they may correspond with events that we call pleasurable.

Moreover, the "simple account of pleasure" seems strangely flat. Surely there is a difference between the pleasure of scratching an itch and that of sipping from a particularly well-made cup of tea. Are we to compare the climbing of Everest with the pleasure of a drug addict's first month of "going cold turkey"? It is not just that pleasure is a complex of different sensations; it is also that, for us, pleasure seems to be something that goes beyond individuals—it is tied up with and embedded in our complex social lives and the judgments we make using language. In part of our culture that supports the distinctions we make in different forms, types, and intensities of pleasure-things not easily mapped onto simple reactions of parts of the brain.

The mistake here then is in misunderstanding pleasure by putting too much weight on the internal experiences. As Nagel (1974) points out, if we reduce different experiences to brain events we lose any way of considering the characteristic differences between how experiences feel. Not all pleasures feel the same. Our shared human life together is based, to a large extent, on understanding those different pleasures and talking about them—you might prefer tea to coffee, but that doesn't mean that your pleasure of drinking tea is the same as my pleasure of drinking coffee. Moreover, some pleasures, such as that of learning to play a musical instrument, can take a long time to come to fruition. The pleasure we get from such a complex long-term undertaking is hardly just a flash in the brain. The sense of achievement and of gradual progress involves a complex sense of self-progression, of forming a sense of who we are and what we can do.

As a contrast, we might look at the public world of managing, talking about, and engaging with pleasure in a public way, one that has developed over thousands of years. There is a rich language we use when describing and talking about pleasure —and we are well tuned for detecting when others are in states of pleasure or its absence. We have a whole library of skills based on differentiating pleasure for others. The language and the tools that we use in thinking about others are often used when we think about ourselves. When we describe how we feel, we use the same language and concepts that others are using—we talk about how our partner feels and about how we feel in the same way. And we use those same concepts to think about ourselves, to train ourselves, and to learn about different feelings and sensations. Our expressions of pleasure are produced from what we learn from others.

4 Pleasure Is a Skill

Let us develop this argument further. If we are to reject the idea that pleasure is simply a mental sensation, what else can we say about it? What form does it take if it is not just in the mind? To answer these questions, we need to analytically step into our shared world of pleasure. This leads us to investigate the acquisition of pleasures from others.

Take what might be thought of as one of the most immediate of pleasures: the consumption of a narcotic drug. Surely here we have pleasure in a particularly physical form: the release and the high. Yet even the stimulus that a drug produces is something that is viewed through and understood in shared life with others. In "*Becoming a marihuana user*", (Becker 1953) draws on interviews with users to describe how becoming a user entails training in a social setting to understand what it is to be high. A marijuana user must first pick up an interest in trying the drug, perhaps from friends who are users. After the first experiment with smoking the drug, there is a period during which the user gets accustomed to doing so in various ways. First, one has to learn a technique for using it effectively. This sort of training is usually done in a group including more expert users. Second, one must learn to recognize the symptoms produced by marijuana and connect them to the smoking. Becker's interviewees testify that they initially had symptoms, such as hunger, that they failed to connect to the drug until fellow users pointed them out. Becker explains:

> I was told during an interview: "As a matter of fact, I've seen a guy who was high out of his mind and didn't know it." I expressed disbelief: "How can that be, man." The interviewee said, "well, it's pretty strange, I'll grant you that, but I've seen it. This guy got on with me, claiming that he'd never got high, one of those guys, and he got completely stoned. And he kept insisting that he wasn't high. So I had to prove to him that he was." (ibid.)

The phenomenon and experience of being high is not necessarily something that is immediately available to a novice smoker, without training or some sort of discussion and explanation of the results. One has to get some tutoring in recognizing what might count as the different effects of the drug, and what could be just the general good feeling of socializing with others. After the effects have been realized, the smoker has to go a little further to become a regular user-he must actually enjoy smoking marijuana. Again it might be argued that surely one immediately and unproblematically enjoys the sensations. But according to Becker's interviewees it is possible to understand the sensations either as enjoyable or as frightening. In order to become a user, a person must therefore learn to enjoy the drug, or at least must come to think that it may produce pleasure in the future. Becker further argues that a person who has learned to use marijuana not only must be able to identify its symptoms but also must believe them to be pleasurable. From this the person can either progress in that "skill," and identify other symptoms and sensations, or revert and lose the abilities. Thus, even experiencing a chemical that is introduced into the body depends on a series of communicative acts "in which others point out new aspects of his experience to him, present him with new interpretations of events, and help him achieve a new conceptualization of his world, without which the new behavior is not possible" (ibid.).

The mere stimulation of the brain and body by a drug is something that must be understood through talking and interacting with others about the effects. It is through interaction with others that the drug becomes pleasurable. This is not to say that a drug couldn't become pleasurable without the presence of other people. We could, of course, use our own previous experiences with others to reflect on something and decide that it was pleasurable. But we might also doubt ourselves.

As with drugs, so with other things that seem to be *automatically* enjoyable. By listening to other people tell jokes, we learn when it is appropriate to laugh. Although the laughing may seem involuntary, it is shaped by the institution of jokes. One must know when to laugh so as not to be a bad joke teller or a poor listener. The jokes other people tell us teach us lessons in what is "funny"—what could be something that you would laugh about. The fine taste of the connoisseur (for example how to differentiate the flavor of a good coffee from the many tastes of bad coffee) is learned not as an individual but in interaction with others. It might seem odd that we need some sort of education to have a particular sensation or experience. We are accustomed to thinking that sensations and experiences are things anyone can have—things granted to us by our being alive rather than by our having "learned" a particular orientation.

Yet it is through our experiences of pleasure and pain with others that sensations achieve their stability and reproducibility. Pleasure becomes "what everyone knows," and we can come to understand how arduous "life projects" might give us pleasure even though they can be difficult and at times unpleasant. Of course, there are some pleasures, perhaps illicit ones, that we will keep to ourselves. Yet in *principle* we could share them with others. Though others might struggle to make sense of them, they could at least try to do so. Moreover, most "private" pleasures aren't strictly private at all. Some such pleasures (shared private pleasures, if you will) are derived from public rejection-pornography is an obvious example. And sexuality is hardly asocial.

What is gained by moving to understanding enjoyment as a public and social phenomenon is a whole new research world-the study of experiences, such as enjoyment, for which we do not have to invent a new language to "extract" what the mental acts are. Humans have been talking about and engaging in those very same experiences for thousands of years, and we need no special tools or apparatus to get at the phenomena. We don't have to put people into expensive brain-scanning tools, nor do we have to spend years in hermit-like contemplation of our inner lives. What is there can be discussed in the very same language used by those who are having those experiences and sharing them with others. The reason we don't need a specialized language is that whatever sense we could ever hope to make of what people do will come from their efforts (alone or in engagement with a researcher) to produce publicly what their pleasures are.

Pleasure in its richest form, then, is something embedded in and produced by the rich cultural and social world we find ourselves in. Understanding pleasure in this way explains some of its curious features. Pleasures can be gone in a second or can extend over years. One can say that something (for example, one's childhood) was pleasurable even though at many points-even most of the time-one may not have been enjoying oneself. The pleasure in something may be revoked if it is later determined to be fake or misunderstood. We can question other people's pleasures; we can even question our own. And, as was mentioned earlier, we can distinguish between pleasures that are dispositions ("I like to do this") and pleasures that are experiences ("I like this").

5 Pleasure Is Ordinary

Questions of criteria lead to our third point—how we can talk about and describe enjoyment. It is important here to consider the role of *experiences*. Focusing on experiences—is the approach that McCarthy and Wright take in their 2004 book *Technology and Experience* (McCarthy and Wright 2004). This serious and successful attempt to engage with enjoyment (along with other experiences), has been influential in the field of design research, particularly for how it has grounded a range of attempts to move beyond descriptions of the properties of artifacts and to understand how it feels to use designed things. The interest in enjoyment here comes from McCarthy and Wright's general interest in examining, and designing for, the experiences that technology gives. Arguing that for purposes of designing things we should refocus our attention on "the felt life," McCarthy and Wright discuss how users experience technology "in action." Rather than optimizing technology for performance, or ease of use, we should focus on experience in the form of our inward feelings.

While McCarthy and Wright enrich the view of experiences, they also take a wrong turn down the path of "philosophicalizing the everyday". This misleads them empirically by making us look for exceptional artistic experiences rather than for the enjoyment we see around us every day. Central to McCarthy and Wright's argument is the concept of the "felt life." This encompasses the emotional quality of technology-our emotional and sensual responses in interaction with technology. They argue that previous theoretical approaches ignore such experiences. Psychological and sociological approaches that have focused on the social and physical circumstances surrounding action and interaction have not paid attention to how different artifacts have roles not only in efficiency or work but also in our felt experiences of technology use. McCarthy and Wright argue that those aspects of social life have been missed as a result of the influences of "beliefs and ideology" (p. 43). Neglecting lived experience, they assert, researchers have focused on what can easily be seen and studied. This leads to a set of questions about exactly what "lived experience" is and how it might be studied.

McCarthy and Wright's argument for the need to focus on *felt life* when designing and conducting research on digital technologies is valuable. It helps us to think about enjoyment as something felt, rather than just something we can measure. It also begins to highlight situations in which feelings affect our attitude toward and our use of technology. Yet at the same time, it struggles conceptually with what are actually everyday parts of our experience and life. For example, the book makes a claim to the value of certain sorts of "authentic" experience. McCarthy and Wright provide two enjoyable "felt life" examples from their own lives: a concert by the jazz musician Courtney Pine and the purchase of a new computer, experiences that they rate as worthy of inquiry. They argue that it is in aesthetic moments such as these-that is, moments in which the role of language is downplayed and sensing is prioritized- that we live fully.

Yet this method starts with somewhat extreme experiences, ones that on the whole have been intellectually validated as "worthy." We see no particular reason why these experiences will be as illuminating. One might as well pick the first few minutes after one wakes up or the feeling of exhilaration one gets when urinating after a long car journey. What we have here are a set of normative judgments—there are certain experiences that intellectuals might consider more valuable, but we see no intellectual justification for specifying 'aesthetic' experiences as those that would unlock the subjective world (or, in our case, let us understand the variety of forms that enjoyment takes).

A related worry is that McCarthy and Wright seem to want to understand felt life not by enhancing and drawing on pre-existing ways of doing that, developed in the course of millions of years of evolution and shared life, but instead they attempt to draw on *philosophical* inquiry. From the start, this approach gets stuck as an intellectual exercise with poor grounding in the felt life. We find little that can direct us toward understanding what is already under our nose—our everyday ways of describing understanding and interpreting experiences, enjoyment included. McCarthy and Wright foreground the need for philosophical discussions about the foundation of salient conceptualizations of the world.

In contrast, we would agree that there is a need for theoretical discussions to focus on experience, that the "turn to practice" has missed out. Yet we are not convinced that this is the time to abandon detailed empirical investigation with a turn to individual introspection guided by philosophy. There is the danger of chasing a new language for feelings when we already have one that works. "We know… without using any specialist methods of research," Ryle writes (1954, p. 55), "whether we enjoyed something or not this morning, and even more generally whether we prefer cricket to football." Something that from a theoretical perspective is difficult or impossible-reading others' minds, is, from a practical point of view, much easier (Watson 1994). Our feelings seem to be accessible to others, as least in the routine sense.

This may not be much of a surprise, since so much of our happiness depends upon the goings on of our shared lives together-without means of working out or assessing others' happiness, at least to some sort of approximation, it would be hard to imagine how our affairs would come to be managed. So what we think of as private is routinely shared with others. This routineness means that we have a rich pre-existing vocabulary on which to rely when seeking to understand pleasure. We are already experts in seeing pleasure in others. This move is the opposite of

McCarthy and Wright's assent to philosophy-we remain in the green rich hills of ordinary pleasure.

6 Pleasure Is Felt

Although we are critical of the above-mentioned moves by McCarthy and Wright, their move to understanding 'felt life' is valuable and productive. This emphasis resists describing enjoyment as something *merely* orderly and methodical—they emphasize that it is also felt, that there are characteristics of the experience, and that we are habitually able to talk about and describe those characteristics. It is not enough to simply describe what people do, where they are, the 'orderliness' of their activity. If we want to understand enjoyment we also have to engage with feeling—the felt life of pleasure.

An example might be illustrative. Family life is one important place where we learn to feel and experience the world, along with its pleasures. In terms of enjoyment and family life, clearly our parents produce the place for not only our first emotional experiences, but also some of the strongest experiences we will ever feel.

Family life has also been an active site of technology research (Harper 2006)—indeed, in our own work (Brown et al. 2007) we have looked at some length at the role that family life plays in technology use. Yet in studying families, technologies, it is all too easy to take the family for granted as an entity-to take the social arrangement or organization of a family as a given. With an eye on the technology, we can lose sight of what is actually going on, and the feelings that are a crucial part of what makes families what they are. In many senses a family is a "work in progress," with at times strenuous work sometimes needed to keep the members together, to keep them in touch, and to maintain a common identity. In short, being in a family relies upon the work of its members to organize, in some recognizable fashion, itself as a social group. And this work is *felt*. One of the most important jobs of the family is to care for each other—to care for others and to be cared for oneself. This could be as simple as a glance seen by another family member, or a joke made in the evening over dinner. These are short but important actions that could have been lost in an interview or an academic study. Our family lives are rich sites of emotions of different sorts, of at times bittersweet feelings and responses. To understand family life, and in turn the technology that is used in those settings we must touch on the feelings that guide us.

Indeed, technology plays an important role in this 'felt life'—keeping in touch, messaging, communicating and sharing. Indeed, technology is often successful in family life not just as a device for providing information, but for how technology can give us reassurances that the fragile institution of the family was holding together, if only for another day.

In our own research we have looked at how different family awareness systems came to be adopted by families to get a sense of what different family members are doing different things and different times. Yet this technology was not about simply communication, but much more importantly for letting family members show their concern for each other. The fears that go hand in hand with parenting were hardly solved, but at least they were soothed a little. Technology provides worries or reassurance about where people were at certain times. This did not radically reform family relations, but rather it worked within the boundaries of existing family practices-to help families do together that makes them a family. A glance at a device that is seen by other family member could speak to those around about as concern, or even worrying or care. By glancing at a device a daughter could show that she was thinking about a younger sibling, or a mother that she was thinking about the well-being of her son. Caring, at least for the families we have worked with, was not a burden, not a problem or a cost, but rather something that brought the families together in shared enjoyment. As Gopnik (2009), when we talk about our children, while we may remark on the cost and the burden of raising them, few of us would seriously describe it as simply a form of labor. Indeed, for many people, their children are one of the best things about their life. The pleasures of family life then—which come of course with all the many burdens also—are clearly felt, and moveover are felt *with* technology.

7   Conclusions

We can summarize this argument with the words of Aristotle: "Happiness depends on ourselves." Addressing the question 'what is pleasure' we have explored a viewpoint on pleasure that makes primary our role in creating, maintaining and understanding what pleasure is.

Our rational in producing this foundation though is not just some conceptual housekeeping. These concepts give a foundation for going out and studying pleasure empirically-going out and looking at the various forms pleasure takes, and why it is organized as it is. Our own bias is for ethnography-a particular set of methods based on participating in and watching different experiences at first hand and recording them in detail. These methods are qualitative, descriptive, and time consuming, but we would argue that they are among the few methods available for getting at the sorts of complexities we find in enjoyable experiences. Doing this sort of empirical work is one way in which we seek to avoid reducing enjoyment to unified categories.

As we turn to technology, we would argue that the story of technology cannot be told without taking seriously the enjoyment that it brings to the world. Here we have tried to sketch out how we might respond conceptually to understanding enjoyment and how we might think of the role it plays in our lives. As we use technology to make ourselves happy, in turn we have made happiness something as diverse as we are.

References

Becker HS (1953) Becoming a Marihuana user. Am J Sociol 59(3):235–242

Bennett MR, Hacker PMS (2003) Philosophical foundations of neuroscience. Blackwell Publishing. http://ingenieria.uao.edu.co/hombreymaquina/revistas/29%202007-2/Resena%201%20%20HyM%2029.pdf

Berridge KC, Kringelbach ML (2008) Affective neuroscience of pleasure: reward in humans and animals. Psychopharmacology 199(3):457–480

Brown B, Juhlin O (2015) Enjoying machines. MIT Press. http://scholar.google.com/scholar?cluster=17785146301824089302&hl=en&oi=scholarr

Brown B, Taylor AS, Izadi S, Sellen A, Jofish'Kaye J, Eardley R (2007) Locating family values: a field trial of the whereabouts clock. Springer

Bruni L (2007) Handbook on the economics of happiness. Edward Elgar Publishing. https://www.google.com/books?hl=en&lr=&id=zCPzDfUlNpwC&oi=fnd&pg=PR4&dq=Handbook+on+the+Economics+ofHappiness&ots=48PRmPKKQZ&sig=207TXWZb1APpSKI1QCTZg0x577g

Gopnik A (2009) The philosophical baby: what children's minds tell us about truth, love & the meaning of life. Random House

Harper R (2006) Inside the smart home. Springer Science & Business Media

Kahnemann D, Diener E, Schwarz N (eds) (1998) Well-being: the foundations of hedonic psychology. Russell Sage Foundation

Layard R (2005) Happiness: lessons from a new science. Allen Lane, London, 3 Mar 2005, 256 pp

Marcuse H (2013) One-dimensional man: studies in the ideology of advanced industrial society. Routledge. https://www.google.com/books?hl=en&lr=&id=eXlTAQAAQBAJ&oi=fnd&pg=PP1&dq=one+dimensional+man&ots=-zvlXy08b0&sig=ZL9rYzSutpxs9nAMXOphyGRQUUE

McCarthy J, Wright P (2004) Technology as experience. The MIT Press. http://www.amazon.fr/exec/obidos/ASIN/0262134470/citeulike04-21

Nagel T (1974) What is it like to be a bat? Philos Rev 83(4):435–450

Rehberg KS (2000) The fear of happiness anthropological motives. J Happiness Stud 1(4):479–500

Ryle G (2015) Dilemmas: the Tarner lectures 1953. Cambridge University Press. https://www.google.com/books?hl=en&lr=&id=95yNCgAAQBAJ&oi=fnd&pg=PR6&dq=ryle+dilemmas&ots=szUnXoQXbH&sig=_yxgmCCL-HOklRrbbO9Dnark-hA

Schroeder T (2004) Three faces of desire. Oxford University Press

Vowinckel G (2000) Happiness in Durkheim's sociological policy of morals. J Happiness Stud 1(4):447–464

Watson Rod (1994) Harvey Sacks's sociology of mind in action. Theory Cult Soc 11(4):169–186

White NP (2008) A brief history of happiness. Wiley. https://www.google.com/books?hl=en&lr=&id=BLEY-Eur_eAC&oi=fnd&pg=PP7&dq=A+Brief+History+of+Happiness&ots=6-eEv75hg6&sig=2MImkDJo_Azez6lv0eyCLOF1aSE

Chapter 5
The (Un)Enjoyable User Experience of Online Dating Systems

Doug Zytko, Sukeshini Grandhi and Quentin Jones

1 Introduction

Online dating systems are used by millions of people around the world to pursue love, sex, friendship, and other goals. Several product features of online dating systems contribute to a seemingly enjoyable and rewarding user experience. For example, the "swiping" mechanism commonly found in many of today's mobile dating apps has been likened to a game (Purvis 2017). Users swipe right to "like" profiles that they find attractive, and swipe left to reject the others. Receiving a match in these apps (i.e. discovering that an attractive user reciprocated a "like") can be an exciting and addictive experience, not unlike winning a trivial amount of cash on a casino's slot machine. Let's pull the lever just one more time, let's view just one more profile.

Despite a barrage of reward mechanisms and gamified features, research into online dating system-use suggests that user experiences with these systems are sometimes anything but enjoyable. Frost and colleagues found that users more often than not "preferred to stay home and watch a movie" than engage in online dating system-use (Frost et al. 2008, p. 54). Several other studies have deepened our knowledge of aspects of online dating system-use that users may find difficult, confusing, anxiety-laden, and stressful (e.g. Ellison et al. 2006; Masden and Edwards 2015; Zytko et al. 2014a), which can contribute to an unenjoyable online dating user experience.

D. Zytko (✉) · Q. Jones
New Jersey Institute of Technology, Newark, NJ, USA
e-mail: daz2@njit.edu

Q. Jones
e-mail: qgjones@acm.org

S. Grandhi
Eastern Connecticut State University, Willimantic, CT, USA
e-mail: grandhis@easternct.edu

© Springer International Publishing AG, part of Springer Nature 2018
M. Blythe and A. Monk (eds.), *Funology 2*,
Human–Computer Interaction Series,
https://doi.org/10.1007/978-3-319-68213-6_5

This chapter frames enjoyment of online dating user experiences through user motivations for online dating system-use. Such motivations commonly revolve around achieving social relationship goals (e.g. committed romantic relationships, casual sex, platonic friendship) (Gatter et al. 2016; Sumter et al. 2017). An advantage that online dating systems provide to users pursuing these goals is an expanded pool of potential partners for one's desired social relationships. We argue in this chapter that user experiences with online dating systems become unenjoyable due to perceived failures and inabilities to reap the benefits of this digitally expanded pool of potential relationship partners. These can include experiences that represent lost or ruined opportunities to attract desirable partners, experiences that instigate doubt regarding one's general appeal to the pool of potential relationship partners, and so on.

The chapter begins by defining and describing prototypical online dating system design, and reviewing motivations for online dating system-use. This understanding of system design and user motivations is then leveraged to dissect dimensions of the online dating user experience that are commonly unenjoyable. The chapter concludes by reflecting on why users may continue their use of online dating systems despite some of their user experiences being unenjoyable.

2   Online Dating System Design

Online dating systems are a type of social matching system, or system designed to recommend people to people for various reasons (Terveen and McDonald 2005). While many online dating systems have the stated aim of recommending users to each other for long-term romantic relationships like marriage (e.g. *eHarmony.com*), systems are increasingly becoming more inclusive of a range of relationship goals—such as short-term dating, casual sex, and platonic friendship—providing system designs that allow users to specify one or more relationship goal to other users.

Online dating systems are accessible as websites and also mobile apps that allow for accurate location aware features (such as "you crossed paths with"). From a system design point of view, today's online dating systems comprise three typical user interface elements regardless of the platform they are accessed on: (1) user discovery interfaces, (2) user profile pages, and (3) private messaging interfaces. *User discovery* is often supported by proprietary algorithms that display system-recommended users on a "matches" page (e.g. *eharmony*), or a "swiping" interface that lets users indicate initial (un)attraction to potential partners recommended with non-proprietary criteria like GPS-based proximity (e.g. *Tinder*). User discovery is also sometimes facilitated with a search function that lets users proactively find others that satisfy objective trait requirements such as for height and age (e.g. *OkCupid*). Users are represented to one another in online dating systems as *profile pages*, which are comprised typically of self-provided information in the forms of profile pictures, dedicated trait fields (e.g. age, height), and open-ended "about me" text fields. *Private messaging interfaces* allow users to engage in dyadic communication through the exchange of asynchronous text-based messages.

3 User Motivations for Using Online Dating Systems

Usage of online dating systems has continually grown through the years, with more than 38% of single Americans having used online dating systems as of 2015 (Smith and Anderson 2015). Uses and Gratifications Theory (U&G) has been applied to study the varying motivations for using online dating systems (Clemens et al. 2015; Timmermans and De Caluwé 2017). Under U&G, the user plays an active role in choosing and using media—sometimes at the expense of other, competing media choices—to satisfy a particular need (Katz et al. 1973). In the context of this paper, we leverage U&G to posit that users actively decide to use an online dating system over other media choices (including other online dating systems) to satisfy particular needs or goals.

While individual motivations or goals for using online dating systems can be numerous (e.g. "to get over an ex" or "to live out a sexual fantasy") (Timmermans and De Caluwé 2017), common motivations and anecdotes of online dating system-use revolve around achieving social relationship goals that require in-person meetings (Couch and Liamputtong 2008; Ellison et al. 2012; Zytko et al. 2014a). Long-term romantic relationships may be the most publicized of these goals. Of couples that married between 2005–2012, more than one third had discovered each other online (Cacioppo et al. 2013), and online dating systems were the most common way that those online couples met. The literature has indicated that use of online dating systems is also motivated by other social relationship goals like finding short-term dating partners, casual sex partners, platonic friendships, travel partners, and local residents who can give advice to travelers (Gatter et al. 2016; Hsiao and Dillahunt 2017; Sumter et al. 2017). In line with the U&G perspective, prior research has indicated that the choice of which particular online dating systems to use (as there are many to choose from) is informed by one's particular relationship goals, such as casual sex (Zytko et al. 2015b).

If motivations for using online dating systems commonly revolve around social relationship goal achievement, then there must be an element of online dating systems that leads users to believe that they can achieve their social relationship goals through online dating system-use. While the designs of online dating systems may vary (e.g. some have proprietary matching algorithms while some do not), the most commonly used online dating systems offer the universal advantage of partner choice and access. Specifically, online dating systems provide users with access to hundreds to thousands of users in their geographic area, and millions of users around the world, whom they can assume are available as potential partners for their social relationship desires. This pool of potential partners, and the air of partner availability implied by their presence in an online dating system, almost certainly trumps the relationship partner pools that users have access to through their social circles, their school or job, and even bars and clubs due to social, geographical, and temporal constraints.

4 What Makes Online Dating User Experiences (Un)Enjoyable?

What do we mean by an unenjoyable or enjoyable user experience? Blythe and Hassenzahl differentiate two distinct forms of enjoyable experiences based on "fun" and "pleasure" [x]. Under this distinction, fun is characteristic of distraction, repetition, and triviality. Pleasure is derived from, among other things, *progression* towards a goal and *anticipation* ("fantasies about activities or objects that are about to happen") (Blythe and Hassenzahl 2005, p. 97). We refer to Blythe and Hassenzahl's conceptualization of pleasure when discussing (un)enjoyable online dating user experiences because common motivations for online dating system-use revolve around social relationship goal achievement and the anticipation or expectation that use of online dating systems can satisfy these goals.

We argue that pleasure derived from online dating system-use stems from the perceived progression towards one's social relationship goals (anticipation of a goal that appears closer to being actualized). Pleasurable user experiences with online dating systems are contingent on the user's perception that the resources they devote to online dating system-use (e.g. time, money, emotional energy) are bringing them closer to achieving their social relationship goals. Conversely, unpleasurable user experiences are ones in which users believe that the resources devoted to online dating system-use are not bringing them closer to achieving their anticipated social relationships. By extension, if an expanded pool of potential relationship partners is what facilitates the initial anticipation of relationship goal achievement through online dating system-use, then unpleasurable user experiences with online dating systems are reflective of perceived failures or the fear of failure to progress towards one's relationship goals in spite of the expanded pool of potential relationship partners.

5 Dissecting Unenjoyable Online Dating User Experiences

In the previous section we argued that unenjoyable online dating user experiences stem from perceived and expected failures to progress towards one's relationship goals in spite of the expanded pool of potential relationship partners facilitated by online dating systems. What particular scenarios of online dating system-use may trigger these perceived failures and thus unenjoyable user experiences? In this section we identify and discuss four such scenarios: conveying one's relationship goals to other users, evaluating user profile pages, crafting one's own profile page, and crafting message content.

5.1 What if the Next One Is Better?: Shopping for Profile Pages

Several scenarios of online dating system-use involve user evaluation: forming impressions of other users as potential relationship partners to determine who is worthy of an in-person meeting. The process of evaluating a given user often begins with the discovery of that user's profile page and the decision to express initial interest in them (e.g. by sending a message or "liking" their profile page).

A user can easily spend hours sifting through an almost-endless sea of profile pages before ever going on an in-person date (Frost et al. 2008). This may appear an advantage to users looking for romance—the ability to search through and select profile pages that appear "just right" can negate awkward blind dates, costly trips to bars and clubs, and the pressure to settle for someone "good enough." However, there is evidence that excessive choice of profile pages can be detrimental to users' romantic relationship goals.

Confronting people with many choices of profile pages can induce choice overload (Iyengar 2010), which may spur them to reject all available choices, or to evaluate choices based on traits that are easiest or fastest to evaluate (González-Vallejo and Moran 2001). Research has demonstrated that as the choice of online dating profiles increases (from 4 to as many as 64), users increasingly adopt a faster evaluation strategy that leverages the traits easiest to evaluate on the profile page like age and height (Lenton and Stewart 2008). They also deviate from their ideal romantic partner preferences (Chiou and Yang 2010; Wu and Chiou 2009; Yang and Chiou 2010), they misremember which traits were listed in particular profile pages (Lenton et al. 2009), and they are less satisfied with profile pages that they express interest in (D'Angelo and Toma 2016).

System design may encourage a hasty and dissatisfying profile page evaluation process. Some online dating systems facilitate profile page discovery with a "browse and search" page, which provides users with search parameters to curate their list of recommended potential partners based on trait fields in profile pages (e.g. "a man over 6'0", between 25 and 32 years old, who does not smoke"). Decisions of which potential partners to select for further evaluation on "browse and search" pages are largely predicated on ideal partner preferences; a phenomenon Heino and colleagues call "relationshopping" (Heino et al. 2010). As they describe (p. 437):

> [...] the ability to filter through thousands of profiles [...] encouraged a shopping mentality, in which participants searched for the perfect match based on discrete characteristics and reduced potential partners to the sum of their parts. Decision making based on these qualities was quite different from offline dating situations in which individuals often get a more holistic impression of the individual, usually taking into account unquantifiable aspects of personality (such as energy level) and interaction (such as chemistry).

"Relationshopping" is detrimental to users' romantic relationship goals for two reasons. One, there is evidence that ideal romantic partner preferences do not predict romantic attraction in-person (Eastwick and Finkel 2008). Two, potential romantic partners that may otherwise be evaluated favorably in later stages of evaluation (e.g. messaging, in-person dates) may go undiscovered or may be prematurely disqualified if their "discrete characteristics" do not resonate with the evaluating user's conscious preferences.

Worse, this "relationshopping" spree of profile page evaluation can become a circular process of dissatisfaction. Finkel et al. (2012) suggest that discovery of multiple potential romantic partners at a time induces an assessment mindset (Kruglanski et al. 2000) in which choices are evaluated against each other (e.g. "is this potential partner more or less attractive than ones that I also just discovered?") rather than solely in regards to one's goal (a locomotive mindset). Finkel and colleagues consider this mindset detrimental to evaluation of potential romantic partners in online dating systems, drawing on romantic compatibility research which demonstrated that romantic relationship partners with assessment mindsets are more critical of their partners and more pessimistic about their relationships (Gagne and Lydon 2001; Kumashiro et al. 2007). Under this line of thinking, an abundance of discovered potential partners may increase online dating system users' tendencies to disqualify a potential partner even if their profile page is considered attractive because another, potentially more attractive user "is a mere mouse-click away" (Finkel et al. 2012, p. 29).

Ultimately, sifting through a sea of profile pages is poised to be an unenjoyable user experience despite the apparent advantages of partner choice. This is because users' perceived progress towards their relationship goals can be stifled by endless thoughts of "what if?" regarding the next discoverable profile page and tendencies to evaluate profile pages using criteria that poorly predicts later attraction.

5.2   Pick Me, Pick Me: Temptations of Deception in Profile Pages

"People have an ongoing interest in how others perceive and evaluate them" (Leary and Kowalski 1990, p. 34).

The previous section highlighted user evaluation as a fundamental theme of online dating system-use. Yet evaluating a user favorably as a potential relationship partner does little to progress one towards their relationship goal if the respective user does not reciprocate interest. As such, users have a vested interest in presenting information about themselves to influence potential partners' attraction to them. This is called self-presentation: the act of managing impressions that other people form about us (Goffman 1978).

The profile page is often the first source of information used for evaluation or impression formation in online dating systems. This means that profile page creation stands to be one of the most vital aspects of online dating system-use for self-presentation. Users experience tension between desires to maximize attractiveness and accurately portray oneself when crafting their profile pages (Ellison et al. 2012;

Hancock et al. 2007). Desires to maximize attractiveness can give way to deception
—deliberately lying about aspects of one's profile page to attract potential partners.
Prior work has shown that users do exaggerate particular traits in their profile pages,
like height, age, and income in text fields dedicated to those traits, and physical
appearance through profile pictures (Hall et al. 2010; Hancock and Toma 2009).
Deception may be the result of users thinking there is no better strategy to attract
potential partners. For example, the lower an online dater's physical attractiveness,
the more likely they are to deceptively present information about their physical
appearance (Toma and Hancock 2010).

What kind of user experience culminates from decisions to deceptively present
in profile pages? After all, deception and lying are generally understood to be
morally wrong, and online daters do not want to be deceived by their potential
partners (Gibbs et al. 2006). Prior work has revealed ways that users rationalize
their deceptive self-presentations in profile pages. Profile pages may be conceptualized as "promises" that self-presented traits in profile pages will not differ too
drastically from in-person impressions (Ellison et al. 2012). Other work has called
this strategic self-presentation—mildly exaggerating traits to the extent that
deception will not be obvious come an in-person date (Hancock et al. 2007). Users
have also justified deceptive self-presentations of malleable traits such as body type
and career through the concept of an "ideal self" (Ellison et al. 2006). This entails
self-presenting traits in the form that a user expects them to be in the future. For
example, a user may exaggerate their job title because they expect to get a pro-
motion or are training for their desired job. Another example from prior work
(Ellison et al. 2006, p. 426): "The only thing I kind of feel bad about is that the
picture I have of myself is a very good picture from maybe five years ago. I've
gained a little bit of weight and I feel kind of bad about that. I'm going to, you
know, lose it again."

The process of profile page creation can be an unenjoyable user experience
because of fears of failure to procure initial attraction in potential partners and then
subsequent rationalization of deceptive self-presentation to avoid this failure. Given
that profile page creation is a requisite step of signing up for an online dating
system, this means initial experiences with online dating system-use are likely to be
unenjoyable. This unenjoyment can mount if users do not receive as much attention
from potential partners as they were expecting, which some users interpret as a sign
that they are universally unattractive to the system's user base (Zytko et al. 2014a).

5.3 Private Messaging: What Do I Say, and When Do I Say It?

Self-presentation does not end with the profile page. Conversation through an
online dating system's messaging interface is an integral step towards a face-to-face
meeting with another user. Messaging interfaces are used not simply to organize

in-person meetings, but to evaluate users beyond their profile page to determine their suitability for an in-person meeting (Zytko et al. 2015a, b). As such, messaging is a stage for further, tailored self-presentation to potential relationship partners.

Messaging interfaces are typically unprompted with no instructions, meaning users can discuss whatever they would like. What should one say in their messages to a user of interest? There is evidence that male users, in particular, experience anxiety over how to craft their messages due to fear of a lack of response and thus failure to attract desirable relationship partners (Zytko et al. 2014a). This may be because male users typically initiate messaging conversations more than women and receive less messages in general than women (Fiore et al. 2010).

Anxiety over what to say in messages can lead to erratic, potentially offensive messaging behavior. Prior work has discussed male users randomly changing their messaging strategies in search of ways to procure the most responses (Zytko et al. 2014b). Such strategies ranged from writing poems to discussing politics to making fun of women's appearance. As one male user reported (Zytko et al. 2014b, p. 6):

> I used to send long paragraphs, but now I send short messages where I try to make fun of the girls. Honestly, I have no idea what's working, I just don't want them to think I'm insecure.

Once male users do procure a response from a female user of interest, the question of when to escalate communication off of the online dating system to the phone or an in-person meeting becomes a new source of anxiety. Female users in a prior study called this a "moment of truth" (Zytko et al. 2014a). They reported discontinuing contact if male users attempted to move communication off of the online dating system too early (thus making them uncomfortable) or too late (thus making them annoyed that the man was not taking a leading role).

In summary, user experiences with deliberating message content can be quite unenjoyable for male users because of potentially ruined/missed opportunities to attract potential partners—the possibility that a user did not respond because of one's choice of message content, and the possibility that they may responded if one chose different message content.

Conversely, receiving message content can be an unenjoyable experience for female users who receive offensive messages that cause them to doubt their value to the potential partner pool or the general quality of male potential partners available to them. As one female user described (Zytko et al. 2014b, p. 2):

> This one guy called me fat and messaged me with a diet plan. I guess he thought it was funny? It made me feel horrible about myself. I didn't log in [to the online dating system] for a couple weeks because of that.

Furthermore, some female users intentionally avoid responding to men to overtly reject them because of prior experiences in which rejected men repeatedly messaged them demanding a second chance or insisting that the female user formed the wrong impression of them (Zytko et al. 2014a). This means unenjoyable user experiences with receiving messages can be quite frequent because male users have little opportunity to learn why their messages do not yield responses and thus how to correct potentially off-putting behavior.

Some users, however, do seek advice from others about how to craft their message content, such as by soliciting free advice from public online communities (Masden and Edwards 2015) or paid advice from online dating coaches (Zytko et al. 2016). These coaches commonly sell advice for crafting messages in the form of prewritten message content that clients can simply copy and paste into their own messages. Online dating coaches do not claim particularly high response rates for their prewritten message content; one coach claimed a response rate of 7–20% (Zytko et al. 2016). While response rates to prewritten message content may seem low, the copy-and-paste strategy enables users to contact hundreds of users very quickly, meaning they can procure at least a few responses with relatively little effort. Of course, the majority of users who do not respond to such copy-and-pasted messages stand to have unenjoyable user experiences due to time wasted reading these messages and doubts over the kind of users they are attracting.

5.4 What Brings You Here?: Expressing and Identifying Relationship Goals for System-Use

The relationship goal perhaps most synonymous with online dating systems is long-term romance (e.g. marriage), but users have adopted online dating systems to pursue a variety of alternative relationship goals (Hsiao and Dillahunt 2017). Prior research has revealed user struggles with conveying and identifying such alternative goals, which can instigate unenjoyable user experiences.

Casual sex is a type of relationship goal that online daters commonly pursue or are open to experiencing (Blackwell et al. 2014; Couch and Liamputtong 2008; Hardy and Lindtner 2017; Zytko et al. 2015a, b). Recent work has indicated that heterosexual men and women sometimes disguise this relationship goal because of the belief that its disclosure may negatively impact their achievement of this goal (Zytko et al. 2015a, b). Work in the psychology domain regarding social stigma around promiscuity supports this user concern (Crawford and Popp 2003). If users are unwilling to elucidate their interest in casual sex, they stand to have difficulty probing for casual sex interest in others. Collectively, this increases the chances of relationship goal misinterpretation, which can culminate in unenjoyable user experiences because of resources (e.g. time, emotional investment) committed to users that are discovered to be incompatible in later online or in-person discourse. A study of online dating coaches revealed that they commonly coach their clients to expect first dates to go poorly in order to avoid the emotional let down of misinterpreted relationship goals and other traits (Zytko et al. 2016). This can be interpreted as a way to temper the anticipated benefits of online dating system-use and thus unenjoyable user experiences when those benefits go unrealized.

Some users have adopted specific strategies for identifying others who are open to casual sex. For example, a subset of male online daters are called "pickup artists" (Zytko et al. 2015a, b). They formulate and sell strategies for how to seduce women for casual sex in online dating and physical world environments. Prior work has indicated that "pickup artists" engage in exhaustive "field research" to develop profile and message content that attracts women open to casual sex (Zytko et al. 2015a, b). This includes a method akin to cold-calling—sending copy-and-pasted message content that implies casual sex interest to hundreds of female users in geographically distant areas to gauge which messages yield the highest response rates. This method enables users to test various system-use strategies without ruining their chances with female users who they are feasibly able to meet. Such a method can trigger unenjoyable user experiences for both message senders and receivers. For one, the repetition of sending mass quantities of messages and recording response rates is likely not an enjoyable user experience for the sender. Indeed, the "pickup artists" who employ the strategy have outsourced the messaging process to avoid a user experience altogether (e.g. developing message automation software and paying others to send messages on their behalf) (Zytko et al. 2015a, b). On the flip side, receiving, reading, and potentially responding to a copy-and-pasted test message costs time that can never culminate into relationship goal achievement.

Aside from romance and sex, users also pursue platonic relationships through online dating system-use (e.g. friendship, travel partners) (Hsiao and Dillahunt 2017). Both same- and opposite-sex users are potential partners for these goals, which effectively doubles the pool of potential partners. However, prior work has indicated that potential platonic partners of the same and opposite sex tend to assume romantic/sexual intent when they are contacted. There is documentation of users with platonic interests having unsuccessful in-person meetings with opposite-sex partners because of assumed romantic intent ("*I think she thought it was a date, but I wasn't attracted to her like that. We never messaged each other again after that*") (Zytko et al. 2014a, p. 60). Other work reported that users stopped contacting same-sex potential partners for platonic connections because of assumptions of homosexual intent by the users they contact (Hsiao and Dillahunt 2017). Ultimately, pursuit of platonic relationship goals can lead to unenjoyable user experiences because of perceived and expected failure to clarify one's relationship goal to the pool of potential partners.

6   Why Do Users Endure Unenjoyable Experiences?

The previous section discussed various scenarios of online dating system-use that can culminate in unenjoyable user experiences, or experiences that spur users to believe that the resources devoted to online dating system-use are not bringing them closer to achieving their anticipated social relationships. This begs the question: why do users not simply discontinue online dating system-use if they perceive a lack of progress towards their relationship goals? There certainly are users who discontinue online dating system-use, either temporarily (Zytko et al. 2014b) or permanently. But users may continue online dating system-use despite frequent unenjoyable user experiences because they are unsure as to why they are not achieving their relationship goals. Could they be using the online dating system "wrong"?

Recall that users solicit advice from others (Masden and Edwards 2015; Zytko et al. 2016) about how to use online dating systems, which suggests that some find themselves at least partially responsible for their lack of relationship goal achievement through online dating system-use. Online dating systems are not like bars, clubs, and other physical-world social settings for relationship partner recruitment where one can observe how others act and how successful others are at soliciting potential relationship partners. Ignorance over other users' behavior and experiences in online dating systems can lead some to assume that unenjoyable user experiences with online dating systems are largely their fault and can be rectified by altering one's system-use strategies. This assumption may be bolstered by system design elements that can make it seem like relationship goal success should be inevitable or just around the corner (e.g. matching algorithms that emphasize supposed romantic compatibility with several other users).

We can leverage Uses & Gratifications Theory (U&G) to pose an alternative explanation for why users continue online dating system-use despite unenjoyable user experiences. U&G was referenced earlier in this chapter to frame unenjoyable user experiences with online dating systems as those that represent failures to progress towards the expected gratifications (i.e. relationship goal achievement) that motivated online dating system-use. Katz and colleagues describe a recursive cycle between gratifications sought and obtained through media-use: needs or desires that are gratified through media-use in turn construct new desires, creating a cycle of sought and obtained gratifications (Katz et al. 1973).

In terms of online dating systems, unenjoyable user experiences may be offset by a recursive cycle of sought and obtained "mini-gratifications" that can re-instill a perception of progress towards one's relationship goals or a general sense of desirability by the system's user base. Several of these small gratifications are emphasized through system design, such as notifications that a user "liked" one's profile page or that multiple users viewed one's profile page, bright icons that highlight a number of unread messages in one's inbox, alerts by the system's matching algorithm of romantic compatibility with newly signed-up users, and so on (see Fig. 1).

If enjoyable user experiences are characteristic of anticipation and progression towards one's relationship goals, online dating system-use can be viewed as a cycle of enjoyable and unenjoyable user experiences that sustain system-use in ways reminiscent of the "carrot and the stick" idiom. Unenjoyable user experiences weaken perceptions of progress towards one's relationship goals, but frequent small gratifications help users endure these experiences and renew anticipation for the "carrot" of relationship goal achievement.

Fig. 1 Small gratifications from online dating system-use such as "matching" with an attractive user in the mobile dating app Tinder can reinstate anticipation of relationship goal achievement and sustain system-use despite other unenjoyable user experiences

7 Conclusion

Several aspects of online dating system design seem primed to facilitate an enjoyable and rewarding user experience. Notifications of user interest like "9 unread messages!" or "20 people viewed your profile!" are commonplace, as are myriad other reward mechanisms. However, research suggests that unenjoyable user experiences with online dating systems can be quite frequent.

In this chapter we framed unenjoyable online dating system experiences through users' relationship goals that commonly motivate system-use. We argued that the large pools of potential relationship partners on offer in online dating systems lead users to believe that relationship goal achievement is possible or probable through online dating system-use, and that unenjoyable user experiences are indicative of perceived or expected failures to exploit this digitally expanded choice of potential partners. Various scenarios of online dating system-use that can culminate in unenjoyable user experiences were reviewed, such as crafting one's profile page and message content to attract potential partners, and detecting alternative relationship goals in other users such as casual sex.

We concluded the chapter by suggesting that users sustain their system-use despite unenjoyable user experiences because of frequent, small gratifications or rewards of system-use that reinvigorate anticipation of relationship goal achievement.

Acknowledgements Some material cited in this chapter (Zytko et al. 2016) is based upon work supported by the National Science Foundation under Grant No. 1422696. Any opinions, findings, and conclusions or recommendations expressed in this material are those of the author(s) and do not necessarily reflect the views of the National Science Foundation.

References

Blackwell C, Birnholtz J, Abbott C (2014) Seeing and being seen: co-situation and impression formation using Grindr, a location-aware gay dating app. New Media Soc 1–20. http://doi.org/10.1177/1461444814521595

Blythe M, Hassenzahl M (2005) The semantics of fun: differentiating enjoyable experiences. In: Funology, Springer, pp 91–100

Cacioppo JT, Cacioppo S, Gonzaga GC, Ogburn EL, Vanderweele TJ (2013) Marital satisfaction and break-ups differ across on-line and off-line meeting venues. Proc Natl Acad Sci 110 (25):10135–10140. https://doi.org/10.1073/pnas.1222447110

Chiou W, Yang M (2010) The moderating role of need for cognition on excessive searching bias: a case of finding romantic partners online. Ann Rev Cyberther Telemed 120–122

Clemens C, Atkin D, Krishnan A (2015) The influence of biological and personality traits on gratifications obtained through online dating websites. Comput Hum Behav 49(August): 120–129. https://doi.org/10.1016/j.chb.2014.12.058

Couch D, Liamputtong P (2008) Online dating and mating: the use of the internet to meet sexual partners. Qual Health Res 18(2):268–279. http://doi.org/18/2/268 [pii]; https://doi.org/10.1177/1049732307312832

Crawford M, Popp D (2003) Sexual double standards: a review and methodological critique of two decades of research. J Sex Res 40(1):13–26

D'Angelo JD, Toma CL (2016) There are plenty of fish in the sea: the effects of choice overload and reversibility on online daters' satisfaction with selected partners. Media Psychol 3269 (May):1–27. https://doi.org/10.1080/15213269.2015.1121827

Eastwick PW, Finkel EJ (2008) Sex differences in mate preferences revisited: do people know what they initially desire in a romantic partner? J Pers Soc Psychol 94(2):245–264. https://doi.org/10.1037/0022-3514.94.2.245

Ellison N, Heino R, Gibbs JL (2006) Managing impressions online: self-presentation processes in the online dating environment. J Comput Mediated Commun 11:415–441. https://doi.org/10.1111/j.1083-6101.2006.00020.x

Ellison NB, Hancock JT, Toma CL (2012) Profile as promise: a framework for conceptualizing veracity in online dating self-presentations. New Media Soc 14(1):45–62. https://doi.org/10.1177/1461444811410395

Finkel EJ, Eastwick PW, Karney BR, Reis HT, Sprecher S (2012) Online dating: a critical analysis from the perspective of psychological science. psychological science in the public interest, vol. 13. http://doi.org/10.1177/1529100612436522

Fiore AT, Taylor LS, Zhong X, Mendelsohn GA, Cheshire C (2010) Who's right and who writes: People, profiles, contacts, and replies in online dating. In: Proceedings of the annual Hawaii international conference on system sciences, IEEE, pp 1–10. http://doi.org/10.1109/HICSS.2010.444

Frost JH, Chance Z, Norton MI, Ariely D (2008) People are experience goods: improving online dating with virtual dates. J Interact Mark 22(1):51–61

Gagne FM, Lydon JE (2001) Mind-set and close relationships: when bias leads to (In) accurate predictions. J Pers Soc Psychol 81(1):85

Gatter K, Hodkinson K, Kolle M (2016) On the differences between Tinder™ versus online dating agencies: questioning a myth. An exploratory study. Cogent Psychol 3(1):1162414. https://doi.org/10.1080/23311908.2016.1162414

Gibbs JL, Ellison NB, Heino RD (2006) Self-presentation in online personals: the role of anticipated future interaction, self-disclosure, and perceived success in internet dating. Commun Res 33(2):152–177. https://doi.org/10.1177/0093650205285368

Goffman E (1978) The presentation of self in everyday life. Harmondsworth

González-Vallejo C, Moran E (2001) The evaluability hypothesis revisited: Joint and separate evaluation preference reversal as a function of attribute importance. Organ Behav Hum Decis Process 86(2):216–233

Hall JA, Park N, Song H, Cody MJ (2010) Strategic misrepresentation in online dating: the effects of gender, self-monitoring, and personality traits. J Soc Pers Relat 27(1):117–135. https://doi.org/10.1177/0265407509349633

Hancock JT, Toma CL (2009) Putting your best face forward: the accuracy of online dating photographs. J Commun 59(2):367–386. https://doi.org/10.1111/j.1460-2466.2009.01420.x

Hancock JT, Toma C, Ellison N (2007) The truth about lying in online dating profiles. In: CHI Proceedings, pp 449–452. http://doi.org/10.1145/1240624.1240697

Hardy J, Lindtner S (2017) Constructing a desiring user: discourse, rurality, and design in location-based social networks. In: Proceedings of the ACM conference on computer-supported cooperative work & social computing—CSCW'17. http://doi.org/10.1145/2998181.2998347

Heino RD, Ellison NB, Gibbs JL (2010) Relationshopping: investigating the market metaphor in online dating. J Soc Pers Relat 27(4):427–447. https://doi.org/10.1177/0265407510361614

Hsiao JC-Y, Dillahunt TR (2017) People-nearby applications. In: Proceedings of the 2017 ACM conference on computer supported cooperative work and social computing—CSCW'17, February, pp 26–40. http://doi.org/10.1145/2998181.2998280

Iyengar S (2010) The art of choosing. Twelve

Katz E, Haas H, Gurevitch M (1973) On the use of the mass media for important things. Am Sociol Rev 164–181

Kruglanski AW, Thompson EP, Higgins ET, Atash MN, Pierro A, Shah JY, Spiegel S (2000) To "do the right thing" or to "just do it": locomotion and assessment as distinct self-regulatory imperatives. J Pers Soc Psychol 79(5):793–815. https://doi.org/10.1037/0022-3514.79.5.793

Kumashiro M, Rusbult CE, Finkenauer C, Stocker SL (2007) To think or to do: the impact of assessment and locomotion orientation on the Michelangelo phenomenon. J Soc Pers Relat 24(4):591–611

Leary MR, Kowalski RM (1990) Impression management: a literature review and two-component model. Psychol Bull 107(1):34–47. http://doi.org/10.1037/0033-2909.107.1.34

Lenton AP, Stewart A (2008) Changing her ways: the number of options and mate-standard strength impact mate choice strategy and satisfaction. Judgm Decis Mak 3(7):501–511. Retrieved from http://citeseerx.ist.psu.edu/viewdoc/download?doi=10.1.1.419.9483&rep=rep1&type=pdf

Lenton AP, Fasolo B, Todd PM (2009) The relationship between number of potential mates and mating skew in humans. Anim Behav 77(1):55–60. https://doi.org/10.1016/j.anbehav.2008.08.025

Masden C, Edwards WK (2015) Understanding the role of community in online dating. In: CHI Proceedings, pp 535–544. http://doi.org/10.1145/2702123.2702417

Purvis J (2017, February 14). Why using Tinder is so satisfying. The Washington Post

Smith A, Anderson M (2015) 5 facts about online dating. Pew Research Center. Retrieved from http://www.pewresearch.org/fact-tank/2016/02/29/5-facts-about-online-dating/

Sumter SR, Vandenbosch L, Ligtenberg L (2017) Love me Tinder: untangling emerging adults' motivations for using the dating application Tinder. Telematics Inform 34(1):67–78. https://doi.org/10.1016/j.tele.2016.04.009

Terveen L, McDonald DW (2005) Social matching: a framework and research agenda. ACM Trans Comput Hum Inter (TOCHI) 12(3):401–434

Timmermans E, De Caluwé E (2017) Development and validation of the Tinder Motives Scale (TMS). Comput Hum Behav 70:341–350. https://doi.org/10.1016/j.chb.2017.01.028

Toma CL, Hancock JT (2010) Looks and lies: the role of physical attractiveness in online dating self-presentation and deception. Commun Res 37(3):335–351. https://doi.org/10.1177/0093650209356437

Wu P-L, Chiou W-B (2009) More options lead to more searching and worse choices in finding partners for romantic relationships online: an experimental study. CyberPsychol Behav 12(3):315–318

Yang M-L, Chiou W-B (2010) Looking online for the best romantic partner reduces decision quality: the moderating role of choice-making strategies. Cyberpsychol Behav Soc Networking 13(2):207–210

Zytko D, Grandhi SA, Jones Q (2014a) Impression management struggles in online dating. In: Proceedings of the 18th international conference on supporting group work, pp 53–62

Zytko D, Grandhi S, Jones Q (2014b) Impression management and formation in online dating systems. In: European conference on information systems (ECIS) 2014, pp 1–10. Retrieved from http://aisel.aisnet.org/ecis2014/proceedings/track12/9/

Zytko D, Freeman G, Grandhi SA, Herring SC, Jones QG (2015a) Enhancing evaluation of potential dates online through paired collaborative activities. In: Proceedings of the 18th ACM conference on computer supported cooperative work & social computing, pp 1849–1859

Zytko D, Grandhi SA, Jones Q (2015b) Frustrations with pursuing casual encounters through online dating. In: Proceedings of the 33rd annual ACM conference extended abstracts on human factors in computing systems, pp 1935–1940

Zytko D, Grandhi SA, Jones Q (2016) The coaches said…What?: analysis of online dating strategies recommended by dating coaches. In: Proceedings of the 19th international conference on supporting group work (GROUP '16). ACM, New York, NY, USA, 385–39

Chapter 6
"My Peaceful Vagina Revolution:" A Theory of a Design

Jeffrey Bardzell and Shaowen Bardzell

> *Therefore turn at least your minds around now;*
> *now they'll begin to speak, and now we'll be pressed to answer.*
> *Hymenaeus Hymen, come! O Hymen Hymenaeus!*
> —Catullus *Carmen* 62

An interesting feature of online media meta-review sites such as Rotten Tomatoes (a meta-review site for movies) is that they feature two independent rating systems: critic and audience scores. It would have been quite easy for a site like Rotten Tomatoes to have developed an algorithm to blend these scores into one. But they did not, implying the difference between the scores is meaningful, and consideration of a few cases bears that out. When the two meta-scores are in agreement, one tends to infer that the two scores are validating one another. For example, Kieslowski's *Three Colors: Blue* has a critic score of 100 and an audience score of 93%, so site visitors probably see the two scores as agreeing that it is an excellent film.

When the two meta-scores disagree, however, something interesting happens. Site visitors are not confused; they do not try to somehow average the two scores together; and they do not simply pick a side and dismiss the other. Instead they most likely draw inferences about the meanings of that difference. Figure 1 shows two such disagreements. For *Sex and the City*, the Tomatometer (i.e., critics' score) was 49% shown with the "Rotten" badge, while the audience score was a fairly high 77%. In contrast, *Summer Hours* has an elite 93% Tomatometer score, with the site's highest badge, "Certified Fresh," and yet a merely solid 69% audience score. We suspect most readers would interpret the differing scores in the first example as suggesting that *Sex and the City* is relatively forgettable as a work of cinema, but nonetheless probably a fun, feel-good movie while it lasts, especially for fans of the TV series on which it was based. And we suspect that many readers would interpret the differing scores in the second example as implying that *Summer Hours* is an aesthetically ambitious, yet slow-paced art film, possibly with a tragic ending, a film

J. Bardzell (✉) · S. Bardzell
Informatics, Computing, and Engineering, Indiana University, Bloomington, USA
e-mail: jbardzel@indiana.edu

© Springer International Publishing AG, part of Springer Nature 2018
M. Blythe and A. Monk (eds.), *Funology 2*,
Human–Computer Interaction Series,
https://doi.org/10.1007/978-3-319-68213-6_6

Fig. 1 When critics and audiences disagree on films

that rewards careful viewing, but which is probably not a feel-good romp. And if a Rotten Tomatoes visitor that evening happens to be in the mood for something more aesthetically challenging, she will probably pick *Summer Hours*, and likewise if she just wants a feel-good stress reliever, she will probably pick *Sex and the City*.

This example suggests that when critics versus audiences evaluate films, they are up to two different activities. Audiences are reporting more or less directly on their own personal experiences. Critics are doing something else, offering some kind of aesthetic evaluation, one that goes beyond the immediate experience of watching— though what, exactly, they are doing is a topic we will come back to. This example also suggests that visitors to Rotten Tomatoes intuitively understand this difference, and if they might not be able to articulate it with academic precision, nonetheless it is obvious to them that the scores should be different and should be interpreted using different criteria.

1   HCI and Criticism

What is at stake in a question like this is the role of viewer response versus a more professionalized interpretative and critical practice to help us understand cultural products, be they films or interaction designs. Yet in the field of human-computer interaction—a primary audience for *Funology* and where we co-authors usually situate our work—a skepticism towards subjective reasoning is still oft-asserted, even after nearly two decades of scholarship within the field emphasizing the roles of interpretation in understanding user experiences and design methods. And if HCI had its own proprietary meta-scoring system, there is no doubt HCI researchers would tend to apply it to the user score. Indeed, the following combination of HCI commitments tends to prioritize user perspectives: prevailing notions of "user-centered design," the methodological centrality of the deployment and user study, and the scientific commitment to objectivity. And in fact a whole tradition of user experience and aesthetic interaction research (e.g., Tractinsky 2012; Hartmann et al. 2008; Law et al. 2009; Hassenzahl 2010) is carried out only with reference to users, as if the researchers themselves had no particular aesthetic opinions or even a modicum of taste themselves. (Of course, they do have aesthetic opinions and tastes —they just suppress them insofar as they act in researchers.)

There seems to be an opportunity for HCI to ramp up its critical practice to complement what we are able to learn from users. That leads to the question of what a critic brings to an interpretation or evaluation than an ordinary citizen viewer/user typically does not. We argue in this chapter that the kinds of critical evaluations offered by critics cover a much wider range of phenomena, and take far more into account, than typical user evaluations. A serious aesthetic critic will read a work with reference to such matters as its location in and contribution to its medium/ genre, the artist(s)'s prior work, the narrative and emotional structures of the work, its pacing and structural economy, its stated and implied politics, the depth and subtlety of its handling of people and ideas, its reception upon release, and similar factors. As for outcomes, criticism typically contributes a very different kind of theory: critics often offer a theory of an individual work. As philosopher of art and art critic Danto (1983) puts it, "To interpret a work is to offer a theory as to what the work is about, what its subject is" (p. 119). In other words, instead of contributing towards a theory of modernity, a critic of T. S. Eliot will often instead produce a theory of *The Waste Land*. Applied to HCI, this line of reasoning would suggest room for thoroughly worked out theories of individual interaction designs.

2   Design Artifacts as Theory/ies

We need theories of works because in both HCI (e.g., Carroll and Kellogg 1989) and design research (e.g., Archer 1995; Frayling 1994), the idea that design artifacts embody theories is prevalent. But as is well known, as a community we struggle to make good use of theories embodied in artifacts. There are many possible approaches to dealing with this issue. For example, Gaver and Bowers (2012) propose "annotated portfolios" as a new discursive mechanism in HCI to fore- ground curated collections of artifacts that are annotated with explanations. This is promising in a number of ways, but it also has two key drawbacks: one is that the annotations are produced by the designers themselves, with no mechanism for anyone else to participate in the sense-making and annotating (Bardzell et al. 2015); and second, an "annotation" is not as robust as a full-blown critical essay, at least as we attempted to define the nature and roles of the essay in HCI (Bardzell and Bardzell 2015).

We propose that HCI should more seriously consider existing scholarly practices that also are organized around the knowledge, theories, values, and concepts embedded in artifacts: film, literature, art history, cultural studies, and so on. What we tend to see in such practices is an aggregation of knowledge that does not so much unfold through generalized theory, as it does through the ongoing con- struction of a collection of exemplars accompanied by critical works theorizing about what those works mean. However controversial the literary canon is, it remains at the heart of literary practice: it is what is taught in literature classrooms, what is written about in literary scholarship, and key to how literature continues to be legitimated in university budgets and state curricular mandates.

Art history, film, and much of design are no different. Consider any book on architecture used in undergraduate design studio education, for example, Ching's (2014) *Architecture: Form, Space, and Order*. This academic book has a higher percentage of image-based content than text-based. It is filled with exemplars, and the exemplars in this case are organized into formal categories: form and space, organization, circulation, proportion and scale. Other books from fashion design to product design on first glance look similar, though their organizational principles might vary to include era, style, geographic region, materials, social class, level of difficulty) as well as catalogues of exemplary elements (e.g., an introduction to fashion will survey different forms of closure, such as buttons, zippers, Velcro, laces, etc.). This collection becomes a living repository of knowledge used for creative inspiration, pedagogy, the construction of elements and categories, and ongoing scholarly debate.

The collection of exemplars—not an organized collection of abstract and verbally articulable theories—comes to organize the knowledge in the discipline. And of course, individual practitioners both master the collection in general and also specialize in a certain area of it. That specialization often entails a more intimate and insightful mastery of a smaller set of exemplars, which the individual uses to gain creative and/or analytical access into new exemplars and/or problem spaces.

It is not hard to see in such a description room for major design concepts, such as Christopher Alexander et al.'s "patterns" (1977). Practitioner's understandings of these are not mediated by a larger abstract theory (along the lines of, say, the theory of evolution, plate tectonics, or big bang theory); the exemplars *are* the theory: in Christopher Alexander's idiom, those patterns collectively form a "language." The more personalized specialization is captured by Schön's (1983) notion of a "repertoire"—a personalized collection of exemplars that an individual designer turns to again and again to help frame and solve similar problems. All of this comes together in the designer's act of problem framing, which entails the construction of one or more ways of viewing a problem space, by proposing relationships between particulars and broader themes (Schön 1983; Dorst 2015). Without going into a full analysis, such a description resembles the practice of constructing a *theory of a work*, as described above (Danto 1983). In short, there exists a robust scholarly practice—interpretation—that is key to many professional and scholarly disciplines, including design. And we believe that HCI has too often outsourced this practice to users—users that are certainly qualified to characterize their perceptions and experiences, but who do not do critique, because they do not offer a theory of the work.

In what follows, we offer an interpretation of a work of design—the Formoonsa Cup—which is a menstrual cup designed in Taiwan. The interpretation—that is, the theory of the work—that we will develop offers an organization structure in which a wide array of highly diverse details—details, for example, like the material out of which it is made, or a political value expressed at the time of its creation, or a disapproving statement by a prospective buyer—come to make sense together, in what a humanist would call an interpretation, Danto might call a theory of the work, and a designer might call a "frame."

3   A Theory of the Formoonsa Cup

The Formoonsa Cup is a menstrual cup inserted in the vagina during a woman's period. Like other menstrual cups, it catches the menses as it is discharged by the uterus, offering a similar function to tampons and pads. Menstrual cups are made of a durable material—rubber and latex in the past, while silicone is most common today—and they can be washed and reused for many years. The menstrual cup was first patented in the United States in 1932, and the first commercially successful cup appeared in the United States in 1987.

The Formoonsa Cup came on the market in Taiwan in 2017. Its name is a play on words: "Formosa" is the Dutch word for "beautiful island" and it was their name for the island when they colonized it in the seventeenth century; "moon" of course connotes (among other things) menstruation. Its Chinese name is also a play on words, though it is somewhat difficult to translate: 月釀杯. The name does not use the characters for "moon" or "period" ordinarily used; instead, it substitutes in a character that sounds like (almost rhymes with) moonlight, but suggests a notion of intentional fermentation (e.g., that of winemaking)—notions of cultivation, or brewing—as if the period is a precious and carefully crafted fluid. The name itself raises several themes that we will develop in what follows. In our interpretation, however, what makes the diverse details of the Formoonsa Cup "hang together" is the product's political objective to reframe and reclaim menstruation as something to look forward to, and more broadly to affirm the value of loving oneself as part of a biological process and often debilitating experience that is culturally construed as repugnant.

Formoonsa Cup is an important design, therefore, not because it is functionally superior to Western menstrual cups, but because it was an ambitious and hopeful design project. By the time it had become a mainstream consumer product in Taiwan, it had raised unquestioned misogynistic cultural tendencies to the public's awareness in a pointed way, and that in turn set in motion democratic processes leading to meaningful social change. The misogynistic cultural tendency it challenged was a taboo against putting anything into one's own vagina, a taboo based on an often unspoken ideal to treat the vaginal cavity as a protected space belonging to one's future husband, including but not limited to a fetishization of hymen intactness and anxieties about vaginas being too loose. The most literal change that Formoonsa Cup led to was the legalization of menstrual cup sales in Taiwan beginning in May 2017. Menstrual cups had not been legalized in part because they were categorized in a stringently regulated category of medical devices, making regulatory changes unusually difficult. Thus, Formoonsa Cup's contributions to this legal change through democratic procedures constituted a remarkable political achievement: "my peaceful vagina revolution" to borrow designer Vanessa Tsen's oft-repeated phrase.

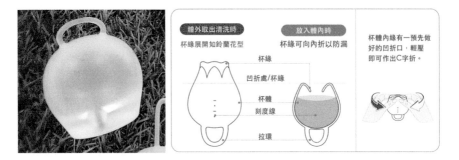

Fig. 2 Photo of the Formoonsa Cup (left) and instructions on its use (right). *Image source* http://www.formoonsacup.com/

3.1 The Formoonsa Cup

Heralded as "The First Cup for Taiwanese Women," the Formoonsa Cup (Fig. 2) is made with medical-grade soft silicone. Unlike pads and tampons that have to be discarded after each use, each cup, like most other menstrual cups, lasts 5–10 years and can be re-used. To use the cup, one squats or sits on the toilet, folds the cup using clean hands, parts the labia, and then inserts it into the vagina. After rotating it a few times to ensure it is firmly in place in the internal walls of the vagina, the cup will then unfold inside. After 4–8 h of use, it can be removed and emptied by gently squeezing the bottom of the cup to release and slide the cup out, again in the sitting position. For each subsequent use during the menstrual period, it needs to be rinsed with water. Between periods, the cup needs to be sanitized using lightly boiled water for 5–7 min. Although functionally the Formoonsa Cup works like other menstrual cups in the market, it is distinctive in many other ways, partly by identifying itself as a menstrual cup "for Taiwanese Women," a claim that we explicate below.

In 2015, Vanessa Tsen came up with the idea for the Formoonsa Cup and began to seek investor funding. She was denied support from the company she worked for, whose CEO happened to be her father, so the thirty year old inventor went to Taiwan's Kickstarter-like crowd-funding platform, Backer-Founder. Within three days she had over US$100,000 in pledges, eventually closing the campaign with US$320,000 raised from 6,361 supporters to manufacture 10,000 cups, more than 300% of her original target.

Not surprisingly, Formoonsa Cup and Vanessa[1] earned a lot of press attention. However, the project was not celebrated as an example of Taiwanese innovation in the new shared economy. Instead, the press focused on its politics, and in particular

---

[1]We note that the founder uses only her first name, "Vanessa", as her professional name. Even when she writes her name in Chinese, instead of using her real name, she strings together Chinese characters that sound like the English "Vanessa." Respecting her preference, we refer to her as Vanessa throughout the chapter.

the questions it raised about women's rights. That Formoonsa Cup could have been featured in both types of stories but was only featured in one of them itself is interesting, and we believe itself a political act. At any rate, raising the funds turned out to be one of the easiest challenges for Vanessa to overcome. What lay ahead for Vanessa and her supporters was a long journey to negotiate with state, legal, medical, and ideological power over the control of women's bodies. This struggle prompted debates, new understandings of, and new legal and medical norms about Taiwanese womanhood.

3.2 Seeing like a Hymen

A major obstacle to any kind of menstrual cup in Taiwan are social taboos about crossing the boundary into the vagina, a taboo that often focalizes on the care and maintenance of the hymen. Most Taiwanese women have never had anything in their vaginas aside from their husbands' penises, nor have they been taught that it is okay to do so. Even as recently as the late 1990s, only 2.1% of Taiwanese women used tampons. Like tampons, menstruation cups are considered by many Taiwanese to be 異物 (a "foreign body," with derogatory connotations). An intact hymen is considered to be a woman's greatest gift to her husband. This cultural more is reflected in contemporary law: prior to 2009 tampon packages were required to carry warnings: "unmarried women should use it with caution," and "consult your GYN before using the product."

As a result, it is widely considered an indecent and/or unethical behavior to put a "foreign body" in there. Although society has begun to change, some of these values die hard. Resistance to change can be seen in Formoonsa Cup user forums and as well as sections of ptt.cc (the largest bulletin board system [BBS] in Taiwan) devoted to the topic of women's health. In such places, one still finds resistance from mothers who feel strongly about safeguarding their daughters' hymens for their future husbands. For example, Vanessa quotes a conversation she had with a mother about the product, *"Why are you promoting products like this? Why do you ask my daughter to put something like Formoonsa Cup in her body?"* (Vanessa 2016). And it's not just the mothers; some husbands also do not support the use of Formoonsa Cup. Vanessa describes her experience at the booth at an expo promoting the Formoonsa Cup: *"I saw a couple walking by the Formoonsa Cup booth a couple of times. The wife showed a lot of interest in Formoonsa Cup and wanted to learn more, but the husband told her 'Don't use something like this' and dragged her away"* (Vanessa 2016). These quotes suggest that the primary caretakers of women's bodies are first their mothers, and then they are transferred to husbands after marriage.

The preoccupation with the hymen is thus not medical, but sociological. The hymen is a signifier of virginity and its accompanying notions of purity, worthiness, and marital eligibility. Many Taiwanese women are taught to be reluctant to use either tampons or menstruation cups in fear of losing their virginity. But as one PTT poster writes,

> Let's return to the question 'whether a virgin can use Formoonsa Cup.' This really should be a question one asks oneself as opposed to turning to others for an answer. If you are really that crazy about the hymen, you should not even ride a bike, because bike-riding is likely to tear it.....As to the question whether the use of Formoonsa Cup might bring discomfort, well, it's your own body, so ask yourself whether it is okay to use it. Who cares about whether others like it or not? We should disregard those chauvinists who insist that it's not comfortable to use FC. They don't have vaginas.... [N]obody can deprive you of your right to choose for your body, disguised in the 'I am only trying to protect you' rhetoric. (https://tinyurl.com/yaawpofb)

The symbolic value of the hymen was so great that it was worth denying women choices about menstrual products: it has been illegal to import, sell, and purchase menstruation cups in Taiwan—a law that only changed in May 2017 thanks largely to the Formoonsa Cup. With menstrual cups illegal and tampons packaged with government warning labels, women have little choice but to use pads. But pads have their own disadvantages. One is forced to get by with bloody undergarments, which introduce various forms of restriction (e.g., no swimming, sex, or ballet) and body anxiety/shame (e.g., did I stain my clothes or the chair? Can others see or smell my period?).

Restated from a Foucauldian perspective (1988, 1995), the body is a surface where rules and regulations, hierarchies, and socio-cultural traditions are inscribed and enforced. The physical nature of the hymen renders the female body subject to examination in a way that has no equivalence in men (i.e., there is no visible change in the penis when a man has sex the first time), and the semiotic nature of the hymen (as a signifier of virginity, purity, and worthiness) renders the female subject to serious forms of social, political, legal, and medical control. But as Foucault is careful to stress, the operations of power are not always so obvious as in the case when the military cracks down on a protest. The loving mother who admonishes her daughter to protect her marital value by caring for her hymen is a social subject who is passing her subject position onto to the next generation; there are no tanks, but the compliant daughter is now a subject of the same regime. Making room for a product like the Formoonsa Cup will require new subject positions.

And indeed, attitudes in Taiwan are changing. As outspoken Taiwanese activist and feminist Wen-Fei Shih wrote on Facebook,

> I'd like to share something I witnessed when I was behind the Formoonsa Cup booth today. A mother began by asking me whether Formoonsa Cup might tear the hymen. She wanted her daughter to wait till after she gets married and has obtained the approval from the husband first before using Formoonsa Cup. .... I find it mindboggling that the mother believes her daughter's vagina belongs to her future husband. It's also incomprehensible that parents believe they have the right to decide whether their daughter can insert something into her vagina. I don't think I am not a virgin because I bleed from using Formoonsa Cup. I am a virgin because I have never had sexual intercourse with a man.... Why is it that women and girls in the West can put different things in their vaginas but we cannot? (Wen-Fei Shih, from her FB post: https://tinyurl.com/ybxyfcf9)

Her post reflects changing views from younger generations. The regime of the hymen was coming to an end.

3.3   Seeing like a Taiwanese Menstrual Cup

The Formoonsa Cup mounted a direct attack on both the laws and the ideologies underlying them. Combining political activism, concrete design choices, an education platform, and a strong communications campaign, Vanessa and her Formoonsa Cup would offer an alternative vision, change the law, and create alternatives for women—not just hygienic alternatives, but also alternative relations to the self and others.

We begin with Vanessa's political activism. Enlisting the help of Taiwanese women's rights activist Wen-Fei Shih, Vanessa petitioned for a new law to legalize the sale of menstruation cups, using join.gov.tw in 2016. Join.gov.tw is an online open government platform in Taiwan that facilitates direct public engagement with different branches and offices in the central government in Taiwan. The platform offers systems and procedures by which citizens can participate in policymaking, demand transparency and accountability from government, and collaboratively leverage open data. Vanessa's petition was seconded by 6150 people, prompting the relevant government branches to respond (in this case the Ministry of Health and Welfare in the Executive Yuan). In spring of 2017, Taiwan's Food and Drug Administration amended the law, and Taiwanese women can now purchase menstruation cups legally.

With the funding secured and changes to health regulations underway, Vanessa set out to design the first Taiwanese menstrual cup. There are two important points here. One is that she set out to design a high quality menstrual cup, choosing medical grade materials and so forth. The other is that she specifies that it will be "Taiwanese." The latter move is unusual. Vendors of menstrual products do not typically frame their products in national terms: "this is a tampon for Canadian women" or "these pads are made for Tunisian women." Most of us intuitively think of hygiene products in ergonomic terms: this is for larger versus smaller women, or for heavy days versus light days, but not: Taiwanese versus Korean women.

But the choice to make the menstrual cup Taiwanese is a recognition that Formoonsa Cup is not addressing itself primarily to an ergonomic problem, but rather to a sociocultural one. And although many cultures fetishizing the hymen as a signifier of virginity and therefore purity, worthiness, and femininity, each region or culture plays that out in its own concrete ways. Vanessa would scope her inquiry to Taiwan. Of course, part of it is ergonomic: women in Taiwan tend to be smaller than Western women, and so the sizing of Western menstrual cups might not work for Taiwanese women. To determine this, Vanessa collaborated with a group of over 60 menstrual cup users in Taiwan (one of them also happened to be an industrial designer).

Trying to find the right cultural connotations to resonate with Taiwanese women, Vanessa sought to create a cute design. She designed the Formoonsa Cup in the form of a lily of the valley (Fig. 2), because the flower has gentle, harmless, and appealing characteristics. A similar design principle was that the cup should be "adorable" and not "intimidating" because as Vanessa put it "Taiwanese girls are cute and are drawn to cute things." Here, she is encouraging women to identify with the Formoonsa Cup, rather than seeing it as a "foreign body" (異物).

Helping women to relate to menstruation in a new way required a learning and education campaign, one supported by design choices and also figures, diagrams, and public appearances. Two design choices reflect the educational agenda of the Formoonsa Cup. First is that the cup itself is made of translucent material, rather than being dyed a solid pink or purple. The translucent material allows a woman to engage with, rather than merely dispose of, her menses. The second choice was to put measuring marks on the side of the cup (visible in Fig. 2): encouraging women to measure their menses. Vanessa explains this choice as follows:

> My hope for Formoonsa Cup is to provide women in Taiwan with a meaningful way to know their own bodies. With the measurement marks on the Formoonsa Cup, you can accurately tell your friend, 'hey, how much did you bleed in this menstruation cycle? I got 78 cc. How about you?' 'I got 125 cc.'... Gone are statements such as 'I bleed too much this time' or 'I hardly bled last time.' Formoonsa Cup as a project promises Taiwanese women a more accurate way to discuss heavy vs. light bleeding during menstruation, and such knowledge cannot be obtained from using pads or tampons. ... Additionally, with Formoonsa Cup, you can actually tell yourself during menstruation that 'you have been taking good care of yourself'... as a woman, loving oneself means you are taking good care of yourself. Only when you do that you can then turn around and love and care for others around you. (Vanessa 2016)

Even as Vanessa tried to prepare women to switch to a menstrual cup, there is still the physical discomfort of putting a foreign object in the body. Beyond detailed anatomical diagrams and descriptions, Formoonsa Cup is also available as a "learning cup"—a tiny cup that is not functional as a menstrual cup but which is designed to help women practice putting it in and taking it out. The learning cup is the one on the left in Fig. 3.

None of this matters, however, if women are construed as hymen caregivers. For advocates of the Formoonsa Cup, this means confronting attitudes toward the hymen directly. As one bulletin board poster put it,

> Who actually cares about the hymen? We do? Or we are worried about the fact that it is important for men? [...] A more important issue is for us to critically reflect what that piece of membrane actually means for us now.... If you for whatever reason want to keep yourself intact, it is your choice, and it is not for me to say otherwise. But if you are worried that your future husband might be upset because you do not have an intact hymen and that he would only marry you if you are a virgin, then please tell him to travel back in time to the century he actually belongs. Clearly 2017 is too dangerous and unsuitable for him. (https://tinyurl.com/ydxdjngb)

The assertion of a new generation with new values also implies a whole new set of subject positions. In other words, not only has the woman changed, when it

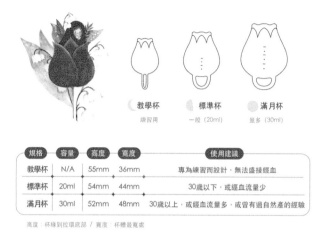

規格	容量	高度	寬度	使用建議
教學杯	N/A	55mm	36mm	專為練習而設計，無法盛接經血
標準杯	20ml	54mm	44mm	30歲以下，或經血流量少
滿月杯	30ml	52mm	48mm	30歲以上，或經血流量多，或曾有過自然產的經驗

高度：杯緣到拉環底部 / 寬度：杯體最寬處

Fig. 3 The "learning cup" and two usable sizes. *Image source* http://www.formoonsacup.com/product

comes to how she cares for her body, but also what constitutes an acceptable husband has changed. Such views are increasingly seen on social media, online boards, and streaming political programs, such as "阿苗帶風向" ("Trendsetting by Miao") hosted by Miao PoYa, a lawyer turned activist.

In one episode of her show, host Miao invited Vanessa to the show to answer the question, "Does Formoonsa Cup destroy hymens?" Vanessa then demonstrated 8 different types of hymen using acrylic models. Afterwards, Miao observed:

> the hymen is only a thin layer of membrane. The act of sexual intercourse does not necessarily destroy the hymen or make the hymen disappear…. The fact that the hymen is intact does not mean that one never has sex…. After all, the membrane is very fragile… just Google it… our bodies have many different kinds of membranes and they don't stay with us forever. Even if you never had sex in your entire life, it does not guarantee that you will forever have an intact hymen…. The fact that you have the hymen only means you have it, it doesn't mean anything else…. (https://tinyurl.com/yatqhl6z)

This talk asserts a merely anatomical status of the hymen, attempting to sever it from questions of virginity and all the meanings attached to that. The changing perceptions about the hymen are happening just in time: an article published in Taiwan Journal of Public Health in 2015 on the use of tampons in the country, the authors surveyed 363 female college students and learned that 49.7% had sexual experiences, and 35.3% have used tampons before. These college students also claimed that the potential risk of tearing their hymens was not a concern (Chang et al. 2015). Instead, lifestyle choices and personal hygiene (e.g., the desire to keep up the exercise regimes, the anxiety over leakage and bodily odor during menstruation, and influence from peers and siblings, etc.) were offered as reasons for tampon use.

The changes are felt as empowering at least by some women. One self-described virgin describes her experiences of the Formoonsa Cup:

Ever since I started using Formoonsa Cup, I noticed I stand taller and feel more beautiful. I often score 100 in exams. The important thing is I am a lot more comfortable and also enjoying playing with [sex] toys now. (https://tinyurl.com/ydacgwok)

And coming back to Vanessa: "I do not want to be sneaky when I menstruate anymore, because menstruation is not a disease."

In her discussion of menstruation, feminist political theorist Iris Marion Young proposes a concept of menstrual etiquette, a concept that concerns "who can say what to whom about menstruation, what sort of language is appropriate, and what should not be spoken" (Young 2005, p. 111). This etiquette creates disciplinary burdens for women, both physically and emotionally. The Formoonsa Cup not only proposed a new way of managing menstruation; it also changed the rules of menstrual etiquette in Taiwan. For that reason, Formoonsa Cup is indeed—however counterintuitive it may sound—a menstrual cup for Taiwanese women.

4   Seeing like a Critical Computing Researcher

We hope it is obvious that everyday users are unlikely to produce a reading such as the foregoing interpretation of the Formoonsa Cup. To finish this chapter, we want to address ourselves to two questions: The first is this: in what way did we offer a "theory" of Formoonsa Cup, or, what exactly is a "theory of a work"? And the second is: "how can such a theory inform HCI?"

4.1   In What Way Did We Offer a "Theory" of the Formoonsa Cup?

Danto's formulation of an interpretation as a theory of a work is slightly unusual; it is more common to see words like "reading," "interpretation," "critique," or "analysis" used. But we believe "theory" is actually a very accurate word. A theory of a work is basically a single account that unifies and explains the salient particulars of the work, not unlike the way that "evolution" offers an account that unifies and explains an extraordinary number of empirical findings. The more variety, and the more surprising the connections among them, the more elegant and interesting the theory is.

Our "theory of the work" might be paraphrased explicitly as follows: The Formoonsa Cup was far more than the creation of a functional industrial product; it was a sociological and political intervention that prompted conversations and led to actual legislative change. More poetically, we suggest that just as the physical Formoonsa Cup has the potential to destroy a woman's hymen, so does the Formoonsa Cup as an ideology destroy the ideology of the hymen fetish, and the latter was required before the former was ever possible. The product—including the Cup itself, the creative process and designer intentions, all the accompanying materials packaging and illustrations, the teaching at expos and public events, the political activism leading to changes in Taiwan's laws, and the nature of the press coverage—can be explained by this theory of the work.

4.2   How Can Such a Theory Inform HCI?

The short answer is that works embody theories in particulars, and in doing so, they model how particulars, by embodying certain theories, can solve problems in future situations.

Philosopher of art Danto (1983) makes the point that when we commit to one statement about a work of art, we immediately also commit to many other statements about the work. If we look at a painting, for example, and say, "that is a mountain," then we are committed to dozens of other statements: "these are trees," "this is in the foreground and that is in the background," "this painting uses 3D perspective," "this is a landscape painting," "these daubs of paint are representational," and on and on. We also attend to some things and not to others, as soon as we have made this claim. We might see one daub of paint as representing a person on a road, but a similar daub of paint in the sky we might not notice at all, merely processing it as a variation of color in the sunset. The lack of representationality of the daubs making up the artist's signature would likely be overlooked.

So it is with the hymen regime in Taiwan. If one accepts that the hymen is connected with virginity, one has also accepted innumerable other statements—about purity, responsibility, the worth of women, the exclusive and non-reciprocated rights of the husband's penis, the relative importance of hymen care versus menstrual pain, and on and on. To reject one statement is to undo all (or many) of them. The Formoonsa Cup project is significant because it is a design that reframed care of the vagina away from *care of the hymen* towards a *care of menstruation*—managing its pain and inconvenience, but also becoming comfortable with it as one's own. This reframing depended on and drove meaningful social change. Perhaps it is likely that another product, another activist, or another designer would have changed Taiwan's antiquated ban on menstrual cups anyway. But in 2017, it was the Formoonsa Cup, and the group of activists it attracted, and a democratic platform that gave them a voice, that led to this change.

If we compare Danto's "theory of a work" to the way that design researcher Dorst (2015) talks about a "design frame," the similarities are hard to miss. In both cases, an interpretative structure is proposed, and accordingly certain details become important while others fade away. Propose a new frame and the same thing happens again, only with different details. An intractable problem in one frame becomes a readily solved problem in another. By connecting Dorst to Danto, we avoid the reification of design thinking as a unique cognitive style proper only to designers and instead connect it to millennia of examples, theories, and methods of criticism. The idea that theories are embedded in artifacts is no longer an epistemological mystery better avoided in appeal to empiricism (i.e., user research); instead, the notion of design objects embodying theories is simply the same sort of thing that art historians, literary critics, and so on have taken for granted and worked from for millennia. Danto's skillful analyses of how artworks become meaningful can support readings of design seeking to unpack design objects' meanings.

For us, staging a dialogue between Dorst's characterization of design framing and Danto's characterization of interpretation (i.e., a "theory of a work") helped us learn to see how the Formoonsa Cup achieved meaning. The cup built on and pushed forward a social movement by integrating into a single work a coherence of specific design decisions, activist movements, consumer demands, sociological changes, and political structures. Our part in offering a theory of the work is to point to that coherence. Its benefit to HCI is to add that work and its theory, its coherence, its framing to the collection of exemplars that the HCI community uses to think through similar problems. We can imagine an HCI designer-researcher interested in using design to pursue gender-oriented social emancipation. The Formoonsa Cup is an exemplar, featuring reusable design patterns and design rationales, available in our community's design repertoire. One way to contribute theory to HCI is to fill out that collection of exemplars, enriched by accompanying theories of them as individual works to make those patterns and rationales more visible and actionable.

References

Alexander C, Ishikawa S, Silverstein M, Jacobson M, Fiksdahl-King I (1977) A pattern language: towns, buildings, construction (Center for environmental structure). Oxford UP, Oxford

Archer B (1995) The nature of research. Co-design Interdiscip J Des, 6–13 (Jan 1995)

Bardzell J, Bardzell S (2015) Humanistic HCI. Synthesis Lectures on Human-Centered Informatics (Series editor: John M. Carroll). Morgan & Claypool Publishers

Bardzell J, Bardzell S, Hansen LK (2015) Immodest proposals: research through design and knowledge. In: Proceedings of CHI2015. ACM, New York

Carroll J, Kellogg W (1989) Artifact as theory nexus: hermeneutics meets theory-based design. In: Proceedings of CHI 1989

Chang C-T, Huang J-H, Wu S-C (2015) An Exploration of tampon use intentions among female college students based on the theory of planned behavior: examining sexual orientation and gender characteristic as effect modifiers. Taiwan J Public Health 34(4):424–436. Retrieved 08/17/2017. http://www.publichealth.org.tw/english/index.asp?HidDID=2

Ching FDK (2014) Architecture: form, space, and order. Wiley

Danto A (1983) Transfiguration of the commonplace: a philosophy of art. Harvard UP, Cambridge

Dorst K (2015) Frame innovation: create new thinking by design. MITP

Foucault M (1988) The history of sexuality, vol. 3: the care of the self (Trans: Hurley R). Vintage Books

Foucault M (1995) Discipline and punish: the birth of the prison (Trans: Alan S). Vintage Books

Frayling C (1994) Research in art and design (Royal College of Art Research Papers, vol 1, no 1, 1993/4). Royal College of Art, London

Gaver W, Bowers J (2012) Annotated portfolios. Interactions 19(4), July + August

Hartmann J, Sutcliffe A, De Angeli A (2008) Towards a theory of user judgment of aesthetics and user interface quality. Trans Comput Hum Interact

Hassenzahl M (2010) Needs, affect, and interactive products—facets of user experience. Interact Comput 22(5):353–362

Law E, Roto V, Hassenzahl M, Vermeeren A, Kort J (2009) Understanding, scoping and defining user experience: a survey approach. In: Proceedings of CHI2009. ACM, New York

Schön D (1983) The reflective practitioner. Basic Books, NY

Tractinsky N (2012). Visual aesthetics. https://tinyurl.com/ya38wznh. Retrieved 08/17/2017

Tseng Vanessa (2016) TED talk on Formoonsa Cup. Retrieved 08/17/2017. https://www.youtube.com/watch?v=tmYa6ubl4Ew

Young I (2005) On female body experience: "Throwing like a Girl" and other essays. Oxford UP, Oxford

Part III
"Ideation"

Chapter 7
Improv for Designers

Elizabeth M. Gerber and Florence Fu

1 Introduction

Theater-based improvisation, or improv, is gaining status as a useful practice for interaction design teams. Demand for improv practice grows as practitioners and instructors realize the effectiveness of using improv to foster productivity. For example, Facebook and Google teams use improv practices when creating digital content for their users and customers, Twitter includes improv classes as a part of their employee's benefits, and Northwestern's and Stanford's Design Institutes teach oversubscribed improvisation classes every semester.

While researchers and practitioners have designed and studied critical team practices for supporting interaction design such as collaboration (e.g. Muller 2003; Simsarian 2003), exploration (e.g. Holtzblatt and Jones 1993; Sutton and Hargadon 1996), generation (e.g. Hartmann et al. 2008; Yeh et al. 2008), experimentation (e.g. Buchenau and Suri 2000; Rettig 1994), and communication (e.g. Buxton 2010; Gerber and Carroll 2012), few have considered how improv practices influence team practices for supporting interaction design (Gerber 2007a, b, 2009, 2013). This chapter explores how theater-based improv reinforces practices needed for collaboration between designers as they develop novel computer human interaction interfaces across diverse industries.

E. M. Gerber (✉) · F. Fu
Northwestern University, Evanston, USA
e-mail: egerber@northwestern.edu

© Springer International Publishing AG, part of Springer Nature 2018
M. Blythe and A. Monk (eds.), *Funology 2*,
Human–Computer Interaction Series,
https://doi.org/10.1007/978-3-319-68213-6_7

2 Design Practice

When working to generate novel interfaces, designers gather together with inter-disciplinary team members for discrete amounts of time to work on a proposed project. With the discipline's growing popularity, researchers and practitioners have sought to understand the practices that support collaborative interaction design, specifically why and when the practices work. When design teams are effective, they are able to immerse themselves and explore the problem area, question pre-viously held conceptions of what is possible, generate new possibilities, commu-nicate their ideas to stakeholders, and generate novel and valuable design solutions. For this to occur, participants must feel safe questioning the status quo, sharing their expertise and knowledge, building on the ideas of others, and expressing their ideas fluidly to various stakeholders (Gerber 2009, 2013).

These collaborative practices are intended to govern behavior and enhance the productivity of interdisciplinary teams and desirability of complex interaction designs. Further, these practices, especially the variety codified and popularized by design firms (e.g. www.ideo.com, www.frogdesign.com) and schools (e.g. segal.northwestern.edu, dschool.stanford.edu) have arguably contributed to the broad diffusion of a collaborative approach to interaction design (Gerber 2009).

But designs teams are not always effective. They can fail to work effectively together in at least as many ways as they can succeed (Gerber 2009). With the increased demand for interaction design in the face of competing resources (e.g. time, money, attention), we need more established approaches for fostering col-laboration, exploration, generation, experimentation, and communication.

Improvisation offers a promising approach.

3 Improvisation Practice

Improvisation is the convergence of composition and execution activity in time (Moorman and Miner 1998). Improvisation stands in contrast to performing pre-existing routines, yet it is not accidental. It is a deliberate creation of novel activity and has been examined across diverse domains including musical and theatrical performance, management, medicine, education, and most recently, interaction design.

In the 1950s, jazz trumpeter Miles Davis introduced modal jazz, a framework whose only rule is to use the scale (Danzico 2010). Although many formalists initially resisted, the improvisation method explored an entirely new sound and style, giving musicians the freedom to rearrange, mix, and play around with the pieces to produce infinite possibilities.

In the 1960s and 70s, theater directors and instructors, Viola Spolin and Keith Johnstone adopted improvisation for drama by designing exercises to encourage actors to perform more spontaneously (Spolin 1963). Johnstone offered simple guidelines such as "Be Obvious," "Fail Cheerfully" and "Show Up" to liberate actors and alleviate the pressure they feel to be original and creative on stage. Acting troupes such as Chicago's Second City and New York City's Upright Citizen's Brigade incorporated these improv guidelines and helped to establish the practice in drama schools and stages ever since.

In the 1990s, improv's reach extended beyond the performing and musical arts into management, medicine, and education. Concerned about the accelerating pace of competition in the global marketplace, management scientists focused on how stored knowledge and skills shape improvisation and the inherent tension between the exploration and execution needed to maintain momentum within a firm (Moorman and Miner 1998). Medical researchers, concerned about the time-consuming formal decision analysis process, sought to understand how medical providers improvised when responding to individual patient needs (Embrey et al. 1990). Similarly, facing limited time for instruction, researchers investigated educators use of improvisation when deciding what and how much to teach when presenting case studies to their students under time constraints. Just as musicians originally reassembled sets of notes into a novel tune, researchers found that skilled improvisers in medicine, education, management can learn to recombine existing routines to create action.

4 Beyond Performance: Improvisation in Human Computer Interaction Design

With the ubiquity of technology and designers' concerns extending beyond usability to aesthetics and user experience (Buchenau and Suri 2000), designers became increasingly invested in user-centered design practices (Norman 2013), a practice characterized by collaboration, contextual awareness, adaptability, and experimentation (Gerber 2007a, b). Interaction designers began to explore how improvisation practices could be used to understand existing and generate new user experiences with complex technological systems (Macaulay et al. 2006). Within this context, designers used improvisational performance to enact different roles and scenarios in simulated environments to understand the related activities and context (Burns et al. 1994; Simsarian 2003). Building small scale, low-fidelity environments effectively reveal both the constraints and opportunities within a user's context. Designers can easily adjust the stage environment or props to test their hypotheses about users' experience interacting with complex systems (Bredies et al. 2010).

In contrast to designers performing as "users," designers have engaged users in improvisational performances through participatory design, a process in which designers interactively create artifacts with users (Buchenau and Suri 2000; Davidoff et al. 2007; Iacucci et al. 2002). Similarly, in the early stages of the design process, designers have engaged in live theatre to design with stakeholders (Vines et al. 2014). Through conversations, designers can understand and communicate, or perform their stakeholders' concerns to domain experts. Designers create a staged environment such as traveling on a plane, seated users on chairs organized in rows, and ask them to hold props and prototypes, while designers improvise the travel experience. As they role-play the scenario, designers and users learn about possible uses for the props and prototypes (Kuutti et al. 2002). Designers then must work with users to craft sustainable and ethical solutions. They can not just look at behavior, but must envision how things should be, and observe what actually happens in real time.

While this research illustrates ways in which interaction designers use improvisation to gain empathy for users and to generate ideas, it primarily focuses on the performance aspects of improvisation and how performance improves usability and experience. Most recently, we investigated how designers used techniques from theatrical improvisation to adhere to the rules of the popular method of group brainstorming. The usefulness of improvisation for brainstorming stems from the similarity of the goals of improvisation and brainstorming, the similarity of the recurrent problems that actors and designers encounter when collaborating, and the distinctness of the ways each have devised to resolve the problems that block the group's performance (Gerber 2009, 2013). The more frequently people practice in many different ways, the better they become.

Building on this previous body of work we examine improvisation through its core practices in interaction design. Specifically, we focus on (1) collaboration (2) exploration (3) generation (4) experimentation and (5) communication.

5   Approach

This chapter compiles reflections from teaching and observing the ways in which participants integrated theater-based improvisation into their design practice. The study builds directly on an initial 5 year exploratory research study (Gerber 2007a, b). The ideas and examples were developed and drawn from a 15 year multi-case study (2002–2017) of approximately 350 practicing designers and 1000 undergraduate and graduate design students using theater-based improvisation to support their work. These cases are based on theater-based improvisation workshops delivered to undergraduate and graduate design students in Northwestern's Segal Design Institute and the Product Design Program and Hasso Plattner Institute of Design ("d.school") at Stanford University and to practicing designers in the enterprise, financial, knowledge management software, e-commerce, consumer products, and medical industries. Workshops varied from two to sixty hours over the course of one day to ten weeks (Fig. 1).

For the practitioners, the improvisation training typically occurred during the launch of a product development effort. Their work demanded collaboration throughout the project's six month to one year long duration. For the students, their improvisation training occurred at various stages of industry sponsored design projects. Student teams worked together for the duration of the project which ranged from one to ten weeks. In most cases, the students had limited to no

Fig. 1 Designers using theater-based improvisation to practice generating novel interaction designs for information services

experience working together prior to the course. This situation is typical for many student design teams and requires students to quickly negotiate ways of working effectively together. Both the practitioners and students were engaged in design projects. Examples of projects included interfaces for enterprise, financial, and knowledge management software, search engines, consumer websites, and mobile devices. To complete this work, practitioners and students worked in teams of two to 23 members. The teams included members with 1 month to 20 years of design experience. Because the student teams were working on industry sponsored projects, like the practitioners, their projects had the potential to be introduced to market if proved viable during user testing.

6 Shared Practices of Improvisation and Design

For each practice, we describe the cognitive basis of the shared practice in interaction design followed by how the improv serves as a supportive role. Table 1 outlines and defines the shared practice. While practices are described as distinct, in truth, there is overlap.

Table 1 Improvisation and design share 5 core practices including collaboration, exploration, generation, experimentation, and communication

Shared practices	Definition
Collaboration	Understanding and working with others to gain inspiration and create new concepts
Exploration	Searching for and making sense of new information
Generation	Creating new ideas within constraints
Experimentation	Testing new ideas to learn what works and what does not work in context
Communication	Sharing information with others to elicit an intended reaction

7 Collaboration

Collaboration involves understanding and working with others to gain inspiration and design appropriate solutions. Collaboration is critical for interdisciplinary teamwork where participants learn reciprocally through discussions across and through differences, eloquently described as "symmetries of participation and symmetries of learning" (Muller 2003; 4). To participate, designers must be present, attentive, and empathetic. Consider the following situation, one designer is in a dispute with another designer about a feature. The designers can fixate on the dispute or they could look and listen for what their partner is offering. By understanding different people's ideas and perspectives, designers can then gain inspiration from a situation and build new ideas together, taking concepts into surprising, unforeseen directions, rather than have single ideas exist independently (Gerber 2009).

Like designers, improvisors strive to "co-create" performances together by combining different people's contributions into a unified concept. Improvisors prioritize the "ensemble" over the individual. They focus on reconciling the needs of individuals to be the star performer with those of the broader group to put on an effective performance. Improvisors advise listening to others for what they need and "making them look good" by reacting to their requests (Ryan 2005). They focus on noticing what they receive from others. Improvisation rests on the egalitarian belief that everyone has the potential to contribute, which is not the same as involving every potential collaborator in a design project at every stage.

During an idea generation session, to develop novel and valuable ideas, designers aim to build on each others' ideas. To build on each others' ideas, all ideas must be heard. To hear ideas, both improvisors and designers must pay attention to each other. With the goal of hearing every idea, only one person should be speaking at a time. If a designer generates an idea while another is talking, she should note the idea on a piece of paper and communicate the idea when she is able. This reduces the likelihood that she will forget her idea, and increases the likelihood that she will focus on the ideas presently being generated in real time.

Similarly, improvisers focus on their colleagues' words and actions so they may be prepared to react and develop a novel and interesting narrative. Instead of rehearsing a specific scene, improvisors rehearse ways of interacting, training themselves to listen carefully and react in such a way that is natural. As such, audience members are convinced their performance has been scripted because they respond to each others actions in expected ways, creating compelling narratives and characters. By intently looking at each other, improvisers can anticipate and are aware if another player is going to speak.

To cultivate this awareness of individual members of the ensemble, improvisers may play "Alphabet." In this practice, the ensemble stands in a circle and attempts to say the letters of the alphabet one by one without interrupting each other. If two actors say a letter at the same time, the ensemble returns to the beginning of the alphabet. The objective is to focus on each other's intention to speak and act rather than to create a shortcut for getting through the alphabet such as going around the circle in order. Interested in raising awareness of other's intentions to contribute, a designer led her team through this activity prior to a brainstorming session. According to the designer, this activity helped her team to transition from focusing on their individual work to their group work. She observed the participants adhering more closely to the rules of brainstorming including holding one conversation at a time—actively listening and responding to what the participants are saying and doing rather than contributing to the group dialogue independently of what is being said and done.

8 Exploration

Exploring involves searching for and making sense of new information in the context of what is known. Designers explore to determine user needs and recognize new opportunities for interaction design. Designers understand users' needs in a real-world context by visiting them where they work and play and challenging their pre-existing assumptions by asking users "Why do you do what you?" (Holtzblatt and Jones 1993). An attitude of curiosity, or willingness to find out why things are the way they are, is core to exploring. Additionally, exploring requires awareness, paying attention to what is happening, both locally and more broadly. Awareness increases when we engage multiple senses to get new perspectives (Dyer et al. 2013). Through meaningful multi-sensory observation and inquiry, designers identify gaps in their understanding, empathize with users, and uncover insights that would not have been otherwise discovered if they were holed up in the studio.

Like design, improvisation requires searching for and uncovering new information to create performances never created before. Improvisors are advised to both "be present" (Ryan 2005) and "challenge convention". By engaging all senses in a particular context, improvisors are exposed to more information. They use sight, sound, smell, and touch to compose and perform in a new way that has not been rehearsed before. By challenging convention, improvisors may create a surprising performance that delights the audience.

Designers practice exploring by playing a modified version of an improv activity we call "Floor as Door." During this activity, designers meander around a space, pointing at familiar objects in the room and labeling them inaccurately out loud. The goal of the game is to point to and uniquely label as many different, but familiar objects as possible. For example, while doing this activity, a team of four designers walks around their studio. The first might point to an electric outlet and label it a "pencil sharpener." The second points to a computer monitor and labels it a "book." The third points to a window and labels it a "painting." And the fourth points to a studio dog and labels it a "bird." The first several labels generated tend to resemble the object at which the designer is pointing. An electric outlet features holes for a plug like a pencil sharpener features a hole for a pencil. The first labels generated also tend to be based on other things the designer is seeing. The designer who is pointing to the computer monitor sees a book on the same desk. As they continue, the designers realize they can't just rely on sight, they must use more of their senses to see familiar objects in new ways. The plant becomes "an ocean breeze" and the smell of coffee becomes "an urgent deadline." During this activity, designers practice being present, using their senses, and questioning existing labels. Designers experienced and modeled for each other the increased awareness and connection with the subconscious when they are able to break free of conceptions about familiar contexts—the desired state for effective exploration. By seeing their collaborators participate in this behavior, they felt safe and encouraged to challenge pre-existing conceptions and explore new possibilities. Like designers, improvisors craft engaging narrative performances by asking questions and revealing new perspectives the audience could not previously see. The best improvised performances happen when the unexpected occurs (Gerber 2009).

9   Generation

Generating involves coming up with new ideas within constraints. Designers create new ideas when combining two from different fields. They must regularly ask "What if we put these ideas together?" This practice demands discomfort precisely because it is about imagining something that has never been done before. And yet this conflicts with the desire to be knowledgeable and not make mistakes. In order to be predictably good at things, we must practice and do the same things over and over again (Hartmann et al. 2008; Yeh et al. 2008). To generate new ideas, designers must learn to relinquish their fear of being judged by others and need to know the answer. Designers are encouraged to explore, or generate ideas free of constraints (e.g. resource allocations, timing, technological restrictions).

Similarly, improvisers generate ideas free of constraints when performing a unique show for their audience. Paradoxically, improvisation teaches that improvisers can more easily generate ideas when they are not trying to be clever or unique, but when they are "being obvious" with their fellow players (Gerber 2009. Designers innovate not when they are trying to be clever but when they are attentively reacting to the needs and opportunities they observe. As they attend to their environment, they are inspired to take what may be "obvious" to them but not necessarily obvious to others.

In addition to "be obvious," improvisors follow shared guidelines for generating ideas with others including "Accept all offers" and "Don't stockpile ideas" (Madson 2013). Designers "defer judgement," "build on other's ideas," "free-wheel," and "go for quantity" in an effort to reduce social inhibition and generate as many ideas as possible (Osborn 1953; Gerber 2009). Improvisation and design recognize both that there are no shortage of new ideas and that while some will be "good," a lot will be "bad." However, by letting imaginations run freely, they are able to learn and bring ideas that wouldn't have existed otherwise. Within these ideas, they can uncover possibilities that allow them to assess and evaluate where to go next.

Of all improvisation guidelines, designers practice generation by readily adopting the classic improv phrase "Yes, and" (Leonard and Yorton 2015). This phrase is based on a central tenet of improvisation—every thought, behavior, and feeling offers a chance to be acted upon (Leonard and Yorton 2015). Rather than hesitating and adhering to their perceived notion about should happen, improvisors must act. The best improvised performances happen without hesitation.

Improvisors offer ideas to each other beginning each offer with the phrase, "Let's." A member of a group replies, "Yes, let's" and offers an additional suggestion that builds on the first. Consider how a design team might use the phrase "Yes, let's" to generate new interaction designs. One designer might say "Let's design enterprise software that is easy to use." Rather than sharing their critiques, the team members reply saying "Yes, let's." Another team member offers a second suggestion to build on the first, "And let's make software that people look forward to using every day at work." The group responds unanimously, "Yes, let's." The team continues until the product and enterprise are defined. While the idea seemed outlandish at first, the team reported the potential viability of certain design elements within the idea. By withholding judgment of each other's ideas and building on the ideas of others, designers and improvisers create new combinations of ideas not previously considered (Gerber 2007a, b, 2009).

Designers devised a variation of this game using props called "Yes, Let's, with props." A pair of designers passed a small wood block back and forth developing multiple uses for the block such as a paperweight and a door stop. The first several ideas generated were ideas that more or less assumed the constraints and typical uses of the object. However, the designers realized that to explore more alternate uses, they had to relinquish their preconceptions of what a wood block can or could be. As they continued the activity, the block became a podium for giant ants and a bed for a mouse. During this activity, designers practiced generating wild ideas. Designers modeled for each other the generation of novel ideas that occurs when they are able to break free from cognitive, emotional, and behavioral bounds of socially shared conceptions of what is possible—the desired state for an effective brainstorm (Thompson and Choi 2006). By seeing their collaborators participate in this behavior, they felt safe and encouraged to develop their own wild ideas when brainstorming.

10 Experimentation

Experimenting involves testing new ideas with users to learn what works and what does not work in context. This is especially critical when working on complex problems with no evident solution. Low-fidelity prototypes, or physical manifestations of ideas made with easy to manipulate materials such as paper and pencil, allow designers to quickly test an interface early in the design process before investing time and money into hi-fidelity, and often expensive, prototypes.

In this process, the designer must have humility, letting go of the assumption that their idea is the best and recognizing that there is always still more to learn (Gerber and Carroll 2012). If a design does not go as planned, a designer must recover and continue forward. Experimenting requires optimism that a solution will be realized (Loewenstein 1994), while working in an unstructured environment and managing work with loosely defined goals. Through this process, designers assume ownership of the project and guide direction (Gerber and Carroll 2012).

Similarly, improvisors are performing without a script, yet they remain optimistic that by committing to the guidelines of improvisation, they will deliver a performance. Improvisers learn that it is not only okay to make mistakes, but also that they should always include it as part of their practice (Leonard and Yorton 2015). However, fixating on mistakes during a performance does not make for a good show. Rather, they focus on moving forward. When people act, they experience more successes and more mistakes. Successes breed confidence and mistakes provide learning experiences to inform future design practice (Gerber 2009).

Within a limited time frame, designers can use improv and stage small scale, low fidelity environments to test experiences to "better understand the constraints and opportunities of the user's context" (Gerber 2009). These constraints provide clarity for both the users, who can focus on the key points of interactions, while designers gain empathy about their user and understand critical needs.

We observed designers adopting and modified a classic improvisation game to practice rapid experimentation. To play "Body Sketch," designers stand in a circle facing each other. The first designer steps into the middle of the circle to offer an idea, such as, "I am a screen." Rather than deliberating as a group, a second designer may jump in, adding something like, "I am a trackpad." A third designer adds a third element to the sketch, such as "I am a keyboard." Now the first designer leaves the circle, taking another designer with him/her. The remaining designer idea is used as the next starting off point, and the activity continues to develop and test these triadic scenes of socio-technical situations. To be sure, some "sketches" work well and the team members all nod with satisfaction. As for other "sketches", even if they do not work well, the activity continues.

11 Communication

Communicating involves sharing information with others to elicit an intended reaction. Reactions may include providing feedback or providing sponsorship and a coalition of supporters for an idea. To elicit a desired reaction, ideas must be crafted and shared in an engaging and compelling way and through different modalities (written, verbal, behavioral). When designers tackle complex problems, they involve a greater number of interdependent stakeholders. The ability to work effectively with others and communicate ideas in a way that excites others about a team's design is critical (Buxton 2010).

Storytelling is one important kind of communication for interaction designers to inform their work and share their ideas (Muller 2003). Stories may be used to trigger conversation, analysis, or feedback, collected from end users to understand new product opportunities, and constructed by designers to stand as proxies for real users and present their concept of what a designed service or product will do, how it will be used, and what changes will occur as a result (Muller 2003; Gerber and Carroll 2012). Designers begin a story by asking stakeholders to "To imagine a time when …" there was a challenge to be overcome. The design solution is the hero of the story, addressing the challenge and changing the future state. Strong stories elicit emotion.

Like designers, improvisors focus on the importance of dialogue and action in creating meaning. Improvisors rely on story structure which starts with establishing the scene, the characters, and the norms and routines. Once established, improvisors present the "tilt" or the unexpected action that changes the day to day routine. The tilt is followed by a series of subsequent events leading to a climax, changed characters and a new reality.

Designers practiced communicating by adapting an improvisation activity called "Half Life." The activity begins when one member picks a familiar product and describes the problem it is trying to address, the product features, and the way in which the product addresses the problem described. For example, he may choose to describe a mobile phone. "People need to be able to communicate with others regardless of location. Mobile phones are portable telephones that can receive and make calls while the user is moving within a designated service area. The phone relies on a cellular network which also supports other services such as text messaging, internet access, digital photography, games, and general computing capabilities. Phones with greater capability are called smart phones. For example, if a person is walking down the street and notices a car accident, a person could call the police on their mobile phone. If a person wanted to commute to the city by train, she could look up the schedule on her phone using Internet access. And so on." While the first member is telling this story, a second member records the time it takes to tell the story. A third member must then tell the story in half the amount of the original time yet still retaining the important details of the story. A fourth member then tries to describe the story in half of that time. They continue until the story can be told in 15 s, capturing the most important elements needed to communicate the story.

12 From Stage to Studio

In this chapter, we expand our understanding of theater based improvisation in interaction design beyond performance to enhance design practices. We reflect on the similarities between improvisation and interaction design practices and how interaction designs use improvisation to pre-empt and resolve interpersonal and personal dynamics that impede successful design practices (Gerber 2009). Specifically, we focus on five core practices: collaborating, exploring, generating, experimenting, and communication.

Our formal training in both improv and design allowed us to work with designers to adapt the techniques in the context of the design process. Over the past 15 years of application, we have adapted 49 popular improvisation exercises from the Improv Encyclopedia for easier adoption by designers not trained in improvisation. Instead of actors, we refer to members. Not feeling that acting was critical, instead of acting, we refer to interacting. Each activity was coded based on shared practice, group composition (new versus established) and size, time to complete, and formation. These activities are not intended to replace successful interaction design practices such as contextual inquiry, paper prototyping, and think alouds, but rather complement. See www.improvforinnovators.org for a complete list of exercises.

13   Guidelines for Adoption

Since engaging in improvisation activities tends to be personally satisfying and enjoyable, it is not surprising that design teams may want to do the activities all day. However, this approach would not be productive for getting the needed design work done, we offer the following guidelines to inform the selection of improvisation activities to foster design practices.

Consider Goals: Prioritize those exercises that help to achieve the most pressing practices needed to prepare you for your next group activity or reinvigorate a group when energy is dragging. Like a coordinated team sport, improvisation and interaction design require concentration and agility (Gerber 2009).

Group Composition: Due to their relative simplicity, some activities are better designed for newly acquainted teams while others work well for established teams.

Time: Practices vary greatly in length. While some only require 1–3 min, others require 10 min or more.

Group Size: Just as design teams range in size so do improvisation activities. While some activities require a small team of 5 or less people, others require larger groups of 10 or more. Pick accordingly.

Space: Designers and improvisors are trained to perform their work wherever they may be—on a stage, at a studio, on a train or even on a bicycle! This is because the people engaged are more important than the space. However, designers must be able to clearly communicate with each other, feel free to move physically, and have limited distractions. For this reason, we encourage stepping outside the work area into a back hallway, a backyard or a empty room.

Leadership: Regardless of which activity is chosen or who chooses the activity, establishing a leader is useful. The leader need not be a great improviser but rather is needed to encourage explicit reflection between the practices of improvisation and interaction design. Leaders can also articulate goals, encourage participation, ease people's concern that acting ability is needed, and keep time. It is helpful to have someone who is starting, ending the activity, and facilitating reflection and application to design work.

Leaders can start exercises with "Let's give this a try." By framing the exercise as an experiment, the consequences for failure are less. Engaging in these activities can feel intimidating to people who are unfamiliar. Offer clear directions so participants can have a better idea of what they might expect. Offer a time limit (5–10 min), so people trying them for the first time feel a sense of control. Aim for 2–3 activities that emphasize a single practice. Doing too many at a time can be overwhelming and feel unfocused.

14 Limitations

As our consciousness of the similarities between improvisation and design practices increases, we are aware of much we do not yet understand about how to best utilize the similarities to produce the most innovative and successful designs. While we do know that improvisation is generally useful when understanding existing user experiences and context (Gerber 2007a, b), exploring and evaluating design ideas (Gerber 2009), and communicating ideas to an audience (Gerber 2007a, b), more research is needed to understand when the interventions are most useful.

Further, we each come into the design studio with values, practices and perspectives that can perpetuate unconscious cultural assumptions and biases in our work—both in design and improvisation. Greater awareness is needed of who is and is not participating and reflection on how this might limit us from designing inclusive technology.

15 Conclusion

Design is inherently interdisciplinary and creative practice informed by computer science, cognitive science, and art and it continues to draw from a range of interdisciplinary practices over time. Interaction designers learned game mechanics from video game designers, ethnographic research practices from anthropologists, and typography from graphic designers. What other creative disciplines might we want we use to inspire interaction design? How might we be inspired by physicists who devise methods related to experimentation of physical laws and theories? Or copy writers who write compelling language to promote products in publications and broadcast media? Or writers who create original written works including scripts, essays, literature, poetry, or song lyrics?

Acknowledgements We are grateful to many colleagues and peers in the interaction design community for their contributions to this paper, whether through example or critique. Specifically we wish to thank and credit Interaction Designer Molly Lafferty helped to adapt the improvisation activities for designers. And deepest gratitude to Patricia Madson Ryan, Dan Klein, and Nicholas Switanek for their continued instruction in improvisation.

Appendix

As with most things in life, competence comes with practice. A few notes on our personal practice.

Elizabeth I have not always been a improvisor or leader of improvisation. My Stanford engineering Professor Rolf Faste introduced me to improvisation professor Patricia Ryan Madson in my early twenties. At first, I was scared to engage, hiding behind excuses of not being an actress, but I soon became convinced that improvisation was not about acting but about a set of mindsets at which I actively wanted to improve. Over the last twenty years, many people have asked me if they can lead these exercises without being a skilled improviser and my answer is always yes. I've seen countless people pick the activity with which they most connect, try it out, and now use them regularly with their teams.

Florence As a student in Elizabeth's class, I've seen my own growth from fear to excitement towards improv. While taking her course on Human Centered Service Design, I remember dreading the morning improv activities and fearfully thinking *What's it gonna be this time?* As someone who can't help but over think, over plan, and fears 'saying something wrong,' these activities have sincerely encouraged me to have fun and embrace ambiguity. As each week went on, I eventually came to class curious, excitedly thinking *What's it gonna be this time?!* Through these activities, I've seen dramatic changes in the way I work and in the dynamics of my design team, when we're all up and out of our seats, engaging with each other mentally, emotionally and physically. I'm eager to continue applying these mindsets to both my life and my design work.

References

Baer M, Oldham R (2006) The curvilinear relationship between experience, creative time pressure and creativity. J Appl Psychol 91(4):963–970

Bredies K, Chow R, Joost G (2010) Addressing use as design: a comparison of constructivist design approaches. Des J 13(2):156–179

Buchenau M, Suri JF (2000) Experience prototyping. In: Proceedings of the 3rd conference on designing interactive systems: processes, practices, methods, and techniques. ACM Press

Burns C, Dishman E, Verplank W, Lassiter B (1994) Actors, hairdos & videotape—informance design. In: Proceedings of CHI 1994. ACM Press, pp 119–20

Buxton B (2010) Sketching user experiences: getting the design right and the right design. Morgan Kaufmann Press

Danzico L (2010) From Davis to David: lessons from improvisation. Interactions 17(2):20–23

Davidoff S, Lee M, Dey A, Zimmerman J (2007) Rapidly exploring application design through speed dating. In: Proceedings of the conference on ubicomp. ACM Press, pp 1069–1072

Diehl M, Stroebe W (1987) Productivity loss in brainstorming groups: towards a solution of a riddle. J Pers Social Psychol 53(3):497–509

Dyer J, Gregersen H, Christensen C (2013) The innovator's DNA: mastering the five skills of disruptive innovators. Harvard Business Press

Embrey DG, Guthrie MR, White OR, Dietz J (1990) Clinical decision making by experienced and inexperienced pediatric physical therapists for children with diplegic cerebral palsy. Phys Ther 76(1):20–33

Gerber E (2007a) Devotion to an innovation process: the case of human centered design. Dissertation, Stanford University

Gerber E (2007b) Improvisation principles and techniques for design. In: Proceedings of the SIGCHI conference on human factors in computing systems. ACM Press, pp 1069–1072

Gerber E (2009) Using improvisation to enhance the effectiveness of brainstorming. In: Proceedings of the SIGCHI conference on human factors in computing systems. ACM Press

Gerber E (2013) Using improvisation to enhance brainstorming sessions. In: Martin RL, Christensen K (eds) Rotman on design: the best on design thinking from Rotman magazine. University of Toronto Press, pp 258–260

Gerber E, Carroll M (2012) The psychological experience of prototyping. Des Stud 33(1):64–84

Hartmann B, Yu L, Allison A, Yang Y, Klemmer SR (2008) Design as exploration: creating interface alternatives through parallel authoring and runtime tuning. In: Proceedings of the 21st annual ACM symposium on user interface software and technology. ACM Press, pp 91–100

Holtzblatt K, Jones S (1993) Contextual inquiry: a participatory technique for system design. In: Participatory design: principles and practices, pp 177–210

Iacucci G, Carlo I, Kari K (2002) Imagining and experiencing in design, the role of performances. In: Proceedings of the second Nordic conference on human-computer interaction. ACM

Kulhan B, Crisafulli C (2017) Getting to "yes and": the art of business improv

Kuutti K, Iacucci G, Iacucci C (2002) Acting to know: improving creativity in the design of mobile services by using performances. In: Proceedings of the SIGCHI conference on creativity and cognition. ACM Press, pp 95–102

Leonard K, Yorton T (2015) Yes, and: how improvisation reverses "no, but" thinking and improves creativity and collaboration–lessons from the second city. Harper Business Press

Loewenstein G (1994) The psychology of curiosity: a review and reinterpretation. Psychol Bull 116(1)

Macaulay C et al (2006) The emerging roles of performance within HCI and interaction design, pp 942–955

Moorman C, Miner A (1998) Organizational improvisation and organizational memory. Acad Manag Rev 23(4):698–723

Muller M (2003) Participatory design: the third space in HCI. Hum Comput Interact Dev Process 423:165–185

Norman D (2013) The design of everyday things: revised and expanded edition. Basic Books

Osborn A (1953) Applied imagination. Scribners, Oxford, England

Oulasvirta A, Kurvinen E, Kankainen T (2003) Understanding contexts by being there: case studies in bodystorming. Pers Ubiquitous Comput 7(2):125–134

Rettig M (1994) Prototyping for tiny fingers. Commun ACM 37(4):21–27

Ryan P (2005) Improv wisdom. Bell Tower, New York, USA

Scott S, Bruce R (1994) Determinants of innovative behavior: a path model of individual innovation in the workplace. Acad Manag J 37(3):580–607

Sengers P et al (2005) Reflective design. In: Proceedings of the 4th decennial conference on Critical computing: between sense and sensibility. ACM Press

Simsarian K (2003) Take it to the next stage: the roles of role playing the design process. In: Proceedings of the SIGCHI conference on human factors in computing systems. ACM Press, 1012–1013

Spolin V (1963) Improvisation for the theater: a handbook of teaching and directing techniques. Northwestern University Press, Evanston, Illinois

Sutton RI, Hargadon A (1996) Brainstorming groups in context: effectiveness in a product design firm. Adm Sci Q, 685–718

Thompson LL, Choi HS (eds) (2006) Creativity and innovation in organizational teams. Psychology Press

Vines J et al (2014) Experience design theatre: exploring the role of live theatre in scaffolding design dialogues. In: Proceedings of the SIGCHI conference on human factors in computing systems. ACM Press (2014)

Yeh R, Paepcke A, Klemmer S (2008) Iterative design and evaluation of an event architecture for pen-and-paper interfaces. In: Proceedings of the 21st annual ACM symposium on user interface software and technology. ACM Press

Chapter 8
Playing with Provocations

John Vines

1 From Design-For to Design-With

In the fifteen years since the first edition of Funology, the notion that designers of digital products, services and applications should take seriously aspects of user experience has become mainstream and part and parcel of the design of new systems. Alongside the mainstreaming of experience centric approaches, we've seen a related diversification in the types of methods and approaches researchers and practitioners draw from when engaging with users. Along with shifts from cognitive to social constructions of users, and from the context of work to all areas of peoples' lives, we've observed a shift from designing-for to designing-with, and indeed designed-by, those who live with and use digital technologies. Of course, participatory and collaborative approaches to design have a long history in the field of human-computer interaction (HCI), with the Scandinavian tradition of participatory IT development and its politics of user empowerment in design having a particularly strong influence. Although initially utilised as methods for supporting the exploration of and inspiration for ideas within design teams, the last decade has seen personas (Cabrero et al. 2016), scenarios (Chin et al. 1997) and probes (Graham and Rouncefield 2008) be appropriated as resources for dialogue and collaboration between designers and research participants. Generative tools (Sanders and Stappers 2012) and low-fidelity techniques such as magic machines (Andersen 2012) have emerged as valuable mechanisms that enable 'non-designers' to work within design teams and speculate on alternate visions of future

J. Vines (✉)
Northumbria University, Newcastle upon Tyne, England, UK
e-mail: john.vines@northumbria.ac.uk

© Springer International Publishing AG, part of Springer Nature 2018
M. Blythe and A. Monk (eds.), *Funology 2*,
Human–Computer Interaction Series,
https://doi.org/10.1007/978-3-319-68213-6_8

111

technologies. While diverse in their uses, a commonality across these techniques is an appreciation that people are experts in their own experiences, that those affected by a new design should be directly involved in design processes, and that design teams should need to take seriously the processes by which they involve users in their design processes. Thus, a critical component of contemporary exploratory design work is not just the design of new products, services, interfaces or systems, but also designing the processes by which people are able to participate in and contribute to design work.

While this participatory turn is positive, from a methodological point of view this raises questions and challenges: how do we promote meaningful discussion with participants around design concepts in workshops?; how do we seed dialogue and discussion with participants without leading too heavily?; how do we create playful, engaging, yet serious opportunities for people to participate in design processes?; and how might we use workshops meaningfully to bridge engagements with different participants, in different stages of a design process? In this chapter I don't have full and detailed answers to these questions, but I will explore them through examples of techniques that collaborators and I have used across a range of participatory design projects over the last half a decade. While the methods we use are diverse, a common feature across them all is the use of provocation as a means for stimulating discussion around issues pertinent to design. Like participatory design, provocation has been used extensively as a technique in HCI. Provocation is a core element of critical and speculative approaches to design (Dunne and Raby 2001, 2013), and artefacts created purposely to de-familiarise habitual settings often intend to provoke reflection and question taken for granted assumptions designers and users bring to particular issues (Bell et al. 2005). There are many examples of HCI projects that have used techniques of provocation to promote critical engagement and reactions from participants, including around subject matters that are hard to articulate or envision (such as imagining proximal futures (Tanenbaum et al. 2016)), around reflection on practices (such as around home energy usage (Raptis et al. 2017)) or indeed around topics that are seen as a little taboo (such as eliciting talk around sex toys (Bardzell and Bardzell 2011)). Using provocation in design can help explore complex topics, to promote debate around design ideas, and help generate new ones. They can also be highly stimulating, engaging, and promote "complicated pleasure" (Dunne and Raby 2001, p. 63, building on Martin Amis). However, there is surprisingly little work out there that discusses how researchers and practitioners might use provocation in participatory and collaborative design activities. In this chapter I'll talk through three approaches to using provocation in participatory design projects, and then draw out four issues for design teams to consider when using aspects of provocation in participatory projects in the future.

2 Three Approaches to Exploring Provocations with Participants

In the following sections I'll discuss three techniques collaborators and I have used across a range of projects: (i) Questionable Concepts, (ii) Invisible Design and (iii) Experience Design Theatre. These techniques share a number of commonalities. They were all grounded in the use of provocation as a starting point for stimulating discussion and debate, and, as we shall see, each used fun, in different ways, as a resource for designing with participants. Another common element was each has been used primarily (but not exclusively) in projects where we examined the implications of digital technology on experiences in later life. Although we collaborated with a diverse range of participants and stakeholders in each of these projects, the primary groups we worked with were older participants, many of whom had limited experience of digital technology (indeed, they were often deeply sceptical and critical of it) and were highly unfamiliar with what design processes might entail. Furthermore, each of the techniques were used to gather qualitative research data that would be valuable for the ongoing analytical and design activities, as well as generating new ideas and concepts along the way.

Beyond these broad commonalities, the techniques are rather different from each other. Each represents design concepts very differently: for instance, in Questionable Concepts the provocations come in the form of illustrations that visually characterise an idea with short explanations of their functions; in Experience Design Theatre the design concepts are not seen, with the focus being on the relationships and interactions between characters whose lives are being affected by a speculated future service; Invisible Design also avoids showing the visual and material form of the design concepts, focusing instead on scripted dialogue between characters, though this dialogue is more fixated on interactions with a technology that is left unseen on screen. Each technique is also intended to have a different function when it comes to gathering input from participants in a design process; Experience Design Theatre is purposely designed for participants to "re-script" engagements between characters and in doing so influence the design of concepts that structure these engagements; the intention of Invisible Design films are to promote discussion among participants in workshops as they make sense of and unpick what's happening on screen, which can be captured and used for analysis later in the project; Questionable Concepts are intended to both provoke conversations in workshops and also offer a space for the illustrations to be scribbled on, amended, or redacted by participants. The various techniques also had different roles in connecting different phases of design activities in projects; both Questionable Concepts and Invisible Design presented an opportunity to convey ideas that were grounded in earlier contextual research in projects, articulated as design concepts; Experience Design Theatre also did this, but at the same time gave the opportunity for using the re-scripted interactions between characters in workshops with other participants later in the project.

The following sections describe each of the techniques in more detail. Following this, I draw out a number of recommendations for researchers and designers who wish to make use of provocation in their own design processes.

2.1 Questionable Concepts

Questionable Concepts (Vines et al. 2012) are sketches, of differing levels of fidelity, and brief textual outlines of design concepts. We call them Questionable Concepts as they are purposely created to be provocative, aiming to promote critique and discussion with participants. A concept might be questionable because of its overtly simplistic response to a complex problem; or because it is "un-useless", like Chindogu designs, in that they solve one problem but cause another; or because they resonate with a problem that people might identify with but the design deals with in a particularly impractical or unnecessarily complicated way.

We initially used the Questionable Concepts approach in a project exploring the design of future banking services for "eighty somethings" (people who are in their eighties or above). In the context of this first project, our intention was to use the concepts as both a way to bridge different activities in the design process (in this case, between initial interviews with participants, our analysis of this data, and a series of design workshops to be conducted with new groups of eighty somethings), and as a means for structuring design activities and promoting discussion with our new participants. The Questionable Concepts used in this first project were therefore intended to illustrate specific issues and problems that had been identified by our first set of interviewees. For example, the Pin Thimble concept (Fig. 1, left) was a proposition where a person places a thimble on their finger and upon recognising the user's fingerprint the gadget displays their Personal Identification Number

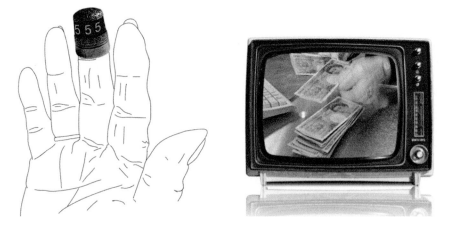

Fig. 1 The Pin Thimble (left) and Cashier TV (right) Questionable Concepts

Fig. 2 Questionable Concept cards

(PIN) for their bank account. This spoke to issues that many initial participants expressed around struggling to recall their PINs or passwords when they needed to, and how many felt compelled to write them down even if they knew this was a dangerous practice. Another proposition, Cashier TV (Fig. 1, right), proposed a direct video link to your local bank or shop so you could watch someone else completing a purchase or withdrawing money on your behalf. This related to how some of our participants were housebound and would often have to ask others to do shopping on their behalf or entrust them with their account credentials to withdraw money from automated teller machines.

For the banking for eighty somethings project we developed the concepts into A6 sized folded cards with four sides (Fig. 2). On the cover of the card was an illustration of the design idea. On the inside of the cover was a brief description of the idea (e.g., "reminds you of your PIN when you need it") and some anonymous quotations from the interviews that relate to the idea. The rest of the space on the cards was dedicated to three prompts for participants to consider and respond to, either through a written response or a doodle, or when in person at one of the workshops. We created 11 concepts in total. The cards were handed out to participants at the end of an initial 'getting to know you' workshop. We asked them to take the pack of cards home and to look through the ideas in their own time. They were asked to complete as many of the cards as they wished, requesting they responded to at least 3 but not to feel compelled to do them all. These three cards they had chosen would then get discussed at the later workshop. Invariably, when it came to meeting again at the next workshop almost all of the eighty somethings had completed all of the cards in the pack, or at the very least had comments and, usually very strong, opinions on them. Through this process we generated a range of diverse data for analysis, including audio recordings of discussions participants had around the concepts, the written responses on the cards, and the various doodles and sketches participants had created.

Fig. 3 Questionable Concepts as fictional service leaflets and brochures, with pastiche banking services for eighty somethings (left) and pastiche fresher student services (right) as examples

We have also used the Questionable Concepts technique in different ways. In the banking for the older old project (where the Questionable Concept cards discussed above were first used) we created more refined versions of the same concepts, developed into pastiche advertising material for the service providers that partnered us on the project (Fig. 3, left). These were used partly as a means of iterating the design ideas within the project team, but also as a way of promoting discussion with our project partners around our concepts, how they might fit in with their current service provision and how they might speak more explicitly to particular groups of their customers. We did a similar activity for a more recent project that explores near-future services to help University students intergrate into University life (Fig. 3, right). Again, we used these pastiche brochures of future student services to invite critique and discussion between students in the early stages of the design process. In many respects, these adaptations of the Questionable Concepts technique relates closely to work conducted by the Near Future Laboratory who have created fictional catalogues of future products and services in projects with IKEA and Stockholm city[1].

2.2 Invisible Design

Invisible designs are short film-based scenarios where two characters discuss technologies that are never shown on screen. Much like Questionable Concepts, the films are created to purposely promote critical discussion and creative dialogue with participants, and are used for generating insights and ideas in the early stages of concept development. Invisible Design draws on a rich history in HCI of using characters and scenarios as part of design processes, and more specifically work that

---

[1]See https://shop.nearfuturelaboratory.com/collections/frontpage/products/ikea-catalog-from-the-near-future as an example.

Fig. 4 Shots from the Invisible Design films 'Smart Money' (left) and 'Panini' (right)

has used film to communicate user needs (e.g., Raijmakers et al. 2006; Newell et al. 2011) or to promote discussion among users around the implications of future technology (e.g., Mancini et al. 2010) (Fig. 4).

The Invisible Design films have been used in a range of different projects, such as exploring future identity management and authentication technologies (see Briggs and Olivier 2009), urban mobility for older people (Lindsay et al. 2012), future smart payment technologies (Briggs et al. 2012) and recommender systems for people with health conditions (Vines et al. 2015). In each case, professional scriptwriters and directors were hired to create a compelling story. In a similar vein to Questionable Concepts, the starting point for the scripts were research materials from prior studies, initial interviews and focus groups that contextualised problems, and initial sketches of ideas and example technologies that related to these. What followed was an iterative process whereby the scriptwriter developed initial drafts of the script, shared these with the research team for feedback, and after several redrafts led to producing and editing the film (a process described in detail in Briggs et al. 2012). While the specific content of the final films differed from one project to the next, there are a number of common qualities across them. First, in each film the focus is on the dialogue between two characters who discuss an object or technology that always remains unseen. Rather like the invisible monsters of classic horror films and the never revealed technologies in many a science fiction book, the viewer is left to imagine and speculate what these designs might look and feel like, and what is occurring with them off-screen. Purposely avoiding showing what the objects might look like was, in part, a reaction to previous experiences colleagues and I had of participants in workshops quickly focusing on the material and aesthetic qualities of prototypes rather than the potential experiences they might promote and issues they might address. Second, the dialogue between the characters in the films would try to convey aspects of the relationship between the characters, portraying one as slightly more informed than the other, and would attempt to bring in humour often based on some political or social commentary. In other words, the script would purposely avoid fixating on talk around the "invisible design" (Fig. 5).

As an example, let's describe the narrative of the film 'Cucumber' which was created for a project exploring recommender services for older people with health conditions. The film opens with an older gentleman—Billy—sat in his armchair

Fig. 5 Shots from the 'Cucumber' Invisible Design film, with Stan and Billy discussing his DIY woes and how Stan gets advice and recommendations from others through a new online service

watching TV. A "knock knock" is heard followed by a voice: "it's only me!". Another older gentlemen—Stan—enters the room holding a bag of shopping. He slams the door and a picture falls off the wall behind Billy. What ensues is a back and forth between the two about how long this has been a problem and whether Billy knows how to fix it. Stan starts talking about a new service he uses to get advice and tips from people to fix all manner of problems. But Billy refuses to accept Stan's advice, assuming that those giving tips would be "cowboys". When Stan explains "he bought a knife sharpener, based on a recommendation", Billy replies with "Aye, from a guy who makes knife sharpeners". After some more discussion, Billy starts to come round to the idea, but Stan says: "You've got to sign up. I can ask for you, but if you want one you've got to sign up. You should. You'd be good man!" Billy sits back in his chair: "Ah well. Mebbes" he sighs. As with other Invisible Design films, humour plays an important role, emphasized here by the film's ending of Stan giving Billy a cucumber, the reason for him popping by in the first place.

The Invisible Design films have been used in design workshops with participants to provoke discussion around the issues and topics they infer. In these workshops, after a brief introduction to the broad topic of the project, the film would be shown and a loosely structured discussion would follow, facilitated by one of the research team. Here, participants would be asked to comment on the film and then to address the Invisible Design associated with it. Typically, discussion transitions from initial sense-making of the film (who the characters appear to be, what they were talking about, what the purpose of the film is) to unpicking the relationship between the characters and addressing the object that is not seen on screen. The discussions participants have around the film are audio recorded and then later transcribed for analysis.

2.3 Experience Design Theatre

Experience Design Theatre (Vines et al. 2014) is a technique where we work with professional actors and theatre directors to engage participants in responding to and 're-scripting' scenes and scenarios related to near-future service designs. This process is conducted multiple times with different groups of participants, after which the input from these different workshops is integrated into a longer story that is performed to new audiences. As with the other two approaches described above, this technique builds on a wealth of work in HCI and user-centred design where theatre has been used as a resource for design work. Brenda Laurel's (1993) classic work examined how interactions with computers could be described in performative terms, while role-play has for a long time been used as a means for design teams, and to a lesser degree participants, to work through possible design concepts. Work directly related to ours is that of Alan Newell and colleagues who have appropriated Boal's forum theatre into the context of design. In Theatre of the Oppressed (Boal 1979), Augusto Boal coined Forum Theatre as a means for involving citizens in discussion and action in support of social change. Forum Theatre would involve the improvisation of a situation by actors, which could then be halted by an audience member who could suggest different actions that might lead to less oppressive outcomes for characters. Newell et al. (2006, 2011) and Rice et al. (2007) have used Forum Theatre inspired approaches to promote discussion in design processes, where scripted scenarios would be performed live by actors (or via pre-recorded short films) and participants would be invited to discuss the scene, focusing on the characters' motivations, the role of technology and how the scene could be changed (Fig. 6).

Our approach deviated from Newell's and others work in several ways. First, as well as the opportunities live theatre and improvisation offer to engage participants in discussion and debate, we wished to explore the live re-scripting and re-drafting of speculated scenarios with them. The actors would begin workshops by performing very loosely scripted situations between characters that related to the service designs that we were exploring. In having these loose starting points, we wanted to provide the space for participants to express their own stories in relation to these initial scenes and have these stories and opinions fed back into new improvisations by the actors. Second, as with Invisible Design, we wanted to explore the experiences that might arise as a result of proposed services and technologies, rather than fixating on functions, aesthetics and the probability of the technology itself. Therefore, the focus of the starting scenes and resulting improvisations was primarily on the interactions and relationships between characters, rather than direct interactions with interfaces and services. Third, through iterating the scenes with participants in workshops, the actors explicitly incorporated participants' voices, stories and experiences in the re-drafted scenes. This was done, in part, to acknowledge to participants that their input had been heard and taken on board. But it also allowed us to carry these stories forward to later workshops with

Fig. 6 Documentation from Experience Design Theatre workshops

different audiences, so that the experiences contributed by participants could be communicated to those participating in later stages of the project.

We explored component parts of the Experience Design Theatre technique in a number of projects, although it was explored in most depth as part of a project where we were exploring the future of later life social care provision. The project involved working with a broad range of stakeholders—care organisations, public funders of care, technology companies developing ICT for care services, and most importantly older people, people in receipt of social care and carers themselves. Collectively we were exploring alternative models for delivering care in communities in the future, particularly models where (younger) volunteers would give up their time to provide informal care. Our project was exploring some of the socio-technical infrastructure and interfaces required to sustain such a service, as well as exploring its general appropriacy and the nature of "rewards" (material or otherwise) that volunteers might receive for doing care work of this kind. Here, the technologies themselves were not particularly provocative—they were primarily envisioned to be tools to help timetable care visits, to share information and data between parties, and to support reward card infrastructures. The provocative elements in this project related to the broader implications of these new care services, the social and political values they implied, and how they fundamentally change the nature of care work and the relationships between carers and carees.

To explore this project context, we divided the Experience Design Theatre process into multiple stages of activity. We conducted a first set of workshops with small groups of different participants—informal carers, young people seeking

Fig. 7 Documentation from the final stages of the Experience Design Theatre process, where participants representing various care organisations responded to theatrical performance of a new care services that was created with potential service users

volunteering activities, healthy older people and people receiving community care services—where we explored the loosely scripted scenes and re-drafted these with participants. After these workshops were conducted, we developed a set of scenes where participant input across multiple workshops was sewn together into one longer performance. We then invited our initial participants back to a follow-up workshop where this was performed back to them for further discussion and iteration. The various scenes were then refined further based on participant input and then presented to a large audience of representatives of various organisations implicated in new services like these (Fig. 7).

The data collected through this process was complex and diverse. As with Questionable Concepts and Invisible Design, sessions were audio recorded for transcription and analysis. The performances were also video recorded, to capture the interactions between the actors as they were being guided and re-drafted by participants. In the later workshops with larger groups, we also provided a myriad of means for participants and audience members to comment on and discuss the performances, including opportunities to question the actors (both in character and out, Fig. 7, left), to write brief notes as the performances were happening (e.g. on table cloths as in Fig. 7, right) and in short break-out activities (Fig. 7, centre).

3 Reflections on Using Provocations in Participatory Design

In the following sections, I conclude the chapter with some brief reflections on how these various techniques have worked in workshops with participants, and draw out some recommendations and guidance for using these and similar techniques.

3.1 Balancing Technological and Social Imaginaries

In Design Noir, Dunne and Raby note that central to critical design is "the suspension of disbelief [...] if the artefacts are too strange they are dismissed, they have to be grounded in how people really do behave" (p. 63). They note that provocativeness here means to design for a "slight-strangeness". Bardzell and colleagues have since noted, however, that defining what is "strange" and designing for a slight-strangeness is actually rather hard (Bardzell et al. 2012). Indeed they highlight a number of attempts in HCI research to develop provocative systems which, when given to participants for evaluation, were found to be neither provocative or interesting enough to promote engagement over any extended period of time. In our cases, it was perhaps through accident rather than intention that some of our ideas appeared to be provocative (in that they received strong reactions from participants, at least); we certainly did not set out knowing precisely how to design for provocativeness, and even looking at the ideas and discussions they stimulated retrospectively it is hard to define why certain ideas worked well. What was clearer though was what didn't work well. Some ideas participants found highly agreeable and perhaps because of this the ideas struggled to promote discussion in the groups. Like Dunne and Raby note, those concepts and ideas that were slightly too believable—the online review system inferred in the *Cucumber* Invisible Design film, and Questionable Concepts like video banking and top-up debit cards—were too close to existing platforms and services and did not explore in any great detail how the social settings of banking or care might be radically altered by these. Those ideas that did produce strong reactions were those that clearly played with combinations of altered technological and social futures, leading to questions and discussions among participants around "how might that work", "how does that change how we do things now" and "what are the values entwined in these ideas". This was particularly evidenced in the work we did on future voluntary care with Experience Design Theatre; the setting of care seemed very familiar to our participants, but the way these were played with to infer reward mechanisms for the 'volunteers' and the interchangeability of carers led to debates from participants about the values implied by such a service, and indeed the politics of the technologies that might underpin them. Similarly, the smart money Invisible Design film worked well in as much as it portrayed dialogue around the everyday experiencing of paying another person, altered via this new proposed intelligent bank note.

A further point to note here is that the mediums by which we engaged participants in discussion had a bearing on the potential to stimulate discussion. The loose illustrations on the cards, the hidden designs in the films, and the improvised scenes of the performances, all presented situations where it was rather clear we were working with things that did not yet exist, and provided space for participants to fill the gaps we had left in the materials. Bardzell et al. (2012) note that it is hard to find examples of provocative technologies that have been successful in engaging participants over extended periods of time. They reflect that more experienced (critical) designers might be better at designing for provocativeness. However, our experiences suggest that creating snap shots of ideas and designs is a productive way to defamiliarize settings and promote discussion on technological and social imaginaries; for us this was through the various sketches of ideas, whereas for much critical design work this is through highly developed proposals of future designs. Perhaps as soon as you try to embody these ideas in working prototypes, to be given to participants for evaluation, you lose the potential to find these productive middle grounds.

3.2 Harness Critique as Creativity

Central to all the participant related activities in these methods was promoting critical discussion around the proposed ideas. Many of the concepts and ideas were purposely created to be whacky, strange and impractical. We felt that from the start participants would critique the designs, but we had not expected the degrees to which they would do so. In the Questionable Concept workshops the participants took great joy in pointing out the flaws in ideas. Sometimes the flaws would be a result of a lack of clarity in the illustrations of the concepts. The Pin Thimble was a particular example where the simple sketch did not communicate well the intended idea of biometrics and an appearing and disappearing display of one's PIN. At other times critique came from unpicking the feasibility of the technology, but not necessarily from the perspective of "how would this work?" but instead "what other problems might an idea like this cause?". Smart money was critiqued not because the technology was unfathomable to eighty somethings, but because it was seen to be making valued relationships between people more technologically determined and reduced opportunities for informality and spontaneous acts of trust between friends and family. Across all the activities using our techniques, participants appeared to take great pleasure and pride in being able to identify flaws in the ideas as presented. Indeed, the researchers were also the target of critique—the flaws in ideas being blamed on our naivety in knowing how care relationships worked, or our poor personal banking and savings practices, or our perceived unwavering fascination with technology as a replacement for human agency and responsibility (Fig. 8).

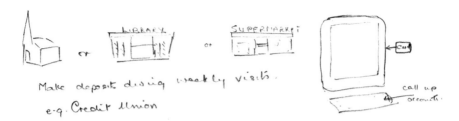

Fig. 8 Examples of sketches of new design propositions from Questionable Concept workshops where participants emphasised ideas that valued social relationships around money, or the importance of designing technologies that fitted their existing practices

Engaging in critique was therefore a fundamental way for participants to find ways into exploring the ideas as presented. However, the criticisms levelled at the ideas were not statements that would end debates; they were often the starting points for continued discussions around the imaginaries presented in the various provocations, the values and politics they implied, and how they related to matters of concern in the present. In the banking for eighty somethings project, the workshops where we used Questionable Concepts and Invisible Design presented opportunities for participants to clearly say "no" to not just our propositions, but to a whole raft of changes they saw happening to banking and finance in the UK. They used the workshops to challenge how the banking sector were, at the time of the research, proposing to withdraw cheques from circulation; they also used our project to challenge the ethics and morals of modern banking, its spendthrift ways, and how this was seen to influence younger generations. As with the origins of co-operative design, where researchers worked closely with workers to challenge the introduction of technologies that would lead to deskilling and impoverished working conditions, critique and saying "no" was a way of challenging the values of new technologies. Saying no though was also an opportunity to critique an existing state of affairs and request a very different type of change. We saw this in the Questionable Concept workshops where new ideas were generated by participants on the back of our provocations that articulated how current services and systems might be better aligned with their values. We also saw this in the theatre workshops where re-drafted scenes would better account for the value of people who give up their time to care for others in the community, and where care was more experience-oriented than task and role driven.

3.3   Fun as a Resource for Design

Some readers might at this point be wondering when this chapter would talk about fun, especially given the nature of the book it lives within! While the research conducted in these various projects was not focused on understanding and designing for 'fun' per se—indeed, the topics examined in most of the projects were considered rather serious matters, potentially dealing with personal vulnerabilities and sensitive information—fun, humour, silliness were critically important dimensions of the techniques when they were put into practice. Blythe et al. (2016) have highlighted the benefits "plain silly ideas" that are partial and ill-formed can bring to exploring, advancing and better defining problem spaces, and this was an explicit feature of the techniques I've described in this chapter. We used humour and silliness in multiple ways across our various provocations. As noted some of the Questionable Concepts were just plain ridiculous ideas intended to provoke strong reactions. It was intended that the Invisible Design films were to have comedic elements to them in order to make the various situations more light-hearted, to develop the characters being portrayed, and to purposely not fixate entirely on the Invisible Design. That the performances in the Experience Design Theatre workshops were either partially scripted or entirely improvised meant there would often be breakdowns of communication between the actors as they responded to each other's actions. Introducing comedic elements, either on purpose or by accident, gave people a way into talking about the ideas. The breakdowns in the performances were typically followed by outpours of laughter and hilarity from participants, and brought down the theatrical walls between the actors and audience. The comedic elements of the films often aided making very serious topics rather more playful, even if what was seen to be "funny" was a matter of taste and some participants failed to get the jokes embedded in the short dialogues. However, without doubt, fun and humour was both a resource for settling nerves and initiating conversations and, over time, became a resource for design. The qualitative and participatory design literature is full of examples of the problems researchers face in having participants simply agree with what researchers produce or say out of politeness, so the absurdity of some of the ideas were an easy way in for our participants to start critique and unpicking ideas.

There is, of course, many dangers of using fun and silliness so explicitly in participatory design processes like these. Researchers can be viewed as trivialising issues, especially when it comes to topics such as identity and security, personal finance, and social care. While for the most part this was not an issue, there were points in workshops where participants became frustrated with what was occasionally noted to be "simplistic" characterisations of people's experiences. I'll conclude by talking about such situations, and other times we have got it wrong, in the following section.

3.4 Getting It Wrong

Building on the last point above, it is easy to get provocation wrong. I noted in Sect. 3.1 that sometimes things we might assume to be provocative are felt to be quite normal by participants and do not adequately play with norms, values and practices to promote debate and discussion. In some workshops we also saw situations where participants were simply unable to make sense of ideas, became disengaged as they did not "get" the provocation, or we felt it necessary to verbally elaborate on the presented ideas. Other times, provocations might be viewed to simply be too absurd or silly, or the use of humour in the concepts themselves (e.g., an Exploding Handbag Questionable Concept, which was intended to deter thieves) or in the dialogues between characters or actors (e.g., comedic elements of the dialogue between the two characters in the Invisible Design films) would fall flat, be misunderstood, or overtly distract participants from exploring the ideas and relationships we were presenting to them. These situations highlight how important it is to not just create carefully produced materials but also to sensitively facilitate discussions, reading the dynamics of the room, and to carefully negotiate a degree of clarity around the purpose of the workshops and the ideas being presented without feeling compelled to apologise for ambiguities or light-hearted elements at the first sign of confusion or frustration. As noted in Sect. 3.2, critique can be a valuable resource for generating new ideas and be inspiration for new forms of change, and often misunderstandings, misreadings and ambiguities can be a great promoter of such critical conversations with participants. Furthermore, it's important to give people space and time to make sense of the ideas being presented to them; Questionable Concepts allowed people to carefully view ideas in their own time at home, while the other techniques have been used as part of projects where we had reoccurring meetings with participants which meant the content of the films or the performances from earlier workshops would continue to resonate and be built upon much further into the projects.

Finally, it's important to note that there is no right or wrong way of creating a provocation; it comes from experience, and very much what makes an idea provocative is dependent on the context being explored and the participants it is being interrogated with. What I have described here are just three examples of using provocation in design processes that I and collaborators have used at different points in time. This is in no way an attempt to claim that these methods are particularly better or more robust than others. There are many other methods and techniques that could be drawn upon, such as extreme characters, pastiches, magic machines, and many more. These various techniques illustrate multiple 'ways in' to using provocation in engagements with participants, and also highlight the great breadth of approaches and techniques designers and researchers can draw upon and make provocation work for them.

Acknowledgements The techniques and projects I have described above have come from a range of projects and are the fruit of collaborations with many fantastic people. Specific thanks must go to Pam Briggs, Mark Blythe, Andrew Monk, Patrick Olivier, Pete Wright, Stephen Lindsay, Paul Dunphy and Tess Denman-Cleaver. The projects described have been funded by a range of sources, including: the EPSRC funded 'New Approaches to Banking for the Older Old' project (EP/H042911/1); the EPSRC Digital Economy theme 'Social Inclusion through the Digital Economy Research Hub' (EP/G066019/1); the Technology Strategy Board funded 'SALT' project (2377-25137); and the EU FP7-ICT 'OASIS' project (215754). Thanks to all the participants across these many projects whom have contributed their time and effort to our research, and put up with our silly ideas and our occasionally strange activities.

References

Andersen K (2012) Making magic machines. In: Proceedings of the 10th European Academy of Design conference

Bardzell J, Bardzell S (2011) Pleasure is your birthright: digitally enabled designer sex toys as a case for third-wave HCI. In: Proceedings of CHI 2011, pp 257–266. ACM Press, New York

Bardzell S, Bardzell J, Forlizzi J, Zimmerman J, Antanitis J (2012) Critical design and critical theory: the challenge of designing for provocation. In: Proceedings of DIS 2012, pp 288–297. ACM Press, New York

Bell G, Blythe M, Sengers P (2005) Making by making strange: defamiliarization and the design of domestic technologies. ACM Trans Comput-Hum Interact 12(2):149–173

Boal A (1979) Theatre of the oppressed. Pluto Press, London

Blythe M, Andersen K, Clarke R, Wright P (2016) Anti-solutionist strategies: seriously silly design fiction. In: Proceedings of CHI 2016, pp 4968–4978. ACM Press, New York

Briggs P, Olivier P (2009) Film as invisible design: the example of the Biometric Daemon. Extended abstracts of CHI 2009, pp 3511–3512. ACM Press, New York

Briggs P, Blythe M, Vines J, Lindsay S, Dunphy P, Nicholson J, Green D, Kitson J, Monk A, Olivier P (2012) Invisible design: exploring insights and ideas through ambiguous film scenarios. In: Proceedings of DIS 2012, pp 534–543. ACM Press, New York

Cabrero D, Winschiers-Theophilus H, Abdelnour-Nocera J, Koch Kapuire G (2016) A hermeneutic inquiry into user-created personas in different Namibian locales. In: Proceedings of PDC 2016, pp 101–110. ACM Press, New York

Chin G Jr, Rosson M, Carroll J (1997) Participatory analysis: shared development of requirements from scenarios. In: Proceedings of CHI 1997, pp 162–169. ACM Press, New York

Dunne A, Raby F (2001) Design noir: the secret life of electronic objects. Birkhäuser, Berlin

Dunne A, Raby F (2013) Speculative everything: design, fiction, and social dreaming. MIT Press, Cambridge

Graham C, Rouncefield M (2008) Probes and participation. In: Proceedings of PDC'08, pp 194–197. ACM Press, New York

Laurel B (1993) Computers as theatre. Addison Wesley, Boston

Lindsay D, Jackson D, Schofield G, Olivier P (2012) Engaging older people using participatory design. In: Proceedings of CHI 2012, pp 1199–1208. ACM Press, New York

Mancini C, Rogers Y, Bandara AK, Coe T, Jedrzejczyk L, Joinson AN, Price BA, Thomas K, Nuseibeh B (2010) Contravision: exploring users' reactions to futuristic technology. In: Proceedings of CHI 2010, pp 153–162. ACM Press, New York

Newell A, Carmichael A, Morgan M, Dickinson A (2006) The use of theatre in requirements gathering and usability studies. Interact Comput 18(5):996–1011

Newell A, Morgan M, Gregor P, Charmichael A (2011) Theatre as an intermediary between users and CHI designers. Extended abstracts of CHI 2006, pp 111–116. ACM Press, New York

Raijmakers B, Gaver W, Bishay J (2006) Design documentaries: inspiring design research through documentary film. In: Proceedings of DIS 2006, pp 229–238. ACM Press, New York

Raptis D, Jensen R, Kjeldskov J, Skov M (2017) Aesthetic, functional and conceptual provocation in research through design. In: Proceedings of DIS 2017, pp 29–41. ACM Press, New York

Rice M, Newell A, Morgan M (2007) Forum theatre as a requirements gathering methodology in the design of a home telecommunication system for older adults. Behav Inf Technol 26(4):323–331

Sanders E, Stappers PJ (2012) Convivial toolbox: generative research for the front end of design. BIS Publishers, Amsterdam

Tanenbaum J, Pufal M, Tanenbaum K (2016) The limits of our imagination: design fiction as a strategy for engaging with dystopian futures. In: Proceedings of LIMITS 2016, article 10. ACM Press, New York

Vines J, Blythe M, Lindsay S, Dunphy P, Monk A, Olivier P (2012) Questionable concepts: critique as a resource for designing with eighty somethings. In: Proceedings of CHI 2012, pp 1169–1178. ACM Press, New York

Vines J, Denman-Cleaver T, Dunphy P, Wright P, Olivier P (2014) Experience design theatre: exploring the role of live theatre in scaffolding design dialogues. In: Proceedings of CHI 2014, pp 683–692. ACM Press, New York

Vines J, Wright P, Silver D, Winchcombe M, Olivier P (2015) Authenticity, relatability and collaborative approaches to sharing knowledge about assistive living technology. In: Proceedings of CSCW 2015, pp 82–94. ACM Press, New York

Chapter 9
Sketching the Polyphonic Design Space of Theme Parks

Abigail Durrant, Michael Golembewski and David Kirk

1 Introduction

Drawing is a fundamental human practice of making sense of the world and communicating ideas. It gives form to acts of imagination and free thought, enabling not only collaboration but also co-creation (Petherbridge 2010). As such, drawing can be fundamental to design practice. We can think about *sketching* as a particular type of drawing activity that helps practitioners refine, explore, and understand complex systems and concepts (Allen 1999). Design sketching is well explored in relation to cognition and expression through action (e.g. Arnheim 1995; Fish and Scrivener 1990; Gedenryd 1998; Schon and Wiggins 1992), and has been open to new interpretation and relevance as computer-related technologies have developed and proliferated (e.g. Gross and Yi-Luen Do 2004; Tovey 1989; Winograd 1996). In the field of human computer interaction (HCI), design sketching practices have been transformed by, amongst other things, new digital tools for making and collaborating (e.g. Scrivener et al. 1994). Creative design techniques have matured for prototyping through sketching as a generative endeavour, in particular to attend to *user experience* (e.g. Buxton 2007; Carroll 2000; Fallman 2003; Moggridge 2007), or to *envision* possible worlds (e.g. Bleecker 2009; Blythe 2014; Kirby 2009; Morrison et al. 2013), or in *visual argumentation and dissemination* (e.g. Gaver 2011, 2012; Gaver and Bowers 2012).

A. Durrant (✉) · D. Kirk
Northumbria University, Newcastle upon Tyne, UK
e-mail: abigail.durrant@northumbria.ac.uk

D. Kirk
e-mail: david.kirk@northumbria.ac.uk

M. Golembewski
Microsoft Research, Cambridge, UK
e-mail: mike.golembewski@gmail.com

© Springer International Publishing AG, part of Springer Nature 2018
M. Blythe and A. Monk (eds.), *Funology 2*,
Human–Computer Interaction Series,
https://doi.org/10.1007/978-3-319-68213-6_9

In this chapter, we consider sketching in an HCI research context that seems apt for inclusion in a book about Funology: the theme park. Specifically, the research project investigated technology and service design opportunities in a UK theme park setting, grounded in experiences of visiting. As members of the project team with backgrounds in art, design, psychology and ergonomics, we present a collective account of ideation and prototyping activities that engaged our project partners and stakeholders. We consider design sketching as a dialogical process (Wright and McCarthy 2005), and describe how we used the medium of sequential art for pictorial expression (Eisner 2008; McCloud 2001) in ways that enabled the complexity of both our setting and our design space to be identified and worked with. We had a precedent for working with this medium to pursue HCI research (Rowland et al. 2010) that informed our approach. Central to our account is a proposition that our sketching process afforded us a deep level of collaborative, empathetic, and speculative engagement with our subject, which we valued as constructive and enjoyable.

In broad terms, the project set out to explore how novel pervasive computing configurations in the theme park could enhance visitor experience. The research was part of a UK-funded academic programme (http://horizon.ac.uk) exploring opportunities for product and service innovation based on new empirical understandings of personal data interactions in people's everyday lives. The project team was interdisciplinary, with other members contributing expertise in computer science and business studies. The theme park was one of a number of settings where we conducted our research, and provided a particular context of cultural engagement to investigate. Our commercial partner operated in the entertainment sector and owned the theme park that became our particular setting. Our approach was experience-centred (McCarthy and Wright 2004, 2015), and involved observational fieldwork and interviews alongside design practice.

Responding to empirical insights and using the medium of sequential art, we—the three authors—developed a picture book (Nikolajeva and Scott 2000) that captured a collection of conceptual designs, depicted in the speculative context of character-driven scenarios (Blythe and Wright 2006; Wright and McCarthy 2005). Our method was informed by the literary theory of Mikhail Bakhtin, after Wright and McCarthy (2005), and Bakhtin's concept of Dialogism (Bakhtin 1981). Specifically, we drew on his conceptualisation of polyphony, a literary device for supporting the expression of multiple authorial voices. The picture book scenarios became a valuable resource, not just for expressing diverse ideas and perspectives within the design space of our interests, but also for fostering dialogue between the research team members, and with our partnering stakeholders, as we will go on to describe.

The project provides an example of a design-based, interdisciplinary inquiry in which the practice of sketching and storyboarding was instrumental to the generation of ideas, collective understanding, and for experiential engagement with the empirical materials. The account of our practice in this chapter has three narrative strands that deliver three interlinked sets of insights. One strand explores the aesthetic experience of sketching as a form of collaborative design inquiry, leveraging

multiple versions of hand-drawn and computer-aided sketches in the development
of ideas. A second strand reflects on the use of sequential art and fictional scenarios
to create and explore a polyphonic design space. Multiple perspectives on theme
park visiting are combined to depict a speculative, character-driven storyworld
(diegesis) that is populated with new design ideas. The third narrative strand
describes how the conceptual design inquiry was pursued in conjunction with
empirical fieldwork, interviews, and evaluations of prototype technology; the
activities were mutually informing. The picture book also enabled a broader conversation with our stakeholders and partners, enabling considerations for strategic
innovation to be elucidated and 'entertained'.

2   The Theme Park

The theme park presents a distinct commercial setting for HCI research. It is a place
with a specific cultural identity, geared around fun, thrill, and memorable experiences, and is routinely experienced across the world (Adams 1991; Cross and
Walton 2005; Jones and Wills 2005). It represents and realises visions of fantastical
and future worlds, inviting its visitors to escape from their ordinary lives. It has also
often represented a space for utopian speculation—the visions of Disney's EPCOT
Center being a notable example (Gennawey 2011), a place where both new interactive technologies and proposals for new ways of living are explored. In this
regard, the theme park is often seen as a space in which the play of possibilities of
new design ideas and technology interactions can come to fruition—a technological
proving ground. Yet despite its 'fantastical' aspects, the park is still a world of
mundane practicalities and orderly interactions, the combination of which gives rise
to a host of intriguing considerations for HCI researchers (Durrant et al. 2012).

Extant studies of leisure settings and practices of *cultural visiting* are extensive
in the humanities and social sciences (e.g. Cross and Walton 2005; Moore 1980),
especially for tourism (Harrison 2001). HCI researchers have explored current
(Brown and Juhlin 2015) and future (Rennick-Egglestone et al. 2016) uses of digital
technology within these settings. Such studies often focus on how technologies
shape practices of visiting. This is particularly notable in Bell's (2002) articulation
of museum spaces as 'cultural ecologies', curated to promote certain kinds of
message and to scaffold certain kinds of experience that will shape visiting as part
of an inherently socio-technical activity extending beyond the museum space itself.
Another conceptual framework that has gained purchase in HCI for studying cultural engagement focuses on how such 'trajectories of experience' (Benford and
Giannachi 2011) may be supported by interaction design.

Many of the HCI studies of visiting practices focus on the roles and use of
mobile technologies (Brown and Chalmers 2003) and mobile 'photoware' (Sarvas
and Frohlich 2011), designed to support engagement with cultural events or
entertainment (e.g. Salovaara et al. 2006; Jacucci et al. 2007a, b), or the capture of
these events for posterity (Durrant et al. 2011a, b). The now-ubiquitous availability

of cameras on smartphones opens up new opportunities for creating photo-based souvenirs (ibid).

The potential role of data as a new form of record within the production and consumption of 'souvenirs' has been a recurrent interest within studies of the 'digital economy' in the UK (Nissen et al. 2014) and within literature on personal informatics (Elsden et al. 2017). The commodification of data has been seen in recent times as a new mass-market ripe for exploitation (Beer and Burrows 2013). This is interesting to consider in the setting of the theme park, as a commercial enterprise that exploits the provision of added value services through its dedicated infrastructure. It is worth noting that our research reported herein was conducted between 2009 and 2011 at a time when there was strategic interest in location based services. Our commercial partner was Merlin Entertainments, and Alton Towers Resort, a theme park managed by Merlin, was the setting for our case. At the time of study, Alton Towers welcomed around 2.8 million visitors per year, forming a resort site with flagship roller coasters set amongst gardens and other attractions, organised around themed zones. At the time of study there was minimal networked infrastructure in place, and the business was keen to speculate on how its souvenir systems could be digitally enhanced.

3   Consuming Experience

Our research at Alton Towers set out to explore how emerging pervasive computing could enhance visitor experiences in the theme park setting. We focused on aspects of *experiential* engagement with the theme park and its consumables. Our obser-vational fieldwork (conducted by Durrant and Kirk) revealed how this park is designed to promote and tightly choreograph 'out-of-the-ordinary' experiences; such experiences were scaffolded and steered by park furniture, landscaping and signage, including maps, queue lanes to rides, and the careful placement of cafes and stalls for consuming refreshments and post-ride souvenirs. Also, individuals within visiting groups used their smartphones and other personal devices to capture and consume their own entertainment and expressions of their visit. This led to the analysis of individual *trajectories* of experience through the park (Benford and Giannachi 2011).

A striking feature of these trajectories was how a 'day in the park', for any given visitor, comprised of a range of experiences that could be positive and negative in emotional and physiological charge. A visit could encompass moments of height-ened sensory awareness like thrill, as well as less memorable or mundane times, or even events that should be forgotten. Diverse affective experiences were expressed *within visiting groups*, too, sometimes about a shared engagement in a ride or another park activity. Such trajectories could be interpreted in narrative terms, and we found value in using Bakhtinian concepts to guide our analysis of participants' accounts. Bakhtin argued that literary devices can be engaged with as a resource for exploring and understanding human experience; his insight has been identified and operationalised for HCI design by Wright and McCarthy (2005), amongst others. In particular his notion of polyphony (1984), taken from the genre of the polyphonic novel, helped us to attend to the 'multivoicedness' (or polyvocality) evidenced in the data. This analytic lens foregrounded the *multiplicity of experiences voiced* within an individual's visit, and visitors' 'reading' and meaning-making around souvenirs, captured in post hoc interviews. Also foregrounded in the data was the *continuing* and *pluralistic* nature of sense making on theme park visiting in groups, and on souvenir consumption.

We were keen to understand how emerging pervasive technologies could become integrated with the aesthetic experiences of visiting. In particular, we were keen to explore how services could be designed to accommodate differing orientations towards 'being entertained' and more actively 'pursuing leisure' (Durrant et al. 2012). We saw an opportunity to develop new kinds of service provision within the park that would support the co-production and consumption of theme park experience by both the park and its visitors. Such service provision could utilise the park's systems as well as the visitors' own personal devices in the *co-creation of souvenirs*.

In the following sections, we describe how we used sketching, scenario development, and related material resources to explore and critically consider some possibilities given form by this design space, of consuming experience in theme parks.

4 Concept Sketching

Our field studies and interviews generated socio-technical insights about how visitors experience a theme park and orient to its technologies and souvenir services (see Durrant et al. 2011a, b). These insights provided the original impetus for the design work, based upon four 'sensitising' considerations. One was to *reconceptualise the materiality of souvenirs* to incorporate digital media. A second was to critically explore what it means to conceptualise the theme park as an *anthropomorphised intelligent park system* with discrete agency for intervening in visitor activities (Fig. 1). A third was for visitors and the park systems to collaborate in the *co-creation of souvenirs*. A final consideration was to explore the notion that visitor interactions may be captured as a digital trail of activity, or '*contextual footprint*', and be leveraged in design. These considerations were initially discussed, critiqued and developed at a design workshop that engaged an interdisciplinary group (c. 20 people) of colleagues and representatives from our industry partnerships. The workshop provided critical perspectives to motivate the next stage of the design process.

Two of this chapter's authors and the designers in our team (Durrant and Golembewski) then used a card-based ideation technique, as detailed in

Fig. 1 Conceptualising the co-creation of souvenirs by the park and its visitors.*Image* A. Durrant

Golembewski and Selby (2010), to build on the workshop outcomes. We made a bespoke deck of cards grounded in the aforementioned considerations. Each suit in the deck was colour-coded to represent a category or axis of thinking, each depicting content about a different feature of the design space—type of park visitor, visitor activities, technologies, avenues for engaging with visitor data (Fig. 2). Different coloured cards were then *combined* to create new design briefs, serving to *think through 'instances' of park activity* and establish meaningful constraints for the designers to respond to. A number of novel conceptual designs were rapidly generated and informally critiqued; these were hand-drawn and rendered in provisional form. It's worth emphasising that these designs were not envisioned as solutions to problems per se, but rather as a means to open up and give form to spaces of possibility and speculation (Dunne and Raby 2013; Gaver 2012). A particular strength of the card technique was that it could be used to explore multiple and alternative design situations (Tohidi et al. 2006). This ideation activity guided the team to consider the full breadth of the design space of possibilities, also generating loose working taxonomies of relevant factors to consider.

Early on in this process we realised that it worked well for us to 'place' or arrange these sketched designs not just in the instances of park activity, but also their inter-relation in the context of an *imaginary theme park setting*, hereafter referred to as The Park. Fictional features and branding were inspired and grounded by the field insights from Alton Towers. Through hand-drawn sketching we imagined how a group of visitors might encounter and interact with each design, and with designs in conjunction with each other (Fig. 3). This led us to create fictional interaction scenarios for our designs, as experienced by imaginary visiting parties. We adopted layout structures from sequential art to develop these scenarios

Fig. 2 Making ideation cards to support concept sketching.*Image* A. Durrant

Fig. 3 Early sketch of characters interacting with a design called Magic Cam in The Park

(McCloud 2001). Sequential art has been increasingly drawn upon in creative and visual approaches to HCI inquiry and project dissemination, in especially in the comic book format, with demonstrable impact (e.g. Dykes et al. 2016a, b; Rowland et al. 2010; Sousanis 2015), and it was a logical extension of our hand-drawn mode. As we considered the physical affordances of collaborative work with sequential art sketches, we arrived at the notion of assembling all of the scenarios in a 'picture book' format (Nikolajeva and Scott 2000). We oriented to the picture book as a developing document made manifest in both digital form and print, that brought the scenarios together into an overarching, polyphonic narrative of 'A day at The Park'.

5 Character-Driven Scenarios

The picture book formed a critical-reflexive tool in an ongoing, iterative process of sketching and sense making that was grounded in the field insights. This process was realised in two ways. First, developing the visual format and storyworld formed a rich exchange of ideas and understandings. Second, after the book was produced it prompted researcher-stakeholder dialogue.

The Park formed a diegesis for the scenarios to unfold in, and was depicted with fictional flagship rides in three imaginary zones, populated with characters visiting in three separate groups. We continued to be inspired by the literary devices of the polyphonic novel in our scenario development. Central to this genre is that the characters and their actions are not determined by their situation, rather they are brought together in dialogue to determine plot action. *Characters thus drive the narrative development of the scenario.* The character creation was inspired and informed by the real-world participants in our previous research, including their behaviours, socio-economic and ethnic backgrounds, and intra-group relations. They formed, in many ways, personas, because they were representative of our research population and the visitor demographic of Alton Towers. But they were also distinct as characters, interacting in ways that were open to the imagination. Significant for our account herein, our engagement with the characters fostered our empathetic connection as designers with theme park visitors-as-technology-users (Blythe and Wright 2006; Wright and McCarthy 2005). We also responded to Nielsen's (2002) critique of personas that emphasises the importance of developing emotionally relatable characters in scenarios.

Let us briefly introduce these characters, visiting The Park in groups. One group comprises three friends: Bill, Bella and Bob (Fig. 4). Bob is the fearful rider; roller coasters make him feel ill. A different group comprises a nuclear family: Jo and

Fig. 4 Characters Bella, Bill, and Bob, depicted as a visiting group of friends

John Jones, and their children Jack (aged 10 years) and Jane (aged 7). Jack is a thrill-seeker and keen Park-goer, whilst his father is fearful of coasters. A third, extended family group is composed of Sam Smith, his father Simon, and grandfather Saul. 'Park' is an anthropomorphised theme park system infrastructure and has considerable agency as a character (Fig. 1), looking for opportunities to engineer fun and serendipitous moments in visitor interaction that promote a certain visiting experience. As these characters developed, we allowed their traits and behaviours to result in certain situations and events, hence referring to character-driven scenarios. For example, the interpersonal dynamics between Bill, Bella and Bob changed during the storyboarding process. The decision to ascribe character names that started with the same letter within each visiting group was made to help readers manage the complexity of the storyworld and enable surface-level as well as deeper levels of engagement—deemed important for differing engagement contexts.

The inquiring function of the scenarios was reflected in the graphical rendering of their content. We set out to create multivoicedness in the narrative (Wright and McCarthy 2005); each scenario was depicted through multiple 'voices' (Fig. 5). First, a visitor perspective captured unfolding visitor interactions and experiences at The Park, through hand-drawn cartoons. Second, a notional 'Park system' perspective depicted a symbolic reading of sensed visitor interaction with Park infrastructure, accompanied by text boxes describing how the Park was making sense of visitor interaction. In developing The Park perspective, we considered how it might interpret and draw inferences from visitor interaction. Third, the researchers' perspective served to ground the fictional narratives in the original field findings, depicted as descriptive notes and, at times providing additional context for the other perspectives. Multivoicedness was further expressed through: (a) the different characters represented within the visitor perspective; (b) the combination of symbolic language and textual account within the Park perspective; and (c) through the multiple voices of the design team from the researcher perspective. This potential of the sequential art medium for expressing different perspectives has also been recognised by Dykes and colleagues, who, in a recent account of a making process that was rendered as a comic, distinguished the lead author's voice through his depiction as a character (2016b).

Our choice of sequential art afforded the presentation of these three perspectives with clarity and comprehensibility (Fig. 6). As the three perspectives developed in conceptual terms, we started to combine hand-drawn paper sketches with computer-generated illustration and formatting, scanning drawings and overlaying and juxtaposing the scans as digital media aided by Adobe Creative Suite applications (Photoshop, Illustrator, and InDesign). In the developing digital layouts, the hand-drawn aesthetic of the visitor perspective became visually distinct from the Park perspective to communicate different 'voices'. The researcher perspective became visually distinct from the perspectives through use of a different colour palette, stylisation and typesetting (Fig. 6). As the graphics developed for the storyboards, we oriented to this researcher perspective as if it formed an interjection or reflective pause-point in the narrative proceedings of the other two perspectives (Fig. 5).

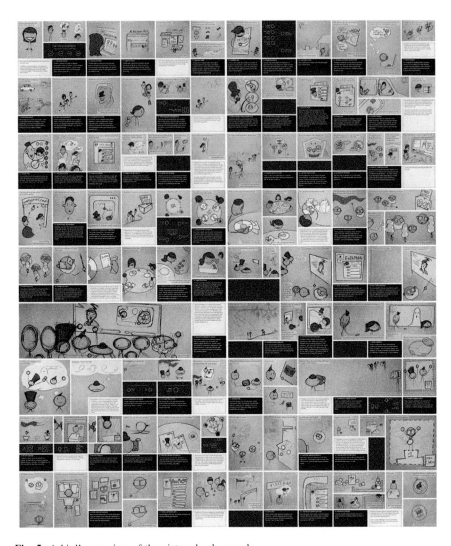

Fig. 5 A bird's eye view of the picture book spreads

As storyboarding continued, we further introduced templates for the layouts, which streamlined our workflow and enabled us to keep focused on the narrative possibilities afforded by interactions in the storyworld. Retaining digital layers, we could rapidly edit text in the storyboards along with other visual elements. The flexible workflow ensured that our working documents remained provisional in form, and open to a degree of revision and redrawing.

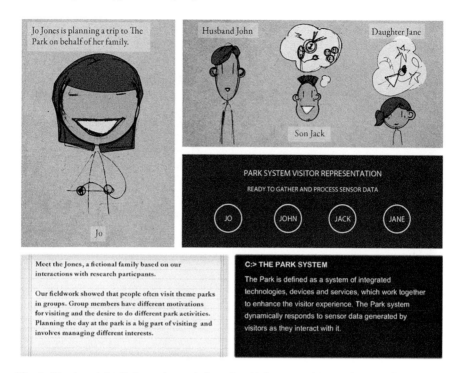

Fig. 6 Storyboard detail shows the rendering of multiple perspectives on the scenario

6 Souvenir Concept Sketches

We now turn to present *extracts* from scenarios that were included in the picture book for this project, to illustrate our account herein. Captured here are three of the eight design concepts and two of the three fictional visiting groups. Scenario content has been adapted for the page formatting of this chapter, and is depicted in indented prose supported by figures. The *visitor perspective is italicised*, the Park perspective is prefixed by a (C: >) command line prompt, and the researcher perspective notes are underlined). For more detail on the designs see Durrant et al. (2011a).

6.1 Daemon Guide

This concept introduces the idea of a customizable virtual agent or bot, providing visitors with a personalized means of interacting with the Park and its systems that is tailored to individual preferences and linked to previous experiences of visiting. The name Daemon references both the entities (daemons) featured in Phillip

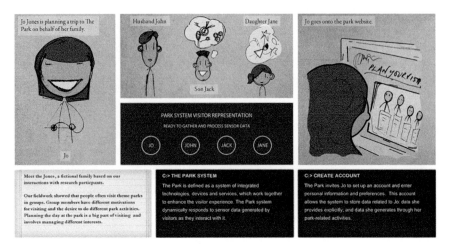

Fig. 7 Example of picture book page layout

Pullmans' 'Dark Materials' trilogy (Pullman 1998), and also the constantly running background tasks found within multitasking computer systems (Fig. 7).

> *Jo Jones is planning a trip to the park on behalf of her family. Jo goes onto The Park's website.*

> Our fieldwork showed that group members have different motivations for visiting. A key challenge when planning a visit was to manage multiple in-group interests and capture individual experiences alongside shared ones.

> C: > CREATE ACCOUNT. Park invites Jo to set up an account and enter personal information and preferences. This account allows the system to store data related to Jo: data she provides explicitly; and data she generates through her Park-related activities. Jo is given a unique identifier, associated with a profile that she creates.

> *Jo is then invited to create a Daemon Guide for her account. Jim-Bug asks Jo about her motivation to visit and who may be visiting with her. Jack already has a Park profile, and a Daemon Guide called Pirate Jones. Jo is invited to link her profile to her son's profile.*

> *Across town, Bill proposes a visit to the Park with his friends, Bob and Bella. Bill is recognised as a frequent Park visitor and already has a Daemon Guide. Bill's Daemon uses Facepage, a social network site, to notify him about a new ride and invites him to visit.*

As the scenario progresses, more aspects of Daemon Guide creation are explored. The storyboards reflect how the social dynamics of visiting and repeat visiting noted in the field findings motivated the technical and logistical aspects of Daemon Guide interaction (Figs. 8 and 9).

> *The next week, the Jones arrive at The Park, followed shortly after by Bill, Bob and Bella.*

> C: > LOCATION SENSING. Upon arrival at the entry gates, position-sensing technologies locate Park's visitors. Park provides them with tailored information - e.g. provide estimated queue times or signpost nearby attractions.

Fig. 8 Storyboard detail depicting the visitor perspective

6.2 Snackshacks

The group consumption of refreshments in The Park provides a context for the social consumption of souvenirs. Snackshacks cater for this. Visitors view, share and triage park-captured media on interactive table-tops whilst consuming refreshments.

C: > DAEMON SUSTENANCE RECOMMENDER

Upon leaving the Heavy Weather Zone, the Jones' family reunite. Whilst Pirate Jack is excitable, John looking pale needs something to eat. Jim Bug suggests that the Jones' visit a nearby Snackshack on site.

Field findings showed how energy levels shaped visitors' experiences over the course of their visit. Sustenance was a key talking point between visitors, and taking time to eat together was a factor in ad hoc planning. It was also found to direct emotional states and social dynamics.

The Jones' sit at a Snackshack table. Virtual placemats magically appear.

C: > SNACKSHACK PLACEMATS: Each visitor's Park activity up to the present time is depicted on a table-top display. Each 'placemat' resembles a pie chart, the size of each

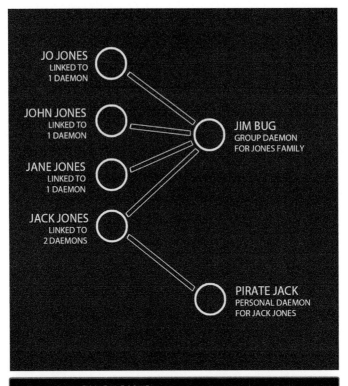

Fig. 9 Storyboard detail depicting the Park perspective

portion representing the relative portion of time spent with a colour-branded attraction since entry.

When Jane presses one of her placemat portions, media captured around the associated ride fans out on the table. Jane laughs at the fearful expression on her father's face, from being in the Heavy Weather Zone. She adds one of his photos to her personal collection.

C: > CENTRALISED MEDIA MANAGER: The Jones' generate a volume of media at The Park on personal mobile devices, which can be automatically uploaded and stored on

Park servers, along with Park-generated media, and then viewed and edited on Park interfaces.

C: > CONTEXTUAL INFERENCE [JANE JONES] Through tracking Jane's recorded contextual data footprint, Park makes inferences about her happiness 'levels'. Jane has queued for several rides but not ridden any. Her profile shows her age. Park alerts the Snackshack staff to order Jane a free gift.

6.3   *Magic Cam*

Magic Cam is an integrated in-park photography system that leverages an extensible network of locative technologies and resident cameras (Park Cams) distributed across the site (e.g. in queues, ride cars). A key interface is a handheld viewfinder called Magic Cam, with a camera selection dial and a shutter button. Magic Cam users can operate the Park Cams and view live video feeds from across this camera network, including those located in inaccessible places such as on rides. It also enables the real-time capture of photos from these feeds (see Fig. 3). Upon hire, the Magic Cam ID is associated with the visiting group.

C: > ID FLAME RIDERS IN QUEUE

Bill, Bob and Bella approach the Flame ride in the Heavy Weather Zone. Bill and Bella want to queue for the Flame Ride, but Bob doesn't want to. He's getting a headache.

In our fieldwork we found that visitors were frustrated when they couldn't see their friends on a given ride, and share in the thrill. This also meant that they couldn't capture the riders using their personal cameras.

Bill gives Bob a Park Magic Cam that he hired upon Park entry, and asks him to take on-ride photos on his behalf. Bob obliges as 'Magic Cam Photographer', whilst Bill and Bella join the queue for The Flame. Whilst in the queue, they look into one of the Park Cams and wave to Bob. Bob sees the Park Cam 1 view through the Magic Cam viewfinder.

C: > PROXIMITY CAMERA VIEWS: Magic Cam senses the Park Cam proximal to group members, in this case, Bill and Bella. It also senses and excludes the photographer, Bob.

Bill and Bella board separate cars on the Flame ride.

Location sensors let Park know which car each rider is occupying. We found that spectators wanted the means to engage with and capture riders' experiences in real time.

By turning the dial on the side of the Magic Cam, Bob can switch between the different views of Bob and Bella. Bob finds these views pretty exciting, although still feels frustrated.

At Flame's scary part, Bob presses the Magic Cam shutter.

C: > MAGIC CAM CAPTURE: When Bob presses the shutter, Flame Cams One and Three simultaneously capture photos of Bill and Bella and send them to the Park servers.

Later on, Bill, Bob and Bella stop for a drink and look over the Flame photos.

We found our real-world visiting groups wanted to review, edit and exchange media captured within them over the course of their park visit. Magic Cam and the Snackshack tables afford this collaborative photowork, and the in-park consumption of photos and video media.

C: > MEDIA ARCHIVING: Data is not kept on Park servers. Park takes note when visitors leave through its gates.

When Bella was looking through her media archive, she came across only two photos of Bob from the day at The Park. In one of the photos he was captured pulling a face at her behind her back. This must've been taken on accident by one of the Park Cams. Viewing this photo made Bella angry and she hasn't spoken to Bob since.

C: > CUSTOM SOUVENIRS: Post-visit, visitors can fashion bespoke souvenirs using online services. Daemons use contextual tags to auto-generate media stories from one or more visits.

7 Engaging Design Through the Picture Book

We've described how the picture book spreads afforded the expression of poly-phony within the character-driven scenarios (e.g. Figs. 10, 11, 12, 13 and 14) and, in turn, within our design space (e.g. Figs. 15, 16, 17, 18 and 19). However, as introduced above, we found the iterative sketching process useful for establishing dialogical understanding between the members of the research team. The sequential art mode helped us acknowledge, maintain, and integrate our personal contributions within our design process. We each identified with particular, complementary, aspects of the process so that *none of us had a distinct authorial voice*. The two designers in the team assumed roles in the production of the storyboards, with one preparing the original hand-drawings (Durrant) and the other developing the digital renderings and inking (Golembewski). The spatial arranging of the storyboards was deeply collaborative. Tangible drawings and print-outs afforded the collaborative

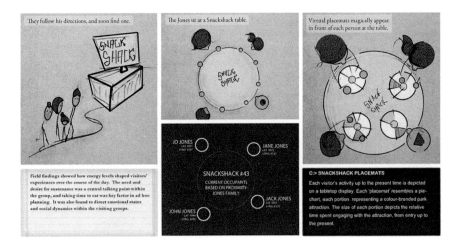

Fig. 10 Picture book spread depicting Park and Snackshack designs

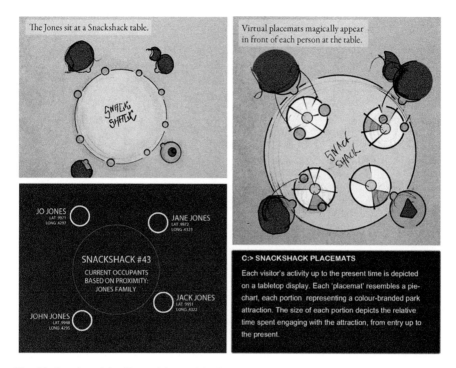

Fig. 11 Storyboard detail: on visitor and Park interpretations of activity

Fig. 12 Picture book spread: on creating and consuming souvenirs in The Park

The Snackshack place mats make visible the different activities people have done on the visit, indicating different individual trajectories of experience through the park. We found that visitors wanted the opportunity to review and share media captured by the park, to create a record of the visit. Visitors wanted to retain access to all the media that was captured from the whole visit, so that they could retain the option to make editorial decisions later on, and with a holistic view.

Fig. 13 Storyboard detail depicting the researchers' perspective

Fig. 14 Picture book spread depicting Park interacting with its visitors

Fig. 15 Picture book spread depicting Park making contextual inferences

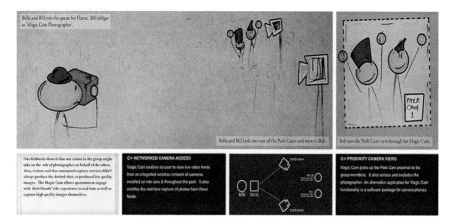

Fig. 16 Picture book spread depicting multiple perspectives from Park and its visitors

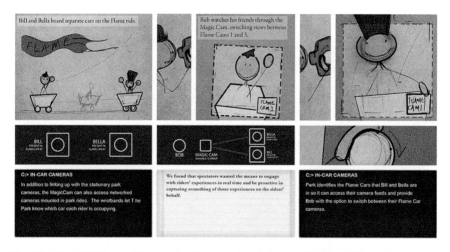

Fig. 17 Picture book spread: on co-creating souvenirs between visitors and The Park

organising of content into frames, storyboards, and page layouts. In this activity we worked through characters' encounters with a given design element, and their interactions with its functional features—whether serendipitously or in pursuit of particular goals. Our workflow encouraged this inquiry whilst ensuring that the storyboards had visual coherence and that the production work was efficient to maximize our time for ideation. The experience of sketching-as-inquiry was constructive because it felt like our process flowed in the layering of material forms and resources that we shared to-hand.

Shifting from our 'researcher' engagement with the book, we may also reflect on how the multi-perspectival visualisations also enabled dialogue with the book's other readers, once it was produced: the members of our wider research team, and

Fig. 18 Book spread detail: on co-creating souvenirs

the project partners and stakeholders. As McCloud (2001) and others including Rowland et al. (2010) and Dykes et al. (2016a, b) have noted, a valued feature of the sequential art form is that it is accessible to multiple levels of reading. The pictorials are widely accessible, and, technical or academic language aside, they communicate effectively at a glance. There is a relatively linear reading of the book, considering the instances of activity as they unfold in time, from page to page. Alternatively, a reader may choose to follow one perspective more closely, or jump across pages and storyboards to analyse a particular design in a different context of

Fig. 19 Interacting with multiple designs in The Park

use. The visual coding of perspectives was intended to aid such alternative readings. As also noted by McCloud (2001), the gaps between the panels in a storyboard become almost as important as the panels themselves; the careful pacing and layout of the designs enabled the inference of storyworld elements between scenes. This was intended to support our readers' imaginative exploration of the design space— as it had supported ours when we developed the book.

The picture book was circulated, in both digital and print form, to others in the wider project team including colleagues working in business and management studies. It was a resource in round-table gatherings, to consider strategic design implications and opportunities for business modelling around new forms of souvenir consumption in the entertainment industries. These colleagues reported finding the book valuable for giving form to near-future design directions, and making 'real' and 'consequential' the concepts and people's interactions with them. Participants described how the characters felt more 'alive' and 'relatable' than personas because of their closeness in representational terms to the individuals who participated in our empirical studies; this was emphasised in the juxtaposition of the researcher voice and the visitor voice in the storyboards. They felt that they could link the empirical instances to possible interactions in the storyworld, in a way that enabled them to anticipate character *reactions*. This also meant that the designs were open to critique and iteration. Indeed, this sense of unfinalisability was also communicated through the graphics; the original hand-drawn sketches were kept visible behind layers of digital rendering, which was picked up in the round-table discussions.

We also presented the scenarios to our partners at Merlin for their consideration. In this context the picture book usefully enabled a level of nuance and complexity about The Park to be communicated to a large number of people in a very short time. Certain features of a system or service could be considered with quite precise technological configuration. The printed book form afforded straightforward navigation and glanceability, aiding group discussion. One tension with presenting the content in book form to industry partners was that it risked elevating the designs to potential product proposals. However, the provisionality of the designs, reinforced by their sketched rendering, implied that design thinking and envisioning in the conceptual space was agile and open to ongoing development, a valued feature of the generative 'research through design' endeavour (Gaver 2012).

8 Picture Book for Prototyping

Having developed the conceptual designs, and critically discussed them, we were interested to select certain features and functions to develop further, prototype, and deploy back in the 'real world' of Alton Towers theme park, for the participants from our original field studies to experience. Based on the wider team and stakeholder engagement, we drew out particularly interesting 'design interactions' and 'instances of activity' to synthesise. We subsequently developed functioning physical prototypes of aspects of an envisioned souvenir service called 'Automics' (see Durrant et al. 2011b for a full account of this prototyping work).

The Automics service itself was based on a smartphone application (for the Android platform) that was designed to support *souvenir co-creation*. The app provided templates for its user(s) to arrange a montage of provided visual media, auto-captured by the existing park systems, and personally captured on the smartphones of visiting groups. The design of this service was only partially prototyped and implemented: critical 'user touchpoints' with functioning features were built at 1:1 scale for field deployment at Alton Towers; other aspects of the service were non-functioning and were staged in the field using a Wizard of Oz method (Martin and Hanington 2012); and the proposed service as a whole, along with its integration into a bigger network of systems and park infrastructure was left as envisioned in the fictional Park of the picture book. The 'big picture' afforded by the book, of The Park as a socio-technical landscape, provided an important scaffold for designing Automics as an end-to-end service, with functioning features to respond to particular instances of activity—intersecting with trajectories of visiting (Durrant et al. 2011b).

A field evaluation of Automics, in turn, provided further experiential insight about potential models of *souvenir consumption* that speak to a digital economy of co-creating data-as-product. At the end of the project with the theme park, these new insights were brought together with the picture book content and the physical prototypes to deliver a set of understandings that were speculative and future-oriented whilst remaining empirically grounded. These deliverables, as an assemblage, offered concrete examples of possible worlds made human-scale, relatable, and remaining open to interpretation.

9   Closing Discussion

In this chapter we have described how a creative design process furthered an interdisciplinary programme of research about cultural visiting mediated by pervasive interactions with personal data and personalised souvenirs. We have reported on a collaboration anchored around the sketching of design concepts embedded in fictional, character-driven scenarios and assembled in a picture book. We explained how the scenarios were developed from empirical insights and used by the project team to advance design understanding. Taken together, the scenarios have helped deliver strategic design implications for shaping socio-economic, infrastructural and human concerns in the theme park and related settings. The scenarios also scoped out a rich design space for novel forms of souvenir creation and consumption.

Returning to our three narrative strands, we consider the value of sketching as inquiry in the sequential art mode. This approach facilitated a rewarding collaborative process which fostered the flow of ideas and the envisioning of multiple perspectives on a set of designs and their potential instances of use. The literary devices we used, in particular those from the polyphonic novel, guided us in composing and speculating on the narrative of the scenarios and the storyworld of The Park. This approach offered us conceptual resources for designing to foster dialogue and provisionality around the creation of characters. Arguably, imaginary users and their interactions brought as much to the shaping of the design space as the real-world research participants did. Moreover, by exploring how non-human characters (e.g. systems) may be given voice and agency (Fig. 12), we expand upon Dykes et al. (2016a) to examine how a system might be made scrutable and accountable within a narrative.

Leading from this and returning to the second and third narrative threads, we have shown how sequential art afforded the visual presentation of multiple perspectives to convey complex human-data interactions—in this case with an extensible theme park infrastructure. We reiterate here that our conceptual designs were positioned as critical, speculative devices to foster ideation around instances of park activity, and possible near futures, by the partners and stakeholders in our research. We highlight the efficacy of the sketching and storyboarding for enabling the development and critique of concepts more rapidly and flexibly than may be possible with physical prototyping (Blythe 2014). We've also demonstrated how this conceptual work can be pursued in conjunction with physical prototyping and empirical studies.

Gaver has critiqued scenario use in HCI design processes, in particular suggesting that it may create a 'unitary vision' that closes down interpretation of how the designs may be used, evaluated, and further developed (Gaver 2011, p. 1553). We argue that by developing characters inspired by our real-world participants, and allowing these characters to ground, inspire and direct the unfolding of design interactions in our storyworld, we could open up and entertain spaces for interpretation and empathetic connection. This potentially suggests a bridging between prototyping and the purely fictive scenarios explored by Blythe (2014) and others.

However, as advocated by Blythe (ibid.), our theme park scenarios did provide a vehicle for the generation of *narrative rather than design solutions*. By juxtaposing and weaving through multiple accounts of sense making, and retaining layers of process and provisionality in our renderings—for example making visible the hand-drawn sketches under the computer graphics, we arguably sustained dialogical engagement around the various interactional elements in the design space and with the voices of those involved in the research.

Through our account herein, we hope to have demonstrated to our readers the rich potential of sketching in the sequential art mode to pursue creative and generative design inquiry. The methodological insights gained from this work have since been applied to another HCI research setting, in continuing efforts to critically explore ways to draw and design in conversation with empirical materials, and leverage the picture book format (Durrant et al. 2015, 2016). In current and future work, we extend modes of inquiry and media into online video presentational platforms, opening up further spaces for dialogical encounter and polyphonic design.

Acknowlegements This research was funded by Horizon Digital Economy Institute (EP/G065802/1). The first author was supported in developing this work by The Leverhulme Trust (ECF-2012-642). We also thank our research partners Merlin Entertainments Group Limited and Picsolve International Limited, and we thank the staff at Alton Towers Resort along with our colleagues at Nottingham University for their participation in the project. We thank those who took part in the related field studies, for having fun at Alton Towers in aid of our research. We give special thanks to Steve Benford for directing the project 'A Day in the Park' and for supporting and discussing the picture book development. Image credits for Figures 3 to 19 inclusive: Abigail Durrant and Michael Golembewski.

References

Adams JA (1991) The American amusement park industry: a history of technology and thrills. Twayne, Boston, MA

Allen S (1999) Points + lines: diagrams and projects for the city. Princeton Architectural Press

Arnheim R (1995) Sketching and the psychology of design. In: Margolin V, Buchanan R (eds) The idea of design. MIT Press, Cambridge, MA

Bakhtin MM (1981) In: Holquist M (ed) The dialogic imagination: four essays (trans: Emerson C, Holquist M). University of Texas Press, Austin, London

Bakhtin MM (1984) In: Emerson C (ed) Problems of Dostoevsky's poetics (trans: Emerson C). University of Minnesota Press, Minneapolis

Beer D, Burrows R (2013) Popular culture, digital archives and the new social life of data. Theory Cult Soc 30(4):47–71. https://doi.org/10.1177/0263276413476542

Bell G (2002) Making sense of museums: the museum as a cultural ecology. Technical report. Intel Corporation

Benford S, Giannachi G (2011) Performing mixed reality. The MIT Press

Bleecker J (2009) Design fiction: a short essay on design, science, fact and fiction. Near Future Lab, vol 29. http://tinyurl.com/jbbhvg4. Accessed 09 Sept 2015

Blevis E, Hauser S, Odom W (2015) Sharing the hidden treasure in pictorials. Interactions 22(3): 32–43

Blythe M (2014) Research through design fiction: narrative in real and imaginary abstracts. In: Proceedings of the SIGCHI conference on human factors in computing systems. ACM Press, pp 703–712

Blythe MA, Wright PC (2006) Pastiche scenarios: fiction as a resource for user centred design. Interact Comput 18(5):1139–1164

Brown B, Chalmers M (2003) Tourism and mobile technology. In: Proceedings of ECSCW'03. Springer, New York, NY, USA, pp 335–354

Brown B, Juhlin O (2015) Enjoying machines. MIT Press

Buxton B (2007) Sketching user experiences: getting the design right and the right design. Morgan Kaufmann

Carroll JM (2000) Making use: scenario-based design of human-computer interactions. MIT Press

Cross GS, Walton JK (2005) The playful crowd: pleasure places in the twentieth century. Columbia University Press, New York, NY

Dunne A, Raby F (2013) Speculative everything: design, fiction, and social dreaming. MIT Press

Durrant A, Golembewski M, Kirk DS, Benford S, Rowland D, McAuley D (2011a) Exploring a digital economy design space in theme parks. In: Proceedings of the second conference on creativity and innovation in design. ACM Press, pp 273–284

Durrant A, Rowland D, Kirk DS, Benford S, Fischer J, McAuley D (2011b) Automics: souvenir generating photoware for theme parks. In: Proceedings of the SIGCHI conference on human factors in computing systems (CHI'11). ACM, New York, NY, USA, pp 1767–1776. https://doi.org/10.1145/1978942.1979199

Durrant A, Kirk DS, Benford S, Rodden T (2012) Pursuing leisure: reflections on theme park visiting. Comput Support Coop Work 21:43. https://doi.org/10.1007/s10606-011-9151-1

Durrant AC, Trujillo Pisanty D, Moncur W, Orzech KM (2015) Charting the digital lifespan: a picture book. Newcastle University, UK. ISBN 978-0-7017-0250-2

Durrant A, Moncur W, Kirk D, Trujillo-Pisanty D, Orzech KM (2016) On presenting a rich picture for stakeholder dialogue. In: Proceedings of design research society conference (DRS 2016). https://doi.org/10.21606/drs.2016.169

Dykes T, Blythe M, Wallace J, Thomas J, Regan T (2016a) RtD comics: a medium for representing research through design. In: Proceedings of the 2016 ACM conference on designing interactive systems (DIS'16). ACM, New York, NY, USA, pp 971–982. https://doi.org/10.1145/2901790.2901821

Dykes T, Wallace J, Blythe M, Thomas J (2016b) Paper street view: a guided tour of design and making using comics. In: Proceedings of the designing interactive systems conference. ACM, Brisbane, Australia

Eisner W (2008) Comics and sequential art. W.W. Norton, London. ISBN: 978-0-393-33126-4

Elsden C, Durrant AC, Chatting D, Kirk DS (2017) Designing documentary informatics. In: Proceedings of designing interactive systems (DIS 2017). ACM, New York, NY, USA. http://dx.doi.org/10.1145/3064663.3064714 (forthcoming)

Fallman D (2003) Design-oriented human-computer interaction. In: Proceedings of the SIGCHI conference on human factors in computing systems. ACM Press, pp 225–232

Fish J, Scrivener S (1990) Amplifying the mind's eye: sketching and visual cognition. Leonardo 23(1):117–126

Gaver W (2011) Making spaces: how design workbooks work. In: Proceedings of the ACM conference on human factors in computing systems (CHI 2011). ACM, pp 1551–1560

Gaver W (2012) What should we expect from research through design? In: Proceedings of the SIGCHI conference on human factors in computing systems. ACM, Austin, Texas, USA, pp 937–946

Gaver W, Bowers J (2012) Annotated portfolios. Interactions 19(4):40–49

Gedenryd H (1998) How designers work. Lund University, Sweden. Retrieved from: http://archive.org/details/HowDesignersWork-MakingSenseOfAuthenticCognitiveActivity

Gennawey S (2011) Walt Disney and the promise of progress city. Theme Park Press, pp xiii. ISBN: 1941500269

Golembewski M, Selby M (2010) Ideation decks: a card-based design ideation tool. In: Proceedings of the 8th ACM conference on designing interactive systems (DIS'10). ACM, New York, NY, USA, pp 89–92. http://dx.doi.org/10.1145/1858171.1858189

Gross MD, Yi-Luen Do E (2004) Three R's of drawing and design computation. In: Design computing and cognition'04, Springer, Dordrecht, pp 613–632

Harrison J (2001) Thinking about tourists. Int Sociol 16:159–172

Jacucci G, Oulasvirta A, Ilmonen T, Evans J, Salovaara A (2007a) CoMedia: mobile group media for active spectatorship. In: Proceedings of the SIGCHI conference on human factors in computing systems (CHI'07). ACM, New York, NY, USA, pp 1273–1282

Jacucci G, Oulasvirta A, Salovaara A (2007b) Active construction of experience through mobile media: a field study with implications for recording and sharing. Pers Ubiquitous Comput 11(4):215–234

Jones KR, Wills J (2005) The invention of the park: from the Garden of Eden to Disney's magic kingdom. Polity Press, Oxford

Kirby D (2009) The future is now: diegetic prototypes and the role of popular films in generating real-world technological development. Soc Stud Sci 40:41–70

Martin B, Hanington B (2012) Universal methods of design. Rockport

McCarthy J, Wright P (2004) Technology as experience. MIT Press

McCarthy J, Wright P (2015) Taking [a]part: the politics and aesthetics of participation in experience-centered design. MIT Press

McCloud S (2001) Understanding comics. William Morrow paperbacks

Middleton D, Brown SD (2005) The social psychology of experience: studies in remembering and forgetting. Sage

Moggridge B (2007) Designing interactions. MIT Press

Moore A (1980) Walt Disney World: bounded ritual space and the playful pilgrimage center. Anthropol Quart 53(4):207–218

Morrison A, Tronstad R, Martinussen ES (2013) Design notes on a lonely drone. Digital Creativity 24(1):46–59. https://doi.org/10.1080/14626268.2013.768534

Nielsen L (2002) From user to character: an investigation into user-descriptions in scenarios. In: Proceedings of the 4th conference on designing interactive systems: processes, practices, methods, and techniques. ACM, pp 99–104

Nikolajeva M, Scott C (2000) How picturebooks work. Garland Press

Nissen B, Bowers J, Wright P, Hook J, Newell C (2014) Volvelles, domes and wristbands: embedding digital fabrication within a visitor's trajectory of engagement. In: Proceedings of the 2014 conference on designing interactive systems (DIS'14). ACM, New York, NY, USA, pp 825–834. http://dx.doi.org/10.1145/2598510.2598524

Petherbridge D (2010) The primacy of drawing: histories and theories of practice. Yale University Press

Pullman P (1998) Northern lights: his dark materials, vol 1. Scholastic Point

Rennick-Egglestone S, Brundell P, Koleva B, Benford S, Roussou M, Chaffardon C (2016) Families and mobile devices in museums: designing for integrated experiences. J Comput Cult Herit 9(2), Article 11, 13 p. http://dx.doi.org/10.1145/2891416

Rowland D, Porter D, Gibson M, Walker K, Underwood J, Luckin R, Smith H, Fitzpatrick G, Good J, Walker B, Chamberlain A, Rennick Egglestone S, Marshall J, Schnädelbach H, Benford S (2010) Sequential art for science and CHI. In: CHI'10 extended abstracts on human factors in computing systems (CHI EA'10). ACM, New York, NY, USA, pp 2651–2660. http://dx.doi.org/10.1145/1753846.1753848

Salovaara A, Jacucci G, Oulasvirta A, Kanerva P, Kurvinen E, Tiitta S (2006) Collective creation and sense-making of mobile media. In: Proceedings of the SIGCHI conference on human factors in computing systems (CHI'06). ACM, New York, NY, USA, pp 1211–1220

Sarvas R, Frohlich DM (2011) From snapshots to social media—the changing picture of domestic photography, 1st ed. Springer, Berlin

Schon DA, Wiggins G (1992) Kinds of seeing and their functions in designing. Des Stud 13(2):135–156

Scrivener SAR, Clark SM (1984) Sketching in collaborative design. In: MacDonald L, Vince J (eds) Interacting with virtual environments. Wiley, Chichester, UK
Sousanis N (2015) Unflattening. Harvard University Press, Cambridge, MA
Tohidi M, Buxton W, Baecker R, Sellen A (2006) Getting the right design and the design right. In: Proceedings of the SIGCHI conference on human factors in computing systems (CHI'06). ACM, New York, NY, USA, pp 1243–1252. http://dx.doi.org/10.1145/1124772.1124960
Tovey Michael (1989) Drawing and CAD in industrial design. Des Stud 10(1):24–39
Winograd T (1996) Bringing design to software. Addison-Wesley, Reading, MA
Wright P, McCarthy J (2005) The value of the novel in designing for experience. In: Pirhonen A, Isomaki H, Roast C, Saariluoma P (eds) Future interaction design. Springer, Berlin

Chapter 10
Playful Research Fiction: A Fictional Conference

Ben Kirman, Joseph Lindley, Mark Blythe, Paul Coulton,
Shaun Lawson, Conor Linehan, Deborah Maxwell, Dan O'Hara,
Miriam Sturdee and Vanessa Thomas

1 Introduction

Fiction has long been important to Human-Computer Interaction (HCI) research and practice. Through familiar tools such as personas, scenarios and role-play, fictions can support the exploration and communication of complex psychological, social and technical requirements between diverse collections of designers, developers and end-users. More recently, HCI and design research has embraced the development and evaluation of make-believe technologies as a way to speculate and study the possible future effects of technological innovation, since it enables us to unpack and understand the implications of technology that does not yet exist. In this chapter we explore the weird relationship between fiction and technology research through the lens of a fictional conference, a playful project that gathered ideas about fiction in research *through* fictional research, and explore the fluid relationship between the real and unreal in HCI.

In terms of practical fictions, there is growing use of fiction to explore implications of emergent technologies. For example, Lawson et al.'s (2015) speculative prototypes of wearable technologies for cats and dogs give insight into unforeseen

B. Kirman (✉) · D. Maxwell
University of York, York, England, UK
e-mail: ben.kirman@york.ac.uk

J. Lindley · P. Coulton · M. Sturdee · V. Thomas
Lancaster University, Lancaster, England, UK

M. Blythe · S. Lawson
Northumbria University, Newcastle upon Tyne, England, UK

C. Linehan
University College Cork, Cork, Republic of Ireland

D. O'Hara
New College of the Humanities, London, England, UK

© Springer International Publishing AG, part of Springer Nature 2018
M. Blythe and A. Monk (eds.), *Funology 2*,
Human–Computer Interaction Series,
https://doi.org/10.1007/978-3-319-68213-6_10

ramifications of this technology on both pets and their owners. The real-yet-fictional objects are known as "diegetic prototypes", in that they belong to a larger imagined story world, as part of what science-fiction author Sterling (2005) terms "Design Fiction". They are fictional objects that "make sense on the page" and help suspend disbelief about the world in which they exist. Although this includes fantastic visions of phasers and space battles, in HCI, these story worlds are usually around plausible near-future scenarios, such as those where drones enforce parking restrictions (Lindley and Coulton 2015b) or dogs can access the Internet (Kirman et al. 2017). Design fiction remains a contested space, including among the authors of this chapter: Blythe (2017) notes that despite often being seen as isolated "objects", they implicitly follow traditional plot patterns, where they don't already explicitly exist as part of wider performances, such as Buttrick's et al. (2014) erotic stories about modems or Elsden's et al. (2017) "speculative enactment" of a fictional datagraphic wedding service. Meanwhile, Coulton et al. (2017) explicitly reject the notion that narrative and plot is central to design fiction and instead cast them as groups of speculative artefacts that, when viewed together, have the ability to invoke and define the properties of artificially constructed worlds.

These fictional prototypes, and the story worlds they represent, carry persuasive, and therefore critical, powers. Just as design fiction is exploited by corporations to sell slick visions of the future, where Iron Man's suit always works and Siri understands even the thickest Scottish accent, it is also used to imagine critical (Dunne and Raby 2013) and adversarial (Di Salvo 2011) futures based on different value structures. Although the term "design fiction" seems to be extremely broad, Lindley and Coulton (2015a, b) argue for embracing the ambiguity of the term as a core feature, in that it defines fiction as provocation for discourse around the desirability of imagined interactions rather than an evaluation of the interaction itself.

It is clear why design fiction is of such great interest to technology researchers in particular, as a field seemingly obsessed with defining proximal techno-utopian futures through idealized gadgetry, from Google Glass to robots in nursing homes. Along with the proliferation of academic papers concerned with reporting studies based on fictional prototypes, two major HCI conferences, ACM GROUP 2016 and NordiCHI'16 both explicitly solicited fictional submissions in tracks dedicated to design fiction research and practice.

2   Play in Research Fiction

Writing stories, building worlds and telling tales are rewarding pastimes, and the "play" of creating fictions gives a special freedom to authors to push boundaries and experiment with unusual scenarios. The fun of world-building has naturally informed the work of design fiction creators, who use this freedom to explore the edges of both what is possible in this form, but also how it is understood as a research method.

For example, Blythe et al. (2016) report on a series of workshops where participants created a range of "silly" prototypes, explicitly in reaction against the perceived "solutionist" (Morozov 2013) stance implicit in much HCI research. Encinas and Blythe's (2016) "author eraser" is described as a tool that removes the names of the authors that made only minor contributions to research papers, a function that only makes sense to an academic audience. Buttrick's et al. (2014) "50 Shades of CHI" is a series of erotic vignettes that are critical of how the HCI research community describes relationships between humans and technology. Baumer et al. (2014) collected a series of vignettes describing fictional futures of HCI, and similarly, Kirman et al. (2013) gave an in-character performance at an academic conference, where they claimed to be robots from the future sent back through time to congratulate HCI researchers on hastening the enslavement of humankind by machines. In a bizarre turn of events, this performance inspired the creation of a young adult science fiction novel that has since been read over a million times (Adams and Moreau 2015; Dalton et al. 2016). All of these examples are interesting in their use of humour, particularly through irony (Blythe and Encinas 2016), to critique the culture and practice of technology research. Rather than aiming outward at a wider audience, these playful "research fictions" use the forms typical of this kind of research—prototypes, papers and presentations—to build fictions that critique values of academic research directly.

Of particular note is the paper "Game of Drones" (Lindley and Coulton 2015a, b), since, in contrast to the previous examples, a deliberate pursuit of plausibility ultimately lead to a deceptive amount of ambiguity (Coulton et al. 2016). The project consisted of a paper and poster, describing an entirely fictional project, submitted to a real conference (ACM SIGCHI Symposium on Computer-Human Interaction in Play 2015). Although papers with unreal content are not uncommon (both in the examples above, and e.g. Lem 1973; Zongker 2006; Mazières and Kohler 2005), this was unique in how it was presented as a real project. The authors have since discussed both the idea of fictional research papers and the reviewer response to the same (Lindley and Coulton 2016a, b).

3 A Fictional Conference

This trajectory of fictional abstracts (Blythe 2014, Blythe and Buie 2014) through fictional papers (Lindley and Coulton 2016a), led the authors of the current chapter to consider the natural next step—a fictional conference containing only fictional research.

Apart from the humour inherent in such an endeavour, which was a prime motivator, it also provided a playful way to collect thoughts and reflections on the fluid understanding of design fiction research from a broad audience who might not be already involved in this kind of work. To this end, we built a conference committee (the current authors) and circulated a Call for Papers soliciting submissions for the Fictional Conference on Design Fiction's Futures (FCDFF).

Throughout the conception and development of the conference, it was important to the committee that this example of design fiction was coherent, rather than a simple joke. As experienced design fiction creators, we were concerned with the texture of the 'diegetic landscape' of the project. Although meaningless due to the conference not existing, committee members took on specific roles typical of conferences (e.g. workshop chair, industry liaison) in HCI, a Website (www.fictionalconference.com) was built and Twitter account (@fictionalconf) created to represent the conference. Although only existing in the virtual world, these two digital artefacts gave the otherwise entirely fictional conference a sense of tangibility in that it involved the same work as organising a real event. In a similar way as the appointment of a committee, one role the online presence played in creating the research fiction was to underwrite the "reality" of the conference to the public.

The Call for Papers (CfP) was designed to follow the style and format typical of HCI conferences. This includes a short description of the conference aims and a list of suggested topics. The call invited submissions for paper or workshop titles. The CfP stipulated that submissions should be accompanied by a list of authors and a list of keywords to help understand the contribution. The decision was made to limit the requirement to submit to titles, author names, and keywords to keep the work required to make a submission minimal, thereby making participation in the project as simple as possible for as wide an audience as possible. There was some discussion amongst the committee about the potential to ask authors to submit an abstract, to help give context for ideas that might not fit in a short title. However, since the typical HCI conference programme only includes a title for each paper, it would be hard to include abstracts inside the programme without undermining the fictional frame of a conference. The restriction to titles also attempted to force the authors to be brief and concise, with mixed success.

The CfP was distributed via several popular HCI mailing lists, including ACM CHI, NordiCHI, CSCW and BCS-HCI. All of these lists receive multiple CFP notifications for a range of conferences and journals each day, so the request sat naturally among the other, presumably real, calls. The CfP was also advertised on social media by members of the conference committee (Fig. 1).

The following pages contain the final programme for FCDFF, built from submissions from research active members of the HCI community and the public. The sections following the programme discuss the response to the call, and the themes emerging from the submissions.

ƒ international fictional conference
on design fiction's futures

Venue: Tlön, Uqbar | Dates: Irrelevant

CALL FOR PAPERS

We are inviting real submissions that describe fictional papers and workshops. Practically speaking, we would like you to submit a title of a fictional paper or workshop, along with a list of authors (authors may be real or fictional), and a set of up to 4 keywords.

We are primarily interested in fictional papers and workshops that reflect on the future of Design Fiction research. From a future perspective, what will the impact of the current increase of interest in fictional research be in the long term? How will it change the research landscape, if at all?

Paper concepts could include (but are not limited to) ideas on:

- Arguments on how to evaluate the quality of Design Fiction
- Reports of Lost Futures and Counterfactual Histories
- Analyses of the relative importance of data and narrative
- Analyses of how fiction is used instrumentally in scientific papers
- Accounts of interaction with fictional systems (fictional user research)

Fig. 1 Extract from FCDFF Call for Papers

Day 0

08:30–17:30	Registration Desk Open (Foyer)	Workshop (Room α)		Workshop (Room ω)	
		Lickable City Workshop: an exploration of the current and future flavours of our urban environments	Ding Wang and Vanessa Thomas	*Utopian design methods for lost futures and imaginary pasts*	Ding Wang, Louise Mullagh, Serena Pollastri, Vanessa Thomas

Day 1

09:00– 09:15	Welcome to FCDFF—[Conference Chairs]			
09:15– 10:00	Opening Keynote—The @_CHINOSAUR: "They're Made Out of Meat: the CHI Community and Me"			
10:00– 10:30	Coffee Break			
10:30– 12:00	Session: Temporal Insecurities (Room z)		Session: (Room o) Evaluation	
	Future of Non-Standard Design: A Philo-Design Fiction	Fictilis Ensemble	Evaluating Design Fictions: From minimal departures to engagement	Tau Lenskjold, Eva Knutz and Thomas Markussen
	Same Old Design Fictions: Rehashing Tomorrows for Today	Tom More	All quiet on the western future, designing nor-colonialist fictions for and with the rest of the world	Dounia Ben Hassen and Alessandra Renzi
	The Challenges of Time Travel: Loss of Granularity with Artifacts From the Future	Leo Frishberg and Charles Lambdin	The downsides of world-building approaches: excluding diegesis from design fiction	Sandy Brown and Mikhail Markovsky
	Almost there: an analysis of phenomena perpetually "increasing," "emerging" and "becoming" in HCI papers over the past 20 years	Conor Linehan	All fiction is fact in the making, if you try hard enough: Deconstructing design fiction to reconstruct design fact	Dhruv Sharma

(continued)

(continued)

Time	Title	Presenter	Title	Presenter
12:00–13:00	Lunch and Networking (Note: Lunch is not provided)			
13:00–15:00	Session: Applications		Session: ???	
	Designing better privacy controls: Implications for the fictional world where personal data preferences matter	Jen Golbeck	It Was the Best of Times, It Was the Worst of Times, We Had Everything Before Us, We Had Nothing Before Us	P. Iglesias
	Haptic Communication in Virtual Reality English Education: 3D Creative Writing	Josh T. Jordan,		Benjamin Durr
	The art of creating fictional cases for experiencing realistic organisational design: a research study on case writer skillsets	Clive Holtham and Charles Mill	Afterfutures and Restructured Temporalities: Serresian Fold as Worldbuilding Design Paradigm for Augmented Immersive Fictions (AIF)	Bodhisattva Chattopadhyay, Josh Tanenbaum, John S. Seberger, Caitlin Lustig and Anita Marie Tsaasan
	Participatory design of military attack drones with stakeholders in developing countries	Conor Linehan	The Rise of San Leodis and Silicon Shore	Imran Ali
15:00–15:30	Coffee Break			
15:30–17:30	Session: Autonomous Fictions		Session: Ethics	
	Lessons from the field of robotics on how to use fiction to sell childish, fantastical, impractical and reckless research agendas	Brian Cox	Ethics for the confrontation of design fictions, when fictitious products meet real people	Gedebor Houston, Amara Mital and Christian Blach
	When bots generate their own speculations: what is left for designers?	Estelle Kery	Celebrity Design Fictioneers: The Chosen Few	Noah Bodie
	I wrote this! Copyright issues of autonomous AI academic papers	Ben Griffin	Blasphemy or prophecy; how design fictions might engage inter-religious dialog to discuss societal futures?	Rasmus Rasmussen and Marcel Picoli-Picolo
	A Study of Users Experience of AI Controlled Sex Toys	Aldridge Prior and Finbarr Saunders	Dr Strangefutures or: How I learned to stopped worrying about ethics committees and love Design Fictions	Alan Hook
18:30–22:00	Drinks Reception: Macondo Lounge			

Day 2

09:00–10:30	Session: Virtual Futures	Session: Reflecting	
	A risk assessment of deep augmented reality	On hope, nostalgia, determinism, and modernity in post-modern design fictions	Susann Wagenknecht, Siri Hustvedt, Bjarne Mädel, Suzanne Treister and Fred Turner
		Ben Griffin	
	A headset for every child: How virtual reality will transform education	Is Design Fiction only available in vanilla?	Steve Todd and Marc Harry
		Lawrence Angelo	
	Using human-generated design and speculative fiction as training data to synthesize criteria (or, in the case of dystopian fiction, counter-criteria) for the automatic generation and evaluation of useful algorithms	The verdict is in: Gamification always works. Looking back at fifty years of unquotable empirical evidence for psychological, behavioral and economic gamification effects	Andreas Lieberoth and Juho Hamari
		Yana Malysheva	
	Refunghi: An Internet Platform to Monitor Refugee Growth in Europe	[Footprints: Breaking Distributed Data Displays To Experience Understanding And Method]	Jon Hook and Marian Ursu
		Enrique Encinas	
10:30–11:00	Coffee Break		
11:00–13:00	Session: Methods	Session: Outsider Design Fiction	
	Using the Anatidae/Non-Anatidae Algorithm to Quantify the Plausability of Design Fictions	A Survey of Design Fiction in Comics and Graphic Novels	Aaron Kashtan, Winsor McCay, Jack Kirby, Will Eisner, George Herriman, Osamu Tezuka, Hergé and Tove Jansson
		Paul Coulton, Joseph Lindley, and Emmett L. Brown	
	Speculative Identities: Fictional characters as diegetic prototypes	On Anticipatory Ethnomethodology: Foundational Relationships Between Ethnomethodology and Design Fiction	Chris Elsden, David Kirk, and Garold Harfinkel
		Mark Dudlik	
	Design as Science. A Complete Mathematical Formulation of Design Theory	Nail Bombs, Pipe Bombs, and Sawn-off Shotguns: the DIY-maker culture, community participation and security	G. Adams and M. McGuinness
		Enrique Encinas	
	Fiction pieces for real species: design fiction as a posture to involve non-humans in discussing near-nature scenarios	Vernacular Design Fiction: Case Study of the Speculative Practices of Nordic Larpers	Jaakko Stenros
		Georges Abitbol, Maurice Vian	

(continued)

(continued)

Time	Session: Looking Forward	Speaker	Session: Bad Fiction	Speaker
13:00–14:00	Lunch and Networking (Note: Lunch is not provided)			
14:00–16:00	Session: Looking Forward		Session: Bad Fiction	
	Design Fiction and Its Discontents: The Future of an Illusion	Chris Elsden and Sigmund Fraud	*Design Fiction Considered Harmful*	Mark Blythe
	While Magic is Symbolism Weaponized, Design Fiction is Merely a Doomed Occupation of Storytelling	Koli Löyly	*Shit just got real: designer as dystopian soothsayer or that time I designed that horrible thing that was deemed commercially viable*	Alan Hook
	Where the Fiction Is: is Bruce Sterling HCI's latest Heidegger?	Conor Linehan	*The Death of Design Fiction—The Definition Deficit in Practice*	Allison Dunne
16:15–16:45	Coffee Break			
16:45–17:15	Closing Panel: *Fictional Fictions and Futuristic Futures*, with Ariadne Oliver, Kilgore Trout, Harriet Vane and Garth Marenghi			
17:15–17:30	Prizegiving and Conference Close			
	Introduction to FCFCDFF—the fictional conference on fictional conferences on design fictions futures			
19:30–end	Conference Banquet: Milliways, the Restaurant at the End of the Universe			

www.fictionalconference.com | @fictionalconf

Fig. 2 Commentators reflect
on how the conference
reminds them of the work of
magical realist author Jorge
Luis Borges

4 Response

The conference call received a good amount of attention on social media, being
shared dozens of times and generating some lively discussion around the unusual
concept. In terms of formal submissions, the conference received 56 paper submissions and 2 proposals for workshops. Submitters had a wide variety of backgrounds. Many were from researchers active in the areas of research fiction and
HCI, however, pleasingly, we also received submissions from freelance designers,
artists and high school teachers.

In addition to these formal responses to the call, we also received other kinds of
submissions including suggestions for fictional locations, as well as logos, typography and suggestions for keynotes and other speakers. This suggests that the
'world building' component of design fiction was seen as an important part of the
practice of creating a fictional conference for contributors as much as it was for the
committee members (Fig. 2).

5 Building the Programme

Faced with a diverse selection of submissions, and their universally "unreal" nature,
the task of translating those into a design fiction artefact—the conference programme—seemed to be a significant challenge for the committee. During a discussion about how to deal with this issue, the concept of a 'Nolan Number'
emerged. This was in reference to Christopher Nolan's (2010) film *Inception*, and
the multiple 'levels of dream world' in the film. The multiplicity of levels in

Nolan's film was reminiscent of research fictions, which sometimes become 'levels deep' in a similar way due to inevitably becoming meta-commentaries upon themselves (e.g. Lindley and Coulton 2016a, b). Some submissions clearly interpreted the FCDFF CfP as an invitation to showcase examples of how design fiction could be applied in the future, e.g. *Haptic Communication in Virtual Reality English Education: 3D Creative Writing*. In contrast, some other submissions were applying design fiction to itself, e.g. *Design Fiction Considered Harmful*. The Nolan Number that we started referring to was a subjective measure of 'how much' a particular title was in fact a design fiction referring to itself. This resurfaced when discussing the closing plenary, a fictional event in a fictional conference, which features a collection of fictional characters noted for creating their own fictions, talking about fiction.

The discussion about Nolan Numbers then lead on to considering a range of other subjective properties that we could attempt to quantify. The conference chairs asked the committee members to rate each paper according to each of a number of emergent characteristics:

- Funniness—a measure of how humorous the intent of the paper, if not the success!
- Plausibility—on a scale between fantastical and realistic/likely
- Enticement—how interesting or compelling the subject to the reader (i.e. would you read this paper?)
- Transparency—How clear the submission topic was to understand
- Nolan Number—how much of a comment on design fiction and FCDFF the submission was.

The categories were chosen because they allowed distinction between papers across broad themes based on the approach they took to using design fiction. This was necessary because the actual topics of the fictitious papers were so diverse.

Sessions were built based on similarity along one or more scales, and also similarity in content. For example, there are sessions on evaluation, methods, meta-commentaries and also a session titled "???" for submissions which shared low scores for transparency, in that even the committee could not discern what the papers were about.

6   Discussion

The aim of FCDFF was to playfully explore perceptions of how design fiction may be used, and how it may evolve, within the context of research. The framing of a fictional conference provides a familiar structure for contributions and makes the work itself an example of research fiction. Through the process of collecting, analysing and grouping submissions we have observed a number of broad themes

Fig. 3 The "official" Twitter account clarifies the fictional nature of the conference

FCDFF'16 @fictionalconf · Mar 24
FAQ:
* YES, we are organising a fictional conference
* This is not a troll, we DO want submissions
* Any Qs, reply or ask

↩   ⇄ 8   ♥ 3   •••

that can help us reflect on issues and opportunities around design fiction in HCI through the practice of research fiction.

6.1   Humour

As discussed earlier in the current chapter, humour is a common aspect of design fiction work in HCI. Although the method is serious, arguably humour makes it more accessible and signifies its queerness (Light 2011) in contrast to mainstream work. Blythe et al.'s (2016) "Silly" prototypes, Encinas and Blythe's (2016) "author eraser", Buttrick's et al. (2014) erotic BDSM fictions about wifi routers and Kirman et al.'s (2013) tale of robot enslavement all use humour as a tool, and this is reflected in the submissions to FCDFF. For example, the workshop on the "Lickable City", or the dark humour of the "pipe bomb" paper that frames terrorists as part of the DIY-maker movement. Other submissions use the opportunity to present sarcastic visions of impossible futures where (e.g.) gamification is proven to "always work" and jab at the positivist bias of HCI through submission of a "mathematical model" of design. Indeed the conference itself was read as a "joke" by enough people on Twitter to warrant an official statement from the conference Twitter account to clarify that the CfP was real (Fig. 3).

The question of why so much design fiction is humorous is interesting. GK Chesterton pointed out that the opposite of funny is not "serious", the opposite of funny is "not funny". Similarly much non-fictional design work is presented in a very grave and somber manner, but as critics including Evgeny Morozov point out, this does not necessarily make it serious. Morozov (2013) popularized the term "solutionism" which he defines as either solving problems that do not exist, or presenting quick technological fixes for complex social, political or environmental problems. For Morozov much of the new and emerging technology produced in Silicon Valley, and some academic HCI labs, is deeply solutionist. Although solutionist technologies are presented with gravity this does not mean that they are not inadvertently funny. Much design fiction uses irony as a defence against solutionism to signal the limits of technological interventions in complex social and political problems.

Fig. 4 Logos submitted to FCDFF

6.2   Form

Some submissions experimented with format, and some disregarded it altogether. The most obvious is perhaps the paper of emojis attributed to a fictional character from British adult comic *Viz*. One unusual submission from Hook and Ursu took the form of a data feed from a specially built tool that generates a constant stream of fictional paper titles. Although it does not work in print it is available online (http://jonhook.co.uk/titler). Other submissions suggested alternative logos and graphic design (Fig. 4), and one suggested the conference location to be moved to Tlön, Uqbar, suggested by contributor Michael Muller, an extremely fitting reference to Borges's (1940) tale of a group of intellectuals who build a fictitious world through reference in books and academic papers (such as this one).

In addition, in thinking of fictional characters to serve as keynotes, we followed a suggestion to contact a "living" fictional character, and were privileged to secure the services of notable figure @_CHINOSAUR, a Twitter character known for its biting commentary on contemporary HCI research and insatiable appetite for meat.

6.3   Confusion

Related to the confusion around the actual reality of the conference was confusion as to what to submit. Although the CfP is clear that the conference was fictional and about how and what should be submitted (a title, list of authors, list of keywords), some contributors had the perception that it was a real conference with a multi-stage submission process. Others submitted abstracts rather than just titles. The committee was contacted by more than one contributor asking if they would need to attend the conference, and if so, what the cost of registration would be. Within the gamut of interpretations of the CfP we saw, occasionally submissions were ambiguous and we too were confused as to whether the contributors had submitted something they thought was real, or whether it was simply dressed up to appear real.

This tendency towards confusion is congruent with Lindley and Coulton's (2016a) assertion that reviewers struggle with papers including elements of design fiction, and is perhaps indicative of a wider issue to do with research fictions. We suggest that the community is not yet entirely comfortable with what research fictions mean or how they should be interpreted.

6.4 Authors

One unexpected aspect was the choice of authors. Although we have seen the inclusion of fictionalized versions of authors in papers (e.g. Kirman et al. 2013) and authors from popular culture fictions (e.g. Lindley and Coulton 2015b; Linehan and Kirman 2017) in design fiction in HCI, we were surprised by the range of approaches taken by contributors. In particular, many papers did not include the name of their creator anywhere in the author lists, and also many papers included real but unaware people as authors of fictional papers. In most cases these people were public figures, however not exclusively. Given this, there is a disclaimer attached to the programme that stated authors, where real, may not in actual fact have any connection to the submission made in their name.

This very inconsistent use of identity created unanticipated problems in terms of attribution. As well as real people wishing to use their real name, we also had authors who wished to have their real name attributed as a contributor, but not linked to their specific contribution, and also several authors who absolutely insisted on not being attributed at all at any point. This messiness has led to a complicated arrangement of authors and acknowledgements in this chapter and in the online programme, which is our best attempt at meeting the various wishes of the individuals who supported this work.

7 Conclusion

In 1968 JG Ballard argued that science had become the largest producer of fiction (Sellars and O'Hara 2012):

> A hundred years ago or even fifty years ago, science took its raw material from nature. [...] nowadays, particularly in the social, psychological science, the raw material of science is a fiction invented by the scientists. You know, they work out why people chew gum or something.

There is a sense in which much social science is speculative in ways that are analogous to fiction. In HCI there is a much more direct use of fiction. Weiser's (1991) Sal Scenarios, produced over twenty-five years ago, were plausible fictions that fairly accurately predicted the world we live in today. The production of scenarios, personas, concept design has been standard practice for HCI since its inception.

Why then the current interest in design fiction and research fiction? Again, research fiction is nothing new and has its antecedents in the nineteen seventies. In 'Imaginary Magnitude', Stanisław Lem wrote fictional reviews of books and conference proceedings about academic fields that do not yet exist (Lem 1973). The conceit is powerful because it is economical, science fiction is a literature of ideas and the review of a book that hasn't been written allows for great economy in storytelling: a synopsis of a plot, an elevator pitch of an idea. The form also allows for a degree of plausibility, pastiching academic house styles, publishing conventions and controversies. The fictional conference proceedings here demonstrate the entangled feedback loop between science fiction and actual real world developments.

The term "design fiction" is resonant, playful and ambiguous in ways that "scenarios" or "concept designs" are not. The difference between a design fiction and a scenario is the potential for rich insights to emerge from possible conflicts. Whether viewed in terms of narrative or world building, the titles and author lists which make up these design fictions certainly evoke conflict. Although the titles here do not have a beginning, middle or end they do suggest conflict. Beneath the strongly playful and humorous tones of the submissions, the very form is agonistic: all academic work is contested whether through verification and falsification as the traditions of scientific method or argument and agonism as in the traditions of social science. The playfulness of the fictitious titles and the fictional conference itself is a vehicle for serious discourse on the futures of design fiction within HCI.

Sterling (2013) points out that there is going to be a lot more design fiction regardless of what anybody thinks of it, simply because it is quick and cheap. But there is also going to be a lot more real prototype development because that too is increasingly quick and cheap. Design fiction allows us to pause before we make and ask why we are doing it and what the consequences of making might be, either intended or unintended.

As research fiction, FCDFF asks us to reflect on our playful journey through the use of fiction in research, to consider how it fits within the ever-growing corpus of HCI work. Indeed, now the conference has published its programme online, it has become as real as the thousands of other conferences who have left the same digital traces—echoes of research that may or may not exist. To paraphrase Encinas and Blythe's (2016) comment on the similarly unclear realness of HCI publications, the FCDFF programme is now just as real as the programme of any other HCI conference.

Acknowledgements We thank all the contributors to the conference for their engagement and contributions: Imran Ali, Bodhisattva Chattopadhyay, @_CHINOSAUR, Sally Jo Cunningham,

Andy Darby, Mark Dudlik, Chris Elsden, Enrique Encinas, Lidia Facchinello, Leo Frishberg, Jennifer Golbeck, Ben Griffin, Estelle Hary, Clive Holtham, Alan Hook, Jon Hook, Jo Iacovides, Josh T. Jordan, Juho Hamari, Aaron Kashtan, Bastien Kerspern, David Kirk, Eva Knutz, Charles Lambdin, Tau Lenskjold, Andreas Lieberoth, Caitlin Lustig, Yana Malysheva, Thomas Markussen, Louise Mullagh, Michael Muller, Serena Pollastri, Søren Rosenbak, John S. Seberger, Dhruv Sharma, Jaakko Stenros, Josh Tanenbaum, Anita Marie Tsaasan, Marian Ursu, Susann Wagenknecht, and Ding Wang.

References

Adams RK, Moreau R (2015) I'm a Cyborg's Pet. Wattpad. Available online: https://www.wattpad.com/story/47397263-i%27m-a-cyborg%27s-pet-girlxcyborg

Baumer EPS, Ahn J, Bie M, Bonsignore EM, Börütecene A, Buruk OT, Clegg T, Druin A, Echtler F, Gruen D, Guha ML, Hordatt C, Krüger A, Maidenbaum S, Malu M, McNally B, Muller M, Norooz L, Norton J, Ozcan O, Patterson DJ, Riener A, Ross SI, Rust K, Schöning J, Silberman MS, Tomlinson B, Yip J (2014) CHI 2039: speculative research visions. In: CHI'14 Extended abstracts on human factors in computing systems (CHI EA'14)

Blythe M (2014) Research through design fiction: narrative in real and imaginary abstracts. In: Proceedings of the SIGCHI conference on human factors in computing systems (CHI'14)

Blythe M, Buie E (2014) Chatbots of the gods: imaginary abstracts for techno-spirituality research. Proc NordiCHI 2014:227–236

Blythe M, Kristina A, Clarke R, Wright P (2016) Anti-solutionist strategies: seriously silly design fiction. In: Proceedings of the 2016 CHI conference on human factors in computing systems (CHI'16)

Blythe M, Encinas E (2016) The co-ordinates of design fiction: extrapolation, irony, ambiguity and magic. In: Proceedings of the 19th international conference on supporting group work, pp 345–354. ACM

Blythe M (2017) Research fiction: storytelling, plot and design. In: Proceedings of the 2017 CHI conference on human factors in computing systems (CHI'17), pp 5400–5411. ACM, New York, NY, USA. https://doi.org/10.1145/3025453.3026023

Borges JL (1940) Tlön, Uqbar, Orbis Tertius, Sur

Buttrick L, Linehan C, Kirman B, O'Hara D (2014) Fifty shades of CHI: the perverse and humiliating human-computer relationship. In: CHI'14 Extended abstracts on human factors in computing systems (CHI EA'14)

Coulton P, Lindley J, Akmal HA (2016) Design fiction: does the search for plausibility lead to deception? In: Proceedings of design research society conference 2016, pp 369–384

Coulton P, Lindley J, Sturdee M, Stead M (2017) Design fiction as world building. In: Proceedings of research through design conference 2017. Edinburgh, UK

Dalton NS, Moreau R, Adams RK (2016) Resistance is fertile: design fictions in dystopian worlds. In: Proceedings of the 2016 CHI conference extended abstracts on human factors in computing systems, pp 365–374. ACM

Di Salvo C (2011) Adversarial design. MIT Press

Dunne A, Raby F (2013) Speculative everything: design, fiction, and social dreaming. MIT Press

Elsden C, Chatting D, Durrant AC, Garbett A, Nissen B, Vines J, Kirk DS (2017, May) On speculative enactments. In: Proceedings of the 2017 CHI conference on human factors in computing systems, pp 5386–5399. ACM

Encinas E, Blythe M (2016) The solution printer: magic realist design fiction. In: Proceedings of the 2016 CHI conference extended abstracts on human factors in computing systems (CHI EA'16)

Inception (2010) [film] Directed by Christopher Nolan. Warner Bros. Pictures, USA

Kirman B, Linehan C, Lawson S, O'Hara D (2013) CHI and the future robot enslavement of humankind: a retrospective. In: CHI'13 Extended abstracts on human factors in computing systems (CHI EA'13)

Kirman B, Lawson S, Linehan C (2017) The dog internet: autonomy and interspecies design. In: Proceedings of research through design conference 2017. Edinburgh, UK

Lawson S, Kirman B, Linehan C, Feltwell T, Hopkins L (2015) Problematising upstream technology through speculative design: the case of quantified cats and dogs. In: Proceedings of the 33rd annual ACM conference on human factors in computing systems (CHI'15)

Lem S (1973) Imaginary magnitude. Wielkość Urojona

Light A (2011) HCI as heterodoxy: technologies of identity and the queering of interaction with computers. Interact Comput 23(5):430–438

Lindley J, Coulton P (2015a) Back to the future: 10 years of design fiction. In: Proceedings of the 2015 British HCI conference (British HCI'15)

Lindley J, Coulton P (2015b) Game of drones. In: Proceedings of the 2015 annual symposium on computer-human interaction in play (CHI PLAY'15)

Lindley J, Coulton P (2016a) Pushing the limits of design fiction: the case for fictional research papers. In: Proceedings of the 2016 CHI conference on human factors in computing systems (CHI'16)

Lindley J, Coulton P (2016b) Peer review and design fiction: great scott! The quotes are redacted. In: Proceedings of the 2016 CHI conference extended abstracts on human factors in computing systems, pp 583–595. ACM

Linehan C, Kirman B (2017) MC hammer presents: the hammer of transformative nostalgification-designing for engagement at scale. In: Proceedings of the 2017 CHI conference extended abstracts on human factors in computing systems, pp 735–746. ACM

Mazières D, Kohler E (2005) Get me off your fucking mailing list. Unpublished paper. Available: http://www.scs.stanford.edu/~dm/home/papers/remove.pdf

Morozov E (2013) To save everything, click here: technology, solutionism, and the urge to fix problems that don't exist. Penguin UK

Sellars S, O'Hara D (2012) Extreme metaphors: interviews with JG Ballard 1967–2008. Fourth estate

Sterling B (2005) Shaping things. MIT Press

Sterling B (2013) Presentation at NEXT13—fantasy prototypes and real disruption. Available online: https://www.youtube.com/watch?v=2VIoRYPZk68

Weiser M (1991) The computer for the 21st century. Sci Am 265(3):94–104

Zongker D (2006) Chicken Chicken Chicken: Chicken Chicken. Ann Improbable Res, 16–22

Part IV
"Approaches and Directions"

Chapter 11
Slow, Unaware Things Beyond Interaction

Ron Wakkary and William Odom

1 Introduction

In this chapter we provide an overview of concepts and methods that have become part of our approach to gain a broader and deeper understanding of the relations between humans and technology. Over the years, our efforts have been to move past the field of interaction design's dominant focus on *human interaction* with technology to develop a design-oriented understanding of *human relations* with technology. In our view, this begins by looking at technology beyond its functional, utilitarian, or instrumental value toward a broader set of perceptions and meanings. This theme is emblematic of a broader shift in interaction design and HCI. The first edition of this book contributed significantly to a trajectory in which designers and researchers see technology as a matter of experiences that are fun (Blythe and Hassenzahl 2003), rich (Overbeeke et al. 2003), embodied (Dourish 2004), somaesthetic (Höök et al. 2016), spatio-temporal (McCarthy and Wright 2004), hedonic (Hassenzahl 2003), reflective (Sengers and Gaver 2006), and ludic (Gaver et al. 2004). However, understanding technology through more than solely a functional lens is only one part of more deeply viewing and inquiring into human-technology relations. We believe it is necessary to also understand people's relations to technology beyond interaction and engineered experiences of technology. In the context of *funology,* we aim to critically and generatively contribute to the investigations of the experiences of technology to go beyond both instrumentalism and interaction. In many respects, interaction, like functionality, is too narrow of a lens for both understanding and influencing people's experiences and

R. Wakkary (✉) · W. Odom
School of Interactive Arts + Technology, Simon Fraser University, Surrey, BC, Canada
e-mail: rwakkary@sfu.ca

R. Wakkary
Industrial Design, Eindhoven University of Technology, Eindhoven, Netherlands

© Springer International Publishing AG, part of Springer Nature 2018
M. Blythe and A. Monk (eds.), *Funology 2,*
Human–Computer Interaction Series,
https://doi.org/10.1007/978-3-319-68213-6_11

relations to technology through design. Interaction is only one form of technology relations that happens explicitly, in present time, and consciously (Verbeek 2015). What about relations to technology that manifest over time, incrementally, knowingly and unknowingly (or somewhere in between) that become part of our everyday lives?

A key goal in our design research has been to take a step toward expanding the notion of *interaction design* beyond purposed manipulations, explicit interactions, and experiential engagements to also include the implicit, incremental and, at times, ambivalent or unknowing encounters and relations that emerge among people, artifacts, and environments. Among the specific questions we ask ourselves as interaction design researchers: how do people relate to and make meaning from a lifetime of digital photos and music? What are design approaches that can enable us to design digital artifacts that will productively contribute to how people knowingly and unknowingly construct and reconstruct the complexities of their everyday lives? And, how can we viably design digital artifacts when people do not fully understand the artifacts that they live with and even rely upon?

These lines of questioning have translated into a series of investigations and explorations of designing for slowness, in which interactions are minimal yet meaningful over time; unawareness, in which interaction design artifacts are lived with but are not designed for people directly; and thingness, in which interaction design artifacts are designed to engage each other rather than people. As a result, we have been constructing a design vocabulary and series of exemplars to enable our approach. To support these investigations, we have developed and adopted a series of methodological commitments including inquiry through artifacts and lived-with experiences that are embodied in the related approaches we have termed material speculation (Wakkary et al. 2015) and research products (Odom et al. 2016).

Our design investigations are inspired and informed by a strand in philosophies of technology known as postphenomenology. Briefly, postphenomenology (Ihde 1993; Verbeek 2005) argues that technologies are mediators of human experiences and practices rather than functional and instrumental objects (Rosenberger and Verbeek 2015; Verbeek 2005). In a postphenomenological relationship between humans and technological artifacts, each mutually shapes the other through mediations that form the human subjectivity and objectivity of any given situation. Design is central to and bound up in a postphenomenological understanding of the world since digital technologies do not come to us in a "raw" form but in a form that is *designed*. In this respect, designed digital artifacts, or in our case *things*, manifest technologies and directly influence the mediation of our experiences and practices.

Our design investigations relate to approaches by interaction design and HCI researchers that have been investigating complex matters of human-technology relations that often involve messy, intimate, and contested aspects of everyday life. For example, Wiltse and Stolterman (2010) view interaction architectures of online spaces as they exist rather than as intended by designers in order to reveal how these spaces mediate human activity. Pierce and Paulos (2009, 2011, 2015) investigate the materializing of technologies as embodied relations within technologies. Odom et al. (2009) explore how functional, symbolic, and material qualities of everyday devices and systems shape the potential for sustained, long-term human-technology relations. Fallman (2011) inquires paradigmatically into the nature of what is considered "good" in design extending philosophies of technology (e.g. Borgmann and Ihde) to HCI in order to examine the potential role of values and ethics as a "new good" in interaction design. Relatedly, Tromp et al. (2011), reflect on the social consequences of mediated relations and argue that designers should make more informed decisions to design for socially responsible behavior. This related research is evidence that technological mediation with respect to design is emerging as an HCI research program. Our research discussed in this chapter aims to contribute to these efforts.

In this chapter, we describe three concepts we have been investigating that include *slowness*, *unawareness*, and *thingness*. With each description we provide one or more interaction design artifacts we designed and deployed in everyday settings as part of these investigations. We see this chapter as a field guide to our recent research rather than a full explication and rationale. For the latter, we suggest readers view the original articles on these works that are cited throughout this chapter. In keeping with the field guide approach, we provide an overview of our emerging design vocabulary that resulted from this work. Our hope is that this chapter will contribute to further refinement and critical testing of these ideas by inspiring other interaction designers and design researchers to mobilize these ideas through the generative discourse of making and researching through design.

2 Concepts of Slowness, Unawareness, Thingness

Slowness investigates the radical slowing down of engagement with digital content and artifacts in ways that dovetail into reflections on and reframing of technologies in everyday life. Unawareness investigates interaction design artifacts designed to be lived with and to enact their respective behaviors without awareness of the needs or demands of a user. Thingness investigates digital artifacts as having a parallel existence alongside us and other things we live with.

Methodologically, we investigate these concepts by way of material speculations (Wakkary et al. 2015). Material speculation is the design of a counterfactual artifact that is experienced and lived with on an everyday basis over time in order to ask research questions. A counterfactual artifact is a realized functioning product or system that intentionally contradicts what would normally be considered logical given the norms of design and design products. More generally, we can also see our counterfactual artifacts as a broader class of artifacts called research products (Odom et al. 2016). A research product is an artifact designed to: drive a research inquiry; have a quality of finish so people engage it as it is rather than what it might become; fit in everyday settings and be lived with over time; and be independent such that it operates effectively when deployed in the field for an extended duration. These methodological considerations embody our commitment to supporting long-term, lived-with experiences of our design artifacts in the service of investigating the complex matters of human-technology relations. Practically this often translates to batch productions of research products for multiple concurrent studies and long-term deployments from six weeks to fourteen months. Reflecting this commitment we also describe a series of research products that, in their design and deployment, mutually explored and informed the concepts.

Articulating each concept as distinct descriptions is useful conceptually as well as rhetorically to communicate our ideas. Yet, it is important to make clear that these concepts are not mutually exclusive rather they mutually inform each other. For example, the temporality of slowness informs the incremental perception and meaning of unawareness. In turn, thingness embodies a temporal presence that may be separate from our own human structuring of time; and unawareness reinforces a different temporal unfolding and a thing-oriented existence for the artifacts. While we describe our interaction design artifacts in relation to particular concepts, it is not surprising that the artifacts could easily be used to describe another or multiple concepts simultaneously.

3   Slowness

We now live in a world where digital technology and systems mediate many aspects of people's everyday lives and experiences. For example, the convergence of social, cloud, and mobile computing have made it easy for people to stay constantly connected and to create, store, and share personal digital content at rates faster and scales larger than ever before. We build on our earlier work (Odom et al. 2012a, b; Odom 2015), which provided strong evidence illustrating that designing technological artifacts that intentionally slow down *interactions* with personal digital content and technologies can open up more unique, diverse, and valued ways of conceptualizing a place for these artifacts as everyday things. The notion of *slow technology* (Hallnäs and Redström 2001) offers promise to positively impact digital overload by envisioning a radically different way that technology could operate in everyday life. Our aim in addressing slowness is that it will bring into focus how technological artifacts shape human relations and interactions in the social and material ecologies they are embedded in, how they contribute to experiences of, for example, digital overload, and how new design strategies can help make digital artifacts more enduring and holistic parts of everyday life.

One example of our approach to understanding slowness is *Photobox* (see Fig. 1). Photobox is a counterfactual artifact designed to theoretically and practically explore how a computational object can critically intervene in experiences of digital overload. Photobox is designed to target digital photos because they are one of the most enduring forms of digital content, and they continue to rapidly proliferate. Three Photoboxes were deployed in three households for fourteen months respectively (see Odom et al. 2014 for more details). Through the long-term field study of Photobox, we wanted to explore how slowing down the consumption of digital photos might

Fig. 1 (left) Photobox in an ensemble of domestic artifacts that were situated with it over time by household members; (right) a cherished selection of photos printed from a participant's Photobox

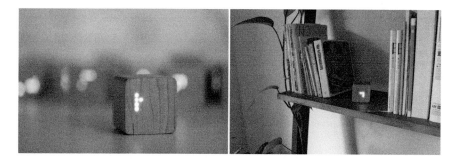

Fig. 2 (left) A Slow Game; (right) A Slow Game on a bookshelf in a home

support experiences of reflection on people's digital materials and also on the Photobox itself as a domestic technology. The two main components of Photobox are an antique oak chest and a Bluetooth-enabled Polaroid Pogo printer (which makes 2×3 in. photos). We decided to use an oak chest as it presents a familiar form with a simple interaction (i.e., it can be opened and closed; things can be kept inside of it and taken out). We decided to use a printer because it produced a simple material form (i.e., a paper photograph) that was open to a range of potential uses.

At the start of each month, Photobox indexes its owner's Flickr archive and randomly prints four or five photos that month. In a similarly random fashion, it selects four (or five) photos and generates four (or five) selected timestamps that specify the print time and date for each photo; at print time, the matching photo is printed. Photobox's computational behavior is designed to make it difficult for the owner to anticipate when it will 'act' next and what might be that action. The computational process never changes. As an interaction object Photobox is extraordinarily simple; the 'interaction' is merely opening the chest to see if a photo from one's past has (or has not) printed.

The simplicity in design of Photobox in part informs the more recent and ongoing *Slow Game* project (see Fig. 2). The project extends our considerations of temporality to the practices of long-term engagement, curiosity, and play. In this project, we design slow interaction into a tangible interactive game that enforces a very low frequency of interaction: one move a day. The game is a small 5 cm^3, with a low-resolution display consisting of 64 tiny white lights that are muted through a thin veneer. The game is based on the classic mobile phone game 'snake', where the player manoeuvres a line that grows in length, with the line itself being a primary obstacle. The game is played by physically rotating the cube, which turns the direction that the snake moves; the user can set the orientation of the next move, but it will only make the move once per day. Whether or not the user interacts with Slow Game, it will continue to slowly advance moves based on its current orientation. Time is represented as slowly moving through the artifact whether or not it is acknowledged or engaged with. The pacing of when a move is made is approximately 18 h; this enables Slow Game to, over time, come in and out of sync with a typical 24-h cycle (see Odom et al. 2018 for more details). By reducing the feedback loop to a frequency of slightly less than once per day, Slow Game challenges our memory, observation and patience. We batch produced fourteen Slow Game cubes, ten of which are currently in long-term field deployments in ten different households.

With respect to slowness it becomes apparent that artifacts and systems may set a pace and time-scale that are irrespective of human norms or expectations. Relatedly, our investigations into human relations with digital artifacts further explore the norms and assumptions of human-centered design through the concept of unawareness.

4 Unawareness

Unawareness centers on the idea of designing objects that intentionally enact their behaviors without requiring or demanding the attention of the people who live with them. These *unaware* objects execute preset computational processes and, in this sense, operate entirely unaware of human presence or actions. They have no explicit output functions to support human interaction and they lack any kind of traditional 'interface' or control mechanisms. Our use of the term *unaware* in this context owes specifically to the fact that these objects are designed to be computationally unaffected by direct interactions. This approach is a counter exploration to the idea of smart technologies and the increasing push to design digital artifacts that vie for our attention and interactions.

Ultimately, the unaware nature of these digital artifacts enables them to be open-ended over time. These qualities are crucial to their ability to foster creative actions and encounters that arise from the negotiation of what and how the human-technology relations are mediated. In earlier empirical work, we saw that mundane objects could be recast as creative resources to be appropriated, for example a chair becomes a coat rack (e.g. Desjardins and Wakkary 2013; Wakkary and Maestri 2007). Through our design research approach we aim to articulate engagements with digital artifacts that go beyond appropriation in ways that are more nuanced and multi-dimensional including creativity that is reflective, mindful, direct, and emergent across complex connections of things and things, as well as things and people.

A good example of unawareness is our *table-non-table* (Odom and Wakkary 2015; Oogjes and Wakkary 2017; Wakkary 2016; Wakkary et al. 2016a). The table-non-table is a stack of paper supported by a motorized aluminum chassis that infrequently moves (see Fig. 3). The paper is common stock (similar to photocopy paper). Each sheet measures 17.5 in. by 22.5 in. with a square die cut in the middle to allow it to stack around a solid aluminum square post that holds the sheets in place. There are approximately 1000 stacked sheets of paper per table-non-table, which rest on the chassis about one half-inch from the floor. The chassis and motors are strong enough to support stacking heavy objects on it including a person sitting or standing on it. The paper sheets can easily be removed and manipulated like any sheet of paper. The table-non-table moves for short periods of movement (5–12 s) once every 6–12 h. The movement pattern is random, yet it stays within an initial radius of less than half a meter square.

Related to the table-non-table is what we call the *Tilting Bowl* (Oogjes and Wakkary 2017; Wakkary et al. 2016b). The Tilting Bowl, as the name suggests is a ceramic bowl that tilts three to four times each day (see Fig. 4). Similar to the table-non-table,

Fig. 3 (left) The table-non-table; (right) the table-non-table in a household intersecting with a cat

Fig. 4 (left) The Tilting Bowl; (right) the Tilting Bowl in an ensemble amongst other things

the movement can easily go unnoticed. Unlike the table-non-table, the Tilting Bowl has a readily intelligible function of holding items. Practically, for all intents and purposes, it is like any other bowl with the exception that it periodically tilts.

The material speculations of the table-non-table and the Tilting Bowl are to investigate the nature and type of computational artifacts that can be shaped and given meaning by people as a matter of living with and performing everyday practices. As counterfactual artifacts, both aim to anticipate unarticulated qualities of technological mediation. Given the subtleties of the experiences table-non-table and the Tilting Bowl we engaged in self-deployments reminiscent of autobiographical design (Neustaedter and Sengers 2012) in which we, the designers, lived with both the early prototypes and final research products for long periods of time. We found that the experience of unaware objects like these is not so much with the direct interactions with the artifact but in the moments the bowl tilts or the table-non-table moves and the ensembles each forms with other artifacts through the course of living with them. Elsewhere we have described these interactions as *intersections* and *ensembles* (Odom and Wakkary 2015; Wakkary et al. 2016a) as we will discuss later in this chapter. The nuanced experiences of incremental and indirect encounters with these unaware objects are hard to both observe and articulate yet it is what makes unawareness as a concept meaningful and distinct. The table-non-table has been deployed to several households from three weeks to five months and continues to be deployed today in ongoing studies giving us a foothold in describing and understanding unawareness. Six Tilting Bowls are currently in long-term deployments among the households of seven philosophers as a form of co-speculation in which we have enlisted the analytical abilities of trained philosophers to help us describe and give more form to the technological mediations of the artifacts (Wakkary et al. 2018b).

5   Thingness

In many respects, human-centered approaches to design aim to close the gap between humans and technologies in order to better serve human needs. Yet, what might a human-centered approach hide with respect to the relations we have with technology? Our approach to thingness asks this question with a positive framing: what might be revealed in the relations we have with technology through a thing-focused approach?

Our investigations of things are informed by postphenomenology (e.g. Ihde 1993; Verbeek 2005) that we discussed at the outset of this chapter, alongside related perspectives in philosophy and technology studies (Baird 2004; Bogost 2012; Harman 2010). With these ideas in hand we view things as non-human technological entities rather than simply artifacts in the physical sense. Further, we understand technology in broad terms in the sense of human made artifacts from hammers to eyeglasses to digital software. While we shift our attention from humans to things, in our design research, we understand that the two are bound up together and mutually shape each other. In short, this means that we cannot understand technologies without humans and conversely, we cannot understand humans without technologies. Thus, while we focus on things we aim to reveal how things are inextricably connected to us.

The *Morse Things* are sets of ceramic bowls and cups (see Fig. 5) that communicate solely to each other over an internet connection (Wakkary et al. 2017). Over time, the conversation of the Morse Things and their degree of connectedness on the network can evolve in degrees of "awareness" on the part of a Morse Thing, from being alone, to being a pair of things, to being a group of things, to being part of a larger network of things. The Morse Things mostly sleep (computationally speaking) and wake at random intervals during the day at least once every eight hours. Upon waking, a Morse Thing will send and receive messages to and from other Morse Things in its set. The messages sent by each Morse Thing are in Morse code and are simultaneously expressed sonically and broadcasted on Twitter. The Morse Things can be used like any other bowl or cup for eating, drinking, and containing items, with the exception that they cannot be put in the dishwasher or microwave. We designed and fabricated six sets of Morse Things each including a large bowl, a medium bowl, and a cup. The form of each Morse Thing is made of ceramics that is shaped around the embedded electronics. This design choice aims to create a fusion of computation and everyday objects; indeed, this is an intention across the series of design artifacts described in this chapter. We deployed the Morse Things in six households in Vancouver for six weeks. Following the deployment, we conducted a workshop to discuss the role of the Morse Things and ultimately the gap between things and people (for more details see Wakkary et al. 2017).

The Morse Things revealed aspects of our relations with things. These include the withdrawal of things from our human understanding and perception in which the non-humanness of things is not a matter that can be ignored or bridged by human-centered design. This withdrawal of things contributes to creating a gap between things and us, in which we live with and rely upon technologies and

Fig. 5 (left) One set of Morse Things; (right) The Morse Things small bowl and cup in a kitchen helping to prepare dinner

artifacts that we can never fully understand. Despite this gap, we have the ability to form attachments with things that are inherently ambivalent and even ambiguous but potentially long lasting and meaningful. As a result, we see the opportunity in Morse Things to design a new type of thing between thing-centered and human-centered technologies in the home.

6 Emerging Design Vocabulary for Going Beyond Interaction

Slowness, unawareness, and thingness are high-level concepts that have enabled us to productively frame design research inquiries aimed at developing a broader and deeper understanding of the relations between humans and technology. Next, we turn to describing a vocabulary that has emerged and been distilled from our research efforts. We understand that up until this point in the chapter we have introduced new concepts to many readers and there is a risk of overwhelming you with yet more new terms with what follows. By way of guidance, we suggest that the three concepts of slowness, unawareness, and thingness are critical high-level ideas we want to communicate. The following vocabulary is intended to support design-oriented research at a more detailed level. In our own work, we have found these terms to be helpful in navigating the challenge of mobilizing the three concepts into concrete, actual design artifacts that embody specific design goals, stances, and research questions. In keeping with our field guide approach, we present the vocabulary in bullet point form and in an abbreviated manner (see the attendant references for more details):

- *Purposeful purposelessness* exemplifies how research question(s) are carefully and precisely crafted into the artifacts that in the use context seem to have no purpose or only an ostensible and weak purpose. This requires purposeful design with a design goal (that should not be confused with a use goal), purposeful crafting of the artifact, and a purposeful aesthetic. Combined together these forms of purpose create a quality artifact that will be accepted into environments alongside other designed artifacts despite not having an obvious purpose or function (Wakkary et al. 2016a).
  The design of the table-non-table is a good example of this as the dimensions, quality of paper and the precision and sturdiness of the motorized aluminum chassis were of exacting and specific requirements despite the opaque function of the artifact. The Slow Game and Morse Things were also designed to exacting requirements and considerations of materials with only the weak purposes of playing a game or containing food in the service of the more carefully crafted research questions related to slowness and thingness.
- *Intersections* refer to people's ongoing incremental encounters with a design artifact in which a modification or transformation may or may not occur. While interaction often involves direct manipulation of an artifact, intersections can

range from experiences of being mindful of the artifact, to subtle uses of the artifact that may be only briefly noticed (or go unnoticed), to piecemeal resituating of the artifact within its physical context. Intersections can be treated as complementary to interaction, but are notably more general in their aim to account for the broader range of known and unknown, incremental and ongoing encounters that unfold with computational and non-computational objects alike in everyday contexts (Odom and Wakkary 2015; Wakkary 2016; Wakkary et al. 2016a; Odom et al. 2018).

The design of the Slow Game and table-non-table offer good examples of intersections: the slow changing lights muted within a wooden cube and the occasional movement of a stack of paper just as easily invite engagement or a momentary glance, or can simply can go unnoticed in the background of everyday life. Our relations with the artifacts continue through time whether we engage with them or not; they may sync more closely with the rhythms of people's everyday lives, or just as easily go unnoticed for numerous days or weeks.

- **Ensembles** manifest through cumulative intersections. As intersections accumulate, qualities emerge that go beyond the individual artifact, often becoming experienced among an ensemble of things and people within their local environment, such as the home. In this sense, the quality of ensembles is comprised through the evolving quality of relational aspects of artifacts, contexts, and human actions. In this way, an ensemble is a dynamic collection of social and material elements within an environment that can become increasingly unique and nuanced over time (Odom and Wakkary 2015; Oogjes and Wakkary 2017; Wakkary 2016; Wakkary et al. 2016a; Odom et al. 2018).

  Our field studies of the Photobox and Tilting Bowl revealed that people began to relate to these artifacts as 'just another thing' in the backdrop of their everyday lives. Photo frames, ceramic vessels, trinkets, books, and various other things were dynamically and unknowingly configured on, in, and around the Photobox and Tilting Bowl.

- **Displacement** refers to the negation of a thing to highlight other less noticed or unarticulated relational elements. When talking about artifacts, understanding an artifact through everything but that artifact helps reveal the relations that are bound up with the artifact. Displacement does not define the thing or its use directly, rather it describes how it relates to the world and the configurations of the world it is a part of. It highlights how artifacts explicitly shape our everyday practices yet remain indirectly present (Oogjes and Wakkary 2017; Wakkary et al. 2017, Wakkary et al. 2018a).

  Our field studies of the Photobox and table-non-table revealed that both of these artifacts became displaced over time. The constellation of other domestic artifacts, spaces, and practices around the counterfactual artifacts became a primary focus over the Photobox and table-non-table. This led study participants to reflect on the relations with other digital artifacts like an Xbox or computers in the case of Photobox or relations with furniture and other household artifacts that occupied the same room as the table-non-table.

- **Withdrawal** Understanding experience from the perspective of things is a difficult task for people. Philosophically speaking, non-human perspectives can be said to "withdraw" from human understanding into a non-human world that we can neither fully comprehend nor articulate (Bogost, Verbeek), which is also evident in our notion of displacement. In addition, non-human worlds are formed in a configuration of materials and performances rather than language (Baird). We refer to this pulling away from our understanding and perception as withdrawal. It is important that while much of the experience of things is beyond our grasp, this perspective is not entirely invisible to us. Rather we establish many commonalities and reliable interactions that form the foundations for the fundamental and ubiquitous relations we have with things (Wakkary et al. 2017, 2018).

  For example, participants in our Morse Things study argued that their 4-year old son could best relate to the Morse Things since he spent his day playing in an imaginary and other world of things. Similarly, a number of participants throughout our workshop compared the Morse Things to pets and teenagers signaling familiar relationships that at times are very unfamiliar if not inaccessible to pet owners and parents.

The emerging design vocabulary is a concrete way in which we are moving away or beyond interaction to investigate in a design-oriented fashion, the complexity and richness of human-technology relations in everyday life.

7 Conclusion and Future Directions

The goal of this chapter has been to offer an overview of concepts, methods, and an emerging design vocabulary that have been central to our collective design research program aimed at developing a broader and deeper understanding of human-technology relations in everyday life. We aim to build on the trajectory of prior works appearing in the funology series that have provided a foundation to understand designing for interaction and to, now, expand beyond it. A focus on human-technology relations provides scaffolding for expanding beyond interaction toward inquiring into the incremental, piecemeal, knowing, and unknowing qualities of consciousness that bind technologies to our everyday lives, and vice versa. We have articulated *slowness, unawareness,* and *thingness* as key higher-level concepts guiding our approach. We then presented the Photobox, Slow Game, table-non-table, Tilting Bowl, and Morse Things as material speculations and research products that concretely ground, embody, and develop these concepts through their actual and lived-with existence. Finally, we described an emerging vocabulary of terms that include *intersections, ensembles, purposeful purpose-lessness, displacement,* and *withdrawal* that have been productive in generatively and critically framing our design inquires. Across this chapter, our aim is not to be prescriptive or conclusive. As the HCI and interaction design communities continue to explore the nature of human relations to technology in everyday life, we hope our work can be seen as a generative framing for supporting these inquiries.

Acknowledgements We thank the many current and past members of the Everyday Design Studio who have contributed to the table-non-table, Tilting Bowl, and Morse Things projects through direct involvement and many discussions. In particular we would like to thank Audrey Desjardins, Sabrina Hauser, Henry Lin, Doenja Oogjes, Markus Lorenz Schilling, and Matthew Dalton. We are also indebted to many other collaborators including Mark Selby for his foundational role in the Photobox project as well as Abigail Sellen, Richard Banks, David Kirk, and Tim Regan; and, Garnet Hertz, Ishac Bertran, Matt Harkness, Sam Beck, and Perry Tan for the Slow Game project; Keith Doyle, Phillip Robbins, Shannon Mortimer, and Lauren Low (from Material Matters at ECUAD) for their invaluable partnership with the Tilting Bowl and Morse Things; and Cheng Cao, Leo Ma, and Tijs Duel for the Morse Things project. SSHRC, NSERC, and 4TU Federation Design United funded parts of the research presented in this chapter.

References

Baird D (2004) Thing knowledge: a philosophy of scientific instruments. University of California Press, Berkeley

Blythe M, Hassenzahl M (2003) The semantics of fun: differentiating enjoyable experiences, pp 91–100. https://doi.org/10.1007/1-4020-2967-5_9

Bogost I (2012) Alien phenomenology, or what it's like to be a thing (2/18/12 edition). University Of Minnesota Press, Minneapolis

Desjardins A, Wakkary R (2013) Manifestations of everyday design: guiding goals and motivations. In: Proceedings of the 9th ACM conference on creativity & cognition. ACM, New York, NY, USA, pp 253–262. https://doi.org/10.1145/2466627.2466643

Dourish P (2004) Where the action is: the foundations of embodied interaction (New edition). The MIT Press, Cambridge, Mass

Fallman D (2011) The new good: exploring the potential of philosophy of technology to contribute to human-computer interaction. In: Proceedings of the SIGCHI conference on human factors in computing systems. ACM, New York, NY, USA, pp 1051–1060. https://doi.org/10.1145/1978942.1979099

Gaver WW, Bowers J, Boucher A, Gellerson H, Pennington S, Schmidt A, … Walker B (2004) The drift table: designing for ludic engagement. In: CHI'04 extended abstracts on human factors in computing systems. ACM, New York, NY, USA, pp 885–900. https://doi.org/10.1145/985921.985947

Hallnäs L, Redström J (2001) Slow technology—designing for reflection. Pers Ubiquit Comput 5 (3):201–212. https://doi.org/10.1007/PL00000019

Harman G (2010) Towards speculative realism: essays and lectures. Zero Books, Winchester

Hassenzahl M (2003) The thing and I: understanding the relationship between user and product, pp 31–42. https://doi.org/10.1007/1-4020-2967-5_4

Höök K, Jonsson MP, Ståahl A, Mercurio J (2016) Somaesthetic appreciation design. In: Proceedings of the 2016 CHI conference on human factors in computing systems. ACM, New York, NY, USA, pp 3131–3142. https://doi.org/10.1145/2858036.2858583

Ihde D (1993) Philosophy of technology: an introduction. Paragon House Publishers, New York, NY, USA

McCarthy J, Wright P (2004) Technology as experience. MIT Press, Cambridge, Mass

Neustaedter C, Sengers P (2012) Autobiographical design in HCI research: designing and learning through use-it-yourself. In: Proceedings of the designing interactive systems conference. ACM, New York, NY, USA, pp 514–523. https://doi.org/10.1145/2317956.2318034

Odom W, Pierce J, Stolterman E, Blevis E (2009) Understanding why we preserve some things and discard others in the context of interaction design. In: Proceedings of the 2015 CHI conference on human factors in computing systems. ACM Press, New York, NY, USA, pp 1053–1062. https://doi.org/10.1145/1518701.1518862

Odom W, Banks R, Durrant A, Kirk D, Pierce J (2012a) Slow technology: critical reflection and future directions. In: Proceedings of the designing interactive systems conference. ACM, New York, NY, USA, pp 816–817. https://doi.org/10.1145/2317956.2318088

Odom W, Selby M, Sellen A, Kirk D, Banks R, Regan T (2012b) Photobox: on the design of a slow technology. In: Proceedings of the designing interactive systems conference. ACM, New York, NY, USA, pp 665–668. https://doi.org/10.1145/2317956.2318055

Odom W, Sellen A, Kirk D, Banks R, Regan T, Selby M, Forlizzi J, Zimmerman, J (2014) Designing for slowness, anticipation and re-visitation: a long term field study of the photobox. In: Proceedings of SIGCHI conference on human factors in computing systems. Toronto, Canada. CHI '14. ACM Press

Odom W (2015) Understanding long-term interactions with a slow technology: an investigation of experiences with future me. In: Proceedings of the 2015 CHI conference on human factors in computing systems. ACM, New York, NY, USA, pp 575–584. https://doi.org/10.1145/2702123.2702221

Odom W, Wakkary R (2015) Intersecting with unaware objects. ACM Press, pp 33–42. https://doi.org/10.1145/2757226.2757240

Odom W, Wakkary R, Lim Y, Desjardins A, Hengeveld B, Banks R (2016) From research prototype to research product. In: Proceedings of the 2016 CHI conference on human factors in computing systems. ACM, New York, NY, USA, pp 2549–2561. https://doi.org/10.1145/2858036.2858447

Odom W, Wakkary R, Bertran I, Harkness M, Hertz G, Hol J, Lin H, Naus B, Tan P, Verburg P (2018) Attending to slowness and temporality with olly and slow game: a design inquiry into supporting longer-term relations with everyday computational objects. In: Proceedings of SIGCHI conference on human factors in computing systems. ACM, New York, NY, USA (in press)

Oogjes D, Wakkary R (2017) Videos of things: speculating on, anticipating and synthesizing technological mediations. In: Proceedings of the 35th annual ACM conference on human factors in computing systems. ACM, Denver, CO, USA

Overbeeke K, Djajadiningrat T, Hummels C, Wensveen S, Prens J (2003) Let's make things engaging, pp 7–17. https://doi.org/10.1007/1-4020-2967-5_2

Pierce J (2009) Material awareness: promoting reflection on everyday materiality. ACM Press, p 4459. https://doi.org/10.1145/1520340.1520683

Pierce J, Paulos E (2011) A phenomenology of human-electricity relations. ACM Press, p 2405. https://doi.org/10.1145/1978942.1979293

Pierce J, Paulos E (2015) Making multiple uses of the obscura 1C digital camera: reflecting on the design, production, packaging and distribution of a counter functional device. In: Proceedings of the 33rd annual ACM conference on human factors in computing systems. ACM, pp. 2103–2112. Retrieved from http://dl.acm.org/citation.cfm?id=2702405

Rosenberger R, Verbeek P-P (2015) A field guide to postphenomenology. In: Rosenberger R, Verbeek P-P (eds) Postphenomenological investigations: essays on human-technology relations. Lexington Books, Lanham, pp 9–41

Sengers P, Gaver B (2006) Staying open to interpretation: engaging multiple meanings in design and evaluation. In: Proceedings of the 6th conference on designing interactive systems. ACM, New York, NY, USA, pp 99–108. https://doi.org/10.1145/1142405.1142422

Tromp N, Hekkert P, Verbeek P-P (2011) Design for socially responsible behavior: a classification of influence based on intended user experience. Des Issues 27(3):3–19. https://doi.org/10.1162/DESI_a_00087

Verbeek P-P (2005) What things do: philosophical reflections on technology, agency, and design (2. printing). Pennsylvania State University Press, University Park, Pa

Verbeek P-P (2015) Cover story: beyond interaction: a short introduction to mediation theory. Interactions 22(3):26–31. https://doi.org/10.1145/2751314

Wakkary R, Maestri L (2007) The resourcefulness of everyday design. In: Proceedings of the 6th ACM SIGCHI conference on creativity & cognition. ACM, New York, NY, USA, pp 163–172. https://doi.org/10.1145/1254960.1254984

Wakkary R, Odom W, Hauser S, Hertz G, Lin H (2015) Material speculation: actual artifacts for critical inquiry. Aarhus Ser Hum Centered Comput 1(1):12. https://doi.org/10.7146/aahcc.v1i1.21299

Wakkary R (2016, October 28). Designing to know: chairs, bowls and other everyday technological things. Inaugural Lecture, Eindhoven, Netherlands. Retrieved from https://pure.tue.nl/ws/files/41066084/Wakkary_2016.pdf

Wakkary R, Desjardins A, Hauser S (2016a) Unselfconscious interaction: a conceptual construct. Interact Comput 28(4):501–520. https://doi.org/10.1093/iwc/iwv018

Wakkary R, Lin HWJ, Mortimer S, Low L, Desjardins A, Doyle K, Robbins P (2016b) Productive frictions: moving from digital to material prototyping and low-volume production for design research. In: Proceedings of the 2016 conference on designing interactive systems. ACM Press, pp 1258–1269. Retrieved from http://summit.sfu.ca/system/files/iritems1/16366/pn0566-Wakkary_rev.pdf

Wakkary R, Oogjes D, Hauser S, Lin H, Cao FC, Ma L, Duel T (2017) Morse things: a design inquiry into the gap between things and us. In: Proceedings of the 2017 ACM conference on designing interactive systems. ACM, New York, NY USA, pp. 503–514. https://doi.org/10.1145/3064663.3064734

Wakkary R, Hauser S, Oogjes D (2018a) Displacement and withdrawal of things in practices. In: Social practices and more-than-humans. Palgrave Macmillan, London

Wakkary R, Oogjes D, Lin H, Hauser S (2018b) Tilting bowl: living with philosophers. In: Proceedings of the 2018 SIGCHI conference on human factors in computing systems. ACM, Montreal (p. in press)

Wiltse H, Stolterman E (2010) Architectures of interaction: an architectural perspective on digital experience. In: Proceedings of the 6th Nordic conference on human-computer interaction: extending boundaries. ACM, New York, NY, USA, pp 821–824. https://doi.org/10.1145/1868914.1869038

Chapter 12
Designing for Joyful Movement

Ylva Fernaeus, Kristina Höök and Anna Ståhl

1 Introduction

Interaction design research has broadened its focus from settings in which people would sit more or less still in front of static computers doing their work tasks, to instead thriving off new interactive materials, mobile use, and ubiquitously available data of all sorts, creating interactions everywhere. These changes have put into question such as play versus learning, work versus leisure, or casual versus serious technology use. As both hardware and software have become mobile—both literally and in terms of transgressing cultural categories—the different social spheres and the rules that they are associated with are changing. Designing for playfulness is no longer confined to leisure time activities, such as computer games or entertainment systems for the home. Instead playfulness design may enter even the most serious work environments, such as control room work (Solsona Belenguer 2015) or learning military tactics (Frank 2012).

The idea of designing for playfulness is not new it has been a topic of research for many years (e.g. De Koven 2013; Gaver 2002; Hobye 2014; Holmquist et al. 2007; Brown and Juhlin 2015; Fernaeus et al. 2012a, b). When turning to the joy of movement as play, there is the New Games movement (Fluegelman 1976), which is concerned with the joy of movement as being more playful and not as structured and measurable as in e.g. sport games. But here, our focus is at the intersection between playful movements and technology through interaction design.

Y. Fernaeus · K. Höök
KTH Royal Institute of Technology, Stockholm, Sweden
e-mail: fernaeus@kth.se

K. Höök
e-mail: khook@kth.se

A. Ståhl (✉)
RISESICS, Kista, Sweden
e-mail: anna.stahl@ri.se

The miniaturization of technology, allows for different forms of mobility. Sensor-nodes connecting wirelessly can be placed onto our bodies or into all sorts of environments. Alongside novel interactive materials such as interactive textiles, these enable a richer palette of interactions. Instead of focusing solely on symbolic, language-oriented, explicit dialogues with interactive technologies, movement-based interactions are on the rise (e.g. Isbister et al. 2011; Simon 2009).

Combining playfulness with movement-based interactions has been at the core of numerous design explorations—both by us, as part of the Mobile Life center, but also by many others. Here we would like to paint a broad picture of movement-based playful design based on some of those design examples. Movement can, for example, be about sensing the movement of a person's body, but it might just as well be the movement of a person's position in geographical space, the movement of a shape-changing object or how moving bodies are tightly connected to and affect social dimensions. We have designed with the joys of moving our bodies actively to achieve a range of different experiences (e.g. Höök 2008; Sundström et al. 2005; Höök et al. 2015; Mentis et al. 2014; Zangouei et al. 2010; Márquez Segura et al. 2013a; Johansson et al. 2011; Tholander and Nylander 2015). But passive interactions such as mirroring bodily processes back to us in evocative, magic shapes have also intrigued us (e.g. Ståhl et al. 2009; Sanches et al. 2010). Another strand of work focused on performance and spectatorship settings (e.g. Montola et al. 2009; Back and Waern 2014). Our designs have touched upon tiny movements as well as full-body experiences (Höök et al. 2015; Márquez Segura et al. 2016; Fernaeus et al. 2012a, b).

The design examples we will engage with can be broadly sorted into four themes:

1. Sensing one's body: the enchanting joys of bellyaches
2. What technology can sense: the magic and mysteries of movement feedback
3. Performance and spectatorship: there is no business like show business
4. Tiny movements and full body experiences: move that body!

2   Sensing One's Body: The Enchanting Joys of Bellyaches

It is an equally obscure point why the corners of the mouth are retracted and the upper lip raised during ordinary laughter. [...] The respiratory muscles, and even those of the limbs, are at the same time thrown into rapid vibratory movements. [...] During excessive laughter the whole body is often thrown backward and shakes, or is almost convulsed. The respiration is much disturbed; the head and face become gorged with blood, with the veins distended; and the orbicular muscles are spasmodically contracted in order to protect the eyes. Tears are freely shed. (Darwin 1872, p. 206)

Reflecting on the above quote, it is obvious that our physical moving bodies can never be quite neglected in the ways we enjoy ourselves. From the thrills of being on a rollercoaster to the simple pleasure of lying down to relax on your sofa.

Proprioception and kinesthesia are two terms that researchers use to describe the personal experience of movement, e.g. how an arm is held in relation to one's torso, or how the whole body is moving through space. Movement-based game platforms such as Wii and Kinect, build on picking up such movements and use it to control games or other applications. There is also a whole range of wearable equipment for sport settings (e.g. Consolvo et al. 2008; Tholander and Nylander 2015; Mueller et al. 2011), which shows that many people like to engage bodily with digital media as well.

Despite all these new technologies, we are not surprised to see that children are still absorbed in ball games, swinging, climbing, hula hooping, making sand castles, or setting up theatre performances. In a study of the unsupervised play in public playgrounds (Jarkiewicz and Fernaeus 2008), tweens made extensive use of their phones as part of playing, but the playing was still highly physical. Cameras and ringtones were used as parts of a hide-and-seek type of game, they recorded music videos and dance performances, but they did not seem to play many games individually on their phones.

That is, designing systems and applications for movement does not have to be complex technology-wise. The design challenge might not be about reading and mirroring the absolute, correct bodily movement, but instead creating space in the interaction for the joy of moving your body. One such example are the games created for the Oriboo platform (Márquez Segura et al. 2013a, b), where the complexity lies in the context and the social setting rather than in the technology. In short, Oriboo is a small, round, portable game console assembled onto a plastic leash. The Oriboo-sphere moves along the leash using a built in motor. It responds to movement as picked up by an accelerometer and a touch screen. The output takes the form of simple sounds and through a small LED-display shaped as two eyes (see Fig. 1). The leash allows for holding the console in different ways, which in turn steers the way the user moves. The idea behind the Oriboo platform was to support vigorous movement—a radical alternative to the traditional game console.

Fig. 1 Left: the Oriboo. Right: children playing with the Oriboo *Photographer* Jonas Kullman

As the Oriboo does not require a connection to a PC or TV, it allows for playing games whenever and wherever you would like to.

When developing game ideas for the Oriboo, we could easily have fallen into the trap of adding more advanced technology to the console to make it more accurate in its estimate of whether the player is performing the "correct" movement according to the game rules. There are two problems with such an approach. First, movement as picked up by the Oriboo on its leash, is very difficult to recognise technically. That in turn risks occasionally providing the wrong feedback to the player—who might lose faith in the system, discarding it (Pakinkis 2011). But second, attempting to stipulate the right and wrong movements, also risks locking in players into performance that has to follow some rigid predefined script. As discussed by Isbister and DiMauro (2011), the improvisation and variation in movement, in particular allowing for social play is a key path to engagement. We therefore changed our design approach, making the simplicity of the technology an asset in our design process. We designed, for example, a social game named Make My Sound (Márquez Segura et al. 2013b). The players, in this case children, are placed in a circle, each with their own Oriboo. They pick one of the players to be the leader. The leader will make movements with the Oriboo resulting in 'music'. The other players mimic the leader's movements and try to create the same music loop. As the Oriboo can only distinguish between slow, fast and jerky movements, mimicking the music loop could be done in many different ways. This openness in the movement repertoire allowed for many different movements that could potentially create the same music loop. The game became open and free and the accuracy of mimicking movements did not become the main purpose. Judging the success or failure of movement perfection was thereby avoided. In addition, the sound and movement of others became the focus of attention rather than the artifact itself (Márquez Segura et al. 2016). What is happening here is that we placed the enactment of the game rules outside the Oriboo. The game rules were instead implemented as part of the context and social setting (Márquez Segura 2016). The players became the judges of whether someone had been able to perform the "correct" movement.

In terms of movement-based joy, the Make My Sound-game touches upon social, bodily play, mimicking, inspiring, loosely following rules, but also bending them to engage in social interactions. The movements are loose, inexact and permissive—all important qualities in creative movement fashioning (Isbister and DiMauro 2011).

One often thinks that a digitally designed experience should lead to nice and comfortable feelings and that physical and emotional pain and discomfort should be avoided. But positive, creative movement interactions where we have total control is but one of the possible experiences we might be designing for. There are also darker, painful or scary movement-based interactions we might orchestrate. The thrill of a rollercoaster ride, the joys of balancing and almost falling down when engaging in Parcour (Waern et al. 2012), or mastering an unruly horse (Höök 2010) are all movement-based playful interactions that touch on being on the edge of our capacity. Maybe one experience cannot exist without its darker counterpart?

Benford et al. (2012) propose crafting carefully and ethically managed uncomfortable interactions as a tool for realizing positive long-term values in interactions (see also Benford et al. 2018). In their system Breathless the user is taken on a ride in an amusement park setting deliberately designed to create discomfort and fear. Breathless consists of a gasmask with breath sensors. The player's breath controls a large powered swing. The rope swing is pulled backwards on inhalation and forward on exhalation. There are always three participants engaged in the experience: the controller, the rider and the voyeur (inspired from a sixteenth century painting called the Swing). The controller's breathing initially controls the ride but halfway through the experience, the rider's breathing pattern takes over the control. After the session the voyeur becomes the rider, the rider becomes the controller and so on. The experience taps into several forms of discomfort. There is visceral discomfort enacted through the choice of material in the gas mask, which is rubbery, close fitting, as well as in the overall claustrophobic nature of not being able to breathe freely. There is also cultural discomfort in the associations we make with the gas masks. Finally, there is discomfort relating to the social setting: playing with who is in control and the intimate experience of isolation and voyeurism.

The "joy of bellyaches" as we choose to name this category, opens a whole space of possible interactions. Playing with visceral, social, cultural limitations and borders, allows us to experiment with what is allowed and what is not allowed in a society (Ferreira 2015). Overstepping boundaries can be done in the safe zone of an orchestrated experience—if we craft it carefully. The active movements, the active engagement, make us committed and involved. But the wonders of movement are not restricted to these full-body active engagements. This brings us to our second category: turning our eyes inwards, discerning the small stimuli, mirrored back to us through technology.

3 What Technology Can Sense: The Magic and Mysteries of Movement Feedback

Many platforms, such as smart phones, tablets and wearable technology, come with sensors that can pick up on movements as well as bio-signals. The sensors may pick up on, for example, radio signals and thereby estimate a person's location within a room indoors, or connect with satellites in the sky to track movement across larger outdoor spaces, or use inbuilt cameras, gyros, compasses sensors or accelerometers to identify small body movements. Bio-sensors worn in bracelets or around your chest can pick up on heart rate, sweating patterns, body temperature, blood pressure and so on. The most common commercial examples of wearable sensors relate to sports performances. Professional as well as amateur athletes use these to reflect over and maximize their performance. For example, by keeping track of heart rate in relationship to e.g. a specific running track, a runner can step by step increase their performance and find their 'optimal' performance (Tholander and Nylander

2015). Sensors can also be used to track and correct the shape of a movement. For example, to ski optimally, your trunk should not move too much sideways as the energy should always be directed forwards.

Most of these sports applications mirror data back to their users so they can improve their performance. The underlying ideal is that of changing your movement until you achieve an optimal performance. The products are made in a material to suit these purposes, to be worn in close skin contact and still be hygienic and sustain sweaty and wet conditions. Injection molded plastic serves these purposes and can be mass-produced, creating for a certain look and feel. The sensors have to be placed in certain locations in order to capture relevant data. The feedback is usually a visual representation in the form of numbers or graphs.

The Nebula (Elblaus et al. 2015) (see Fig. 2) is one example of an alternative and more playful way to mirror bodily movement. The Nebula is a studded cloak that aims to explore the expressive potential of wearable technologies, in the form of a garment. When wearing the Nebula-cloak, a soundscape is generated from your movement. When different clusters of studs get in contact with one-another, data is captured, in turn creating a soundscape. The aim here is neither to measure certain bodily movement nor to give precise, correct, feedback to the user. Instead the Nebula lets movements and soundscapes co-evolve. The usage encourages exploration. The designers describe the Nebula as an alternative means for self-expression through sound and body. The wearer of the cloak becomes aware of the subtleties of their everyday movements. This awareness can be developed into a more playful, expressive or performance-like usage.

The name, Nebula, was chosen to describe how the metal studs on a black fabric could be likened with star fields or nebulas. The Nebula is made out of a carefully

Fig. 2 The Nebula

chosen fabric, where the properties of the textile have been taken into account for its abilities to transfer movements in a preferred way. That is, not only to make it work functionally, but also for the fabric to flow in a certain way, and thereby create a strong visual identity. The soft and delicate fabric is juxtaposed to the large number of metal studs. The studs are positioned in clusters in areas where the fabric produces folds and wrinkles i.e. close to the arms. This positioning of the studs takes into account the characteristics of the fabric. The affordances of the materials chosen, fabric and metal studs, were used in the design of the data capture. The designers behind the Nebula discuss the importance of having both the competences of fashion design and interaction design in the team, to make these two strands equally important, one example of this is the studs that are both aesthetically important and at the same time the central technology in the interaction. On the backside of the fabric each stud is connected by conductive thread, but the studs are divided into two categories: active or receptive. Both the amount of activity as well as the position of the activity of the studs is taken into account when producing the expressive sound that the garment creates.

The Nebula is one example of a future strand in wearable technology that is not aimed at improvement of performance, but the enjoyment of wearing it and experiencing the interaction. It allows for your everyday movements to be mirrored in a different modality—shaping and being shaped by the soundscape generated. An interesting property of many of these mirroring technologies is how difficult it is to make sense of and sketch with invisible data streams and their relationship to our movements. The design work behind the Nebula was as much concerned with the sensing as with ways by which those sensor readings can be mirrored in ways that are aesthetically interesting and meaningful. This kind of work lead to new insights regarding the design features in the intersection of interactive materials and design professions (like fashion design), both how they can be handled by users, and how designers may make use of them, in forming meaningful relationships between physical manipulation, movement, digital media and aesthetic properties. In our examples, the modality and design of the 'mirror' portraying user data back to them, might be more or less evocative. Compared to the sports applications, the Nebula opens a stronger aesthetic experience in the movement. It focuses on the joy of generating sounds by movement, encouraging creativity and curiosity in the moment, while the sports applications create for different forms of joy—a reflective, long-term engagement to achieve set goals and feelings of accomplishments.

All of these mirroring design concepts fit well into what we have named 'Affective Loops' (Höök 2008). When mirroring movements, our designs are not only telling users a story about themselves, but also actively influencing our experience of ourselves—in the moment and over time. The word emotion originates from the French emotion in turn derived from movoir: to set in motion, to move feelings. This gives us a hint of what emotion is—an experience that is both constituted as movement and at the same time sets us in motion—makes us act. The joy of these mirroring designs lies very much in their ability to set us in motion, making us act, making us feel what we might not have felt before.

The joys of moving are restricted to engaging with our own body, turning our eyes inwards or exposing ourselves to movements. We also use movement to 'perform' socially. The joy of performing and viewing others' movements belongs in our third category, which we turn to next.

4   Performance and Spectatorship: There Is No Business Like Show Business

The enjoyment of moving is not only about experiencing yourself, seeing yourself, but also lies in being seen by others. As we move through public spaces, we, in a sense, perform. We walk in certain ways, we dress in certain ways, we put on certain facial expressions to show emotion and mood appropriate to the setting at hand. These performances can be highly enjoyable (or sometimes oppressive or negative).

Accessories, clothes, wearables, interactive textiles and other visible interactions on or around our bodies, are nowadays part of our public performance. Their mere physical appearances can be used to get attention, performing as indicators of one's own sense of taste or lifestyle, or as a trigger for conversations. For example, Juhlin and colleagues have been exploring how interactive technologies can be made part of our fashion statements (Juhlin et al. 2013; Juhlin and Zhang 2011). In particular, they were interested in how to make so-called shape-changing interfaces. A shape-changing interface will alter its physical form in response to the environment. Juhlin and colleagues designed a set of 22 imaginary mobile phone forms (see Fig. 3). The idea was that in the future, the mobile phone would be able to change shape to harmonise with the rest of our outfits. Based on studies of what matters to fashionistas, each of the 22 mobile forms displayed different cuts and drapes. Previously a lot of effort had been put into what shape changing interfaces could portray, but in this study, the focus was on how it might be used. The intention was not that everyone should be pleased by the samples, but to show how Scandinavian fashion design could inspire new interactive designs. In the study participants got to pick from the samples provided to match different outfits. Outfits are often changed more than once a day, you can wear something for work, something else when spending time at home, and yet another outfit for going out in the evening. The participants in the study were creative with the placement of the selected samples, they imagined wearing them as rings on their fingers, on their shoes and as brooches. It is interesting to imagine how such a small accessoire-like object through tiny movements constantly creates a new shape and thereby a novel expression. In turn when worn, in combination with a selected outfit, it becomes a social statement.

Another study of social performance through movement is Back's study of street performance spectators (Back 2016). To a street performer, the urban street landscape is malleable and can be changed from being a place where you run errands

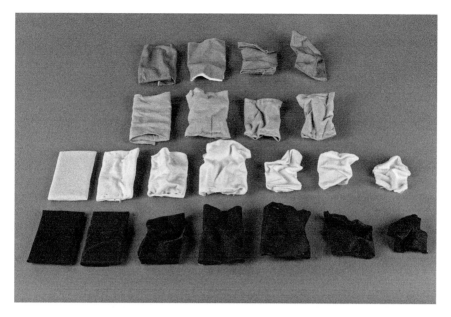

Fig. 3 The 22 imaginary mobile phone forms *Photographer* Anneli Sandberg

and do window-shopping, into a scene for a performance. The study focuses on how the street performers engage with their audience and make them comfortable with the unexpected situation of suddenly watching a show. The act of becoming a spectator, especially for the first ones that stop by, is uncomfortable. The street performers therefore learn a set of tricks to make this transformation possible and interesting enough for the spectators to dare to engage.

Figuring out how to coordinate crowds through movement and attention is interesting when designing for interactions in public spaces. Street performers get their audience's attention step by step: first they draw attention in a non-threatening way, taking a step back to avoid further interaction if it is unwanted, then they approach again, catching attention through talking or doing tricks, this time followed by direct contact through questions, and then once direct contact is established, there is an invitation to watch the show. They talk about this process as 'catching an anchor'. These insights were later used in designing children's playgrounds where landscape architecture designers worked together with interaction designers to create inviting playgrounds where open-ended social play through outdoors, full body movement were core (Waern et al. 2015).

While these two examples (street performance strategies and shape-changing objects as fashion statements) might at first seem unrelated, they, in a sense, build on the same desires. Movement of objects and movement of crowds are immediately attractive to us. We find joy in watching others—their fashion statements, their movements, their interactive behaviours. We find joy in following the crowd, occasionally being invited to partake in activities.

In summary, engaging movement goes beyond individual experience. When designing for public engagement, such as creative self-fashioning or performances between performer/system and spectator/users, the joys of movement may provide fertile grounds for creative processes.

5 Tiny Movements and Full Body Experiences, Move That Body!

When interaction design research initially approached movement-based design, the term "full body interaction" was used to differentiate them from just any interaction. As you sit in front of your computer, you are always moving. Your eyes and hands are moving, the blood is flowing through your body, muscles in your arms and shoulders are activated, and as you sit in your chair, you adjust your position every now and then. What do we mean by movement-based interaction and how is it different from any interaction? To us, it has become a qualitative shift from a predominantly symbolic, language-oriented design stance, to an experiential, felt, aesthetic stance permeating the whole design and use cycle. By putting movement, experience and aesthetics first, our design processes and outcomes have been fundamentally altered. In our work in the Mobile Life centre, we have designed for all sorts of movements: big, physically intense movements, as in sports (Nylander and Tholander 2012) or gesture-based games (Zangouei et al. 2010), very, very small movements, barely discernible to the human eye (Höök et al. 2015), touch (Mentis et al. 2014), dance (Johansson and Tholander 2011), bio-feedback (Sanches et al. 2010; Ståhl et al. 2009), or breathing (Höök et al. 2016). We have based our design work on various studies, such as horseback riding (Höök 2010), nature walks and the sensation of interacting with frozen water (Ståhl et al. 2017). While big movements seem obviously connected with designing for the joy of movement, in our experience equally joyful and immensely satisfying interactions arise when engaging with small and light effort movements. The 'beauty' and joy of such experiences shifts design attention from visual, symbolic, beauty and interaction (which is where most interaction design has been focused). To get at the aesthetics of other senses, we found inspiration in the somaesthetics theories—the aesthetics of the living body or soma (Shusterman 2008).

Two examples of somaesthetic design from our work are the Soma Mat and Breathing Light (Höök et al. 2015, 2016). In short, the Soma Mat (see Fig. 4) uses directed heat stimuli to subtly guide your attention to different body parts. These heat elements follow the instructions from a pre-recorded lesson. For instance, when the person in the lesson asks "How is your body touching the floor right now—where is your right heel today? Left heel? Is there any difference in how they contact the floor?" the mat heats up underneath your right heel and then your left heel. The warmth comes on slowly and leaves slowly. The mat can be used as a support for guided meditation, for body scanning or simply for relaxation. These

Fig. 4 The Soma Mat and the Breathing Light

guided sessions bring participants on a journey of body exploration. When we designed the Soma Mat a lot of effort was put into making the heat subtle enough to not distract, but strong enough to be perceptible. When the right tempo, heat intensity, and interaction with the vocal instructions were found, the experience became intensely pleasurable. The same balance in the material choices was applied for the non-digital materials in the design as well, like using "memory foam" of a certain thickness to create an embracing feeling, without hindering your ability to move freely.

The Breathing Light (see Fig. 4) is a lamp that follows your breathing, creating an ambient light dimming in cadence with the inhalation and exhalation of breath. The light becomes a tool for reflection on your breathing patterns without pulling your attention away from your inner experience. Lying underneath the lamp with your eyes closed, breathing in and out, creates a space for winding down. The space underneath the lamp creates for an enclosed space around your upper body, with an atmosphere intended to make you feel safe and embraced. The string curtains hanging from the lamp, create peaceful patterns with slow responsive movements when touched. During the design process, we put substantial effort into finding the right subtleness, the right timing, intensity, and warmth of the light, to achieve an intimate correspondence between the breathing and the light dimming. We wanted the light to be perceived as an extension of the body—not as a separate device calling for attention.

In common with these particular somaesthetic experiences was their origin in small, conscious, repetitive movements reinforced by subtly designed digital technology. But somaesthetic design is also relevant to all sorts of movement-based design. If you, for example, want to engage the arms in mirroring dance moves in a game, you need to first turn your eyes inwards to really experience what your arms can do. An arm is much more than a limb that can be bent. It is an intriguing, complex system, interconnected through the tiniest fibers and fascia to your shoulders, your trunk, your breathing, your whole body—it is a never-ending treasure open to design explorations. A somaesthetic design process allows you to discern all those sensations of the arms moving and their aesthetic potential in your design process.

In summary, there is an infinite space of bodily experiences we can engage with when designing with movement. Full body engagement can in fact be anything from small, hardly discernible movements to big, vigorous, exertions of your limbs, trunk and whole body. Rhythm, proprioception, deep engagement with all our senses, and explorations of the aesthetics these may bring, are keys to engaging in these new forms of design. It entails paying attention to interactions that are not necessarily based on visual, symbolic, language-oriented, interaction models.

6   Conclusion

What experiences of joy may arise from bodily engagements with technology? Going over the design examples above, we hoped to provide a rich array possibilities: the very small movements as well as the big vigorous ones, the painful as well as the pleasurable, performer as well as spectator experiences, the creative self-fashioning as well as the onlooker experience, the inwards-looking as well as outwards-facing social experience, and so on. The joys of movements are endless, as are the possible applications.

With the four themes, we also wanted to point to possibilities that might move us beyond the first design ideas that pop into our minds when discussing playfulness and movement-based design. With the first theme we aimed to show how the joy of movement does not have to be entirely controlled "inside" the system. Instead the technology can be made into one part of the social and bodily context in which the playful interactions are enacted. With the second theme, we wanted to remind ourselves that movement can be mirrored in many modalities and forms—not only as graphs and numbers. The Nebula cloak allowed us to see how movement can shape interaction and vice versa when we step outside the "measure-and-tell"-formula that is, perhaps, the most prevailing interaction form for sensor feedback of bodily movement. The third theme aimed to remind us that movements are not confined solely to moving limbs, but can also be shape-changing interfaces or the movements of whole crowds. The fourth theme aimed to remind us of all the senses that can be involved and that full body interaction does not necessarily only mean big movements of the limbs.

Above we cited Darwin describing all the movements involved in laughing. Without these movements, the experience of joy would not be the same. Or to quote Sheets-Johnstone:

> Without the readiness to act in a certain way, without certain corporeal tonicities, a certain feeling would not, and indeed, could not be felt, and a certain action would not, and indeed, could not be taken, since the postural dynamic of the body are what make the feeling and the action possible. (Sheets-Johnstone 1999)

As we attempt to set the scene for joyful experiences, orchestrating their initiation, peak and end, engaging with movement as the basis for design is core. The moving body is the basis of what we are, what we can and will experience—whether we engage in small or big movements, engaging our whole bodies, or only certain muscles or nervous system reactions. To design for fun requires an intimate understanding and engagement with our senses and our movements—*feeling the joy*.

References

Back J (2016) Designing public play: playful engagement, constructed activity, and player experience. Doctoral dissertation, Institutionen för informatik och media

Back J, Waern A (2014) Codename heroes–designing for experience in public places in a long term pervasive game. In: 9th International conference on the foundations of digital games (FDG), 3–7 Apr 2014, Ft. Lauderdale, FL

Benford S, Greenhalgh C, Giannachi G, Walker B, Marshall J, Rodden T (2012) Uncomfortable interactions. In: Proceedings of the SIGCHI conference on human factors in computing systems. ACM, pp 2005–2014

Benford S, Greenhalgh C, Giannachi G, Walker B, Marshall J, Tennent P, Rodden T (2018) Discomfort — The Dark Side of Fun. In: M. Blythe and A. Monk (eds) Funology 2, Vol 2. Springer, Dordrecht, pp 215–130

Brown B, Juhlin O (2015) Enjoying machines. Mit Press

Consolvo S, McDonald DW, Toscos T, Chen MY, Froehlich D, Harrison B, Klasnja P, LaMarca A, LeGrand L, Libby R, Smith I, Landay JA (2008) Activity sensing in the wild: a field trial of Ubifit garden. In: proceedings of CHI, 1797–1806

Darwin CR (1872) The expression of the emotions in man and animals, 1st ed. John Murray, London

De Koven B (2013) A player's philosophy. The well-played game. MIT Press, Cambridge

Elblaus L, Tsaknaki V, Lewandowski V, Bresin R (2015) Nebula: an interactive garment designed for functional aesthetics. In: Proceedings of the 33rd annual ACM conference extended abstracts on human factors in computing systems. ACM, pp 275–278

Fernaeus Y, Holopainen J, Höök K, Ivarsson K, Karlsson A, Lindley S, Norlin C (2012a) Plei plei paperback, Nov 1

Fernaeus Y, Jonsson M, Tholander J (2012b) Revisiting the jacquard loom: threads of history and current patterns in HCI. In: Proceedings of the SIGCHI conference on human factors in computing systems. ACM, pp 1593–1602

Ferreira P (2015) Play as freedom: implications for ICT4D. Doctoral dissertation, KTH Royal Institute of Technology

Fluegelman A (1976) The new games book

Frank A (2012) Gaming the game: a study of the gamer mode in educational wargaming. Simul Gaming 43(1):118–132

Gaver W (2002) Designing for homo ludens. I3 Magazine, 12, pp 2–6

Hobye M (2014) Designing for homo explorens: open social play in performative frames. Faculty of Culture and Society Malmö University

Holmquist LE, Höök K, Juhlin O, Waern A (2007) Mobile life: a research foundation for mobile services

Höök K (2008) Affective loop experiences—what are they? In: International conference on persuasive technology. Springer, Berlin, Heidelberg, pp 1–12

Höök K (2010) Transferring qualities from horseback riding to design. In: Proceedings of the 6th Nordic conference on human-computer interaction: extending boundaries. ACM, pp 226–235

Höök K, Ståhl A, Jonsson M, Mercurio J, Karlsson A, Johnson ECB (2015) Cover story somaesthetic design. Interactions 22(4):26–33

Höök K, Jonsson M, Ståhl A, Mercurio J (2016) Somaesthetic appreciation design. In: Proceedings of the 2016 CHI conference on human factors in computing systems. ACM, pp 3131–3142

Isbister K, DiMauro C (2011) Waggling the form baton: analyzing body-movement-based design patterns in Nintendo Wii games, toward innovation of new possibilities for social and emotional experience. In whole body interaction. Springer, Berlin, pp 63–73

Isbister K, Rao R, Schwekendiek U, Hayward E, Lidasan J (2011) Is more movement better?: A controlled comparison of movement-based games. In: Proceedings of the 6th international conference on foundations of digital games. ACM, New York, NY, USA, pp 331–333. http://doi.org/10.1145/2159365.2159429

Jarkiewicz P, Fernaeus Y (2008) September. In the hands of children: exploring the use of mobile phone functionality in casual play settings. In: Proceedings of the 10th international conference on human computer interaction with mobile devices and services. ACM, pp 375–378

Johansson C, Tholander J (2011) Exploring bodily engaging artifacts among golfers, skaters and dancers. In: Whole body interaction. Springer, London, pp 75–85

Johansson C, Ahmet Z, Tholander J, Aleo F, Jonsson M, Sumon S (2011) Weather gods and fruit kids-embodying abstract concepts using tactile feedback and whole body interaction. In: 9th International computer-supported collaborative learning conference, CSCL, Hong Kong, 4–8 July. International Society of the Learning Sciences, pp 160–167

Juhlin O, Zhang Y (2011) Unpacking social interaction that make us adore: on the aesthetics of mobile phones as fashion items. In: Proceedings of the 13th international conference on human computer interaction with mobile devices and services. ACM, pp 241–250

Juhlin O, Zhang Y, Sundbom C, Fernaeus Y (2013) Fashionable shape switching: explorations in outfit-centric design. In: Proceedings of the SIGCHI conference on human factors in computing systems. ACM, pp 1353–1362

Márquez Segura E (2016) Embodied core mechanics: designing for movement-based co-located play. Doctoral dissertation, Department of Informatics and Media

Márquez Segura E, Moen J, Waern A, Onco Orduna A (2013a) The oriboos going to Nepal: a story of playful encounters. In: Proceedings of the 8th ACM/IEEE international conference on human-robot interaction. IEEE Press, pp 411–412

Márquez Segura E, Waern A, Moen J, Johansson C (2013b) The design space of body games: technological, physical, and social design. In: Proceedings of the SIGCHI conference on human factors in computing systems. ACM, pp 3365–3374

Márquez Segura E, Turmo Vidal L, Rostami A, Waern A (2016) Embodied sketching. In: Proceedings of the 2016 CHI conference on human factors in computing systems. ACM, pp 6014–6027

Mentis HM, Laaksolahti J, Höök K (2014) My self and you: tension in bodily sharing of experience. ACM Trans Comput Hum Interact (TOCHI) 21(4):20

Montola M, Stenros J, Waern A (2009) Pervasive games: theory and design. Morgan Kaufmann Publishers Inc.

Mueller FF, Edge D, Vetere F, Gibbs MR, Agamanolis S, Bongers B, Sheridan JG (2011) Designing sports: a framework for exertion games. In: Proceedings of the SIGCHI conference on human factors in computing systems. ACM, pp 2651–2660

Nylander S, Tholander J (2012) Tactile feedback in real life sports: a pilot study from cross-country skiing. In: The seventh international workshop on haptic and audio interaction design 23–24 Aug 2012, Lund, Sweden, p 35

Pakinkis T (2011) Kinect has 'some real problems'—Molyneux. But has a sense of "freedom and emotion". Computer and videogames.com. Available from http://www. computerandvideogames.com/308315/kinect-has-some-real-problems-molyneux/

Sanches P, Höök K, Vaara E, Weymann C, Bylund M, Ferreira P, Peira N, Sjölinder M (2010) Mind the body!: designing a mobile stress management application encouraging personal reflection. In: Proceedings of the 8th ACM conference on designing interactive systems. ACM, pp 47–56

Sheets-Johnstone M (1999) Emotion and movement: a beginning empirical-phenomenological analysis of their relationship. J Conscious Stud 6(11–12):259–277

Shusterman R (2008) Body consciousness: a philosophy of mindfulness and somaesthetics. Cambridge University Press

Simon B (2009). Wii are out of control: bodies, game screens and the production of gestural excess (SSRN Scholarly Paper No. ID 1354043). Social Science Research Network. Retrieved from http://papers.ssrn.com/sol3/papers.cfm?abstract_id=1354043

Solsona Belenguer J (2015) Engineering through designerly conversations with the digital material: the approach, the tools and the design space, KTH Royal Institute of Technology, xvi, 89 p

Ståhl A, Höök K, Svensson M, Taylor AS, Combetto M (2009) Experiencing the affective diary. Pers Ubiquit Comput 13(5):365–378

Ståhl A, Tholander J, Laakoslahti J, Kosmack Vaara E (2017) Being, bringing, briding—three aspects of sketching in nature. In: Proceedings of DIS'17. ACM

Sundström P, Ståhl A, Höök K (2005) eMoto: affectively involving both body and mind. In: CHI'05 extended abstracts on human factors in computing systems. ACM

Tholander J, Nylander S (2015) Snot, sweat, pain, mud, and snow: performance and experience in the use of sports watches. In: Proceedings of the 33rd annual ACM conference on human factors in computing systems. ACM, pp 2913–2922

Waern A, Balan E, Nevelsteen K (2012) Athletes and street acrobats: designing for play as a community value in parkour. In: Proceedings of the SIGCHI conference on human factors in computing systems. ACM, pp 869–878

Waern A, Back J, Pysander ELS, Heefer CJ, Rau A, Paget S, Petterson L (2015) DigiFys: the interactive play landscape. In: Proceedings of the 12th international conference on advances in computer entertainment technology. ACM, p 46

Zangouei F, Gashti MAB, Höök K, Tijs T, de Vries GJ, Westerink J (2010) How to stay in the emotional rollercoaster: lessons learnt from designing EmRoll. In: Proceedings of the 6th Nordic conference on human-computer interaction: extending boundaries. ACM, pp 571–580

Chapter 13
Discomfort—The Dark Side of Fun

**Steve Benford, Chris Greenhalgh, Gabriella Giannachi,
Brendan Walker, Joe Marshall, Paul Tennent and Tom Rodden**

1 Introduction

For many of us, the notion of 'fun' conjures up visions of experiences that are amusing, pleasant, entertaining, playful—perhaps even frivolous. Rides, games, shows and perhaps even the experience of visiting an art gallery can embody these senses of fun, providing amusing and momentary distractions from the toils of life. And yet, such experiences often have a darker side to them. Thrill rides such as roller coasters may be scary and physically demanding. Games routinely involve us in pretending to commit unspeakable acts such as butchering others. And the works we encounter in theatres and galleries may challenge, confront and even outrage us. So perhaps fun is not so frivolous after all? Maybe fun inevitably encompasses a 'dark side' as a vital, even necessary, part of the entertainment. Encountering and passing through suspense, horror and fearful anticipation may be a key part of the fun. They are perhaps also, as some have argued, part of a wider a human need to fill the void created by the 'civilising process' (Elias 1982) as minds and bodies that evolved to confront danger must now inhabit an increasingly safe and sanitised world? Or perhaps these experiences are important ways in which us humans come to reflect, learn, raise and confront our fears and taboos and ultimately, achieve self-realisation?

This chapter explores how such questions apply to the challenge of interaction design. How might HCI researchers and UX practitioners respond to the dark side of fun? Can they, and indeed should they, include it in their armoury of design techniques? We ground our exploration in the concept of 'discomfort', considering

S. Benford (✉) · C. Greenhalgh · B. Walker · J. Marshall · P. Tennent · T. Rodden
The Mixed Reality Laboratory, The University of Nottingham, Nottingham, UK
e-mail: Steve.Benford@nottingham.ac.uk

G. Giannachi
Centre for Intermedia, The University of Exeter, Exeter, UK

© Springer International Publishing AG, part of Springer Nature 2018
M. Blythe and A. Monk (eds.), *Funology 2*,
Human–Computer Interaction Series,
https://doi.org/10.1007/978-3-319-68213-6_13

how different forms of discomfort might be employed and encountered in interactive experiences. Our approach has been one of performance-led research in the wild (Benford et al. 2013), a practice-led methodology in which artists (with technical support from researchers) create new cultural works, that are then studied 'in the wild' as they tour, and ultimately where reflections across multiple works generates so-called intermediate design knowledge such as guidelines, heuristics, taxonomies and 'strong concepts' (Höök and Löwgren 2012). What follows is an example of the latter stage: a reflection across a series of cultural works that have arisen from collaborations between professional artists and HCI researchers, in which we draw on examples of interactive rides and performances to illustrate various kinds of uncomfortable interactions. We focus on two artists as examples—Blast Theory and Brendan Walker—who embody different approaches to designing with discomfort and whose works have been previously widely reported in HCI journals and conferences. An overview and discussions of many of their works can be found in (Benford and Giannachi 2011) while the artists' own descriptions of their works, including links to documentary videos can be found at their websites.[1]

2 Blast Theory

Blast Theory are a UK-based artists group based with a longstanding history of creating interactive performances that employ mobile, virtual environment and other interactive technologies to place the audience at the centre of the work. They have collaborated with the Mixed Reality Laboratory since 1996 and co-authored a series of papers that have appeared in HCI venues since 2001.

Their works are often politically and ethically challenging, directly confronting audience members with contemporary ethical dilemmas and forcing them to the point of adopting positions with respect to and taking decisions about these. They are masters of subtly employing various forms of discomfort to achieve this.

Desert Rain (Koleva et al. 2001) (Fig. 1) tackled the topic of the First Gulf War and its portrayal in the media of the time as making this the first 'virtual war'. The work took the form of a computer game in which six players at a time were sent into a shared virtual world on a mission to locate six targets. This virtual world was in turn, embedded into a physical set that included a briefing room in which they received a military-style briefing and a final chamber in which they watched video interviews with their targets—a soldier who served in the war; a soldier who was bedridden at the time of the war and watched it on television; a peace worker who helped establish a peace camp on the Iraqi-Saudi border in 1990; a journalist who was in Baghdad on the night the air war started; an actor who played a soldier in a TV drama about the war; and an actor who was on holiday in Egypt at the time. At the centre of *Desert Rain*, was the 'rain curtain', a projection screen made of a fine

---

[1]http://www.blasttheory.co.uk/ and http://thrilllaboratory.com/.

Fig. 1 Desert Rain (Blast Theory)

water spray that first performers and subsequently audience members, would physically traverse so as to encounter one another or move on to the next element of the set. The intense nature of the briefing, visceral experience of the rain curtain and surprising encounters with performers was disconcerting for many. However, this was greatly heightened by Blast Theory's deliberate tactic of engineering situations in which teams would have to decide whether to abandon a team member in order to complete their mission. This involved surreptitiously moving avatars from a special control interface behind the scenes so that groups would become separated, disorientated and lost. *Desert Rain* toured to over ten cities worldwide to be experienced by thousands of participants, and to critical acclaim that highlighted it as being a thought provoking and challenging experience.

Blast Theory's subsequent 2003 work *Uncle Roy All Around You* explored such ideas further. Thematically, the work explored matters of surveillance, trust and privacy, engaging participants with the question of whether they should place their trust in remote strangers or systems that appeared to be monitoring their movements from afar. Structurally, like *Desert Rain* before, the work drew on the language and form of computer games, though rapidly subverting the sense of 'fun'. Players on the streets of a city, armed with handheld computers (this was before smart phones) were sent on a mission to find Uncle Roy, a mysterious figure whose identity and motivations were unclear (Fig. 2). Highly ambiguous clues led them on a journey through the city streets in search of Roy's office. Online players then contacted

Fig. 2 Uncle Roy All Around You (Blast Theory)

them with offers of help along the way. This second group of players were logged into a virtual model of the same city over the Internet, which they could explore to find out information such as the location of the office, see the positions of street players, and exchange messages with them. Eventually, most street players found their way to Uncle Roy's office, a real office in a building that they were invited to enter, look around and also fill in a postcard offering a promise of help to an unknown stranger should they be called upon during the next year. From here, they were directed to a phone box where they received a call that guided them to a waiting car and, for those who chose to get inside, a short ride and a face-to-face interview with an actor who asked them questions about the nature of trust. In turn, online players got to observe a street player in the office through a hidden surveillance camera before being invited to make a similar promise to help a stranger in future. Following their experiences, pairs of street and online players who had made promises received postcards from one another with their personal contact details. Like *Desert Rain* before it, the performance deliberately induced discomforting moments of being lost, confused and alone in the city. It also played heavily on the sense of being surveilled while provoking decisions about whether to trust unknown strangers—or indeed the system as embodied by the shadowy Uncle Roy. Finally, the work cast doubt on the relationship between the fictional and the real, inviting participants to enter real places and at one point to follow a stranger (Benford et al. 2006).

Our third example from Blast Theory, *Ulrike and Eamon Compliant*, was also a locative experience designed for the city streets (Tolmie et al. 2012). Originally commissioned for the 53rd Venice Biennale in 2009, the piece led participants on a journey through the city alleyways and waterways (though was subsequently adapted for several other cities) while simultaneously undertaking a parallel journey through the life of one of two convicted terrorists, Ulrike Meinhof of the Red Army Faction or Eamon Collins of the Irish Republican Army. In the words of the artists[2]: "The project is based on real world events and is an explicit engagement with political questions. What are our obligations to act on our political beliefs? And what are the consequences of taking those actions?" At the start of the experience, participants were invited to choose to follow the story of either Ulrike or Eamon. They then undertook a walk through the city, guided by a series of voice messages delivered over their mobile phones that also narrated the life story of their chosen person. As the walk unfolded, so these messages subtly shifted in tone, increasingly demanding compliance from the participant while implying that they were being watched closely, for example requiring them to sit on a particular bench or pause and perform a certain gesture while crossing a bridge. The messages also began to address the participant as if they were the character while offering ambiguous warnings as to whether they might really wish to continue. Eventually, participants encountered an actor who led them to a room to be interviewed about how far they would go to fight for their views (Fig. 3). As they left, they were offered a brief opportunity to pause and look back through a one-way mirror to see the next participant being interviewed. While *Ulrike and Eamon Compliant* shared many commonalities with its predecessors, it is perhaps most notable for the careful design of its instructions, which conveyed the narrative, information about navigating through the city, how participants should behave in public and also set the mood and tone of the experience in a way that was both ambiguous and discomforting (Tolmie et al. 2012).

3 Brendan Walker

The works of artist, TV presenter and "Thrill Engineer" Brendan Walker take a quite different form and provide a distinctive perspective on the nature of fun. Walker's works have consistently explored the nature of thrill as a balance of sensations and emotions in which participants move between fear and suspense and elation and euphoria. Operating under the banner of Thrill Laboratory his works have frequently been inspired by, and occasionally set within, the fairground as a site of entertainment, fun and hedonistic experience, but also one with a sense of

---

[2]http://www.blasttheory.co.uk/projects/ulrike-and-eamon-compliant/ (verified 14th July 2017).

Fig. 3 Ulrike and Eamon Compliant (Blast Theory)

otherworldliness and potentially darkness. His works also draw inspiration from the domain of science, with participants often taking on the role of subjects in unusual physiological experiments. His collaboration with the Mixed Reality Laboratory dates back to 2007 and, as with Blast Theory's, has generated a series of research papers that have appeared in HCI venues.

Fairground: Thrill Laboratory was an early example of using digital technologies to augment the experience of extreme amusement rides (Schnädelbach et al. 2008). Staged at the Science Museum's Dana Centre in London, Walker hired a series of rides over several nights as the basis for public experiments. Riders, selected from the audience, were kitted out with a wearable telemetry system that captured close-up video and audio as well as heart rate, galvanic skin response and accelerometer data. This data was streamed live to a series of visualisations and then discussed by experts including ride designers, psychologists and sensor experts in front of an audience. Subsequent experiences deployed similar telemetry systems on roller coasters (Fig. 4), in horror mazes in two major national theme parks in the UK, at screenings of horror films (Reeves et al. 2015) and even as part of marketing campaigns (Reeves 2011). While roller coasters, horror mazes and similar attractions naturally play with discomfort—fear, suspense, extreme physical sensations—Walker's innovation was to extend this to spectators, enabling them to vicariously tune into close up views of a rider's experience throughout an entire ride rather than just meeting them at the end when they emerge, often euphorically, from the experience (even if only the relief of having survived). This spectating could be an

Fig. 4 Oblivion: Thrill Laboratory (Brendan Walker)

uncomfortable experience in and of itself, for example when watching loved ones go through a ride while being impotent to act or offer comfort.

A series of further works turned to questions of how physiological data captured from data could be used to make the rides themselves interactive. How might one create a tight human-in-the-loop experience in which the ride both responded to but also directly actuated the human's physiological responses? An early experiment was the *Broncomatic*, a breath-controlled rodeo-bull ride (Fig. 5). As is typical with such rides, the underlying movements of the bronco would gradually get more extreme over time in attempts to throw off the rider. However, in this case, the rider's breathing (as measured from a chest strap monitor) added a layer of further movement—breathing in yawed the ride one way while breathing out yawed it the other—while riders also scored points every time they breathed. So breathing scored points while also making it harder to stay on board, with the physical movements of the ride also influencing exertion and hence breathing. The ride proved to be popular, being experienced by thousands of riders at Nottingham's National Videogame Arcade. An early study revealed how it could deliver powerful moments at which riders experienced heightened awareness of their physiological state and imminent lack of control, for example when sitting on top of the ride, holding their breath in attempts to control it, but also gradually "running our of air" (Marshall et al. 2011a). Inspired by such moments, Walker's next work, *Breathless*, took the form of a breath controlled swing ride in which riders' breathing cased the physical actuation of a swing (Fig. 6). In this case, breathing was sensed by a respiration sensor built into a rubberised gas mask, the wearing of which potentially induced further feelings of discomfort due to its temperature, smell, sense of claustrophobia and isolation as well as its wider cultural associations (Marshall et al. 2011b).

Fig. 5 The Broncomatic (Brendan Walker/MRL)

Fig. 6 Breathless (Brendan Walker)

The swing itself is an interesting device inducing visceral sensations and also involving a loss of control—one cannot instantly stop a swing once moving and there is a sense of becoming dangerously out of control if one swings very high. This led to a further work that explored playground swings as an interaction device. In *Oscillate* the riders of a playground swing also wore a virtual reality headset (Fig. 7), with their view of the virtual world gradually becoming increasingly decoupled from their kinaesthetic sensation of swing, for example as the floor dropped away, leading to the idea of new kinds of visual-kinaesthetic-experience that play with the coupling of otherwise of these two senses to create powerful sensations (Tennent et al. 2017).

4   Uncomfortable Interactions

A common these running through both artists' works is their frequent use of discomfort in interactive experiences. While this can take many forms, from the physical discomforts of running out of breath while riding the Broncomatic and wearing a rubberised gasmask on Breathless, to the more emotional discomforts of being lost and disorientated in a strange city or being interviewed as if a terrorist,

Fig. 7 Oscillate (Brendan Walker)

being uncomfortable appears to be part and parcel of all of the experiences that we described.

In unpacking this further, the first question to consider is why employ discomfort at all? In considering this question, Benford et al. (2012) identify three primary reasons. First, feelings of suspense that engender a somewhat uncomfortable sensation of fearful anticipation may be a key aspect of delivering *entertainment*, from roller coasters to thrillers and horror films (Goldstein 1999). Entertainment may also be derived from extreme and potentially uncomfortable physical sensations, from being spun and thrown around, to falling, to the effects of exertion, with the latter forming the basis of new forms of entertainment such as exertion games (Mueller et al. 2011). Second, we may pass through discomfort on the road to *enlightenment*. Engaging audiences with challenging themes such as terrorism and warfare may demand an appropriate framing of the experience, one that sets a serious tone while avoiding the risk of trivialisation. As a mainstream example, Daniel Liebeskind's architectural designs for both the Jewish Museum in Berlin and the Imperial War Museum in Manchester feature challenging architecture that is designed to put the visitor off balance, visually but also physically through the design of sloping floors and irregular angles. Blast Theory's framings in *Desert Rain* and *Ulrike and Eamon Compliant* strive for a similar effect, but through the very precise use of instructions. Experiencing discomfort—even pain—has also long been a tradition in various religions and mystic routes to self-enlightenment. Finally, passing through discomfort together can facilitate *social bonding*. Many groups have initiation ceremonies while even relatively familiar experiences such as riding a roller coaster can be part of a shared rite of passage as reported in Durrant et al.'s (2011) study of families visiting theme parks.

Having explored the "why" of the matter, we turn to the "how"—in what ways are the artists generating uncomfortable interactions in these experiences? While discomfort might potentially take diverse forms, we argue that four primary types are present in the experiences considered above: cultural, visceral, control and intimacy.

Cultural discomfort concerns the themes and meaning of the experience, that is the cultural resonances and potentially dark associations and feelings that it triggers. Thus, experiences may generate discomfort by confronting topics that participants find difficult or that are in some way taboo. A particular tactic is to draw on the form of computer games—a popular source of fun—to place participants in a position of having to take difficult decisions. *Ulrike and Eamon Compliant* for example, tackles the subject of terrorism and ultimately requires participants to decide what they might fight for in similar ways. *Desert Rain* addresses the First Gulf War and requires participants to decide whether to leave a colleague behind. *Uncle Roy All Around You* engages with terrorism and demands decisions about whether to enter strange offices and cars and eventually whether to make an offer of help to an unknown stranger. Moreover, uncomfortable cultural associations may extend beyond the content of the experience to the form of the interface itself. For example, gas masks may invoke chilling associations with, or even direct memories of, warfare, civil unrest for some, or perhaps alternatively of bondage play for others.

Of course, such cultural discomforts are far from new. Nor are they restricted to interactive works. The cultural acceptability of material that is considered adult, difficult or vulgar has provided a significant (and continually shifting) boundary for discomfort. While traditional media such as books and films have long dealt with such material, for example in horror and erotica, interactive works may raise the level of discomfort by requiring users to directly take moral decisions and resolve dilemmas. There is a longstanding tradition of discomforting audiences during performances. Bertold Brecht argued that theatre should contain some level of Verfremdung (alienation), leading to unease or discomfort by encouraging the audience to look at something or someone from another's perspective, thus raising political awareness of social and power structures (Brecht 1993). A series of landmark performances have pushed the boundaries of discomfort. Marina Abramović's *Rhythm O* (1974) was a six-hour performance in which the audience were encouraged to use a series of objects on Abramović's body including a gun, bullet, pocket knife, axe, and matches. While the physical discomfort in this case was primarily experienced by the performer, emotional discomfort was experienced by the audience, whether they acted on Abramović's instruction or simply observed others. More recently, the artist Stelarc (who has given a keynote at the annual CHI conference) created a series of performances in which audience members observe his suspended body being moved and controlled by machinery and stimulated with electric shocks (Smith 2007).

A second type of discomfort that is evident on our examples is **visceral** which refers to those aspects that most directly arise from physical sensation, ranging from the unpleasant feel of physical materials, to strenuous and stressful movements, to experiencing pain. One direct tactic for generating such discomfort is to design wearables and tangibles to feel unpleasant to touch, hold and wear. The physicality of the gas mask used in *Breathless* provides a striking example of this tactic. Gas masks are uncomfortable to wear, with a tight physical fit and overpowering smell. They quickly become hot and fill with sweat that drips down the face in a disturbing way while their restricted visibility can be disorientating, especially as the eye-holes have a tendency to fog. The process of donning wearable technologies may heighten anticipation as reported by Schnädelbach et al. (2008) and may also require intimacy with others as we consider later. Thus, in general, designers may wish to choose materials that are rough, tight, prickly, sweaty, or otherwise physically unpleasant. Similar ideas can be seen in previous interfaces such as *The Meatbook* an artistic tangible interface that involved manipulating raw meat (Levisohn et al. 2007). A quite different tactic is to encourage strenuous physicality. Thrill rides routinely place unusual physical stress on the body through the experience of high G-forces and movements such as inversions, rolls, suspensions and drops. In a different vein, the work *I Seek the Nerves Under Your Skin* required participants to run increasingly fast—to the limits of their physical ability and comfort—in order to listen to a poem (Marshall and Benford 2011). A more extreme and especially challenging, tactic for creating visceral discomfort is to cause pain as seen in game controllers that deliver electric shocks as part of the fun, for example during 'reaction time' games. This typically involves causing mild and transitory rather than chronic pain, without causing significant or lasting physical damage.

While cultural and visceral discomfort are staples of traditional fun experiences, from theatre to rides, our next two forms of discomfort much more directly relate to interactivity and hence to central concerns within the field of Human Computer Interaction.

The question of **control** lies at the heart of HCI's agenda. Conventional wisdom argues that the locus of control should be internal to the user (Shneiderman 1987), that is, that the human should, generally, be in charge of the interface rather than it controlling them. Consequently, perturbing this most fundamental of relationships in HCI can induce a degree of discomfort. In short, participants may become uncomfortable when required to give or control, and perhaps also when required to assume an unusually high degree of control. One tactic is to demand that humans surrender control to the machine. An element of the thrill of rides such as *Breathless* lies in giving up control to the machine; being strapped in and unable to withdraw no matter what follows. Interactive experiences open up new possibilities by giving the user partial control, or as we see with the *Boncomatic*, perhaps inexorably leading them to a crucial tipping point at which they lose control—and also become self-aware of their imminent lack of control. Inspired by the notion of ambiguity in interaction design (Gaver et al. 2003), a further possibility is to emphasise the frustration inherent in unpredictable control and surprising system responses, while the reverse approach of overly precise control may also create discomfort through extreme compliance. A second tactic lies in surrendering control to other people. Theatrical performances invariably involve surrendering control to the performers, which may engender uncomfortable feelings of helplessness and disempowerment. This is a familiar tactic from everyday performance such as when comedians pick on audience members. *Ulrike and Eamon Compliant* induces a far deeper surrendering of control, with participants increasingly complying (and signalling their compliance) with actors' instructions throughout. Our final tactic for control is to require participants to take greater control than normal. There is discomfort to be had in assuming greater control of others as this may invoke feelings of power, responsibility, capriciousness or mischief. As an example, Walker's *Breathless* requires participants to control others as well as being controlled, while *Uncle Roy All Around You* invites online players to assume responsibility for those on the streets of a remote city.

The final form of discomfort that we consider here concerns **intimacy**. The emergence of computer-mediated-communications and then social media over recent decades has placed the inter-personal at the heart of HCI's agenda. As with control, it is therefore possible to generate uncomfortable interactions by systematically distorting social relationships via computers. Intimacy with others may be especially fertile ground. Various HCI papers have promulgated intimate interactions, describing prototype interfaces that aim to enable emotional connectedness and relief from stress or anxiety (Haans and Ijsselsteijn 2006), wellbeing or even sexual fulfilment (Bardzell and Bardzell 2011). However, intimacy also brings much scope for engineering discomfort by distorting social norms. Our first tactic here is to deny the comfort of intimacy by isolating people from the support of friends and loved ones, leaving them alone in unfamiliar environments. Not only is such isolation disturbing, but it also naturally focuses participants inwardly on their own feelings (self-intimacy). *Ulrike and Eamon Compliant* employs this tactic in its solo exploration of the city. The gas mask in *Breathless* delivers localised isolation, anonymising participants and reducing their ability to communicate with others, focusing them instead on the sensation of their own breathing. A second tactic is to force intimacy with strangers which can be very uncomfortable. The final one-to-one interview in *Ulrike and Eamon Compliant* is an especially challenging moment in which the anonymity of being in a large audience is suddenly stripped away and reactions are laid bare in a one-to-one encounter. Taking a step further, the performance *Mediated Body* required participants to physically touch the performer's body (Hobye and Löwgren 2011). Another possibility is to employ surveillance and voyeurism. HCI has seen a growing discussion about the role of the spectator in interactive experiences (Reeves et al. 2005) and this too can be exploited as a source of discomfort. One approach is to emphasise the sense of vulnerability inherent in being surveilled, especially by unseen observers as seen in *Ulrike and Eamon Complaint* and *Uncle Roy All Around You*. Finally, there is discomfort to be found in watching others, for example the helplessness experienced when watching loved ones experience a thrill ride of horror maze without being able to offer succour, a distortion of conventional intimacy by making it unidirectional. The converse is the illicit thrill of voyeurism exploited by *Ulrike and Eamon Compliant* when participants are invited to look through the one-way mirror at the next participant being interviewed.

5 The Appropriate Use of Uncomfortable Interactions

One danger with our argument thus far is that is could be seen as a carte blanche for creating unpleasant interactive experiences, appearing to give designers a licence—backed up with a set of tactics—for doing bad things to users in the interests of what *they* (the designers) think might be fun. We therefore now consider how to ensure that uncomfortable interactions are brought into play in an appropriate and sensitive way so as to ultimately benefit rather than harm people.

The first key point is to restate that uncomfortable interactions are being deployed in pursuit of higher-level goals, specifically those of entertainment, enlightenment or sociality. We suggest that, if used appropriately, they can become a useful aspect of the design of emotional user experiences (McCarthy and Wright 2007). It is important to be clear about the goals of a cultural experience and how any engineered moments of discomfort contribute to these in a proportionate way. Second, is to ensure the skilful application of discomfort. The examples reported above were all delivered by professional artists with many years of training and experience, typically working under the umbrella of a recognised cultural venue. Just as one cannot expect to design other aspects of an interface to an satisfactory standard for public use without extensive experience, so one should not expect to able to create successful uncomfortable interactions. Beginners beware!

Our next point is that discomfort is often resolved within a experience. It is an aspect of experience that is often passed through and typically successfully resolved. Suspense in a narrative is typically resolved. The climb and initial drop of a roller coaster lead to a smoother ride followed by a happy disembarkation. Interaction designers need to consider the entire journey of trajectory through discomfort, including designing in moments of catharsis as it is resolved while also providing people with the resources to tell stories about their experience afterwards as demonstrated by the work of Durrant et al. (2011) and their prototype system for capturing and sharing stories about roller coaster rides. Of course, some experiences may leave the user with a feeling of disquiet as a way to provoke further reflection and enable interpretation which is a key element of experience artworks (Sengers and Gaver 2006), but in this case, we argue, designers should consider what channels are available to people to subsequently express and share their thoughts and views.

This leads us onto a final consideration, that of the wider ethical framework under which these kinds of experiences are delivered. This is a complex issue and one that has been discussed at length by Benford et al. (2015) when considering the wider ethical implications of HCI's turn to the cultural. This discussion emerged from a process of gathering artists, curators, venues, media organisations and researchers together for a three day workshop at which they introduced and reviewed a series of interactive artworks that appeared to push the boundaries of established research ethics in areas such as consent, withdrawal and the treatment of data, especially where the projects served as both cultural works and research projects simultaneously. From out of this discussion emerged a call for a new approach to ethical engagement for grounded in Responsible Research Innovation rather than established approaches that have been adapted from those used for medical research. This RRI approach emphasises engaging a wider variety of stakeholders—including the 'users' of cultural experiences—in ethical debates throughout the process, including giving them channels through which to express their views afterwards. In short, if the goal is to provoke interpretation, then there needs to be appropriate means of allowing people to express and share this and take part in wider ethical and societal debates.

6 But Is It Fun?

So we end by returning to our initial theme, and indeed to the overarching theme of this book: what does this discussion in this chapter contribute to our wider understanding of 'fun' as a useful concept for HCI (Monk et al. 2002)? To be devil's advocate for a moment, one could argue that the kinds of artworks described above are not really fun experiences at all. After all, is it reasonable to consider Blast Theory's *Ulrike and Eamon Compliant* to be a fun experience? Perhaps not and perhaps there is a useful interpretation of fun (within HCI at least) that could be constrained in scope to only those experiences that are amusing, frivolous and even lightweight. On the other hand, Blast Theory's works clearly draw on the cultural form of games that meet this more lightweight definition of fun, but then ambiguously blur their boundaries, using discomfort to undercut the sense of fun with deeper and darker themes. Blast Theory's works may be effective in part because they are fun at some level—or at least directly index fun. Walker's works more clearly ground themselves in the world of amusement. The amusement park after all, is surely the essence of 'fun', a modern day and grown up playground for both children and adults. Even here we see how elements of discomfort heighten the intensity of experience and how playing with fundamental properties of interaction such as control and sociality can be part of this. It therefore appears to us that fun and discomfort are in fact closely bound up and that designers who wish to deliver interactive fun can both create and naturally expect some uncomfortable moments along the way. Our intention in this chapter has been to explain why this might be done, how it can achieved in interactive systems, but also to emphasise the importance of doing so in an appropriate and responsible way.

Acknowledgements This research was funded by the Engineering and Physical Sciences Research Council through the *Living With Digital Ubiquity* (EP/M000877/1) and *From Human Data to Personal Experience* (EP/M02315X/1) projects.

References

Bardzell J, Bardzell S (2011) Pleasure is your birthright: digitally enabled designer sex toys as a case of third-wave HCI, CHI'11, Vancouver. ACM, pp 257–266

Benford S, Giannachi G (2011) Performing mixed reality. MIT Press

Benford S, Crabtree A, Reeves S, Sheridan J, Dix A, Flintham M, Drozd A (2006, April) The frame of the game: blurring the boundary between fiction and reality in mobile experiences. In: Proceedings of the SIGCHI conference on human factors in computing systems. ACM, pp 427–436

Benford S, Greenhalgh C, Giannachi G, Walker B, Marshall J, Rodden T (2012, May) Uncomfortable interactions. In: Proceedings of the SIGCHI conference on human factors in computing systems. ACM, pp 2005–2014

Benford S, Greenhalgh C, Crabtree A, Flintham M, Walker B, Marshall J, Tandavanitj N (2013) Performance-led research in the wild. ACM Trans Comput Hum Interact (TOCHI) 20(3):14

Benford S, Greenhalgh C, Anderson B, Jacobs R, Golembewski M, Jirotka M, Farr JR (2015) The ethical implications of HCI's turn to the cultural. ACM Trans Comput Hum Interact (TOCHI) 22(5):24

Brecht B (1993) The modern theatre is the epic theatre. In: Willett J (ed) Brecht on theatre. Methuen, pp 33–42

Durrant A, Rowland D, Kirk DS, Benford S, Fischer JE, McAuley D (2011, May). Automics: souvenir generating photoware for theme parks. In: Proceedings of the SIGCHI Conference on human factors in computing systems. ACM, pp 1767–1776

Elias N (1982) The civilising process. Pantheon

Gaver W, Beaver J, Benford S (2003) Ambiguity as a resource for design, CHI'03. ACM, pp 233–240

Goldstein J (1999) The attractions of violent entertainment. Media Psychol 1:272–282

Haans A, Ijsselsteijn W (2006) Mediated social touch: a review of current research and future directions. Virtual Reality 9:149–159

Hobye M, Löwgren J (2011) Touching a stranger: designing for engaging experience in embodied interaction. Int J Design 5(3)

Höök K, Löwgren J (2012) Strong concepts: intermediate-level knowledge in interaction design research. ACM Trans Comput Hum Interact (TOCHI) 19(3):23

Koleva B, Taylor I, Benford S, Fraser M, Greenhalgh C, Schnadelbach H, vom Lehn D, Heath C, Row-Farr J, Adams M (2001) Orchestrating a mixed reality performance, CHI'01, pp 38–45

Levisohn A, Cochrane J, Gromala D, Seo J (2007) The Meatbook: tangible and visceral interaction. In: Tangible and embedded interaction (TEI'07). ACM

Marshall J, Benford S (2011) Using fast interaction to create intense experiences, CHI'11. ACM, pp 1255–1265

Marshall J, Rowland D, Egglestone S, Benford S, Walker B, McAuley D (2011a) Breath control of amusement rides, CHI'11, April 2011. ACM

Marshall J, Walker B, Benford S, Tomlinson G, Rennick Egglestone S, Reeves S, Brundell P, Tennent P, Cranwell J, Harter P, Longhurst J (2011b, May) The gas mask: a probe for exploring fearsome interactions. In: CHI'11 Extended abstracts on human factors in computing systems. ACM, pp 127–136

McCarthy J, Wright P (2007) Technology as experience. MIT Press

Monk A, Hassenzahl M, Blythe M, Reed D (2002) Funology: designing enjoyment, CHI'02 EA, 2002

Mueller F, Edge D, Vetere F, Gibbs M, Agamanolis S, Bongers B, Sheridan J (2011) Designing sports: a framework for exertion games, CHI'11. ACM, pp 2651–2660

Reeves S (2011) Designing interfaces in public settings. Springer, Berlin

Reeves S, Benford S, O'Malley C, Fraser M (2005, April) Designing the spectator experience. In: Proceedings of the SIGCHI conference on human factors in computing systems. ACM, pp 741–750

Reeves S, Martindale S, Tennent P, Benford S, Marshall J, Walker B (2015) The challenges of using biodata in promotional filmmaking. ACM Trans Comput Hum Interact (TOCHI) 22(3):11

Schnädelbach H, Rennick Egglestone S, Reeves S, Benford S, Walker B, Wright M (2008) Performing thrill: designing telemetry systems and spectator interfaces for amusement rides, CHI'08. ACM, pp 1167–1176

Sengers P, Gaver W (2006) Staying open to interpretation: engaging multiple meanings in design and evaluation, DIS'06. ACM, pp 99–108

Shneiderman B (1987) Designing the user interface

Smith M (2007) Stelarc: the monograph. MIT Press

Tennent P, Marshall J, Walker B, Brundell P, Benford S (2017, June) The challenges of visual-kinaesthetic experience. In: Proceedings of the 2017 conference on designing interactive systems. ACM, pp 1265–1276

Tolmie P, Benford S, Flintham M, Brundell P, Adams M, Tandavantij N, Row Far J, Giannachi G (2012, May) Act natural: instructions, compliance and accountability in ambulatory experiences. In: Proceedings of the SIGCHI conference on human factors in computing systems. ACM, pp 1519–1528

Chapter 14
Reorienting Geolocation Data Through Mischievous Design

Ben Kirman, Conor Linehan and Shaun Lawson

1 Introduction

It is hard to overstate the impact mobile phones have had on our daily lives. In 2003, when the first Funology collection was published, mobile phones still had keypads, polyphonic ringtones and screens that could barely fit a hashtag. Nowadays, phones have moved past mere voice communication, and are essential tools to help us understand how unfit and unhealthy we are (Boulos et al. 2014), to help us arrange casual relationships ("Grindr" 2009; "Tinder" 2012), to measure how good we are in the sack in those relationships (Alptraum 2016), to bring us beer (Wetherspoon 2017) after our relationships fail, and in making sure we get a ride home to spend another night alone (Rayle et al. 2014). Phones have become our closest confidants; they go with us everywhere and are entrusted with our most private secrets.

It is not just phones themselves that have developed, but their surrounding infrastructure of GPS, 4G data transfer, WiFi, Bluetooth and other technologies have also matured. This infrastructure supports the development of the types of location-based, contextually aware, applications long foretold in ubicomp research. Sal, from Weiser's (1991) famous vision of our ubiquitous future, is indeed now able to use her tablet to remotely spy on her colleagues.

We argue that geo-location services rarely take advantage of the potential richness of the converged capabilities of these devices and data sources. The most

B. Kirman (✉)
University of York, York, England, UK
e-mail: ben.kirman@york.ac.uk

C. Linehan
University College Cork, Cork, Republic of Ireland

S. Lawson
Northumbria University, Newcastle upon Tyne, England, UK

© Springer International Publishing AG, part of Springer Nature 2018
M. Blythe and A. Monk (eds.), *Funology 2*,
Human–Computer Interaction Series,
https://doi.org/10.1007/978-3-319-68213-6_14

common function of geolocation data is in navigation; either egocentric "sat nav" applications, or to get cheeseburgers cycled to our location at a moment's notice. It is rare that a geolocation service engenders a unique social or emotional experience outside of burgers and bedrooms. Locations are still largely considered, by developers of these services, in terms of their physical positions and the nearby business opportunities, rather than as places with complex social, psychological, and physical properties. Moreover, we are concerned about the effect these crude, non-critical, or unthinking application designs have on our established relationships with these places and our interactions within them. As argued in literature on pervasive games (Márquez Segura 2016; Montola et al. 2009) and embodied interaction (Dourish 2004a), these complex contexts of interaction are vital in understanding implications of mobile interaction.

In this chapter, we present a playful exploration of these messy contexts through a series of experimental prototypes. We frame our work as research through design; each design prototype takes a different look at the way mobile phones mediate our relationships both the places we pass through, and with other people. The benefit of this approach is in uncovering affordances of these relationships, revealing the properties of geo-location as a socio-physical material with which we can design experiences.

2 Mischief Making as Research

Theorists have devoted much attention to the novel possibilities opened up by ubiquitous computing (e.g. Dourish 2004a, b). Indeed, location-based services have long been considered the great leap forward in the interaction between humans and computers, allowing us to seamlessly access relevant information and tools, as needed, as we move through the environment carrying out every-day tasks. However, until recently, the reality of location-based services rarely matched the ambition of the theorists. Applications were clunky, flaky and frustrating. They certainly did not blend seamlessly with, or disappear into, our lives. The current generation of smartphones, combined with modern mobile communication infrastructure, allows us to creatively explore geo-location data as a material that we encounter in 'the everyday.' We are now in a position to explore not only what geo-location services can do, but also what they mean, and what kinds of novel experiences they can facilitate.

As research through design (Frayling 1993), the objective of the work in this chapter is not to collect empirical evidence about users or specific services, but rather, through practice of prototyping, prodding and poking, to begin to understand the social, psychological, cultural and physical properties of geolocation data. Taking Löwgren and Stolterman's (2004) perspective of software as material, we seek to understand the unexpected affordances of geolocation data as a material with which to design and craft experiences. Accordingly, our practice focuses on studying artefacts situated in the real world rather than in the user experience lab.

Through design practice, we are able to discuss emergent social, cultural and physical properties of geolocation data that may be overlooked through other approaches.

We have also taken a purposefully subversive and playful approach to our design work. Although not explicitly political as in DiSalvo's (2012) "adversarial design", our work attempts to use play to renegotiate, critique and subvert expectations around data and spaces. This approach is grounded broadly in "critical design" (e.g. Dunne and Raby 2001; Bardzell and Bardzell 2013), however the interactive prototypes are intentionally ambiguous (Gaver et al. 2003), inviting open play and exploration around the themes, in contrast to the sharper critique offered by critical games (Flanagan 2009). We propose this is a kind of "mischievous design" that uses playfulness to push the edge of acceptability without malicious intent (Kirman et al. 2012b). As we will discuss, this mischievousness is brought forth in various ways through the prototypes, but includes asking users to visit crime-ridden areas, behave suspiciously in airports, and wantonly insult strangers. This approach invokes humour, but the humour of discomfort and deviance rather than parody or pastiche.

Each of the four projects has emerged from our ongoing research through design (see Zimmerman et al. 2007; Frayling 1993) practice around designing for mischief. Typically initially inspired by an encounter with a specific context (airports in Blowtooth), or service (crime data in FearSquare), we engaged in ideation work. The content differed across projects—sometimes we designed game mechanics, sometimes we wrote fictional commercials—but always focused on the interaction of people with context-specific data. We fixated on ideas that we found, as designers, felt humorous or uncomfortable. Secondly, we developed a functioning prototype and released it into the real world, presented as real services rather than explicitly critical pieces. This kind of presentation we found key to engaging users, to avoid the danger of being "dismissed as art" (Dunne and Raby 2013).

Thirdly, we undertook a process to understand and capture reactions to the prototype—varying from formal user studies, to scraping social media. Fourthly, we engaged in a reflective process, based on the study, which sought to understand exactly what social norms the prototype subverted. We always found that prototypes commented on, criticised, and explored the properties of geolocation data in ways that we had not considered through the design process. Moreover, through the course of these four projects, we began to build a more coherent understanding of the various social, psychological and physical properties of both the data and the spaces it represents.

In the following sections we describe four of these mischievous prototypes, providing some context for the design and discussing reflections and reception for each.

3 Blowtooth

Airports are curious as places we subject ourselves to, for a few hours, in exchange
for leaving them. Rather than a destination, they are a place of transition—a classic
example of Foucalt's "heterotopia" (Foucault and Miskowiec 1986)—a liminal
other space, neither here nor there. They are "non-places" (Auge 1995; Crang 2002)
that we share with thousands of other travellers, strangers who are temporarily close
but soon to be separated by oceans. Ballard (1997) observed that their architecture
is "designed around the needs of their collaborating technologies", each part of the
physical layout of an airport is simply one step in a line for the next part of a bigger
process; thus "everything is designed for the next five minutes".

However, airports also have a layer of social complexity that sets them apart
from train stations and ferry ports. As we pass the concrete barricades and razor
wire, we know we are in an authoritarian space (Kellerman 2008) watched and
studied by a multitude of cameras, scanners, and sniffer dogs. Bizarrely, they are
part prison, part shopping mall. Airports repeatedly come under scrutiny regarding
security, personal privacy and freedom of movement. For example, Donald
Trump's executive orders in early 2017 led to a whole raft of uncomfortable
experiences for those arriving in US airports.

It is this curious intersection of a strongly defined physical space with complex
social and behavioural rules that make airports a fascinating place for thinking
about design, and indeed what attracted us to working with this space. In response,
we created Blowtooth—a game where players are challenged to smuggle contraband through real airport security.

The game is split into two parts—first, during check-into their flight, before
passing through security, players are asked to "hide" contraband on fellow passengers. When approaching likely "patsies", players are asked to choose from a list
of illicit items (e.g. 101 ml of liquid, unlicensed cheese, etc.) to conceal. The player
is invited to explore this space in order to locate suitable patsies before moving onto
the next phase.

Once the player is happy they have concealed enough contraband, they can pass
through airport security without worry. After all, the patsies now carry all the risk
associated with smuggling the illicit items through the security checks. Once
beyond the security check, the second part of the game commences. Players are
tasked with reclaiming all the items they have hidden by finding the patsies in the
departure lounge. Given the enclosed space contained beyond the security check-
points, there are limited places the patsies may be. There is also time pressure, as
patsies start boarding flights and leaving the airport, unknowingly carrying their
illicit cargo. Although the game does not formally "end", the player's success at
recovering their contraband is recorded and compared against other players at other
airports on the website (Fig. 1).[1]

---

[1]www.blowtooth.com.

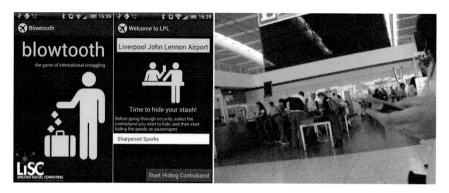

Fig. 1 Blowtooth interface and play at Heathrow

Obviously, no actual contraband is involved. It is implemented (see Kirman et al. 2012a) to use the Bluetooth sensor inside phones to build the list of proximal devices (within ~ 10 m). Although it involves unaware travellers, this only includes people who have set their devices to advertise their presence, and no information other than a hashed identifier is (temporarily) used by the application. This is the same approach many airports and shopping centres use to silently track movement of visitors through those spaces (e.g. Bonner et al. 2010; Bullock et al. 2010).

Blowtooth is an interesting design experiment because of the social and physical geography of the space where it exists. It can only be played in airports, both literally, in that the game code verifies this, and figuratively, in that it does not make sense anywhere else. It is specifically built for both the physical geography of airports, but also the transient nature of its occupants and their devices. In the social aspects, as Moore (2014) observes, Blowtooth relies on surveillance. The game invites anxiety by asking the player to "misbehave" in this heavily watched authoritarian space, but also asks the players to engage in surveillance of their fellow passengers. Although not doing anything "wrong", Blowtooth is provocative in how it asks players to reflect on the weird social rules of the airport as a non-place, and behave in ways that run contrary to these rules, through its subversive narrative.

In terms of the physical location, the design uses the procedural aspects of the air travel experience as core features of the game—the limited space of the departure lounge, and the need to wait, ensures that the player has both the time and space needed to locate their victims. It is tightly predicated on both the specific ways we move around in an airport and the international uniformity of this space. While it can only make sense in an airport, it does so in every airport because all airports are the same place.

In summary, this project demonstrates how through very simple uses, geo-location technology, combined with a coherent understanding of social and psychological nature of a specific environmental context, extremely memorable experiences can be created. In other words, the power of these services is derived not from the cleverness of the application, but from the social rules of the environment itself—and in directing people's attention towards those rules.

4 Feckr

Where Blowtooth is designed for specific kinds of spaces, it is also interesting to consider how our social experiences change across different places. With Feckr, we aimed to explore both the social geography of spaces and the invisible nature of geo-location data. The concept is a refinement of Casey's (2011) MobiClouds social tagging project, whose users made short descriptive messages ("tags") to surreptitiously describe specific spaces. In Feckr, rather than tagging spaces, users are able to save tags about the people they are near at that moment. All tags are permanently saved to a global database. The main Feckr screen shows an updated list of the tags that have been applied to people near your current location (see Fig. 2).

The app is presented as a playful way to vent frustration at annoying situations where it may not be socially appropriate to make a real intervention. For example, sitting near someone with poor hygiene on the bus, or sitting next to a boring

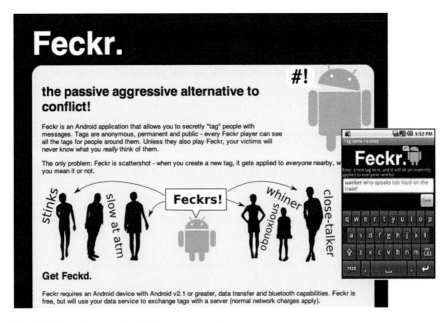

Fig. 2 Feckr website and mobile tagging interface

academic at a conference dinner. Rather than stabbing them with a dessert fork, Feckr invites you to slyly tag them as "boring wankers".[2] The key twist is this act is scattershot—it is not only applied to the individual that raised your ire, but simultaneously applied to all individuals nearby. These tags are only visible to Feckr users, so may never be seen by your target, but are nonetheless permanent. Although the scattershot tagging might seem unfair, over time, tags associated with an individual come to represent the sum of the social environments they have encountered.

It is explicitly social, in that it can only be used in environments where there are other people, but importantly, the tags are communal and not associated with their author. The tags are never displayed as a list of specific individuals with specific tags, but rather a list of tags that have been applied to people near to you at that moment. In this way, users see a cloud of tags that change as they pass through different environments, reflecting their transition through changing social geography along with their change in physical space. As with Blowtooth, Feckr uses the Bluetooth sensor to determine proximal devices, and no personally identifiable information is stored.

The most pleasurable part of using the application is not the act of tagging, which is nevertheless cathartic, but the discovery of tags created by other users applied to the people who you find yourself amongst. In this way, it is revealing how the patterns of tags change around different spaces. This of course reflects the subtle ways our social experiences of moving about the world change between different contexts and situations, but also a shared frustration about the changing social rules and expectations. This is a common experience, and while users understand it can be frustrating they also (hopefully) recognise the reason for those rules existing. The application knowingly points out this tension in its tagline ("the app for closet sociopaths").

In summary, Feckr provides a mischievous exploration of the invisible-and-yet-permanent nature of the social signals carried by geo-location data. Augmented reality apps have long been heralded as useful, due to their ability to "reveal" information related to an environment or device. Feckr poses the question; who gets to decide what information is associated with which location? If normal users are allowed to create that geo-located information, what does that mean for our social interactions? And, if we allow users to "create content" (Kirman et al. 2012b), can we expect people to do anything other than call each other wankers?

---

[2]Real tag a user contributed to the Feckr database.

5   Fearsquare

The desire by governments, technologists and researchers alike to create digital civics applications and platforms (Olivier and Wright 2015) has resulted in the public release of a bewildering amount of so-called "Open Data" that was previously only available to government agencies. The well-meaning intentions of this appears to be the flimsy hope that developers and visualization experts will be able to find hidden unforeseen uses for, and patterns in, these datasets that gives insight into societal trends, needs and wants.

Unfortunately, much of the data has limited usefulness, as councils gleefully avail to us their logs of dog shit data (e.g. City of York Council 2016). However, in 2011 in an astonishing move of openness and foresight the UK government decided to release comprehensive monthly updates of geo-located crime data in a format easily accessible to developers (Home Office 2010). The developer community and the general public briefly went crime data crazed (e.g. Brown 2011), and crime applications rapidly popped up without consideration of problematic implications (Dewey 2014).

Not wanting to be left behind, we released our own app that worked with location-sharing service Foursquare, and mashed users' recent locative "check ins" with the crime data to show them "how dangerous their lives were". For instance, if someone had checked into their local pub, our app, Fearsquare, showed them how many assaults and robberies had taken place there recently. We built automated league tables to show which the most crime-ridden places were, and which users visited the most, supposedly, risky places.[3] Nearly 2500 people registered their Foursquare accounts with our app within the first few months of its release and it was reported widely in media around the world (Garbett et al. 2014) (Fig. 3).

We began this project with the intention of exploring whether it was possible to make Open Data sets more personal and relevant to people. When stumbling upon the blank façade of the UK's data.gov.uk website, the average person would quickly return to their filter bubbles on Facebook. However, our design thinking with Fearsquare was that if you could show people, through their own social media feeds, the violent crime committed in their local kebab shop, this makes the otherwise blandness of data as relevant, captivating and visceral as the shop's spicy sauce. It is known that people are drawn to stories about crime and breaking the law (Glassner 2010); typically however there is a distance (literally and metaphorically) between us and the crimes seen on TV and in the newspapers; and in the UK and Ireland, we are actually very unlikely to be witness to, or victims of, crime. The context collapse that is made possible by social media and Open Data allowed us to

---

[3]Curiously one of the most dangerous places to check into was a local Police station somewhere on the south coast of England; we assumed that this was due to a local plod entering their own postcode by default each a crime was reported, rather it being reflective of some kind of "Assault on Precinct 13" type incident that may have (repeatedly) occurred there.

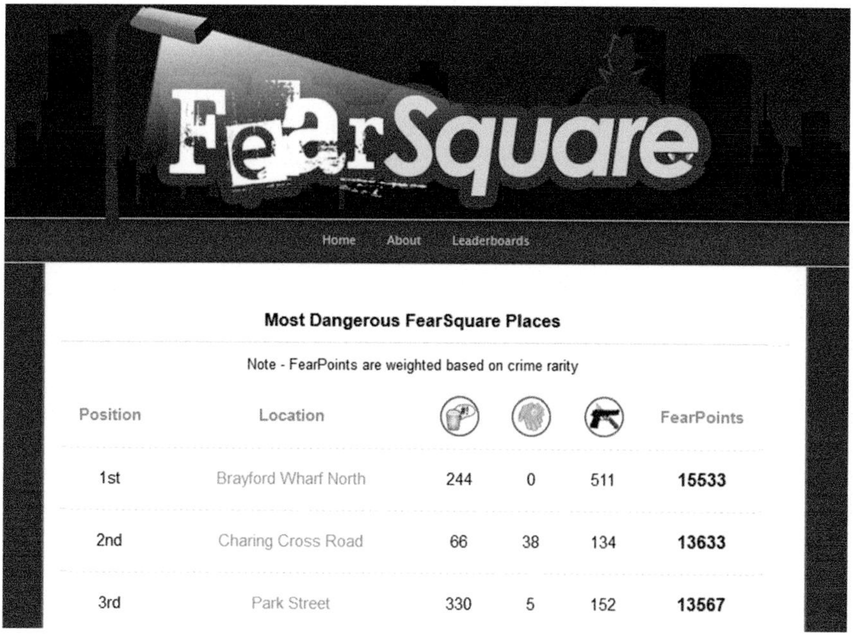

Fig. 3 Fearsquare scores places based on reported crime

eliminate this distance and bring real instances of crime 'home' to our users, making the crime data much more personal.

We observed that the enhancing locations with crime data could make people think differently about those places and the subjective experiences they had had there. Not just in terms of unusual reflection of the spaces, but what the data "means". For example, through its playful leader-board of users who have visited the areas with most crime, Fearsquare implicitly encourages its users to try to visit more "dangerous" places in order to climb the ranking. However, even the most crime-ridden areas, in the UK at least, are still crime-free the vast majority of the time. For example, in most cities, we discover that areas around nightclubs and shopping centres tend to have higher incidents of reported crimes. The publication of this information could quite conceivably cause people anxiety in visiting those locations. However, this increased crime is largely a function of the popularity of these spaces—more people means more crime, but also means more of every other type of social interaction. The subtlety of this point is entirely absent in the data as presented by the government. In essence, Fearsquare criticises the data that is absent from these supposedly "open" and "objective" geo-location based services and the consequences of presenting over-simplified visualisations of geo-location data.

6   GetLostBot

We rely on our phones a lot in terms of generating new experiences—perhaps finding an activity to occupy feral children, or to discover a new band in a vain attempt at seeming cool—aided by recommender services that have learned about our taste. In a similar way, Facebook and Twitter automatically filter our news feeds to show us stories likely to be important to us, based on our relationships and past interactions (Backstrom 2013).

With this in mind, we were curious about what the opposite of a recommender system might be. These systems tend to be quite safe—boring but accurate suggestions are preferred over exciting but incorrect ones. For example, a music app recommending "The Beatles" is likely to be correct but ultimately unexciting. GetLostBot is our attempt to interrogate this, by monitoring your movement around spaces, and giving unsolicited advice to visit new places when you fall into a "routine". For example, if you seem to be going to the same cheap restaurants and bars too frequently, GetLostBot will intervene and suggest somewhere else for you to visit.

There are a few twists to this central function. First of all, the way GetLostBot chooses new locations is purposefully divergent from typical recommendations—it does this by using the Foursquare API to gather recommendations, then removing these from the pool of potential suggestions. It then checks where the user and the user's friends have been and also subtracts these results from the set. A final selection is made randomly to a similar place (e.g. if at a bar, it will suggest another bar) within 1.5 km that is, therefore, neither recommended nor recommendable.

Secondly, GetLostBot is curious since it has no interface, and is rather a responsive system based on the behaviour of users. Users don't interact with it—it interacts with them. Once users register for GetLostBot, with a single click on the website, it quietly monitors the locations they visit, only generating interventions when the user falls into a routine. A "routine" is arbitrarily defined by us as having not visited a new place in the past week. The interventions, as suggestions for new places to visit, are sent to the user via email or Twitter (see Fig. 4).

The final important twist is the way suggestions are presented. Users are never shown the name of their destination, but are instead given a web link to walking directions provided by the Google Maps API. In this way, users are given a "treasure map" and asked to take the application on trust, and not pre-judge the suggestions that it generates. Of course, since they don't know where they are going, they also don't know when they have arrived. This is an intentional attempt to bring ambiguity into the journey that is supposed to be about discovery and exploration.

This framing of "exploration" is important. We call it a "serendipity generator", in that it aims to help users become more adventurous and engage with a spirit of discovery. It seems that this idea is very compelling—we all like to think we are adventurous spirits—and as a result GetLostBot gained some notoriety through its selection as a finalist in the Guardian's Dream Factory competition (Dunlop 2011),

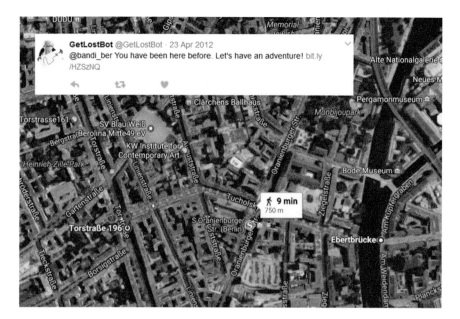

Fig. 4 Example of GetLostBot challenge tweet and walking directions

where it received tens of thousands of votes, and came second place, in the national competition for innovation. It was also covered widely in the press, including being featured in popular science magazine New Scientist (de Lange 2012), and a later book (Brooks 2016), praising how it represented a revival of the nineteenth century urban "flâneur" (Shaya 2004).

Despite the critical acclaim, when speaking to users about their experiences, a slightly different picture emerged. The responsive aspect of GetLostBot received the most criticism. People were unhappy that they would get challenges at strange times, rather than at a time they felt they were ready to explore. In addition, since GetLostBot monitors all movement, often it would make suggestions to change routines that users felt they could not change. For example, a participant who worked in a hotel complained that GetLostBot would regularly suggest they skip work and go to the pub. Another user complained that they received challenges at church every Sunday. GetLostBot insisted that they should visit a mosque instead.

These complaints are fascinating because they nearly all referred to the explicit design of the application. In other words, they were complaining about exactly the features they signed up for. It seems that, although people register for the application with a spirit of exploration, when it actually comes time to put that spirit into practice, they are often reticent.

It is this dissonance between the "idea" of how people engage with different spaces and the reality that is the most interesting aspect of GetLostBot. It is highly personal and reflects only the individual's particular experience of moving through the world. Indeed, since it purposefully excludes locations that friends have visited, it is arguably anti-social. Perhaps it should not be surprising users feel affronted when the application accuses them of, essentially, being boring. All the same, just as it is uncomfortable to be reminded of your drunken exploits the morning after a big night, it is uncomfortable to be reminded that your self-identity as an explorative and exciting soul is often not wholly honest. In this way it uses evidence of your own behaviour to demand some personal reflection.

7 Discussion

In this chapter we have presented four prototypes that mischievously explore the design space around locative and social mobile interaction. We have chosen these prototypes to serve as examples of different ways we can interrogate and build with geolocation data as a design material, since they approach that material from different but complementary perspectives. Through our ongoing design practice, we have come to consider location and sociality as the two key dimensions of how designers can understand "places". In addition, we conceive that both dimensions can be understood in both singular and plural aspects—for example, we can see location as both specific places (such as airports) and the transition between many different places, and we think of sociality in terms of how an individual interprets the social aspects of their environment and also how groups of people do collectively. The matrix in Fig. 5 shows how the prototypes fit into this perspective.

This understanding is not necessarily comprehensive; however we have found it valuable in guiding design since it aligns with the sensing apparatus available on mobile devices. In terms of place, we consider Fearsquare and Blowtooth concerned with specific spaces—this is obvious in an airport, but also how we consider specific locations differently through the lens of crime data. Contrariwise, both GetLostBot and Feckr and more transitory, and about the composite patterns reflected in our movement. In terms of sociality, both Fearsquare and GetLostBot are both very personal, since they are concerned directly with the user's individual perception of places, where Blowtooth and Feckr are centred on the social dynamics and psychology of spaces.

Fig. 5 Matrix of position and sociality

		Place	Transient
Personal		Fearsquare	GetLostBot
Social		Blowtooth	Feckr

Throughout this chapter, we have talked about a guiding attitude of "mischievous design" that has informed most of our work, both here and more broadly. This simple perspective has become an important strategy in helping direct work optimised for engaging academic and public audiences in how (e.g. mobile) systems are designed, and the implications of those design decisions and the wider technology upon which they are built.

The foundation for this is in critical design, a loose but important field within design research centred on provocation, in that it is explicitly situated against what Dunne and Raby (2001, 2013) call "affirmative design", or designing to "maintain the status quo" (ibid). Critical designs aim to "make consumers more critical about their everyday lives" (Bardzell and Bardzell 2013), using artefacts to expose ideas about the future. However, critical design has itself been criticised for failing to recognise its own privilege (Prado and Oliveira 2014), and, despite the audience being "consumers", having a conceptual opaqueness and focus on aesthetics that "normalizes a pretentious taste regime" (Tonkinwise 2015). Our perspective is that, with a few exceptions, critical design is simply not fun.

In the spirit of "Funology", we see mischievous design as critical design with an accessible edge—engaging through its playfulness, yet still offering critique through its function. As Blythe and Wright (2004) discuss, this playfulness allows the extension of established work to "encompass enjoyment". This is hopefully clear in all of the prototypes we have described, which, although Blowtooth is the only "game", all reflect a strong streak of playfulness.

This mischievous approach has been successful in drawing attention to this work. As mentioned, GetLostBot was featured by New Scientist, the Guardian and others. Fearsquare was picked up by news organisations across the world, including the New York Times and Wired, and proved especially popular in Brazil. The controversial aspects of Blowtooth attracted similar press coverage and continues to make the authors extra nervous every time they fly.

8 Conclusions

In this chapter we have reported on a series of projects that use a mischievous design attitude to explore the affordances of the geolocation capabilities of modern mobile devices. We are concerned by the relatively dull way that such capabilities are used by contemporary applications. In particular, how applications confuse position with place, and fail to take advantage of the tremendous potential of the entwined social and physical landscape.

We present four projects in this space—Fearsquare, Blowtooth, Feckr and GetLostBot. Each project takes a different perspective on location data, from Blowtooth's concentration on the specific environment of the airport, through to GetLostBot's subversion of location-based recommender systems. Together the prototypes demonstrate a range of alternative and critical twists that problematise the readily available functionality on modern mobile devices. In particular, we document the "mischievous design" attitude with which each was developed, that uses irreverent playfulness as guiding principles in building engaging work that helps explore location data as a material of design, and offer critique of the way this material is used in contemporary applications.

Acknowledgements Special thanks to Andrew Garbett, the lead developer on Fearsquare, and all other former members of the Lincoln Social Computing Research Centre (LiSC).

References

Alptraum L (2016) Apple's health app now tracks sexual activity, and that's a big opportunity, Motherboard Oct 23rd 2016. Retrieved from: https://motherboard.vice.com/en_us/article/apples-health-app-now-tracks-sexual-activityand-thats-a-big-opportunity

Auge M (1995) Non-places: an introduction to an anthropology of supermodernity. Verso, London

Backstrom L (2013) News feed FYI: a window into news feed. Facebook. Retrieved from: https://www.facebook.com/business/news/News-Feed-FYI-A-Window-Into-News-Feed

Ballard JG (1997) Airports: the true cities of the 21st century. Blueprint, Sept 1997. http://www.utne.com/politics/homeiswherethehangaris

Bardzell J, Bardzell S (2013) What is critical about critical design? In: Proceedings of the SIGCHI conference on human factors in computing systems. ACM, pp 3297–3306

Blythe M, Wright P (2004) Introduction: from usability to enjoyment. In: Funology. Kluwer Academic Publishers, pp 13–19

Bonner BB, Hjelm CT, Jones TA, John EOI, Perkins DB, Sunrise R&D Holdings LLC (2010) Method of tracking the real time location of shoppers, associates, managers and vendors through a communication multi-network within a store. U.S. Patent 7,739,157

Boulos MNK, Brewer AC, Karimkhani C, Buller DB, Dellavalle RP (2014) Mobile medical and health apps: state of the art, concerns, regulatory control and certification. Online J Public Health Inform 5(3)

Brooks M (ed) (2016) Chance: the science and secrets of luck, randomness and probability. Hachette, UK

Brown M (2011) Police.uk shows street-by-street data on local crimes. Wired UK, 1st Feb 2011. Retrieved from: http://www.wired.co.uk/article/police-uk-crime-maps

Bullock D, Haseman R, Wasson J, Spitler R (2010) Automated measurement of wait times at airport security: deployment at Indianapolis International Airport, Indiana. Trans Res Rec: J Trans Res Board 2177:60–68

Casey S (2011) Tagging amongst friends: an exploration of social media exchange on mobile devices. PhD thesis, University of Lincoln, UK

City of York Council (2016) Fixed penalty notices—dog fouling [Dataset]. Retrieved from https://data.gov.uk/dataset/kpi-fpn01a

Crang M (2002) Between places: producing hubs, flows and networks. Environ Plann A 34: 569–574

de Lange C (2012) Let's get lost: apps that help you wander to happiness. In: New scientist, August 2012. Online: https://www.newscientist.com/article/mg21528792.500-lets-get-lost-apps-that-help-you-wander-to-happiness/

Dewey C (2014) The many problems with SketchFactor, the new crime Crowdsourcing App that some are calling racist. Washington Post, 12

DiSalvo C (2012) Adversarial design. MIT Press

Dourish P (2004a) Where the action is: the foundations of embodied interaction. MIT press

Dourish P (2004b) What we talk about when we talk about context. Pers Ubiquit Comput 8(1): 19–30

Dunlop H (2011) Honda and guardian news & media team up to find cultural engineers of the future. Retrieved from: https://www.theguardian.com/gnm-press-office/gnm-team-up-with-honda

Dunne A, Raby F (2001) Design noir: the secret life of electronic objects. Springer Science & Business Media

Dunne A, Raby F (2013) Speculative everything: design, fiction, and social dreaming. MIT Press

Flanagan M (2009) Critical play: radical game design. MIT press

Foucault M, Miskowiec J (1986) Of other spaces. Diacritics 16(1):22–27

Frayling C (1993) Research in art and design. RCA Res Pap 1:1

Garbett A, Wardman JK, Kirman B, Linehan C, Lawson S (2014) Anti-social media: communicating risk through open data, crime maps and locative media. In: Proceedings of HCI Korea (HCIK'15). Hanbit Media, Inc., South Korea, pp 145–152

Gaver WW, Beaver J, Benford S (2003) Ambiguity as a resource for design. In: Proceedings of the SIGCHI conference on human factors in computing systems. ACM, pp 233–240

Glassner B (2010) The culture of fear: why Americans are afraid of the wrong things: crime, drugs, minorities, teen moms, killer kids, mutant microbes, plane crashes, road rage, & so much more. Basic books

Grindr (2009) Grindr—Gay Chat, meet & data [Mobile application software]. Retrieved from: https://www.grindr.com/

Home Office (2010) ASB incidents, crimes and outcomes [Dataset for England, Wales and Northern Ireland, 2010–2017]. Retrieved from: https://data.police.uk/about/

Kellerman A (2008) International airports: passengers in an environment of 'authorities'. Mobilities 3(1):161–178

Kirman B, Linehan C, Lawson S (2012a) Blowtooth: a provocative pervasive game for smuggling virtual drugs through real airport security. Pers Ubiquit Comput 16(6):767–775

Kirman B, Lineham C, Lawson S (2012b) Exploring mischief and mayhem in social computing or: how we learned to stop worrying and love the trolls. In: CHI'12 Extended abstracts on human factors in computing systems. ACM, pp 121–130

Löwgren J, Stolterman E (2004) Thoughtful interaction design: a design perspective on information technology. MIT Press

Márquez Segura, E. (2016) Embodied core mechanics. Designing for movement-based co-located play. Uppsala Studies in Human-Computer Interaction 3. Department of Informatics and Media, Uppsala, 174 pp

Montola M, Stenros J, Waern A (2009) Pervasive games: theory and design. Morgan Kaufmann Publishers Inc.

Moore K (2014) The passenger and the player: blowtooth and the subversion of airport space. Media Fields J 1:8

Olivier P, Wright P (2015) Digital civics: taking a local turn. Interactions 22(4):61–63

Prado L, Oliveira P (2014) Questioning the 'critical' in speculative & critical design. Medium. Retrieved from: https://medium.com/a-parede/questioning-the-critical-in-speculative-critical-design-5a355cac2ca4

Rayle L, Shaheen S, Chan N, Dai D, Cervero R (2014) App-based, on-demand ride services: comparing taxi and ridesourcing trips and user characteristics in San Francisco University of California Transportation Center (UCTC). University of California, Berkeley, United States

Shaya G (2004) The Flaneur, the Badaud, and the making of a mass public in France, circa 1860–1910. Am Hist Rev 109(1):41–77

Tinder (2012) Tinder [Mobile application software]. Retrieved from: https://www.gotinder.com/

Tonkinwise C (2015) Just design. Being dogmatic about defining speculative critical design future fiction. Medium. Retrieved from: https://medium.com/@camerontw/just-design-b1f97cb3996f

Weiser M (1991) The computer for the 21st century. Sci Am 265(3):94–104

Wetherspoon (2017) Wetherspoon order and pay [Mobile application software]. Retrieved from: https://www.jdwetherspoon.com/pubs/order-and-pay-app

Zimmerman J, Forlizzi J, Evenson S (2007) Research through design as a method for interaction design research in HCI. In: Proceedings of the SIGCHI conference on Human factors in computing systems. ACM, pp 493–502

Chapter 15
From Evaluation to Crits and Conversation

Mark Blythe, Jonathan Hook and Jo Briggs

1 Introduction

Soon after the launch of the iPhone the British artist and printmaker David Hockney began sending his friends pictures he had made using painting and drawing apps. One of these friends was the writer and art critic Martin Gayford, he received an iPhone drawing of a sunrise over the East Yorkshire town of Bridlington (Gayford 2007). The file on Hockney's iPhone was identical to the one that he sent to Gayford which prompted him to ask—where was the original? Hockney sent many images to his friends and they in turn could share them until London was awash with "original Hockneys". While Hockney's limited edition etchings and lithographs attract high prices, this new method of production and dissemination challenged the notion of a limited edition "print run" because digital files can be reproduced with no diminution of quality at virtually no cost. These kinds of reflection led to a project called "Digital Originals" where we conducted several studies of practicing artists and developed an app called Repentir. The app was given to a variety of people in technology shows, art shows, and private viewings. Responses were mixed but this chapter argues that a straightforward evaluation of such an app would not tell us very much. Computing technologies are now a part of almost every aspect of human activity, evaluation methods which developed when computers were confined to the workplace are no longer enough. This chapter reflects the move from usability evaluation to arts based techniques like polyphonic evaluation, "crits" and finally conversation.

M. Blythe (✉) · J. Briggs
Northumbria University, Newcastle upon Tyne, UK
e-mail: mark.blythe@northumbria.ac.uk

J. Hook
University of York, York, UK

© Springer International Publishing AG, part of Springer Nature 2018
M. Blythe and A. Monk (eds.), *Funology 2*,
Human–Computer Interaction Series,
https://doi.org/10.1007/978-3-319-68213-6_15

2   The Varieties of Evaluation

When HCI was focused on the workplace evaluation was relatively straightforward. Metrics like ease of use, ease of learning, time on task and accuracy were observable, measurable and comparable. All nice and scientific. But as computing technology seeped from the workplace into the home and then our pockets the goals of the technology became less clear. Workplace computing is almost by definition task and problem focused and the goal of design is to discover optimal solutions to clearly defined problems. Increasingly computing technology is being applied to "wicked" problems where complex vested interests struggle over limited resources and there is no technological quick fix. When technology is conceived as a "silver bullet" it can result in "solutionism" a phrase coined by Dobbins (2009) and popularized by Morozov (2013) as quick fixes for complex psychological, social or environmental problems. It can also take the form of solutions to problems that do not really exist.

In HBO's satirical comedy *Silicon Valley* the protagonists develop "SeeFood" a "Shazam for food", where you take a picture of what you're about to eat and it tells you what it is. The show exaggerates current trends in Silicon Valley and academia but it does not exaggerate much and SeeFood is a perfect example of solutionism. Clearly there would be many usability issues with an app like SeeFood, would it be accurate? How long would it take? Would it be easy to use? But these questions would not answer larger questions like—is it stupid?

Kaye and Sengers (2007) provide a historical overview of the five phases of evaluation in HCI: evaluation by (1) engineers, (2) computer scientists, (3) cognitive psychologists, (4) HCI professionals and (5) evaluation for experience. There is now a body of work that has adopted theory and methods drawn from the humanities, cultural studies and critical theory to address, aesthetics, user experience and politics e.g. (Bardzell 2009, 2010; Blythe and Robinson 2008; Blythe et al. 2006, 2008, 2010; Boehner et al. 2005). Despite the apparent plethora of methods available, evaluation remains problematic (Lindsay and Jackson 2012). Kusunoki and Sarcevic (2012) found that much evaluation work still draws on traditions such as usability testing, heuristic evaluation and cognitive walkthrough. Vermeen et al.'s statistical overview of UX methods (Vermeeren et al. 2010) reveal many methods to be minor adaptations on standard usability tools which do not account for UX factors. Alternatives to standard evaluation metrics include: considering the ways that participants articulate their values [i.e. "Value Centered Design" (Friedman 2006)]; operationalizing values to provide a focus for evaluation (Cockton 2008) and; having designers articulate a particular use quality, so that they can evaluate against it (Greenberg and Buxton 2008). These approaches specify values as the basis for success or otherwise, irrespective of the nature of the values being evaluated.

"SeeFood" might be evaluated in terms of the values of the team developing it. Silicon Valley regularly mocks the rhetoric of the tech industry by having all of their start ups claim that they are "making the world a better place by…" doing whatever it is that they happen to be doing. So the values behind SeeFood could be portrayed as idealistic and utopian. At another moment in the series one character explain that their business model is not really about making apps at all but rather selling stock. Making money might then be a criteria for evaluation and this would rest on popularity. But other approaches take more critical views.

Interdisciplinary criticism calls for a mix of evaluation methods including critical analysis drawing on traditions from the Arts and Humanities (Blythe et al. 2006). Bill Gaver describes "polyphonic evaluation" which draws on multiple perspectives and voices, to argue that conflicting responses can be valuable (Gaver 2007). Bardzell's (2009) "interaction criticism" proposed a mapping of critical theory and aesthetics for interaction design. Greenberg and Buxton (2008) considered usability evaluation as harmful. They argued that it is always possible to find some metric by which a new design might be judged superior to another one. And focusing on usability allows researchers to disregard usefulness. They advocated art school traditions like the "crit" where experienced and new designers respond to work in progressive stages of iteration. The approach suggests that prototypes might be considered as sketches. The drift of all of these approaches is away from binary judgments towards nuanced discussion.

3 Mixed Reviews

Repentir is a mobile app that allows users to take photographs of an oil painting and then rub the image to reveal previous versions of the work right back to the pencil drawings and the blank canvas beneath (Hook et al. 2013). The app was developed with the British artist Nathan Walsh, who took photographs of his urban realist oil painting, portraying the street in New York "*23 Skidoo*" (Fig. 1).

Repentir was presented at Walsh's solo exhibition at a prestigious New York gallery. The gallery's Associate Director, who negotiated sales, said gallery visitors reacted with "wows", and thought the app was amazing. One of the gallery's co-owners was less impressed; he said that this kind of thing had been done in the sixties when artists documented works-in-progress and re-presented them in sequence, on slides or film.

There was another show featuring large scale copies of the work at a small gallery in the English city of York where prints of 23 Skidoo were put on sale to raise funds for a charity that supports new artists (Fig. 2). The app worked just as well with the large-scale facsimiles as the original oil painting. Again, the response was overwhelmingly positive; visitors reacted with "wows" and enthused that the app was "*magical*".

The gallery owner, declared in a webzine video that the app "*revolutionized the way you look at work, it undresses the painting*". After the show, he estimated that the number of prints sold was triple what he would have expected to sell normally.

Another Walsh painting *TransAmerica* was shown with the Repentir app at a CHI conference in Paris. Hiroshi Ishii, a pioneer of Tangible User Interfaces approached the second author and gently informed him that his group had done

Fig. 1 The Associate Director of the gallery demonstrating the app with 23 Skidoo in New York

Fig. 2 Interactive prints exhibition in York

something similar a few years back, making X-ray imagery of hidden traces under the surface of a painting.

The artist's friends and collectors also viewed Repentir in Walsh's studio. While the fellow realist artist David Finnigan was delighted with the app, collectors were less impressed, spending little time with it; one said it added little to the painting. While 23 Skidoo was still in progress, Martin Gayford visited Walsh's studio. Gayford is best known for his work on David Hockney and for his book describing the experience of sitting for the painter Lucien Freud. Gayford reflections on Hockney's iPhone drawings had inspired the project and the researchers were thrilled at the serendipitous opportunity to meet him. For Gayford the app was "interesting" but, as will be discussed in later sections, his responses were nuanced and not easily reducible to approval or disapproval metrics.

Reviews of the Repentir prototype were then mixed. As a group of researchers endeavoring to understand Repentir and how its design and reception might inform the practices of others, what were we to make of this?

4 Process

Over eighteen months the research team met up regularly with Walsh. Initial visits involved biographical interviews and observations of the painter's practice. Walsh's work is sometimes described as "photo-realist" although he prefers "urban realist". Walsh uses photographs for reference but he does not use projection or tracing techniques, a point that he has emphasized in interviews:

> People have said … 'OK, that's a big photograph … or - it's somehow been constructed on a computer. There's a digital print or some sort of underhand process there… It's just hard work and graft … It's a process and there's been a big progression over the last 5 or 10 years.

Walsh works six days a week from a small windowless studio.

During initial visits he was working on the painting in Fig. 3. The first author observed that early stages of the work produced beautiful effects, sometimes lost during the painting process leading him to suggest experimenting with forms of documentation. For example, a "slow print" might play the painting back in real time over the time it took to make the painting (Briggs and Blythe 2012). Walsh was enthusiastic: "*I like the idea of shifting things away from a print that just duplicates what we've already got … [or] a poorer version of the actual painting.*" Following discussion a camera was set up in the artist's studio and Walsh would take a photograph of his work at the end of each working day.

Fig. 3 Nathan Walsh work in progress and painting

5 Developing Repentir

The slow print idea was developed further with the second author suggesting an app: *"where you click on an area ... and it shows you images where painting had been happening"* (Jon). Walsh was again enthusiastic suggesting the name *"Pentimento, which is under-drawing"*. As this was trade marked he later suggested "Repentir", another art history term indicating evidence of painters' corrective work (Fig. 4).

Walsh proceeded to document his next painting "TransAmerica" in eighty images used to develop an app. Users of Repentir take a photograph of an area of Walsh's painting which they can then explore in detail, by rubbing or scrolling to expose the under-workings.

Fig. 4 Repentir TransAmerica

6   Comment and Crits

Initial responses to Repentir were for the most part positive. The camera icon was self-explanatory, the screen featured a slider, which most users dragged to reveal previous stages; while the rub function seldom required explanation. Walsh was very enthusiastic about the app as he relayed to artist Finnigan: *"I thought it was great fun ... I like the rub function because it just seems...you're touching something and you're somehow closer to the process of making the picture"*. Finnigan, was also positive: *"It's an aspect of seeing a picture visually that, it's quite unique."* Mel, one of the New York gallery co-owners was likewise enthusiastic, and reported positive responses from visitors: *"My client the other day, his immediate reaction was, 'Wow, look at all the detail.'"* When asked about the general reaction in New York the Associate Director, Ella relayed reactions like: *'Wow! Oh my God! Wow! How do you do that?'* She suggested, *"for a museum, this is amazing, it's a great educational tool"*. Martin Gayford was also positive if still guarded in his response: *"I think looking underneath the surface it works well, it's quite an interesting proposition."* Overall then, there were positive reactions from artistic, commercial and critical perspectives.

There were however negative comments about both Walsh's work and the app. As previously noted, Leon the other co-owner of the gallery was not impressed saying that this kind of thing had been done before: *"Raphaella Spence, about ten years ago ... the publicity people for Chrysler put a camera in her studio and every five minutes...it snapped a picture. And they made that into a film."* He noted that artists often sent him photographs of works in progress which he used to help generate sales.

Although Walsh's collectors did not entirely dismiss the app they were not impressed and did not engage with it for long: *"It's a good idea but it doesn't work as well as I would expect it to,"* said one. These collectors were responding to an early iteratino of the app that involved waiting for image downloads, and reduced image quality due to the painting being so large—a serious barrier to engagement. Artist Finnigan, while positive, also complained about image quality: *"I keep taking blurred pictures."* Ease of use was impaired by the long processing time and image quality though these were addressed before the New York show. These problems would perhaps be the main focus of a traditional HCI evaluation. But for us they are amongst the least interesting findings with more general discussions more valuable than binary judgments.

As prices of Walsh's work had risen in line with moves to more prestigious galleries, collectors of his paintings now operated at a level of wealth that Walsh found hard to comprehend, something with which he was not entirely comfortable: *"You think we're playthings for the jet set? It's kind of true, isn't it?"* For Walsh, part of the appeal of Repentir was making his work accessible without simply making photographic prints. The fact that the app would work on a postcard as well as the original surprised Mel, the gallery co-owner who joked, *"You just fucked up my whole marketing campaign!"* But for Walsh the app made the work *"inclusive"*.

The art market for Walsh's work is small and specialised. The exposure of labor was crucial for both the galleryists and the artist. Walsh's process is "*very labor intensive. To even make four paintings a year, it requires that level of commitment*". For Walsh and Mel (the gallery co-owner) the app provided value as "*evidence of the effort*", or as fellow-artist Finnigan put it, the "*veracity of the effort*". For the artists the quantity of labor was directly related to quality, prices and marketing. In response to some dismissive comments which were posted on the Huffington Post board Walsh posted: "*I make paintings based on photographic information .... What I don't do is project, Photoshop, grid, trace, or stitch together the information I gather.*" Walsh does not decry other methods but the painstaking nature of his craft is an important part of the narrative that informs its value.

Gallery co-owner Mel, suggested that the app was also of practical value: "*I don't know how people think these paintings get done... You know it's so complex... So it really, really helped justify the prices!*" Marx noted in Capital: "the correctness of the law discovered by Hegel... that merely quantitative differences beyond a certain point pass into qualitative changes" (Marx 1906). Quality increases as more time is spent on a product; ultimately then quality is a quantitative value. While it is no accident that the art which artists produce is known as their "work", the notion of work is disrupted when technology is, or is suspected to be involved. Digital technology is met with suspicion in this space because, as Gayford points out, it "*fundamentally disrupts the art world*". As in the recording industry, previous models of monetization through reproduction have been upturned and new mechanisms must be sought.

Leon, the gallery co-owner made it clear to the first author that he didn't have long for an interview. The first author then decided to risk offence by asking about money and what kinds of people visited the gallery.

> Leon: That's a bad question, everybody in the world comes through and some people buy. Now there's not a lot of this to go round, there's people that respect disciplines, imagery…
> Mark: The people that are buying must be fairly well off though?
> Leon: Yes, of course. But not nearly as well off as the people that spend $50 million dollars on a Basquiat.

Leon claimed that serious collectors did not care how an image was made or if it would increase in value:

> The only time anybody buys a work in this gallery is because they like what it looks like, they can afford it, and they're going to buy it. If I get any indication that it's investment or the decorator told them to have it, they can't buy from me.

Although Leon stressed that art collectors were ordinary and the act of collection nothing to do with money, both Walsh and Finnigan noted that they were producing art for the very wealthy. This made the artists uncomfortable, and also the researcher. It might be possible to construct an argument for the app's economic value from sales, but we would argue that the app was of greater value as a prop for reflection.

7 Conversation

When Gayford visited Walsh's studio he played with the app for a few minutes but remained to engage in a wide ranging discussion for around 2 hours. Much of this discussion concerned the use of digital technology in the production of art. Both Walsh and Gayford insisted there was nothing wrong with using technological aids in painting. Gayford pointed out that if Vermeer did use a camera obscura this makes his achievement no less astonishing. And yet, anxiety about technological intervention was threaded throughout the discussions. Walsh: "*It's just something that I battle against, I have to kind of say, 'Yes, you could project a photograph [but] I'm trying to create a ... completely different space which is independent of the photography*". However, both Walsh and Finnigan in another conversation named another realist painter who they strongly suspected of printing photographs onto canvases then painting on top of them to make them look like paintings. One buyer had asked this artist for "work in progress" shots and, when the request was refused, had asked for their money back. Much of the anxiety around technology then concerned deception. Any technique was acceptable so long as it was not "underhand". The suspicion of technology was ultimately a question of legitimacy.

Gayford noted that it is the universal aspects of digital images that people don't like: "*It doesn't have that difference in texture .... An oil painting is rather good in the digital age because it is still completely impossible to reproduce the actual object, so you've still got unique value in one thing*". For Gayford the value of seeing an original work is practical: there is no better cadmium blue than cadmium blue paint: "*you can't print that color.... So nothing that you see on a screen or a piece of paper is going to correspond to the cadmium blue which is the paint on a canvas*". The notion of an original in this sense is material. However high resolution a screen may become there will always be limitations to digital reproduction and meanwhile value in the tangible original (Fig. 5).

Materiality was also important in the printmaking Gayford had observed. He described Lucien Freud: "*looking at them* [prints] *very hard, he was saying, 'No, there's a tiny mark there [...] What he said to me was buyers of artists' prints are really, really fussy people*." This demand for material perfection perhaps indicates that the immateriality of digital media fundamentally limits its appeal to collectors. Walsh was deeply concerned with creating nuanced texture across the painting surface—with some areas much smoother than others: "*Within a photograph, there's no hierarchy of different marks, it's just one sort of surface*". Mark-making was part of what defined the painting as beyond photographic reproduction. While Repentir presented a flat image on glass, the action, of rubbing through the layers, echoing the gestures of the artist, perhaps added richer experience and texture than

Fig. 5 Martin Gayford in Walsh's studio

if encountering a static reproduction, despite the surface of the device always being uniform.

Many discussions around the app were concerned with other potential uses. At the CHI conference some delegates thought the designed system too simple and pointed out that AR offered far greater functionality, suggesting access to interviews, preparatory drawings and so on. As Walsh noted "*Techs want it to be all singing, all dancing.*" One CHI attendee suggested that it might be used with Old Masters where preparatory sketches were in existence. Gayford suggested "*digital conservation, where… you could look at what's underneath the varnish…*". There is much controversy over art restoration and many claim that some damages the work. Until recently the colors in Michelangelo's Sistine chapel were much darker and while some claim this is how the artist would have conceived it, others argue it is the invention of restorers. While rubbing down to the canvas with the Repentir app Gayford joked: "*this is what art restorers claim they don't do*". A more serious point was that alternative versions of color within an app could show what different types of restoration or varnish removal might look like.

But Gayford also related the app to wider art historical and philosophical concerns: "*I think you've hit on an interesting area. I think this history of pictures is actually deeply involved with time because until the invention of movies, which is pretty recent, the picture, from the moment of drawing bison on cave walls, is a way of freezing time. The image you got was a sort of compact version of reality in which you could dwell on things. You looked at a frozen image for a long time, instead of a constantly changing world. And you have new techniques and you can play around with it more.*" Gayford pointed out that the way an image was displayed drastically affected the amount of time it was possible to look at it. Before reproduction, a painting in a gallery could be looked at for no more than a few hours. But crowded scenes by painters like Breugel repay multiple views, before reproduction this was only possible in private collections. At the same time, he argued, paintings compressed time: the repeated observations of an artist over hundreds of hours are compressed into the single glance of a viewer. In a sense, Repentir allowed an unpacking of those repeated, hidden observations, returning time to the moment frozen in the final image.

Gayford made a comparison to Picasso's performance paintings, made by painting on glass in front of a film camera: "*the drawing at the end isn't particularly interesting, but Picasso quickly worked out that it's the way you got there which is kind of interesting.*" The value in exposing process with Repentir was partly commercial, in that it exposed labor and skill, but for Gayford there was also artistic value in that sometimes process can be more interesting than the final static image.

8   So What?

Responses to Repentir were mixed. For the artist and his community of practice it was a very interesting development, valuable in demonstrating their often-disputed work practices. For the New York galleryists the app was primarily of utilitarian value in demonstrating the labor and time involved to justify high prices. For technology developers, like Ishii, this instantiation of AR was nothing new. And while the app helped to justify prices and make sales, within an art market in service of the "one percent" this was an anathema to the politics and values of both the artists and researchers. The app tripled sales of reproductions in the York gallery, and raised money for the charity, but this was a miniscule fraction of what the original painting sold for. Meanwhile for the critic Gayford, the app was an interesting work in progress.

What then constitutes success or a failure? From the gallery's point of view we might argue Repentir was a successful sales gimmick. From the collector's point of view it added little; perhaps detracting from the magic of the final image. From the technologist's point of view we might note that the idea itself and the technological implementation are not very original. From the critic's perspective we might claim it was a successful resource for historical and critical reflection and discussion.

9   Parallax Views

As Research through Design becomes more common practitioners have begun to ask fundamental questions about its epistemology. Zimmerman and Forlizzi argue for the importance of rigor in research design and methodology (Zimmerman and Forlizzi 2008). Gaver argues for openness and points out that designs are not repeatable in the way that scientific findings are (Gaver 2012). The reasons for making prototypes are also debatable. Academic researchers are not engaged in market research or developing prototypes and evaluating responses to judge whether mass production would be a good idea or not. Both Gaver and Zimmerman and

Forlizzi agree that the point of a research prototype is not market potential but rather contribution to knowledge.

Though most of us would concede that we are not going to solve the world's problems with an app our discourse is firmly based in binary judgements around success or failure. This is rarely the most interesting aspect of Research through Design. Repentir was a success for some and not for others. So what? The reason that such an argument is unsatisfactory is perhaps because evaluation itself is an inappropriate perspective. Over the last ten years the funding criteria for university research emphasizes "impact" stated explicitly as a contribution to the economy, society or culture. Repentir arguably supported sales in the existing art market, so it can be judged a success. But many artists and critics find the existing art market morally reprehensible; Repentir supported this so it is a failure. In this sense Repentir was a success and a failure for the same reasons.

For the philosopher and cultural critic Slavoj Žižek there is no contradiction in the act of interpreting an artifact in opposing or paradoxical ways. Žižek frequently draws on the notion of the Parallax view. Parallax refers to the phenomena where objects appear differently depending on the perspective from which they are viewed. Although we cannot see ourselves looking, our gaze is an integral part of every observation. Žižek illustrates the point with reference to the Moebius strip and the curved space that bends in on itself:

> We do not have two perspectives, we have a perspective and what eludes it, and the other perspective fills in this void of what we could not see from the first perspective. [27:29]

Repentir then can be seen as a success and a failure at the same time. The reality is not found in the agreement between two perspectives. Rather one perspective supplies that which the other cannot. This is not then postmodern relativism, where all perspectives are equally valid, rather it is an insistence on the reflexive inclusion of the observer in any observation. Evaluation strategies that seek simple answers for success and failure metrics are inappropriate for Research through Design.

Attempts to widen HCI's approach to evaluation have encouraged multiple and competing accounts that have shifted the emphasis from binary judgments to nuanced assessment and analysis. But it should be noted that it is not just HCI that has a problem with evaluation. As G. K. Chesterton remarked of the modernist project itself, the position can be characterized as: "Let us not decide what is good, but let it be considered good not to decide it" (Chesterton 1928). He extends this logic in this succinct formulation: "let us not settle what is good, but let us settle whether we are getting more of it" (ibid.).

It is not enough to ask if we did what set out to do, we must also ask if we should have even tried it in the first place.

Acknowledgements This work was supported by the EPSRC grant Digital Originals EP/I032088/1.

References

Bardzell J (2009) Interaction criticism and aesthetics. In: Proceedings of CHI'09. ACM, New York, pp 2357–2366

Bardzell S (2010) Feminist HCI: taking stock and outlining an agenda for design. In: Proceedings of CHI'10. ACM, New York, pp 1301–1310

Blythe M, Robinson J, Frohlich D (2008) Interaction design and the critics: what to make of the "weegie". In: Proceedings of the 5th Nordic conference on human-computer interaction: building bridges (NordiCHI'08)

Blythe M, Bardzell J, Bardzell S, Blackwell A (2008) Critical issues in interaction design. In: HCI 2008, culture, creativity and interaction design. Liverpool, 1–5 Sept 2008

Blythe M, McCarthy J, Light A, Bardzell S, Wright P, Bardzell J, Blackwell A (2010) Critical dialogue: interaction, experience and cultural theory. In: Proceedings of CHI'10 Atlanta, Georgia, USA, 10–15 April 2010. CHI EA'10. ACM, New York

Blythe M, Reid J, Wright P, Geelhoed E (2006) Interdisciplinary criticism: analysing the experience of Riot! A location sensitive digital narrative. Behav Inf Technol 25(2):127–139

Boehner K, DePaula R, Dourish P, Sengers P (2005) Affect: from information to interaction. In: Proceedings of critical computing: between sense and sensibility. ACM, New York, NY, USA, pp 59–68

Briggs J, Blythe M (2012) No oil painting: digital originals and slow prints. In: Slow technology: critical reflection and future directions. Workshop paper in conjunction with DIS'12

Chesterton GK (1928) Heretics. John Lane

Cockton G (2008) Putting value into E-valuation. In: Law E, Hvannberg E, Cockton G (eds) Maturing usability. Springer, London

Dobbins M (2009) Urban design and people. Wiley

Friedman B (2006) Value sensitive design and information systems. In: Zhang P, Galleta D (eds) Human computer interaction in management information systems. Armonk, New York

Gaver W (2007) Cultural commentators: non-native interpretations as resources for polyphonic assessment. IJHCS 65(4):292–305

Gaver W (2012) What should we expect from research through design. In: CHI 2012, pp 937–946

Gayford M (2007) A bigger message: conversations with David Hockney. Thames & Hudson

Greenberg S, Buxton B (2008) Usability evaluation considered harmful (some of the time). In: Proceedings of the SIGCHI conference on human factors in computing systems (CHI'08). ACM, New York, NY, USA, pp 111–120

Hook J, Briggs J, Blythe M, Walsh N, Olivier P (2013). Repentir: digital exploration beneath the surface of an oil painting. In: CHI '13 extended abstracts on human factors in computing systems (CHI EA '13). ACM, New York, NY, USA, pp 2947–2950

Kaye J, Sengers P (2007) The evolution of evaluation. In: CHI 2007. ACM Press

Kusunoki D, Sarcevic A (2012) Applying participatory design theory to designing evaluation methods. In: Proceedings of CHI 2012. ACM, pp 1895–1900

Lindsay S, Jackson D, Ladha C, Ladha K, Brittain K, Olivier P (2012) Empathy, participatory design and people with dementia. In: Proceedings of CHI 12. ACM, pp 521–530

Marx K (1906) Capital. Charles H. Kerr & Company, Chicago, pp 337–338

Morozov E (2013) To save everything click here: technology, solutionism and the urge to fix problems that don't exist. Penguin books

Vermeeren APOS, Law EL-C, Roto V, Obrist M, Hoonhout J, Väänänen-Vainio-Mattila K (2010) User experience evaluation methods: current state and development needs. In: Proceedings of Nordichi 2010, Reykjavik, Iceland, 16–20 Oct 2010

Zimmerman J, Forlizzi J (2008) The role of design artifacts in design theory construction. Human Computer Interaction Institute. Paper 37

Žižek (2009) The parallax view. MIT Press. Cambridge Massachusetts, London England

Part V
Funology 1

How to Use Funology 1

Mark: Ah, someone is reading our book.

Andrew: So they are, quick, say something interesting!

Mark: What?

Andrew: Tell them what a great book this is! We need to get their attention and keep them reading! Quick!

Mark: Oh … uhm, I can't think of anything to say now. Can't Pete do it; he'd put it much better than I could.

Peter: What's going on?

Andrew: We've got a reader and we're introducing the Funology book.

Mark: Snappy title! I thought of that.

Andrew: That's as may be, but it doesn't really say what it's about does it?

Peter: Well, the book is about the move in Human Computer Interaction studies from standard usability concerns towards a wider set of problems to do with fun, enjoyment, aesthetics and the experience of use. Traditionally HCI has been concerned with work and task based applications but as digital technologies proliferate in the home -

Andrew: Gah! Shut up! That sounds really dull! This book is supposed to be about enjoyment! Can't you say something that makes it sound like fun? Where's Kees?

Kees: I'm just relaxing over here. It's very hard work editing a book you know. Andrew, you need to take things easier.

Andrew: But the reader -

Kees: Yes, yes, but the reader can see from the contents page that we have an interesting collection here. For a long time now people in the field have been talking about expanding the concept of usability, even people like Jakob Nielsen -

Andrew: He's in the book!

Kees: Yeah, sure the web guru is here, and even Nielsen, who has been associated with a "no frills" straight usability approach, has been thinking about *engaging* the user.

Peter: The community has been asking questions about enjoyment for some
            time and we're now at the stage where we have a critical mass of work
            providing answers. We've seen quite a lot of ideas in this area coming
            through at the Computers and Fun workshops at York over the last four
            years –

Andrew: My idea, those, you know.

Peter: And then the Funology workshop at CHI last year which this collection
            is based on.

Kees: Yeah, I think the collection maps the field pretty well… but it isn't fun.

Andrew: Gah!

Mark: Well no-one expects an analysis of humour to make them laugh do they?
            We should be telling them "how to use the book" and skip the boring
            bits.

Andrew: There aren't any boring bits! Each sentence is a glittering jewel!

Mark: Well that's as may be, but nobody has to read all of it, that's why it's in
            sections.

Andrew: Three very *exciting* sections! The first is ***theories and concepts***. HCI has
            always been a magpie discipline and here we have a range of positions
            borrowed from a number of fields: anthropology, sociology, psychology,
            literary and cultural studies.

Peter: This section will be of most interest to people who want answers to
            questions like, what's wrong with standard usability approaches, what is
            "user experience", what do we mean by enjoyment, play, fun and is it
            possible to design user experience at all? This kind of theoretical -

Kees: Bullsh -

Andrew: Challenging and stimulating discussion! With each chapter more inter-
            esting than the last will appeal to -

Kees: People with too much time on their hands? No, I'm kidding. It will
            appeal to … uhm …

Mark: People with an interest in understanding the psychological, social and
            philosophical problems inherent in the study of enjoyment and the
            design of enjoyable experiences…. Did that sound alright? I think I
            might go and lie down now.

Peter: And then more practically we have the ***methods and techniques*** section.

Kees: Yeah, not so many of those though.

Peter: No there aren't and I think this might reflect the field. As a relatively
            new area of interest there aren't that many HCI techniques for looking at
            enjoyment that have proven to be useful. So this section begins with
            adaptations of fairly standard usability approaches and moves towards
            more innovative methods.

Andrew: And then in the final section we have a series of case studies -

Kees: A collection of neat ideas.

Andrew: Oh, it's much more than that! Each of the ***case studies*** reflects on the
            problems raised in the previous two sections and tells the story of how

the theoretical problems were addressed in practice, what methods were used, what they produc-

Kees: Yeah, it's a collection of neat ideas. If you are a designer and you want to be inspired – go there straight away.

Mark: Well I think that's the preface taken care of don't you?

Peter: Yeah, what's next?

Kees: Lunch?

Mark: No it's the foreword isn't it?

Kees: What's the difference between a foreword, a preface and an introduction?

Peter: The introduction is longer. It talks to the ontological problems that the publisher wanted us to address doesn't it?

Andrew: [*remark deleted*]

Mark: What?

Andrew: [*further remarks deleted*]

Kees: Well this was fun, let's do it again some time.

Andrew: Are you taking the piss?

Peter: How's that reader doing? Are they looking engaged?

Mark: I'm not sure, it's difficult to see from here.

Chapter 16
Introduction to: Funology 1

Mark Blythe and Peter Wright

These papers are published largely as they appeared in the 2003 edition. Chapter authors were invited to write a note explaining how things have developed since 2003 or setting the context in which the 2003 chapter was written. These notes are included at the start of each chapter. We would draw your attention to the long note added by Wright and McCarthy to Chap. 20 which in many ways sets the scene for the whole book.

Sadly two of our authors have died since 2003, Kees Overbeeke (Chaps. 17, 37 and 38) and John Karat (Chap. 26). The relevant chapters include notes about these authors and their considerable contributions to HCI. We were unable to make contact with two sets of authors in the 2003 volume and so we do not have permission to republish them.

As well as the chapters from the 2003 edition we have included the introductory material from that edition. The chapters were divided into three sections: theories and concepts; methods and techniques, and case studies in design.

1 From Usability to Enjoyment: Book Introduction, as Printed in the 2003 Edition

All I want to do is have a little fun before I die

Says this man next to me out of nowhere

M. Blythe (✉)
Northumbria University, Newcastle upon Tyne, UK
e-mail: mark.blythe@northumbria.ac.uk

P. Wright
Newcastle University, Newcastle upon Tyne, UK

© Springer International Publishing AG, part of Springer Nature 2018
M. Blythe and A. Monk (eds.), *Funology 2*,
Human–Computer Interaction Series,
https://doi.org/10.1007/978-3-319-68213-6_16

It's apropos of nothing

All I Wanna Do: Sheryl Crow.

This book is about enjoyment and human computer interaction (HCI). This may seem like a relatively straightforward topic but humans enjoy so many things that it very quickly becomes tangled and messy. Are we talking about entertainment and play? There's a clear link to technology but games aren't the only applications we enjoy. What about work? People can enjoy that too and a good interface can make a task more enjoyable, but isn't that just usability? Are we talking about cute interfaces or that awful winking paperclip in windows? Is this all about aesthetics? It's possible to enjoy a beautiful Web page so long as it doesn't take half an hour to download. Or are we talking about pornography, isn't this the main way an awful lot of people enjoy interacting with their computers?

And what is enjoyment anyway? Is it an experience, is it an emotion, a sensation, a perception, is it a state of mind, is it a state of being? There is an ontological problem inherent in addressing enjoyment and its associated terms. These terms can be organised fairly simplistically by degrees of intensity: satisfaction, gratification, pleasure, joy, euphoria and so on. The settings in which these states commonly occur can also be organised: work and play, games and entertainment, and so on. But this does not answer the question—to what do these states refer—sensations, emotions, perceptions?

These ontological questions are thousands of years old and many literatures address them. This introduction will provide a very broad sketch of some of the areas of literature relevant to these questions, briefly outline previous work in HCI relevant to the project of extending the concept of usability and finally indicate the differing approaches taken by some of the contributing authors in this book.

1.1 Ontological Problems and Relevant Literatures

Almost every philosopher who ever philosophised has speculated on how and why we enjoy, or take pleasure, in certain things. For Plato pleasure was the absence of pain; in the *Phaedo* a chain is removed from Socrates' ankle and he remarks on the pleasure of relief. For Aristotle, pleasure was caused by the stimulation of the senses through action, in the *Nicomachean Ethics* he is able to explain pleasures that involve no absence of pain such as novelty: when something is new the mind is active and stimulated, the next time it is encountered the mind is less aroused so there is less pleasure in the novelty (Aristotle 2002). In the Confessions of St Augustine pleasures are largely "unlawful" or "awful" unless they are pleasures in the contemplation of God. Historians of philosophy argue that Aristotelian and Christian views of pleasure set the parameters of thought on the subject up until the time of Descartes (Honderich 1995). The Cartesian distinction between self and world, observer and observed paved the way for the measurement of pleasure and Enlightenment philosophers argued that it was possible to do just that. Jeremy

Bentham claimed that pleasures could be judged by intensity and duration and that these could be meaningfully represented on a scale and so measured in a "hedonic calculus" (Honderich 1995). However, Wittgenstein (1953) argued that when we measure some behavioural correlate of enjoyment, we are not getting at the thing itself—the experience; meaning cannot be measured it has to be grasped. Freud famously argued for the existence of a pleasure principle (and later a death principle) as a motivating force for human action that could not necessarily be known by the conscious mind. Questions on pleasure and enjoyment, then, have a long history. Western philosophy offers no coherent approach and nor should we expect it to. Nevertheless it is possible for these literatures to inspire work in HCI; although the current technological challenges are new the fundamental ontological questions are very old indeed.

Physical and social scientists have also created a large body of work that relates to enjoyment. Neurologists have discovered "pleasure centres" in the septal region of the brain which when electrically stimulated produced enjoyable feelings; when animals were wired up so that they could press a lever and administer this stimulation themselves they did so for hours ignoring food, sex and every other need (Gregory 1987). Pleasure then can be regarded as a physical response of the nervous system. But in the twentieth century the apparently simple question—what is an emotion—provoked vexed and contentions debates in psychology. The debate can be broadly characterised as between two schools: the physical and the cognitive. The physical model of emotion, first established by William James, held that "our feeling of (bodily) changes as they occur IS the emotion" (cited: Gregory 1987). The cognitive model suggested that emotion was a decision-making and evaluative process. During grief, for example, we make evaluative decision about the loss, its severity, its permanence and so on. Later, other positions emerged which combined the physical and cognitive aspects of emotion. John Dewey and others suggested that discrepancies between our expectations and the state of the world produce visceral events and that evaluations of these discrepancies dictate whether the emotion is positive or negative. On a roller coaster for example our expectations about the direction and speed at which we move are disrupted producing a visceral response, this is evaluated in terms of how safe we feel and we love or hate the ride accordingly (Gregory 1987). Dewey took the strong position that experience and sense making are relational processes, which, when decomposed into their constituent parts, simply disappear (Dewey 1934). Although chronological accounts of competing theories can suggest a seamless development of coherent thought, it should be noted that there is still considerable debate in these areas.

There are other large bodies of social and anthropological literature on enjoyment. The sociology of leisure is almost as large as the sociology of work. Play has been seen as one of the most important and fundamentally human of activities. Perhaps the best-known study on the subject of play is *Homo Ludens* by Johan Huizinga. In it he argues that play is not only a defining characteristic of the human being but that it is also at the root of all human culture. He claimed that play (in both representation and contest) is the basis of all myth and ritual and therefore behind all the great "forces of civilised life" law, commerce, art, literature, and science (Huizinga 1950). Although Huizinga and others make clear the importance of play to the development of civilisation, until recently, little was known about how and why we do it. Piaget argued that the child at play "repeats his behaviour not in any further effort to learn or investigate but for the mere joy of mastering it" (cited, Gross1996: 639). He divided play into three stages: mastery play (practice play involving repetitive behaviour) symbolic play (fantasy and role playing) and play with rules (structured games). Mastery and skill development form a point of connection to the work of Csikszentmihalyi who offers one of the few theories of intense or peak experiences in his account of "flow". After studying diverse groups engaged in self motivating activities like rock climbing Csikszentmihalyi identified the euphoric feeling of "flow" as a common characteristic of their experiences (Csikszentmihalyi 1975). He was also able to identify the conditions necessary for this feeling to occur. Such models of experience suggest a great many implications for the design of enjoyable products and this work has been drawn on by several of the authors in this book.

There is a further body of literature relevant to the study of enjoyment to be found in the arts and humanities. Perhaps the most famous literary movement to focus on pleasure was that of the aesthetes in the late nineteenth century. The protagonist of Huysman's novel *A Rebours* pursues pleasure so exhaustively that he devotes weeks of study and an entire chapter of description to the subtle differences between scents as he attempts to create a new perfume. The development of enjoyable or pleasurable products and applications can form curious bedfellows. Computer science departments and industrial developers alike are beginning to employ artists to work with programmers. The Surrealist and Situationist art movements of the nineteen sixties have been drawn on by HCI researchers. The work of literary and art critics is also proving, perhaps surprisingly, useful. Concepts such as dialogism drawn from the work of the Russian literary critic Bakhtin have been used to reason about on line shopping, Dewey's theories of aesthetics and the co-construction of meaning between the artist, contemplator and art object, have been adapted to consider the enchantments of such technologies as mobile phones. Insights have also been drawn from film criticism and other branches of cultural studies. There is, perhaps, a degree of similarity between HCI and literary and cultural criticism. The HCI specialist is not necessarily a programmer just as the critic is not necessarily a writer. HCI can be seen as a specialised form of reading, where an application or programme is the object of study rather than a static text. Literary and cultural studies are, perhaps more than any other discipline, concerned with enjoyment and pleasures of a very profound kind; it may be for this reason that a number of the authors in this collection draw on these traditions.

1.2   The Limits of Traditional Conceptions of Usability

In many respects then, the field of human computer interaction is a late-comer to the study of enjoyment. Traditionally, HCI has been concerned with work and work systems, however enjoyment has become a major issue as information and communication technology have moved out of the office and into the living room. Understandings of user concerns derived from studies of the world of work are simply not adequate to the new design challenges. At work we are paid to interact with computers, in the home our motivations are different. Some domestic activities are task based and look very much like work, for example, cleaning and shopping. Clearly efficiency and effectiveness are equally important in the design of technologies to support, on-line shopping for instance, but even here these are not the only important considerations. It is increasingly acknowledged that work tasks are performed better if they are enjoyable. The distinctions between "work" and leisure" and "tool" and "toy" have been challenged by new approaches to design. Further, many activities in the home are not task related at all: they are leisure activities. Where is the task in listening to a piece of music or looking at a family photo album? Of course the activation and control of media can be thought of as tasks and it is even possible to argue that the task in a leisure activity is to relax; but an entirely task based focus is clearly inappropriate.

It is argued throughout this book that traditional usability approaches are too limited and must be extended to encompass enjoyment. Of course HCI has always been concerned with satisfaction. Indeed "usability" is defined as "the effectiveness, efficiency and satisfaction with which a product is used "(ISO 9241-11). But satisfaction is a relatively narrow term; it is an aspect of the question—does this work? In practice the satisfaction element of usability testing often amounts to investigating whether the product frustrates users or not. It is primarily concerned with the prevention of pain. Since the time of Aristotle this has been a limited view of pleasure. If we are attempting to design enjoyable applications there are further questions to be asked.

There have been many attempts in HCI to put enjoyment into focus (Monk et al. 2002). In the early nineteen eighties Malone published a heuristics for designing *enjoyable* user interfaces (Malone 1984). Four years later Caroll and Thomas (1988) proposed game-like, metaphoric cover stories for standard process control jobs as a possible means of addressing boredom and vigilance problems inherent in routine tasks. In the early nineteen nineties Laurel's (1993) *Computers as Theatre* argued that engagement in computer mediated activity is as much about emotional and aesthetic relations as it is about rational and intellectual ones. It is worth noting that this book offered an important warning, Laurel argued that software cannot be made enjoyable with the introduction of gratuitous game-like features. If a student must solve a maths problem before they can play a game in a piece of educational software then either the game or the maths problem is superfluous: both most be shaped in a "causally related way" (Laurel 1993: 74). Two years after Laurel's seminal work, Sherry Turkle (1995) explored the social meaning of computers, the culture of computing and its impact on our sense of self in the age of the Internet. Toward the end of the nineties Donald Norman challenged designers to follow three axioms: simplicity, versatility and pleasurability (Norman 1998). A year later Patrick Jordan's ground breaking book *Designing Pleasurable Products* (Jordan 2000) explored theoretical models of pleasure drawn from anthropology to make concrete recommendations to product designers in terms of aesthetics and ergonomics. Recently applications have emerged which attempt to make even serious work based activities more enjoyable. Dennis Chao's PSDoom, for example, is a Unix process manager that adapts the popular first person shoot 'em up DOOM as the user interface (Chao 2001) and a good example of Laurel's causally related enjoyment.

The move from theories and concepts to design is never an easy one whether as a research activity or as a practical application. Bannon (1997) talks of HCI as dwelling in the "great divide" between the social and the technical sciences. This volume is concerned with theories of experience and enjoyment which originate not only from social sciences but also from the arts and humanities and in some senses we have created for ourselves an even larger, albeit more colourful divide. One of the concerns of many of the 'human scientists' (psychologists, sociologists etc.) who dwell in the great divide has been to develop theories and methods that make the 'important things' of human activity visible and to find ways of 'translating' such information into a form that is usable by designers. It is significant that HCI began as a partnership between computer scientists and cognitive psychologists; one of the main attractions of cognitive psychology was precisely its underlying metaphor of the human as an information processor. This greatly reduces the translation problem since at least in principle, both user and system are modelled in the same framework of concepts. The problem however, is the limitations on what such a theory of human activity makes visible. In this volume, several authors allude to the limitations of cognitive science when it comes to dealing with the affective, and many of the chapters represent attempts to understand what needs to be made visible and how we ought to theorise experience.

1.3 The Breadth of Approaches

Given the ontological uncertainties outlined above, this book provides an overview of where the HCI community, or a part of it, is in terms of theories and concepts, methods and techniques and case studies. The contributions in the theories and concepts section draw on a very wide literature: computer science, psychology, sociology, philosophy, history, literary and cultural studies. The methods and techniques section begins with adaptations of traditional usability approaches to satisfaction and moves on to more innovative methods. The case studies presented in the final section were chosen to provoke and inspire researchers, designers and product developers. Each section is preceded by a brief introduction which summaries the contributions and provides a road map to the section.

The "absence of pain" model of enjoyment can be thought of as a standard usability approach to pleasure; if an application does not frustrate the user then it is more likely that using it will be enjoyable. Contributions in this book from Pagulayan, Nielsen and also Karat and Karat show how standard usability tests can be adapted to focus on enjoyment. Psychological accounts of pleasure inform contributions by Brantdzaeg et al., Desmet and Hassenzahl. An emphasis on the physical rather than purely cognitive aspects of enjoyment can be found in chapters by Overbeeke et al., Hummels et al. and Wensveen et al. Anthropological and social approaches to questions of enjoyment inform chapters by Sengers, Reed, and Blythe and Hassenzahl. The intricacies of social practices are analysed and deconstructed in chapters by Dix, Sykes, Rizzo and Falk in order to generate design implications for technological developments. The influence of art can be seen clearly in the chapter by Hull and Reid, which reflects on a collaborative project with artists, Anderson et al., in the use of techniques inspired by the surrealist movement to develop research methods and by Holmquist et al. in the influence of the artist Mondrian on a particular application. Work drawing on literary and film studies can be found in the chapters by Wright and McCarthy on understanding user experience and also in Braun in the development of an interactive story engine. A concern with making serious or work-based activities more enjoyable can be found in many of the contributions and particularly in Hohl et al. and Rosson and Carrol.

To return to the questions at the beginning of this introduction—are we talking about work, play, games, entertainment and aesthetics, or what? The answer is yes to all of the above, including the—or what. The subject of pornography however will have to wait for another book.

We have not, as editors, attempted to impose a particular theoretical perspective or engineer the appearance of a coherent approach amongst the authors. In fact many of the authors have radically different approaches, where one attempts to measure another tries to grasp. We believe that this breadth of approaches and subjects reflects the development of a relatively young discipline in a dynamic field of enquiry.

2   Theories and Concepts

The introduction to this book questions the limited scope of traditional conceptions of usability. Conventionally, designers are mainly concerned with concepts such as ease-of-learning, low-level ease-of-use and task fit. This has arisen because HCI has been primarily concerned with developing methods and concepts for the design of products to support work. Now that the designers of information and communication technology are turning their attention to the home new concerns come to the fore. Aesthetic attractiveness and enjoyment have their place in the design of products for work but are not of primary importance. In the home, where people are not paid to use the technology, they suddenly become critical. But, as the introduction to this book also suggests, there is no well thought out theory of enjoyment. This first section of the book is to suggest some ways in which the concept of enjoyment may be developed in a way that facilitates the design of enjoyable products.

It is instructive to examine how early concepts in HCI, such as ease of learning were developed. HCI is a great borrower of ideas. It started as a collaboration between psychologists and computer scientists. Concepts and methodologies were borrowed from these disciplines and woven into theories and design methods. For example, psychologists contributed ways of measuring ease-of-learning. Computer scientists contributed abstract system properties such as "mode" and "reversibility" that govern ease-of-learning. In the process the concepts borrowed were subtly changed, to the point they are no longer be recognised by the parent discipline. Thus a psychologist would be very critical of the procedures commonly used to assess usability and many computer scientists would consider that these abstractions lack mathematical formality. They have however lead to the design of better, that is more usable, systems. HCI continues to borrow from: ergonomics, human factors, sociology and anthropology. Further, as we shall see in the chapters in this section, the need identified above to develop the concept of enjoyment suggests a whole new range of borrowings.

The first two chapters examine the problem of designing products that facilitate an enjoyable user experience and take from a classic design perspective. In **Let's Make Things Engaging** *Kees Overbeeke* calls for a less overtly cognitive approach to design. He argues that for a design to be engaging and hence enjoyable it should be physical showing us "the works" and providing clear affordances for action. The point is illustrated by a number of novel designs for digital artefacts with very physical forms and through some putative design principles. In **The Engineering of Experience** *Phoebe Sengers* goes further and questions whether the analytic approach inherent in engineering can be applied to the design of enjoyable products at all. She calls for a holistic approach to design. She rejects the task-based approach to design that can be traced back to Taylor and the deskilling of work and proposes instead a more culturally inspired multidisciplinary approach. She also presents examples of products that fulfil the characteristics she is calling for. By describing the process by which they were created she illustrates the interdisciplinary process she believes is necessary.

These two chapters by Overbeeke and Sengers with their emphasis on concrete examples of design and the process by which they were created could almost be thought of as being anti-theoretical but really they are not. They set a clear agenda for the remaining chapters in this section by illustrating how the problem of designing for enjoyment is radically different from the problem of designing for ease-of-use or task fit. The remaining chapters in this section present different frameworks for thinking about the design of user experience and enjoyment borrowed from a variety of sources.

In **The Thing and I: Understanding the Relationship between User and Product** *Marc Hassenzahl* makes a basic distinction between pragmatic and hedonic attributes of products. His framework is based around a model of user experience. Design features, including content and presentation, lead the user to perceive the product as having a certain character leading in turn to consequences such as appeal. In this way he is able to characterise the relationship between task oriented pragmatic concerns and the hedonic concern of enjoyment. In **Making Sense of Experience** *Peter Wright, John McCarthy and Lisa Meekison* develop a parallel framework. This can be seen as an extension of Sengers' manifesto for a holistic interdisciplinary approach. Four relational elements of experience are identified: emotional, sensual, compositional and spatio-temporal. The process by which a user experience arises is described in terms of sense making through the activities of anticipation, interpretation, reflection, assimilation and recounting.

In their chapters Hassenzahl, also Wright and McCarthy, have ambitious goals. They seek to provide an overarching framework for thinking about user experience as a whole. The next three chapters are more modest in their aims. In **Making it Fun: Lessons from Karasek** *Petter Brandtzæg, Asbjørn Følstad and Jan Heim* draw parallels between Karasek's model of what makes work a rewarding experience (e.g. varied demands and decision latitude) to the design of technology for fun. In **Having Fun on the Phone: the Situated Experience of Recreational Telephone Conferences** *Darren Reed* points out that enjoyment is often a social experience and as such can be described from the point of view of the group rather than the individual. He draws on Goffman and Bateson's idea about play and illustrates the approach by describing the roles for laughter in a conversational analysis. In **The Enchantments of Technology** *John McCarthy and Peter Wright* draw on Boorstin's account of how film enchants the viewer.

Finally, the chapter **The Semantics of Fun** *Mark Blythe and Marc Hassenzahl* examine the semantics of the word "fun" and other words such as "pleasure" and "enjoyment". As in any new field there is as yet no agreed set of terms and this chapter considers problems of definition. They find that "fun" has certain connotations of distraction and frivolity that distinguishes it from other forms of enjoyment.

In these chapters we can begin to see how ideas from various disciplines may be adapted to our purpose of designing for enjoyment. Ideas have been identified from psychology (Csikzentmilhalyi), literary theory (Bakhtin), art history (Dewey), sociology (Goffman), anthropology (Bateson), film (Boorstin) and work design (Karasek). It will take time to adapt the original concepts into a coherent framework. The next step in this process is to refine the concepts by using them in practical guides for design. The second section of this book illustrates some attempts to do just that.

3   Methods and Techniques

The idea of designing for enjoyment rather than simply designing to reduce frustration is a relatively new one for HCI. Usability evaluation methods have tended to concentrate on identifying usability black spots rather than beauty spots. But now it is no longer adequate just to avoid bad experiences, we have to find methods for designing good ones. As we noted in the introduction to this book, the move from principles, theory and concepts to practical methods and techniques has traditionally been quite a tricky one, involving as it does so many different interests, concerns, constraints and viewpoints. There would seem to be two possible fronts on which to advance. The first is to extend our existing armoury of user-centred design methods. The second is to do what HCI often does and borrow some ideas from disciplines new to HCI such as graphic design, art and literary theory. It's not surprising then, that the papers in this section provide examples of both of these ways forward.

The first three chapters in this section approach the problem by extending traditional user-centred design approaches. In his chapter entitled **Measuring Emotion**, *Pieter Desmet*, describes the development of a psychological instrument for measuring emotional responses to products. The chapter is a good example of a classic psychometric approach to measurement. Key factors relating to emotion are first identified and refined down to a small set of measurable attributes. One of the interesting questions that Desmet asks is whether a product that is enjoyable in one culture will also be enjoyable in another. Consequently, he develops a *culture-free* test by using animated faces as stimuli. Using this tool, Desmet is able to show how different products engender different emotional responses in subjects.

In their chapter entitled **That's Entertainment!** *John and Clare-Marie Karat* are concerned to find out whether the Web has to be interactive to be entertaining. They show us how a traditional mix of focus groups, interviews, prototype evaluation and questionnaires can be used to design and assess the entertainment value of cultural Web sites. They conclude that while the Web is an interactive medium it is not exclusively so. Users can be satisfied by watchable experiences on the Web too and a major factor in entertainment is *who* we are entertained by not how much control we have.

In their chapter entitled **Designing for Fun**, *Randy Pagulayan et al.* support Nielsen's argument that you need to get usability right before your product can be fun. Their work is concerned with the rapid evaluation cycles required to test games for Microsoft's *Xbox*. Like Karat and Karat's paper, their approach adapts traditional usability techniques to allow them to test various aspects of their games. They argue that by extending current usability design methods, games designers can get a handle on fun and improve the entertainment experience. Through a series of real product design examples, they show how standard usability tests highlighted problems which when solved, improved users' experiences.

In contrast to the first three chapters that extend traditional user-centred design methods, the next two chapters borrow from the disciplines of art and literary theory. In their chapter entitled **Playing Games in the Emotional Space**, *Kristina Andersen et al.* take on a rather daunting design challenge namely, how to ease the sense of privation felt by people who are emotionally close but physically distant from one another. In trying to solve this problem they set out on a fascinating journey, exploring what it means to be emotionally close. Inspired by the Surrealist art movement they develop a set of games that move people into the *emotional space* they wish to explore. They argue that fantasy games are, like art and poetry, valuable instruments for triggering sincere emotional responses in artificial situations. Once they have moved their participants into the emotional space they use what they call a *survey* to elicit responses from users about what it feels like and how technology might help, but it is certainly a survey with a difference! In contrast to Desmet, their approach is to focus on the *felt life* of individuals and what these feelings mean to them. By getting participants to consciously reflect on their feelings, Andersen et al. uncover what they call a universe of specific symbols and meanings that intimates share and which create a shared, but private, alternative emotional space.

In his chapter entitled **Deconstructing Experience**, *Alan Dix* provides us with a method for analysing experience. Taking as his starting point the literary concept of deconstruction and reconstruction, Dix shows us how a piece of poetry can be analysed in terms of resonances, dissonances and paradoxes at a number of different levels. He argues that this is precisely what strikes us about poetry. He then shows us how principles of reconstruction can be used to produce similar but novel poetic lines. Then he makes the interesting move of arguing that science can be regarded as a kind of deconstruction while design can be seen as a kind of reconstruction. He shows us how these same literary principles can be used to analyse graphic design problems and to deconstruct everyday experiences and reconstruct them in new media. What Dix is offering us then, is a simple but powerful technique for analysing and designing for experience. His point in all of this is not only that literary theory may have a lot to tell us about design but also that in an age when digital media change at such a rate, there is a real need to understand how experience can be re-mediated.

And finally, the last two chapters of this section take an approach somewhere in between the two extremes. In some ways they use traditional HCI concepts and methods but they apply them to artistic and imaginative activities, and they bring together engineers, designers and artists in interesting ways. These two chapters also fall naturally together because they are both concerned with designing for children. In their chapter entitled **Designing Engaging Experiences with Children and Artists**, *Richard Hull and Jo Reid*, show us some inspired design concepts for ubiquitous applications. They bring together low-cost technologies to provide interactive multimedia experiences in a number of settings and show how it is possible to deliver engaging experiences with artful combinations of simple functionality. They argue that engaging experiences share in varying combinations, a number of features including *self-expression, social interaction, bonding, sharing,* *drama* and *sensation*. In order to iteratively design the tools and explore user experience they adapt participatory design techniques. In particular they bring engineers, artists and children together to work as full design partners. One outcome of their experiences as designers is a provisional model of consumer experience that they use to situate their products and to identify future design concepts.

In their chapter entitled **Building Narrative Experiences for Children through Real Time Media Manipulation: POGO world**, *Antonio Rizzo et al.* describe their approach to building POGO world. Where Hull and Reid gave us a lightning tour of their experiences of participatory design with artists and children over a range of products, Rizzo et al. focus on describing in detail how they went about modelling and providing tool support for one pedagogic activity. They argue that narrative construction of experience is a central pedagogical technique in European schools. Through an analysis of children at a number of schools they developed a model of this activity based on Vygotsky's conception of the *cycle of creative imagination*. Using this framework, Rizzo et al. describe how they developed and evaluated a set of novel interactive tools the aim of which is to augment the more familiar media of pencil and paper. In so doing the tools re- mediate the experience of narrative construction. One of the interesting points made by Rizzo et al., is that before the introduction of POGO world teachers saw traditional desktop technology as a potential risk to successful narrative activity. In contrast, POGO world by transforming those same activities into something more, positively enhanced the teachers' and children's experiences and perceptions of technology. Surely it is this kind of transformation which is at the heart of what designing for enjoyment should be about.

4   Case Studies in Design

The first section of this book (now Part 6) offered a range of theories and concepts of use in the consideration of enjoyment in human computer interactions. The second section outlined a number of possible methods for research and development ranging from adaptations of standard usability techniques to more innovative approaches. This section is a collection of case studies concerned with the development of particular applications. They are instances of designers and researchers taking up the challenges of designing for enjoyment as outlined throughout the book.

The first two case studies are concerned with making traditionally serious endeavours, more enjoyable. **From Usable to Enjoyable Information Displays**, is concerned with an innovative form of public information display. *Ljungblad et al.* describe an "informational art" display, based on the work of Piet Mondrian, that presents a weather forecast with squares and colours changing as temperature and conditions vary. A prototype was tested at a University campus and was generally received with enthusiasm. The authors conclude that while such informational art would be inappropriate to display train times or other information where accuracy and readability were graver concerns than aesthetics, there are a range of possible applications for this kind of engaging display. In **Fun for All**, *Rosson and Carroll* explore the use of simulations to promote collaboration and community across generations. The authors consider not only ways of making cross-generational programming more enjoyable but also identify the aspects of the simulations which the participants enjoyed most. They found that the boys were most likely to enjoy the game-like elements of simulations and speculate that gender may predict the enjoyment of different simulations better than age. Creating simulations that are enjoyable but do not trivialise community issues remains an ongoing design challenge.

The next two case studies are all, in one way or another, concerned with narrative and, to an extent, the supernatural. In **Deconstructing Ghosts**, *Sykes and Wiseman* consider the "fun of fear" and report findings from two experiments that attempted to find out what was scary about the allegedly haunted vaults of Edinburgh. They created a computer-participants in the real and virtual spaces with fascinating results. By deconstructing the space in this way they suggest aspects of the visual scene and associated narratives that produce effects of fear with clear implications for designers of video games. In **Interfacing the Narrative Experience**, *Jennica Falk* considers the elements of live action narrative game play that have yet to be captured in virtual gaming environments. Falk's work is concerned with extending virtual game space into the physical world and she illustrates this idea with real world tangible interfaces for multi user domain adventure games. By contrasting on and off-line role play gaming she produces a set of design implications for producing more compelling and immersive virtual environments.

The next three case studies focus on engaging users in interactive experiences. In **Whose Line is it Anyway**, *Blankinship and Esara* describe "talkTV" an ingenious piece of software that allows users to search the text embedded in TV programmes to provide subtitles for the hard of hearing. The particular application of this search engine allowed Star Trek fans to locate and cut and paste scenes from the series in order to construct their own mini-films. The authors took a prototype to a Star Trek convention and found that the "trekkies" who used it had "hard fun" with it. In **The Interactive Installation ISH**, *Hummels et al.* introduce the concept of resonance— the extent to which a product resonates with the user by connecting to cognitive and emotional skills, personal history and aesthetic sensibilities. They then describe their interactive sound handling (ISH) installation featuring a number of innovative image and sound handling devices. The chapter ends with an evaluation of the products based on user testing which support claims for the importance of rich interactions. And finally **Fun with Your Alarm Clock**, is an engaging piece of writing about an engaging technology. *Wensveen and Overbeeke* describe an alarm clock that gauges and responds to the mood of the user. The design problem and the solution are illustrated with character driven scenarios in which the authors attempt to make getting out of bed a little easier for their long suffering heroine Sophie. We thought it appropriate to end the book with this design for making everyday life a little more enjoyable.

References

Aristotle (2002) Nicomachean ethics (trans: Rowe C). Oxford University Press. Augustine S (1960) In: Sheed FH (ed) The confessions of St Augustine. Sheed & Ward, London and New York

Bannon LJ (1997) Dwelling in the great divide: the case of HCI and CSCW. In: Bowker GC, Star SL, Turner W, Gasser L (eds) Social science, technical systems and cooperative work. Lawrence Earlbaum Associates

Caroll JM, Thomas JC (1988) Fun. SIGCHI Bull 19:21–24

Chao D (2001) Doom as an interface for process management. Paper presented at the CHI. Csikszentmihalyi M (1975) Beyond boredom and anxiety: the experience of work and play in games. Jossey Bass Publishers, San Fancisco

Csikszentmihalyi M (1975) Beyond boredom and anxiety: the experience of work and play in games. Jossey Bass Publishers, San Fancisco

Dewey J (1934) Art as experience. Capricorn Books, New York

Gregory R (ed) (1987) The Oxford companion to the mind. Oxford University Press, Oxford and New York

Gross RD (1996) Psychology: the science of mind and behaviour, 3rd edn. Hodder & Stoughton, London

Honderich T (1995) The Oxford companion to philosophy. Oxford University Press, Oxford and New York

Huizinga J (1950) Homo Ludens: a study of the play element in culture. The Beacon Press, Boston. Huysmans J (1998) A Rebours (Against Nature) (trans: Mauldon M)

Jordan P (2000) Designing pleasurable products: an introduction to the new human factors. Taylor and Francis

Laurel B (1993) Computer as theatre. Addison-Wesley, Reading

Malone TW (1984) Heuristics for designing enjoyable user interfaces: lessons from computer games. In: Thomas JC, Schneider ML (eds) Human factors in computer systems. Ablex, Norwood, pp 1–12

Monk AF, Hassenzahl M, Blythe M, Reed D (2002) Funology: designing enjoyment. In: CHI2002, Changing the World, Changing Ourselves. Extended Abstracts. pp 924–925. Plato (1993) Phaedo (trans: Rowe CJ) Cambridge University Press

Norman DA (1998) The invisible computer. MIT Press, Cambridge

Turkle S (1995) Life on the screen: identity in the age of the internet. Phoenix, London

Wittgenstein L (1953) Philosophical investigations. Blackwell, Oxford

Part VI
"Theories and Concepts"

Chapter 17
Let's Make Things Engaging

**Kees Overbeeke, Tom Djajadiningrat, Caroline Hummels,
Stephan Wensveen and Joep Frens**

Kees Overbeeke (1952–2011)

Prof. Dr. Kees Overbeeke was appointed full professor at Eindhoven University of Technology (TU/e) for Intelligent Products and System Design in the Department of Industrial Design in May, 2006. Kees Overbeeke studied psychology at the Katholieke Universititeit Leuven (1974). After working there he moved to the Faculty of Industrial Design Engineering at Delft University of Technology where he earned his Ph.D. (1988) in spatial perception on flat screens. He headed the Form Theory group as an Associate Professor until his move to the Department of Industrial Design of TU/e in 2002. During the academic year 2005–2006 he was invited as the Nierenberg Chair of Design at Carnegie Mellon's School of Design in Pittsburgh. At TU/e he headed the Designing Quality in Interaction group until September 2011.

Kees was one of key figures who introduced the HCI community to Industrial Design. He was dreaming of the impossible, trying to rebalance thinking and doing, to connect ethics and aesthetics, and to educate new kinds of students who are able to combine design, science and engineering in their work. And he succeeded. The imaginative designs that he and his students presented are still influential to this day. As co-editors of the 2003 Edition with Kees, we remember the real sense of fun he brought to everything he did. As well as being an editor of the 2003 Edition, Kees was an author on Chaps. 23 and 24.

2003 Chapter

K. Overbeeke · T. Djajadiningrat · C. Hummels (✉) · S. Wensveen · J. Frens
Eindhoven University of Technology, Eindhoven, Noord-Brabant, The Netherlands
e-mail: c.c.m.hummels@tue.nl

© Springer International Publishing AG, part of Springer Nature 2018
M. Blythe and A. Monk (eds.), *Funology 2*,
Human–Computer Interaction Series,
https://doi.org/10.1007/978-3-319-68213-6_17

1   Introduction

Technology and electronics have given us many positive things. However, the appearance of and the way we interact with products have changed consequently resulting in a less engaging relationship with products towards the end of the 20th century (Hummels 2000; Overbeeke and Hummels 2013). Machinery withdrew to the background and control by means of buttons and icons became prevalent (Fig. 1). The physicality of the machinery became an unnoticeable means to deliver the goods.

People all have senses and a body with which we can respond to what our environment affords (Gibson 1986). Why, then, do human-product interaction designers not use these bodily skills more often and make electronic interaction more tangible (Fig. 2)?

And, as humans are emotional beings, why not make interaction a more fun and beautiful experience? We believe that the physicality of the product should be reinstated, to restore engagement. Fun, as such is not the issue, engagement is. This contribution focuses on those neglected aspects of human-product interaction.

2   What Is Wrong?

Many products are designed by people not trained in product design. The resulting products reflect their maker's training. Psychologists make products that are very "cognitive" (or instruct designers to do so). Software engineers design interfaces that resemble the logic of programming. Cooper (1999) has made a convincing analysis of the latter phenomenon. As a solution, he proposes to get away from "technological artefacts whose interaction is expressed in terms in which they are constructed" (p. 27).

Furthermore, everybody claims to take man, not technology, as his starting point. The talk is all about user-centred design. But what does this mean? This faith

Fig. 1 Buttons and icons stand between the user and the machine's functionality

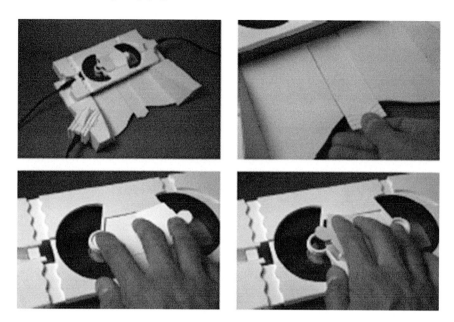

Fig. 2 The action of the user opens up the functionality of this video player. Starting top-left clockwise: the cassette remains visible whilst in the machine, pulling a ribbon triggers eject, and fast-forward/reverse becomes intuitively clear through a toggle placed between the tape reels (Djajadiningrat 2004)

is often professed but seldom applied. We think that user-centred design should be interpreted as design, which shows respect for people as a whole. For the sake of analysis, people's skills, which are used when interacting with products, may be considered on three levels: cognitive skills, perceptual-motor skills and emotional skills. In other words: knowing, doing and feeling; the wholly trinity of interaction (Overbeeke et al. 2002). In the 20th century, research on human-product interaction, however, has concentrated on cognitive skills. Products have become "intelligent", and intelligence has no form. Design research, quite naturally, turned to the intelligent part of humans and thus to the science of cognition to find answers. This has resulted in interface design placing a heavy burden on human intellect. For example, designers start grouping and colour-coding related functions, adding displays with an abundance of text and icons, and writing logically structured manuals. And many design methodologies also suffer from the "logical" disease. Emotions are narrowed down to fun and fun becomes a glued on quality. Products smile at you. I'm not dying to open a bottle of good white wine with a smiling corkscrew. When addressing emotions many designers take a Walt Disney approach and, by doing so, sidestep the real issue: addressing emotions in an adult way. In our opinion the design of electronic products has got stuck as a result of this rather cognitive approach, which neglects the user physically and emotionally. We think that an approach, which mainly addresses the rational and quantifiable human skills, simply does not cut it.

3   How Do We See It?

Users are not interested in products; they are in search of challenging experiences. Therefore the designer needs to create a context for experience, rather than just a product. He offers the user a context in which he may enjoy a film, a dinner, cleaning, playing, working, *with all his senses*. Efforts on improving usability focus on making things easier. However, there is more to usability than ease of use. A user may choose to work with a product despite it being difficult to use, because it is challenging, seductive, playful, surprising, memorable or even moody, resulting in enjoyment of the experience. No musician learnt to play the violin because it was easy. Bringing together 'contexts for experience' and 'aesthetics of interaction' means that we do not strive for making a function as easy to access as possible, but for making the unlocking of the functionality contribute to the overall experience.

Usability is generally treated separately from aesthetics. Aesthetics in Industrial Design appears to be restricted to making products beautiful in appearance. As the ease of use strategies do not appear to pay off, this has left us in the curious situation that we have products, which look good at first sight, but frustrate us as soon as we start interacting with them. We think that the emphasis should shift from a beautiful appearance to beautiful interaction, to engaging interaction. And this should not be a glued on quality. Beauty in interaction is the core, the starting point of interaction design.

This calls for a re-think of product design from the ground up. Design should be given back to designers, as a part of a multi-disciplinary team. Products should elicit the user to engage with them through their physicality. Fun can result from engagement, but is not a goal as such. Design is not about the smile on the product, it is about the smile in the user's heart.

4   How Do We Do It?

But what should designers do once design is given back to them? How can designers open up the products functionality while engaging the user in a beautiful way? The answer to this question is multi-layered and as yet incomplete. In several publications we touched upon parts of the answer. In the first part of this section we mention two, starting from the more general level of a context for experience to the level of design rules of thumb for augmenting fun and beauty. In the second section we give a few examples, as we believe actions speak louder than words.

4.1   Trying to Answer the Question

In her Ph.D. thesis Hummels (2000) makes a strong case for engagement as a means to augment fun and beauty. She argues that the shift towards involvement during interaction means that the designer's emphasis should be placed on a beautiful, engaging interaction with a product. Consequently, the focus shifts towards the aesthetics of interaction. In general one could say that the aesthetics of interaction is the sense of beauty that arises during the interplay between a user and a product in their context. What creates this sense of beauty? Why do some products resonate with a user, while others do not? She believes that five aspects are essential to evoke this sense of beauty. She considers the following five aspects essential.

1. Functional possibilities and performance of the product

   A proper functioning product forms the basis of the aesthetics of interaction. A product that does not do what it is supposed to do, will never allow the user to get intimate and experience the beauty. Spiffy solutions that work well can smooth the way for intimate interactions.

2. The user's desires, needs, interests and skills (perceptual-motor, cognitive and emotional)

   A contextual design approach is based on the experience of the individual. A product may resonate with one person, whereas another person may be indifferent to it. The user's character, skills, needs (short-term and long-term), mood, etc. determine the value of the interaction for an individual.

3. General context

   Although a designer is not able to control the general context in which a person will use his product, this context can influence the experiences of the user when interacting with the product.

4. Richness with respect to all the senses

   Aesthetic interaction requires richness that covers all the senses. Not only does it refer to richness in visual aspects of the product, but the wealth and subtlety of auditory, olfactory, flavoury, tactile and kinaesthetic aspects during interaction, are at least as important to achieve a beautiful interaction and an engaging experience. This richness bears on feed-forward as well as feedback (see below). Moreover, designers need to exploit the range and diversity of design solutions to evoke or intensify the range of feelings (although they can never enforce a specific experience).

5. Possibility to create one's own story and ritual

   Each product tells a story about the user and the relationship between them as it evolves from the moment of purchase onwards (Djajadiningrat et al. 2002).

Intimacy with a product can be enhanced when the product stimulates the user to create his story and rituals during usage. A product should be an open system, which is not an open book, rather a tempting means for exploration and interaction. Due to the advancing digital technology, intelligent products can adapt to the user and actively help to create a never-ending story.

These are very general aspects: they do not tell the designer what exactly he has to do to realize them when designing. Therefore, a few years ago we published a pamphlet with 10 rules to augment fun and beauty in interaction design (Djajadiningrat et al. 2000a, b). These 10 rules do not constitute a guide to "good" design however, and we did not mean to provide one.

Here are the 10 rules:

1. Don't think products, think experiences.

The designer needs to offer the user a context in which he may enjoy a film, dinner, cleaning, playing, working with all his senses. We talk of creating a context for experience rather than just an experience, because we cannot impose a particular experience on a user, who is bound to explore the design in his manner. A design should offer the user the freedom for building his or her experiences.

2. Don't think beauty in appearance, think beauty in interaction.

Usability is generally treated separately from aesthetics. Aesthetics in product design appears to be restricted to making products beautiful in appearance. As the ease of use strategies do not appear to pay off, this has left us in the curious situation that we have products, which look good at first sight, but frustrate us as soon as we start interacting with them. We think that the emphasis should shift from a beautiful appearance to beautiful interaction, of which beautiful appearance is a part. Dunne (1999) too talks of 'aesthetics of use': an aesthetics which, through the interactivity made possible by computing, seeks a developing and more nuanced cooperation with the object—a cooperation which, it is hoped, might enhance social contact and everyday experience.

3. Don't think ease of use, think enjoyment of the experience.

Current efforts on improving usability focus on making things easier. However, there is more to usability than ease of use. A user may choose to work with a product despite it being difficult to use, because it is challenging, seductive, playful, surprising, memorable or rewarding, resulting in enjoyment of the experience. No musician learnt to play the violin because it was easy. Bringing together 'contexts for experience' and 'aesthetics of interaction' means that we do not strive for making a function as easy to access as possible, but for making the unlocking of the functionality contribute to the overall experience.

4. Don't think buttons, think rich actions.

The controls of the current generation of electronic products, whether physical or screen-based, require the same actions. By increasing the richness of actions, controls cannot only be perceptually differentiated, but also motorically. Here again the goal is not differentiation for differentiation's sake, but the design of actions, which are in accordance with the purpose of a control.

5. Don't think labels, think expressiveness and identity.

Not only do current electronic products themselves look highly similar, their controls, whether physical or screen-based, also are often hard to tell apart. This has made it necessary for controls to be labeled with explanatory texts and icons, which are either illegible or unintelligible, regardless of whether they are physical or screen-based. We think that instead designers should differentiate between controls to make them look, sound and feel different. More importantly though, this differentiation should not be arbitrary. The 'formgiving' should express what purpose a product or control serves. This would require a replacement for the current aesthetic with rows of identical controls which so heavily relies on repetition as a means to a achieve a unified and aesthetically pleasing whole, for which the expression of the individual controls are sacrificed.

6. Metaphor sucks.

The use of metaphor has become commonplace in both HCI and product design. 'We could use a such and such metaphor' is an often-heard statement. We think the usefulness of metaphor is overrated. When trying to describe a design in absence of the thing itself it may be necessary to rely on metaphor. But this does not necessarily mean that whilst interacting with the product the user understands the design through one single, consistent metaphor. Gentner and Nielsen (1996) and Gaver (1995) also point out the limits of perfect fitting metaphors. The challenge here is to avoid the temptation of relying on metaphor and to create products, which have an identity of their own.

7. Don't hide, don't represent. Show.

Current product design has a tendency to hide the physical components, even those that are highly informative to a product's operation. A choice is made in favour of an alternative representations rather than physical manifestation.

For example, a videotape becomes completely hidden inside a video recorder when inserted and is then represented on a display (Fig. 3). In photocopiers paper is put inside drawers so that we need sophisticated displays to tell us which paper format lives where. It is the designer's task to make these last remaining physical hold-ons visible and make optimal use of them in the interaction process.

8. Don't think affordances, think irresistibles.

Both the HCI and product design communities have borrowed the term affordances from perception-psychology and have hooked onto mainly its structural aspects whilst neglecting the affective aspects. We lament this clinical interpretation

Fig. 3 First the tape is hidden completely inside the machine, to be then represented on a display

of affordance. People are not invited to act only because a design fits their physical measurements. They can also be attracted to act, even irresistibly so, through the expectation of beauty of interaction.

9. Hit me, touch me, and I know how you feel.

We may slam doors in anger, chew a pen or write with it frantically, sip our coffee or gulp it down in haste. If we design products, which invite rich actions, we can get an idea about the user's emotions by looking at these actions (Wensveen et al. 2002).

10. Don't think thinking, just do doing.

HCI methodologies often separate the cognitive, verbal, diagrammatic and abstract 'thinking' design phase from the visual, concrete, 'doing' phase, and emphasize the former. In product design, 'doing' is seen as equally valid as thinking and as beneficial to the design process even in the very early stages. Handling physical objects and manipulating materials can allow one to be creative in ways that flow diagrams cannot. In the design of the physical, knowledge cannot replace skills. You can think and talk all you want, but in the end, the creation of contexts for experience, the enjoyment and the expressiveness require hands-on skills.

4.2 Examples

Keeping the last of the 10 rules in mind let's turn now our attention to the examples.

In his graduation project, Frens used new methods to explore aesthetics, interaction and role (Djajadiningrat et al. 2000a, b). One of these methods is designing for extreme characters. Designers create products for fictitious characters that are emotional exaggerations. This helps to expose character traits which otherwise remain hidden. For example, Frens used an hedonistic, polyandrous twenty-year old woman as an extreme character. This choice of character required Frens to come up

Fig. 4 Appointment manager (top-left). Public and private mode (top-right). With rotating ring around the top screen (bottom-right). Boyfriend profiles (bottom-left)

with an appointment manager which allows the woman to maximize the fun in her life and which supports her in juggling appointments with multiple boyfriends who may not know of each other. In his final design, Frens aimed to achieve aesthetics of interaction by treating hardware and on-screen graphics as inseparable. The user navigates through time by means of a rotatable ring, which sits around the top screen (Fig. 4).

The appointment manager of the polyandrous woman makes use of five circular screens, which fold up in a fan-like manner. To support the woman in her polyandrous behaviour, the fan is usable in two modes. In the first mode, which is called public mode, all the screens are folded in and only the top screen is visible. This is the mode, which she can use without worries while amongst other people. In the second mode, called private mode, the screens are folded out. In this mode the woman can check upon sensitive information.

Through the playful positioning of the screens, the woman can rate and compare her boyfriends on a fun profile with issues such as dining, shopping, partying, sex etc. The appointment fan fits the twenty-year-old's attitudes. It helps her maintain her hedonistic lifestyle by remembering attributes of boyfriends and allowing her to adjust these through an uncomplicated, playful interface. The dual modes allow her to use the device in public without disclosing the details of her agenda, satisfying her special need for privacy.

The direct coupling between the rotation of the ring and the flow of characters over the screen makes for a beautiful interaction. Through the positioning of the

Fig. 5 An alarm clock
(design: De Groot and Van de
Velden)

multiple screens, the woman can rate her boyfriends on various issues such as shopping, dining, sex etc. in a playful manner. These aspects of the design show respect for the user's perceptual-motor skills, not only from a structural but also from a fun point of view.

The next example (Fig. 5) is an alarm clock from a student exercise. The alarm clock consists of two parts, a base station and an alarm ball. The alarm ball is used to set the wake up time and consists of a display strip flanked by two rotating semi-spheres. If the left hemisphere of the alarm ball is turned while holding the display strip, the hour of the waking time is adjusted. If the right hemisphere is rotated, the minutes are adjusted. The size of the ball and the way it matches the recess in the base station afford picking up and the two halves afford rotation. But more importantly, the positioning of the halves adjacent to the hour digits and the minute digits, informs the user of what he will adjust.

The alarm clock can sense the distance between the base station and the alarm ball. The further the user moves or throws the alarm ball from the base station, the louder, the more aggressive and the more insistent the waking sound may be in the morning. The closer the alarm ball is placed to the base station, the softer and more gentle the waking sound will be. Here it is both the appearance and the actions that are carriers of meaning. Throwing the ball to the other side of the room is a different action from placing it just to the side of the base station and can thus have different consequences. This is also consistent with the actions the user has to carry out to silence the alarm clock. The further the alarm ball is away from the base station, the more of an effort he has to make to find it, to pick it up and to place it over the speaker to muffle the sound. Here again the fit of the alarm ball to the recess and the idea of covering the loudspeaker inform the user of the consequences of his action. The user's actions thus become carriers of meaning and influence the alarm's behaviour.

If the left hemisphere of the alarm ball is turned while holding the display strip, the hours of the waking time are adjusted (top Fig. 6). If the right hemisphere is rotated, the minutes are adjusted (bottom). The alarm clock can sense the distance

Fig. 6 Turning of the hemispheres

between the base station and the alarm ball. The further the alarm ball is placed away, the more insistent the sound will be in the morning. The user's actions thus become carriers of meaning and influence the alarm clock's behaviour.

5   Conclusions

This chapter summarises a position that has been developed over many years and a number of projects. Our work can be thought of as a manifesto for design. Our arguments are deliberately provocative. For too long psychologists have led designers to make overly cognitive designs. We repeat: design should be left to designers! Too often fun is a "glued on" property and interfaces smile. Enjoyment should not be an afterthought and fun does not have to be cute. In order to design enjoyable products we must design for engagement on every level and the physicality of products must be restored, product design must address the user's action potential and capacity to appreciate sensory richness. Products must elicit rich interaction from the user. In this way, not only the functionality but beauty and fun in interaction are opened up. And there is more. Rich physical interaction offers even more possibilities. Products might 'read' the user's emotions and react to it in different ways. (See for example Wensveen and Overbeeke, Chap. 24 in this book).

We believe our approach frees products from clumsy interaction and opens ways to beauty and fun. From a product design perspective, the appearance of interactive products can no longer be considered as arbitrary. A tight coupling between action and appearance in interaction design is necessary. Appearance and interaction need to be designed concurrently.

Acknowledgements Part of this chapter was published earlier in Djajadiningrat et al. (2000a, b). The Frens project was conducted in collaboration with Bill Gaver of RCA London. Most of the work reported was done when all authors were affiliated to the Delft University of Technology. Almost 15 years after the intional publication, we are still exploring and developing engaging designs, although our scope has expanded to product-service systems and large socio-technical systems. Out of respect for Kees Overbeeke, we haven't introduced any new work and refer you to more recent publications to see how we have continued his legacy. Kees was dreaming of the impossible, trying to rebalance thinking and doing, and many of his ideas have been adopted. Kees, your legacy will stay in our hearts, minds and our designs.

References

Cooper A (1999) The inmates are running the asylum. Indianapolis: SAMS. McMillan, Sams

Djajadiningrat JP, Gaver WW, Frens JW (2000a) Interaction relabelling and extreme characters: methods for exploring aesthetic interactions. In: Proceedings of DIS'00, designing interactive systems. ACM, New York, pp 66–71

Djajadiningrat JP, Overbeeke CJ, Wensveen SAG (2000b) Augmenting fun and beauty: a pamphlet. In: Mackay WE (ed) Proceedings of DARE'2000. Helsingor, pp 131–134

Djajadiningrat JP, Overbeeke CJ, Wensveen SAG (2002) But how, Donald, tell us *how*? In: Macdonald N (ed.) Proceedings of DIS2002, London, 25–28 June 2002, pp 285–291

Djajadiningrat JP, Wensveen SAG, Frens JW, Overbeeke CJ (2004) Tangible products : redressing the balance between appearance and action. Pers Ubiquit Comput 8(5):294–309

Dunne A (1999) Hertzian tales: electronic products, aesthetic experience and critical design. RCA CRD Research publications, London

Gaver WW (1995) Oh what a tangled web we weave: metaphor and mapping in graphical interfaces. In: Adjunct proceedings of CHI'95, pp 270–271

Gentner D, Nielsen J (1996) The anti-mac interface. Commun ACM 39(8):70–82

Gibson JJ (1986) The ecological approach to visual perception. Lawrence Erlbaum, Hillsdale, NJ

Hummels CCM (2000) Gestural Design Tools: prototypes, experiments & scenarios. Doctoral dissertation. Delft University of Technology

Overbeeke CJ, Djajadiningrat JP, Hummels CCM, Wensveen SAG (2002) Beauty in usability: forget about ease of use! In: Green WS, Jordan PW (eds) Pleasure with products: beyond usability. Taylor & Francis, London, pp 9–18

Overbeeke CJ, Hummels CCM (2013) Industrial design. In Soegaard M, Dam RF (eds) The encyclopedia of human-computer interaction: the interaction design foundation, 2nd edn. Aarhus, Denmark, pp 237–328. Available online at http://www.interaction-design.org/books/hci.html

Wensveen SAG, Overbeeke CJ, Djajadiningrat PJ (2002) Push me, shove me and i know how you feel. Recognising mood from emotionally rich interaction. In: Macdonald N (ed) Proceedings of DIS2002, London, 25–28 June 2002, pp 335–340

Chapter 18
The Engineering of Experience

Phoebe Sengers

Author's Note, Funology 2

My first experience in HCI was at the CHI 2002 workshop that led to this book. The Funology workshop was a propitious place to begin, because the people involved were asking questions that would soon change the field.

At the time, HCI had an overwhelmingly task-focused orientation; work was its assumed proper domain. While ethnography had made a headway at CHI, the mainstream was cognitive, behavioral, and computational. But new conversations were developing. Dourish's newly published *Where the Action Is* laid out how interaction is inherently socially framed (Dourish 2001). Gaver was developing designs that radically challenged HCI's task-focused orientation (e.g. Gaver and Martin 2000). Soon, McCarthy and Wright would publish *Technology as Experience*, which leveraged philosophy to reframe issues in interaction from 'use of technology' to 'living with technology' (McCarthy and Wright 2004). Blythe would present ever more radical uses of literary theory to create new possibilities for design (e.g. Blythe 2004). Höök would develop new design strategies for creating open-ended emotional engagement (e.g. Fagerberg et al. 2004). Taylor and Swan would publish home studies that raised questions about technology and gender politics (e.g. Taylor and Swan 2005). A few years later, Bardzell and Bardzell would publish "Interaction Criticism," making humanistic criticism salonfaehig in HCI (Bardzell and Bardzell 2008). And these are only a few of many landmarks in what became a rich conversation on user experience, the role of technology in everyday life, the politics of design, and interaction in the wild, drawing not only from existing methods in HCI but also from product design, critical theory, and beyond.

P. Sengers (✉)
Information Science and Science and Technology Studies, Cornell University,
Ithaca, USA
e-mail: sengers@cs.cornell.edu

© Springer International Publishing AG, part of Springer Nature 2018
M. Blythe and A. Monk (eds.), *Funology 2*,
Human–Computer Interaction Series,
https://doi.org/10.1007/978-3-319-68213-6_18

In this chapter, I argued that there were limitations to taking task optimization as the guiding principle for both the topic of HCI and its methods. I described heuristics for approaching design for experience which took into account its messy situatedness and which left the final authority for experience in the hands of users. I argued that engineering of systems had to be coupled with humanistic analysis of the cultural situation of design. Now, 15 years later, these ideas are commonplace—not because of this chapter, but because of the conversations it took part in.

Looking back, it is remarkable to see how this chapter already contained the early germs of the ideas that later became my research group's major contributions to that conversation. It contains the core idea of reflective design (Sengers et al. 2005), i.e. using design and evaluation to support and encourage critical reflection by designers and users. It also lays out principles for affect as interaction (Boehner et al. 2005), i.e. supporting emotional interaction through technology design by placing users, rather than computers, as the final authorities on how they feel.

But I am also now struck by a significant shortcoming of this chapter: its one-sided, simplistic explanation of how Taylorism and the drive for efficiency have shaped contemporary life. I have since come to realize that the cultural impact of technology is much more complicated. Williams's auto-ethnography of persuasive fitness technology, for example, showed that users are certainly influenced by the narratives of optimization that drive its design; but they also fall in love with their data, argue against it, behave irrationally about it, use it to actively construct desired selves, and otherwise make it more interesting than its designers could ever have imagined or perhaps wished (Williams 2015). The empirical reality of how users come to live with technology, our group is finding, is much more complex and fascinating than the cleaner, simpler technological visions that motivated their design.

So I return to this chapter with an empirical, rather than theoretical, understanding of modern labor. I recognize the imperatives of efficiency and control it explores not as absolute and straightforward demands, but as dreams and visions invoked differentially in particular times and places. When thinking of the engineering of experience, my group now asks, "Whose experience?" and "Who is engineering?" The answers are different for white-collar and blue-collar professionals, for rural and urban dwellers, for residents of Silicon Valley and Jamaica or Iceland. The chapter ended with a call to "learn to love complexity and speak its language." And this is what, after 15 years, we still think is worth doing, and we are still learning to do.

2003 Chapter

A deep shift in Western culture has occurred in the last 200 years. We have moved from lifestyles in which work, play, and other forms of experience are inextricably intertwined, to one in which most people separate their work life from a private (and often less societally valued) life of fun and play. Engineering has played a central role in this bifurcation, fulfilling a cultural desire to engineer human experience for optimal functionality. The result has been a great increase in our material comforts, coupled with a harried, frenzied lifestyle for many. In this chapter, I will argue that designing systems to support rich, meaningful, and pleasurable human experiences requires moving away from the model of engineering experience and towards an interdisciplinary approach to computing, in which technology design is intertwined with philosophical and cultural analysis.

1   Fun Is the Dregs of Engineering Experience

The history of the industrial revolution is a story of the gradual optimisation and rationalization of work. Over the last two centuries, work has gone from an integral part of daily life, to something which is bought and sold per hour and engaged in standardized ways. Craftspeople were collected into factories, their work was split into pieces along a production line, some steps of the production line were taken over by machines, and gradually craftspeople became tenders of rote machinery, engaged in soulless work.

This shift is epitomized by the work of the efficiency expert Frederick Winslow Taylor, who in the early twentieth century developed the system of scientific management or 'Taylorism.' Taylorist engineers maximize the efficiency of human labour by observing workers, analysing their movements, and developing a script for the 'one best way' to achieve their work tasks, in the process eliminating all unnecessary or wasteful motions. After the development of the assembly line, which rationalized and optimised machine labour in the production process, the last source of inefficiency in factories was human labour. Businessmen were naturally eager to find ways to reduce this inefficiency, a task which Taylorism solved.

After Taylorist analysis, a worker is told not only the steps to take in order to fulfil a task, but also what order to do those steps in and exactly how to move in order to minimize waste in their work. Because of the mindless, rote nature of Taylorized work, the quality of experience of work is reduced. Rote labour causes both repetitive stress injuries and rebellious, unhappy workers. Offsetting this reduction in experience is a drastic increase in its efficiency and productivity. Because of these great increases in efficiency, Taylorism took the business world by storm. The impact of Taylorism on Western, especially American, culture can hardly be underestimated. It is still felt through later, less extreme manifestations such as ergonomics and time management. Despite the problems of Taylorism, many of us have remained with a model of work in which experience is engineered for maximum efficiency and minimum pleasure. We have also imported these models to the home: to-do lists, appointment calendars, and a clutter of chores regiment our home lives and attempt to ensure that we are as efficient at home as we are at work.

Engineering work leads to a bifurcation of experience. As Blythe and Hassenzahl argue in this volume, if work, on the one hand, maximizes efficiency at the cost of pleasure, we balance out in our free time by engaging in fun: maximizing pleasure and minimizing task achievement. Many of us spend 8–10 h days working efficiently and unhappily, then race home for a mindless evening in front of the TV or Playstation. In the post-industrial West, and especially in America, we have split experience into two: whereas life could be a steady stream of work intermingled with pleasure, we have disengaged the two, often preferring to lavish 'serious' attention only on the first.

2 Computer Science Is Computational Taylorism (but Doesn't Need to Be)

A similar split and imbalance has occurred in computer science. Taylorism is, at heart, simply engineering applied to human behaviour; hence it is no surprise that computer scientists tend to approach work processes the same way as a Taylorist. We break complex processes down into simple steps, we figure out optimal procedures for each work step, and we eliminate wasteful steps and problems.

This process is most clearly seen in Artificial Intelligence, in which both classical planning and the newer behaviour-based approaches attempt to engineer experience by increasing the efficiency and optimality of algorithms and to maximize their functionality (Sengers 1998, 2004). But we see similar emphases in human-computer interaction (HCI). On the one hand, it has a strong emphasis on work-related tasks and increasing the efficiency of their execution. On the other hand, it often focuses on rationalized and optimised techniques to understand and engineer human experience even when the goal is fun.

Engineering is the correct approach to take when there is a well-defined task to be solved. But designing systems that open a space for new kinds of experience is not an engineering task per se. Instead, one must consider the technical challenges to be overcome in the context of the kinds of cultural and social meaning that the system may take on and the ways in which users may choose to interact with it. This necessitates a shift from a pure, task-oriented engineering approach to an interdisciplinary approach that combines socially-oriented approaches such as the social sciences or literary and cultural studies with more traditional human-computer interaction and computer science. Such hybrid approaches are becoming popular both within HCI and in the media art community (see e.g. Ehn 1998; Wilson 2002).

3 Think Beyond Both Work and Fun

The pendulum between work and play is beginning to swing in the other direction. The recent interest in 'fun' as manifested by this volume is important in opening up an understanding of some of the unstated work-related assumptions underlying HCI methods. Funology will necessitate fundamental rethinking of some HCI approaches and the development of new techniques that are less about efficiency and more about quality of experience.

Nevertheless, Funology is not enough. Rather than continuing the bifurcation of experience into work versus play (traditional HCI vs. Funology), as a culture we need to consider systems that take a more integrative approach to experience. This may mean on the one hand systems like that of Hohl, Wissman and Burger in this volume that combine work-related task achievement with pleasurable experience. More fundamentally, it also means that we need to explore the vast and utterly neglected territory of possible systems that are really *neither* work *nor* fun. Such systems may support reflection by users on their lives and activities; they may give users new ways to experience the world; they may make cultural comments in the form of interactive artworks. These systems are neither directly task-related, nor intended simply to entertain. They have a serious point, but they may bring their point across in a playful manner. Examples of such work include Bill Gaver and Heather Martin's conceptual information appliances (Gaver and Martin 2000), which explore the role of technology in our everyday lives; Tony Dunne and Fiona Raby's electronic furniture (Dunne and Raby 2001), which provide people with different ways to sense and respond to activity in the electromagnetic spectrum; and Simon Penny's Petit Mal (Penny 2000), an artwork exploring the nature of artificial agents through a gangly and not very bright robot whose complex and graceful physical activity is almost entirely triggered by human bodily interaction. What these systems have in common is not a desire to engineer experience, but to build thoughtful artefacts that create opportunities for thinking about and engaging in new kinds of experiences. We need to shift from engineering experiences—whether work or fun—to designing them, using principles that draw on both technology design methods and social and cultural analysis.

4   Some Experiences Designing Experiences

In this section, I will describe some experiences in designing systems that are intended to support richer and more meaningful notions of human experience than those traditionally used in computer science by using a broader, interdisciplinary approach combining computer science with cultural analysis. My first work in this area was in designing Artificial Intelligence (AI) architectures for interactive computer characters. Traditionally, AI focuses on activity in the world as problem-solving rationality. The goal for autonomous agents is often to behave optimally rationally in approaching some goal. For interactive computer characters, this focus is problematic, since characters do not need to be particularly smart or rational, instead needing to project emotion and personality in a way that is understandable to users. In the Industrial Graveyard (Fig. 1), I explored how to create agents, not as rational problem-solvers, but as experienced by human users (Sengers 1999). Users observe the antics of a discarded lamp in a junkyard, while controlling the behaviour of its unsympathetic overseer. The technology is based on narrative psychology, which argues that humans interpret activity by organizing it into narrative. I support human interpretation of character action by providing

Fig. 1 The hero of the
Industrial Graveyard

visible cues for narrative interpretation of agent behaviour, most notably through transitions between behaviours that connect them by expressing the reason for the behaviour change to the user.

With the Industrial Graveyard, I started out being interested in how human experience was represented in agents; but in the course of building the system, I began to realize that what was central was the way in which the *user* experienced the system. The next system I worked on explored ways to generate engaging user experiences. With a team of 5 researchers led by Simon Penny, I explored the construction of physical experiences in virtual reality (Penny et al. 2001). In Traces, users' body movements generate 3-dimensional Traces which share their physical space, and with which they can interact (Fig. 2).

In traditional VR systems, the body is an afterthought, left behind when the headset is put on. The goal of Traces was to develop a kind of VR installation where it is possible instead to have strong bodily experiences. Traces is an

Fig. 2 Concept of Traces:
user movement through space
leaves behind 3-dimensional
Traces

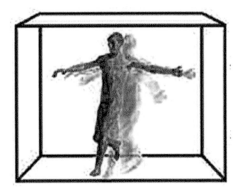

Fig. 3 A user (body model shown in black) moving through Traces leaves behind tracks of physical movement (grey). In the CAVE the voxel model is not shown; instead, the user's own body leaves behind colourful 3-dimensional Traces in the space surrounding him or her. Users are surrounded by the trace they create

installation for the CAVE VR display, a small room onto whose walls 3D images are projected. When users enter wearing 3D glasses, they have the illusion of being surrounded by virtual objects in real, physical space, while they can still look down and see their own bodies. In Traces, vision cameras detect the movement of users, allowing them to leave behind and interact with Traces of physical movements that seem to surround them (Fig. 3).

Gradually, the Traces become more autonomous, turning into "Chinese dragons" which flock together and sense and react to users' physical movements (Fig. 4).

Traces was installed at Arts Electronica '99, where users leapt, ran, skipped, did cartwheels, and came out of the CAVE sweating. Users had strong reactions to the Chinese dragons; though the dragons were not particularly intelligent, they seemed strongly alive and present to human users because they shared the same physical space. With Traces, it became clear that physical interaction and shared physical space with (embodied) users is a way to create meaningful, powerful experiences.

The Influencing Machine (Sengers et al. 2002) explores the human experience of affective computing, or computational systems that recognize, reason about, or can express emotions (Fig 5). It was developed by the author, Rainer Liesendahl, Werner Magar, and Christoph Seibert at the MARS Exploratory Media Lab as part of the EU SAFIRA project. In the Influencing Machine installation, users enter a small room, on one wall of which childlike drawings are being created in real-time, accompanied by an abstract soundscape. In the middle of the room, they discover a wooden mailbox, into which they can put art or coloured postcards. By choosing

Fig. 4 A user (body model shown in black) being chased by a set of Chinese dragons (grey). In the actual experience in the CAVE, the user does not see their voxel model, but only sees Chinese dragons sharing their physical space and responding directly to their physical movement

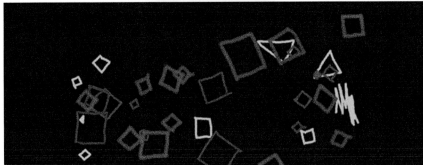

Fig. 5 Input and output of the Influencing Machine

postcards, they can change the "mood" of the drawings and sounds as they are being created. Users explore the postcards, asking themselves what the picture means to them, and exploring what it means to the machine. With the Influencing Machine, we came up with ways to engineer *enigmatic* experience: the interaction is deliberately open- ended and open to interpretation, yet through the interaction of postcards, graphics, and sound, we can create experiences which have concrete meaning for many users. The Influencing Machine was formally evaluated by Gerd Andersson, Pia Mårtensson, and Kristina Höök, who developed new techniques for non-task-oriented evaluation for this project, most notably by using groups of users and recording their conversations in order to better understand the nature of user experience of the system.

These three systems are all examples of critical technical practices (Agre 1977), or practices of technology development which incorporate a cultural, critical component. In all three cases, we built on an analysis of what was missing in the cultural assumptions about human experience that were unconsciously built into previous technology. The Industrial Graveyard twists the notions of optimality, correctness, and action-selection inherent in many algorithms for autonomous agents. Traces alters the assumption of bodilessness behind many VR applications. The Influencing Machine plays off of the assumption in many affective interfaces that "affect" is something to be extracted through surveillance or skin contact, and instead places the user's own choices at the centre of affective interaction. I believe building rich, meaningful experiences will require not just engineering competence but also cultural analysis, design, and art perspectives.

5 How to Design Experience

A pure engineering approach suggests that one can understand human experience by building formal models of it—the traditional approach taken by computer science. In AI, for example, we build conceptual models of people, implement these in code, and run them, in the hopes of better understanding what human experience is like. In HCI, we similarly often build cognitive models of users, allowing software to reason about what users may be experiencing by comparing their behaviour with expectations of human behaviour built into the cognitive models. In many ways these attempts to make human experience computational mimic the efforts of the Taylorists, as we try to clean up, formalize, and organize what is an inherently messy and perhaps fundamentally incomprehensible phenomenon. In my own work as an AI researcher, I became frustrated by the fact that my clean, beautiful models of behaviour always seem to miss the point—they can somehow never generate the complexity and richness of natural behaviour of humans or animals.

The perspectives of the arts and humanities also suggest the futility of trying to formally represent experience. Many humanists and artists feel that complexity, messiness, ill-definedness, and enigmatics are fundamental to the nature of human experience, and that therefore all clean and formal models fundamentally distort that nature. Winograd and Flores's rejection of AI (Winograd and Flores 1986), for example, is fundamentally based on this point. If this is the case, then how can we design systems that can create rich, meaningful, and complex experience for users? I believe we must do so by realizing that we cannot fully represent experience within the software, and instead try to set up more nuanced relationships between (internal, formal, clean) code and (external, messy, complicated) experiences. More concretely, I suggest the following (nonexhaustive) set of heuristics:

- *Instead of representing complexity, trigger it in the mind of the user.*

Instead of trying to contain the complexity of user experience in formal structures such as user models, one should focus on shaping the *actual* (not modelled) experience of the user, which will hopefully be much more complex than its internal, logical representation. One way to do this is to focus on the user's strength: an ability to engage in complex interpretation using a vast amount of cultural background knowledge. By focusing on how users react, rather than on the internal content of the software, a simple computational artefact can be used to communicate a rich and complex idea. In the Industrial Graveyard, I take advantage of the user's 'narrative intelligence' by providing 'hooks' that support narrative understanding of agent behaviour. The agent architecture is structured to support symbolic, narrative interpretation, rather than internal optimality, efficiency, and completeness. Systems built using this heuristic have behaviour that is internally simple, but appears complex to users thanks to the complexity of human interpretation.

- *Instead of representing complexity, bootstrap off it.*

Human behaviour is rich, complex, messy, and hard to organize into rules and formal models. This insight can be used to create rich, complex, messy, and subtle *computational* behaviour with little computational cost simply by driving it directly from human behaviour. For example, in Traces the motion of the Chinese dragons is based on simple rules that respond to human movement. Because human movement is complex and the dragons are responding to it in real time, the movement of the dragons is similarly complex. Unlike the previous heuristic, systems built using this heuristic truly have complex behaviour—but only because they are driven by complex input.

- *Think of meaning, not information.*

Computers care about information. Humans care less about raw data than they do about what information means to *them*. Focusing on meaning instead of information in the design of computational objects means that we adapt to user experience of information rather than to its internal representation. With the Influencing Machine, for example, we tried to move away from the standard affective computing model in which 'emotion' is fundamentally a unit of information to be extracted, manipulated, and communicated, to one in which user interpretation of emotionally valenced postcards, graphics, and sound is central, with the internal informational representation of emotion playing only a supporting role.

6   The Engineering of Everyday Life, or Where's the Fun?

How do these heuristics extend to work outside of the museum and beyond AI? One domain in which non-engineering approaches are clearly needed is in everyday life in the home. In current discussions in HCI, the move of computation from desktops and factories into the home and everyday life is considered motivation to alter the efficiency- and task-oriented approaches on which HCI has largely concentrated in favour of fun- and pleasure-based approaches. Underlying this argument is an assumption that Taylorist models may be appropriate for work, but that they do not apply to the home. Yet the historical record makes clear that Taylorist models are central to current home life in the West, especially in America—to the detriment of our quality of life (Bell and Kaye 2002).

Already at the turn of the century, the popularity of Taylorism and the model of factory production which was rapidly and fundamentally changing the American way of life led to attempts to adapt Taylorism to home life (Strasser 1982). Home economist Christine Frederick was perhaps its most ardent proponent; she proposed that housewives engage in motion studies to minimize the amount of movement they spent on household chores such as washing the dishes or doing the laundry. This attempt to adapt Taylorism to the home ran into resistance, for several reasons. First, many home tasks, such as minding children, are not amenable to being fully engineered and controlled. Second, there was no clear reason why housewives needed to be so efficient. Focusing purely on efficiency causes a great reduction in the experience of work, and many housewives saw no need to get more done at the cost of less pleasure and a more unnatural work process.

The situation today has changed. As Ralph Keyes notes,

> At one time the home was considered a refuge from work pressures. Now its inhabitants March to a businesslike beat. The pace at home has become little different from that at work. It calls for huge calendars on the kitchen wall, constant cross-checking of everyone's schedules, and sophisticated use of complex telephone systems so everyone can stay coordinated.... The tempo of the office and much of its paraphernalia – datebooks, Rolodexes, phone systems, computers, even faxes – have invaded the home. (p. 141)

In fact, especially for two-career families, efficiency and engineering an optimal task schedule are as important at home as at work. What role will HCI play in this? Will we continue the engineering approach, building domestic technology that will allow harried families to cram yet one more activity into a busy schedule, alternating with stress reduction through mindless fun? Or will we design experiences for users that counteract these cultural forces, developing an alternative vision of what home life could be like?

7 Don't just Engineer—Learn to Love Complexity and Speak Its Language

Building computational artefacts that support rich and meaningful human experience requires a variety of perspectives to be combined. Engineering, including technology and algorithm design, is essential in order to be able to turn the vision of an interaction into a functioning system. Traditional and newly created human-computer interaction techniques are needed to support the fine-tuning of interaction and to evaluate the effect the system may have on users. But 'engineering' truly rich experiences requires more of system designers than just technical skills. System designers also need to understand and design for the ways in which user experience exceeds our abilities to formalize it. They can't just love their code; they must learn to love the complexity of user experience as well and be conversant in it. This suggests the incorporation of practices like cultural studies, anthropology, speculative design, surreal art, culture jamming, story-telling, cultural history, sociology, improvisation, and autobiographies, which have found ways to address and understand the complexity of human experience without needing to create complete and formal models of it. Most importantly, it means recognizing the role, not just of fun but of serious *play* as a form of opening the conceptual space for designing, building, and interacting with the new systems with which we will share our lives.

Acknowledgements The Industrial Graveyard was supported in part by an ONR Allen Newell Fellowship. My work on Traces was supported by a Fulbright fellowship. The Influencing Machine is part of the EU SAFIRA project. The systems described here were built at Carnegie Mellon University, the Center for Art and Media Technology (ZKM), at the German National Information Technology Research Center (GMD), and Cornell University.

References

Agre PE (1977) Computation and human experience. Cambridge UP, Cambridge
Bardzell J, Bardzell S (2008) Interaction criticism. In: CHI '08 Extended abstracts on human factors in computing systems (CHI EA '08). ACM Press, NY
Bell G, Kaye J (2002) Designing technology for domestic spaces: a kitchen manifesto. Gastronomica 2(2)
Blythe M (2004) Pastiche scenarios. Interactions 11(5):51–53
Boehner K, DePaula R, Dourish P, Sengers P (2005) Affect: from information to interaction. In: Proceedings of the 4th decennial conference on critical computing. ACM Press, NY, pp 59–68
Dourish P (2001) Where the action is. MIT Press, Cambridge
Dunne A, Raby F (2001) Design noir: the secret life of electronic objects. August/Birkhaeuser, Basel, Switzerland
Ehn P (1998) Manifesto for a digital bauhaus. Digital Creativity 9(4):207–216
Fagerberg P, Ståhl A, Höök K (2004) eMoto: emotionally engaging interaction. Pers Ubiquit Comput. 8(5):377–381
Gaver Bl, Martin H (2000) Alternatives: exploring information appliances through conceptual design proposals. In: Proceedings of the CHI 2000 conference on human factors in computing systems. ACM Press, NY, pp 209–216

McCarthy J, Wright P (2004) Technology as experience. MIT Press, Cambridge

Penny S (2000) Agents as artworks and agent design as artistic practice. In Kerstin D (ed) Human cognition and social agent technology. John Benjamins, Amsterdam

Penny S, Smith J, Sengers P, Bernhardt A, Schulte J (2001) Traces: embodied immersive interaction with semi-autonomous avatars. Convergence 7(2), Summer

Sengers P (1998) Anti-boxology: agent design in cultural context. PhD Thesis, Carnegie Mellon Department of Computer Science

Sengers P (1999) Designing comprehensible agents. In: 1999 international joint conference on artificial intelligence (IJCAI-99). Stockholm, Sweden

Sengers P (2004) The agents of McDonaldization. In: Payr S, Trappl R (ed) Agent culture. Lawrence Erlbaum Associates, pp 3–19

Sengers P, Liesendahl R, Magar W, Seibert C, Müller B, Joachims T, Geng W, Mårtensson P, Höök K (2002) The enigmatics of affect. In: 2002 conference on designing interactive systems. ACM Press, NY

Sengers P, Boehner K, David S, Kaye J (2005) Reflective design. In: Proceedings of the 4th decennial conference on Critical computing. ACM Press, NY, pp 49–58

Strasser S (1982) Never done; a history of american housework. Pantheon Books, NY

Taylor A, Swan L (2005) Artful systems in the home. In Proceedings of the SIGCHI conference on human factors in computing systems (CHI '05). ACM Press, NY, pp 641–650

Williams K (2015) An anxious alliance. In: Proceedings of the fifth decennial aarhus conference on critical alternatives. Aarhus University Press, pp 121–131

Wilson S (2002) Information arts: intersections of art, science and technology. MIT Press, Cambridge, MA

Winograd T, Flores F (1986) Understanding computers and cognition. Ablex, Norwood, NJ

Chapter 19
The Thing and I: Understanding the Relationship Between User and Product

Marc Hassenzahl

1 Introduction

We currently witness a growing interest of the Human-Computer Interaction (HCI) community in *user experience*. It has become a catchphrase, calling for a holistic perspective and an enrichment of traditional quality models with non-utilitarian concepts, such as fun (Monk and Frohlich 1999; Draper 1999), joy (Glass 1997), pleasure (Jordan 2000), hedonic value (Hassenzahl 2002a) or ludic value (Gaver and Martin 2000). In the same vein, literature on experiential marketing stresses that a product should not longer be seen as simply delivering a bundle of functional features and benefits—it provides experiences. Customers want products "that dazzle their senses, touch their hearts and stimulate their minds" (Schmitt 1999, p. 22). Experiential marketing assumes that customers take functional features, benefits, and product quality as a given.

Even though the HCI community seems to embrace the notion that functionality and usability is just not enough, we are far from having a coherent understanding of what user experience actually is. The few existing models (e.g., Logan 1994; Jordan 2000) of user experience in HCI that incorporate aspects such as pleasure are rare and often overly simplistic. In the present chapter, I will propose a more complex model that defines key elements of user experience and their functional relations. Specifically, it aims at addressing aspects, such as (a) the subjective nature of experience per se; (b) perception of a product; (c) emotional responses to products in (d) varying situations. It is a more detailed and further developed version of a research model, I previously presented in Hassenzahl (2002a). I view it as a first step towards a better understanding of how people experience products and a valuable starting point for further in-depth theoretical discussions.

M. Hassenzahl (✉)
Ubiquitous Design/Experience and Interaction, University of Siegen, Siegen, Germany
e-mail: marc.hassenzahl@uni-siegen.de

© Springer International Publishing AG, part of Springer Nature 2018
M. Blythe and A. Monk (eds.), *Funology 2*,
Human–Computer Interaction Series,
https://doi.org/10.1007/978-3-319-68213-6_19

2   A Model of User Experience

Figure 1 shows an overview of the key elements of the model of user experience from (a) a designer perspective and (b) a user perspective.

A product has certain *features* (content, presentational style, functionality, interactional style) chosen and combined by a designer to convey a particular, *intended* product character (or gestalt; Janlert and Stolterman 1997; Monö 1997). A character is a high-level description. It summarizes a product's attributes, e.g., novel, interesting, useful, predictable. The character's function is to reduce cognitive complexity and to trigger particular strategies for handling the product. When individuals come in contact with a product, a process is triggered. First, people perceive the product's features. Based on this, each individual constructs a personal version of the product character—the *apparent* product character. This character consists of groups of *pragmatic* and *hedonic* attributes. Second, the apparent product character leads to *consequences*: a judgment about the product's appeal (e.g., "It is good/bad"), emotional consequences (e.g., pleasure, satisfaction) and behavioural consequences (e.g., increased time spend with the product). However, the consequences of a particular product character are not always the same. They are moderated by the specific usage situation. In the following, each key element is discussed in detail.

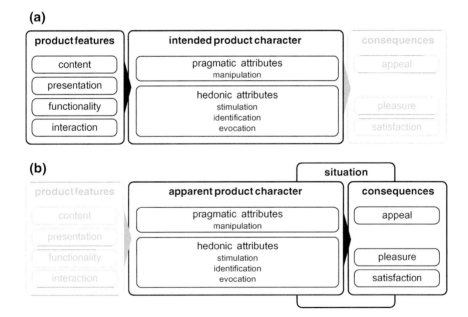

Fig. 1 Key elements of the model of user experience from **a** a designer perspective and **b** a user perspective (for details refer to text)

2.1 From the Intended and Apparent Product Character to Consequences

A product designer "fabricates" a character by choosing and combining specific product features, i.e., content, presentational style, functionality, interactional style. However, the character is subjective and only *intended* by the designer. There is no guarantee that users will actually perceive and appreciate the product the way designers wanted it to be perceived and appreciated. For example, a product with a specific screen layout intended to be "clear" will not necessarily be perceived as "clear." A suitable design process must assure that an appropriate product character is selected and that this character is properly communicated to the user (see Hassenzahl 2002b). For online banking, for example, an appropriate character may consist of attributes such as "trustworthy", "sober", and "clear". The features (e.g., tone of voice, screen layout, colours, news ticker) have then to be chosen and combined by the designer according to the character to be communicated.

When users are confronted with a product, a process is triggered: First, an apparent product character is constructed. It is a user's personal reconstruction of the designer's intended product character. Second, the fit of the apparent character and the current situation will lead to consequences, such as a judgment about the momentary appealingness of the product, and emotional or behavioural consequences.

People *construct* the apparent product character based on the particular combination of product features and their personal standards and expectations. A personal standard most likely consists of other objects the product can be compared to. Variations of the character *between* individuals can be explained by differing standards. The apparent character can also change *within* a person over time. This change is due to increasing experiences with the product. For example, a product that was perceived as new and stimulating in the beginning may lose some novelty and ability to stimulate over time. Conversely, with increasing experiences products originally perceived as unusable may become more familiar and, thus, might be perceived as easier to handle. To date, not much is known about how perceptions of products will change over time. However, the specific way of change, i.e., direction and rate may be an integral part of a product's character extended over time.

Using a product with a particular product character in a particular situation has certain emotional and behavioural *consequences*. In some situations, for instance, to be novel is appreciated in a product; in others it can be neglected or even unwanted. Depending on the situation, character attributes become more or less relevant. The value of a product can be expressed by the user as judgments of appealingness or may manifest itself as emotions (see Sect. 2.3). Compared to perceptions, consequences do vary more strongly because of their embedding into a particular usage situation. Consider, for example, an automated teller machine (ATM) designed to be highly understandable. To achieve this, the designers divided the process of receiving money into a number of small steps. If they got it right, you will perceive the ATM as highly understandable. The first time you try to get money from the

ATM, you will certainly value this attribute. It will add to you satisfaction. Now imagine yourself being more experienced with the ATM or even under time pressure. The succession of small steps slows down interaction, and although you still perceive the ATM as understandable, this attribute is not relevant at the moment. It may even frustrate you. In this example, an individual's (your) appraisal of the ATM strongly varies (from satisfaction to frustration) because of the particular usage situation, whereas the perception of the ATM as understandable remains relatively stable.

The whole process of perceiving and constructing the character and experiencing consequences will *always* take place, no matter how insufficient the available information about the product seems to be. A study using the Repertory Grid Technique to gather product characters, for example, showed that people make far reaching inferences about quality and behaviour of online banking Web sites on the basis of simple screen shots or very short interaction sequences (Hassenzahl and Trautmann 2001). However, the outcome of the process, i.e., the inferences made about the product character and the resulting consequences, may change with growing knowledge and experience of the product. This also implies that the process is repeated over and over again.

In the following section, I will take a closer look at two universal groups of attributes that define the product character and the underlying human needs they address.

2.2   Product Character: Pragmatic and Hedonic Attribute

An apparent product character is a cognitive structure. It represents product attributes and relations that specify the co-variation of attributes. It allows inferences beyond the merely perceived. For example, a product with a simple user interface may also be thought of as easy to operate, although the user has no actual hands-on experience. But what groups of attributes can be distinguished? This is best answered by considering the major functions of products: They enable people to manipulate their environments, to stimulate personal development (growth) and to express identity. Moreover, a product can provoke memories and, thus, has a symbolic value.

2.2.1   Pragmatic Attributes: Manipulation

Manipulation of the environment requires relevant functionality (i.e., utility) and ways to access this functionality (i.e., usability). I call this group of product attributes *pragmatic*. Typical pragmatic attributes of software products are "clear", "supporting", "useful" and "controllable". A pragmatic product is primarily instrumental. It is used to fulfil externally given or internally generated behavioural goals. If, for example, somebody asks you to drive a nail into a wall to put up a picture, you use a tool to do so. From a pragmatic perspective, the only requirement for the tool is that it can in principle be used to drive in a nail and that you are able to figure out how to do so.

2.2.2 Hedonic Attributes: Stimulation, Identification, and Evocation

All other remaining product attributes I subsume as *hedonic*. I have chosen this term for two reasons: first, it is meant to highlight that hedonic attributes and the underlying functions of the product strongly differ from pragmatic attributes. Whereas pragmatic attributes emphasize the fulfilment of individuals' behavioural goals, hedonic attributes emphasize individuals' psychological well-being. Second, the American Heritage Dictionary of the English Language defines something that is hedonic as "of, relating to, or marked by pleasure". Thus, "hedonic" expresses my belief that the functions and attributes it subsumes are strong potentials for pleasure—much stronger than pragmatic functions and attributes. Typical hedonic attributes of software products are "outstanding", "impressive", "exciting" and "interesting".

The hedonic function of products can be further subdivided into providing stimulation, communicating identity, and provoking valued memories.

Stimulation. Individuals strive for personal development, i.e., proliferation of knowledge and development of skills. To do so, products have to be *stimulating*. They have to provide new impressions, opportunities, and insights. McGrenere (2000), for example, found in a study on "bloat" (i.e., "creeping featurism") in Microsoft's Word that on average only 27% of the available functionality was used. However, only 25% of the participants (13 of 53) wanted to have unused functionality entirely removed. I argue that these unused functions are viewed as future opportunities for personal development. They are not needed to fulfil current behavioural goals, but nevertheless wanted for future perfection of the way current goals are accomplished or for future generation of entirely new goals. Thus, functionality that is used and works well will be perceived as pragmatic, whereas functionality not *yet* used but interesting will be perceived as hedonic. The stimulation provided by novel, interesting or even exciting functionality, content, presentation or interaction style will also indirectly help goal fulfilment. It may raises attention, compensates for a lack of motivation to fulfil externally given goals, or facilitates new solutions to problems.

Identification. Individuals express their self through physical objects—their possessions (Prentice 1987). This self-expressive function is entirely social. Individuals want to be seen in specific ways by relevant others. To be socially recognized and to exert power over others is a basic domain of human motives (Schwartz and Bilsky 1987). To fulfil this need, a product has to *communicate identity*. For example, personal homepages can be used to present the self to others. Borcherding and Schumacher (2002) found that students who believed that others hold unfavourable opinions about them, such as a lack of humour and few social contacts, presented more information about family and friends and humorous links on their homepages. In this case, the possession—a personal homepage—is deliberately shaped to communicate an advantageous identity. In general, people may prefer products that communicate advantageous identities to others.

Evocation. Products can *provoke memories*. In this case the product represents past events, relationships or thoughts that are important to the individual (Prentice 1987). For example, souvenirs are a whole product category that provides only symbolic value by keeping memories of a pleasant journey alive. Mackenzie (1997) presented the example of wine collectors, who may appreciate the wine in their cellar because of the aroma *and* the memories and effort attached to each single bottle. A more technology related example might be the trend to play vintage computer games. What do they provide? Definitely neither complex game play nor striking graphics. Their value comes from triggering memories of the good old days, when these games were exciting and kept people captive for hours.

To summarize, a product may be perceived as pragmatic because it provides effective and efficient means to manipulate the environment. A product may be perceived as hedonic because it provides stimulation, identification or provokes memories. Reconsider the example of driving in a nail. From a pragmatic perspective you prefer a tool that allows driving in the nail without much effort. You decide to buy a hammer. From a hedonic perspective you may buy a certain brand that communicates professionalism to others. Or you buy a whole set of tools instead of only a hammer. Although your current goal is to drive in a nail, you anticipate that do-it-yourself may become your new, most exciting hobby. Or you prefer to use an old hammer your mother once gave you as a present. Using it reminds you of the pleasant hours you spent with her as a child in her workshop. What you actually prefer to do depends on what is relevant to you. You then decide for the product which character suggests realization of your needs (i.e., the cheapest hammer that works, a professional hammer, a hammer and other tools, the hammer your mother once gave you).

2.2.3 ACT and SELF Product Characters

I view pragmatic and hedonic attributes as independent of each other. In combination they are the product character. If we take into account that peoples' perception of pragmatic and hedonic attributes can be either weak or strong, four types of product characters will emerge (see Fig. 2). Notice, that products can be pragmatic or hedonic for different reasons. For example, a tool of a certain brand may be

Fig. 2 Product characters
emerging from specific
combinations of pragmatic
and hedonic attributes

hedonic because this tool communicates professionalism to relevant others (i.e., communicates identity). Other tools may be hedonic because they are an innovation, which stimulates its user to do exciting new things.

The combination of weak pragmatic and weak hedonic attributes is simply unwanted. It is a character implying a product that is neither able to satisfy pragmatic nor hedonic needs of potential users. The combination of strong pragmatic and strong hedonic attributes signifies the desired product. An uncompromising combination of both is the ultimate design goal. Most likely, both attribute groups will be not in balance. I call a primarily pragmatic product (i.e., "strong pragmatic/weak hedonic") an ACT product and a primarily hedonic product (i.e., "weak pragmatic/strong hedonic product") a SELF product.

The ACT product is inextricably linked to its users' behavioural goals. As already stated above, goals vary. They can be externally given by others or internally generated by the individual. Moreover, they can be of different importance to the user. Depending on the actual status of goals the appealingness of an ACT product varies (i.e., the importance of pragmatic attributes is decreased). Imagine: to reduce commuting time you bought a downright pragmatic car instead of going to your office by train. Unexpectedly, shortly after you purchase somebody offers you a new, cheap and attractive apartment only a five-minute walk from your office. You accept and suddenly your new car is not as appealing as before, because the main behavioural goal you have meant to fulfil with the car ceased to exist.

On contrary, the SELF product is inextricably linked to users' self, e.g., their ideals, memories, and relationships. If, for example, the car you have bought had been a luxurious sports car that not only stimulates your senses but also communicates success to others, the move would not have decreased the car's appeal. The appreciation of SELF products is much more stable than the appreciation of ACT products, because the probability that individuals change what they require from a product to satisfy their self is much lower than the probability that behavioural goals change. Moreover, the bond between a SELF product and its user should in general be much stronger than the bond between an ACT product and its user. Only when the behavioural goals accomplished with the ACT product are of high personal relevance a strong bond between an ACT product and a user can be expected.

This emphasizes the importance of hedonic attributes. Only products, which provide at least some opportunities for being related to the self, are likely to be truly and stably appreciated.

2.3   Consequences: Satisfaction, Pleasure and Appealingness

Experiencing a product with a certain character will have emotional consequences, such as satisfaction or pleasure. They are momentary and take the usage situation into account. Note that these consequences (i.e., satisfaction, pleasure, appeal) are viewed as outcomes of experience with or through technology (see also Wright McCarthy and Meekison elsewhere in this book).

Human-Computer Interaction regards satisfaction with a product as a major design goal (e.g., ISO 9241-11). However, its definition as a "positive attitude towards the product" (ISO, 1998) remains superficial. Moreover, attitudes differ from emotions in several aspects. Ortony and Clore (1988, pp 118, see Desmet and Hekkert 2002 for a further application of Ortony et al.'s theory) define satisfaction as being pleased about the confirmation of the prospects of a desirable event. In other words, if people hold expectations about the outcome of using a particular product and these expectations are confirmed they will feel satisfied. In contrast to satisfaction, joy or pleasure requires no expectations. It is defined as being pleased about a desirable event per se (Ortony et al. 1988, pp. 86). The more unexpected the event is, the more intense will be the pleasure. In other words, if people use a particular product and experience desired deviations from expectations, they will be pleased.

In practice, one is likely to experience combinations of satisfaction and joy. To give an illustrative example, consider software for playing MP3 music files. You expect that it supports you to manage the files on your computer hard disk by giving you an easy possibility to generate and save play lists. Indeed, the software provides this functionality and you feel satisfied whenever you use it. Moreover, you unexpectedly discover that it is possible to produce standard audio compact discs from the play list by only one click. You are pleased about the unexpected benefit you discovered. Satisfaction is linked to the success in using a product to achieve particular desirable behavioural goals. Pleasure is linked to using a product in a particular situation and encountering something desirable but unexpected.

If a product is able to trigger positive emotional reactions it is appealing. Appealingness is a group of product attributes such as good, sympathetic, pleasant, attractive, motivating, desirable, and inviting. Appealingness weights and integrates perceptions of product attributes by *taking particular situations (i.e., contexts) into account*. For example, individuals may consider an ACT product as appealing because the goals achievable by the product are of high relevance to them in a particular situation. However, other individuals (or even the same individual) can consider the same product as less appealing, maybe because people were rather interested in communicating a favourable identity to others than achieving behavioural goals. In short, appealingness integrates experiences with and feelings towards a product in a particular situation into an evaluative judgment.

In practice, I argue that particular product characters will render some emotional reactions more likely. ACT products emphasize fulfilment of behavioural goals. This can be interpreted as an expectation, which—given that goals had been reached—is more likely to lead to a positive expectation-based emotion, namely satisfaction. With an ACT product, pleasure may additionally be experienced if expectations about goal achievement (e.g., ease of achieving a goal) are excelled. SELF products are used to fulfil psychological needs rather than behavioural goals. Because of the weak connection to goals and expectations about fulfilling these goals, these products are more likely to lead to a positive well-being based emotion, namely pleasure. Satisfaction will only play a role, if hedonic functions are explicitly called for and expected, for example, if a person buys a product to impress a particular other person and is successful in doing so.

The susceptibility of emotional reactions and the judgment of appealingness to variation caused by situation is an argument for separating *potentials* for consequences (i.e., the product character) from the *actual* consequences—the former is simply more stable and, thus, more reliable. Furthermore, in a product design process it is important to know why users judge a product as appealing, pleased by or satisfied by and thus one should rather focus on the product character and the usage situations than the consequences. However, this is not meant to imply that appealingness and the emotional reactions are unimportant. Both will certainly affect future use of the product.

2.4   Situation: Goal and Action Mode

I have repeatedly stressed the importance of different *situations* for understanding both judgments of appealingness and emotional reactions. A usage situation combines the perceived product character with a particular set of aspirations, such as specific behavioural goals or need for stimulation. Obviously, these situations can be quite diverse, which poses a serious problem for predicting emotional reactions or appealingness in particular usage situations. As a solution to this problem, I propose to focus on the mental state of the user by defining different *usage modes* (see Hassenzahl et al. 2002). Specifically, I distinguish a goal and an action mode[1] (see Fig. 3 for an illustration).

Usage *always* consists of behavioural goals and actions to fulfil these goals. In *goal mode* goal fulfilment is in the fore. The current goal has a certain importance and determines all actions. The product is therefore just "a means to an end".

---

[1] Usage modes were inspired by Apter's reversal theory (Apter 1989). In the present Chap. 1, however, use the term "action" instead of Apter's term "activity" to avoid a potential confusion with "activity theory".

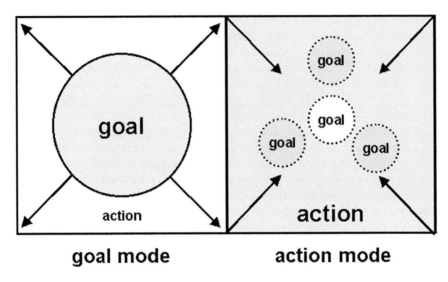

Fig. 3 Goal and action mode (inspired by Apter 1989)

Individuals try to be effective and efficient. They describe themselves as "serious" and "planning". Low arousal is preferred and experienced as relaxation. If arousal increases (e.g., because of a usability problem that circumvents goal fulfilment), it is experienced as mounting anxiety (frustration). In *action mode* the action is in the fore. The current action determines goals "on the fly"; the goals are "volatile". Using the product can be an "end in itself". Effectiveness and efficiency do not play an important role. Individuals describe themselves as "playful" and "spontaneous". High arousal is preferred and experienced as excitement. If arousal decreases (because of a lack of stimulation) it is experienced as increasing boredom.

The particular usage mode is triggered by the situation itself. If, for example, your boss wants you to do an important task that must be finished within two hours, you most likely will be in goal mode. In contrast, if there is not much to do at the moment and you start exploring the new software you just got, you are more likely to be in action mode. In principle, I view usage modes as psychological states and every product can be experienced in either state. The perception of a product character as primarily pragmatic or hedonic will not be influenced by usage modes. However, appealingness and emotional reactions depend on the product's momentary fit to the usage mode. Thus, usage modes become the moderator between the product character and consequences. Usage modes can be chronic, i.e., a part of an individual's self-concept, too; to be in a particular usage mode becomes a stable personal trait.

To conclude, usage modes are certainly a more helpful distinction between ways of approaching a product than the classical "tool" and "toy" or "leisure" and "work"—dichotomy. The advantage lies in the emphasis on the fact that each product, irrespective of whether it is a computer game or a word processor, can be used in both modes.

3   Summary and Conclusion

User experience encompasses all aspects of interacting with a product. Its psychological complexity cannot be underestimated. First of all, user experience is subjective. Consequently, actual experiences with products may considerably differ from experiences intended by the designer. Experiences vary between individuals because of different personal standards. In addition, they vary between situations and they may change over time. Products have a character that suggests a capability to manipulate the environment, to stimulate, to communicate identity or to provoke memories. The first capability is pragmatic, i.e., inextricably tied to internally generated or externally given behavioural goals. The other three are hedonic, i.e., tied to individuals' self and their psychological well-being. People value products on the basis of how they satisfy needs in particular situations. As a consequence, products have a certain appealingness and cause emotional reactions. Different emotional reactions may be distinguished: Satisfaction may be related to the fulfilment of expectations (i.e., behavioural goals), whereas pleasure may be related to the unexpected.

Approaches to user experience in HCI lack theory and empirical investigation. It seems important to better understand user experience itself, its determinants and situational/personal mediation *and* to validate this understanding. So far, several studies tested key elements of the model (e.g., Hassenzahl et al. 2000; Hassenzahl 2002a) and used the concept of "hedonic" attributes in product evaluation (e.g., Kunze 2001; Sandweg et al. 2000; Seifert et al. 2001).

I view the benefit of the suggested preliminary model of user experience as two-fold: First, designers may better understand how people perceive and value objects. Second, it allows operationalisation and measurement of key elements. Both will inform design and lead to better, more satisfying and more pleasurable products.

Acknowledgements I would like to thank Annette Amon, Kai-Christoph Hamborg, James Kalbach, Sara Ljungblad, Andrew Monk, Jürgen Sauer and Peter Wright for their helpful comments on earlier drafts of this chapter, and Katrin Borcherding for bringing the importance of the distinction between potentials for pleasure/satisfaction and their actual realization to my attention.

References

Apter MJ (1989) Reversal theory: motivation, emotion and personality. Routledge, London, New York

Borcherding K, Schumacher M (2002) Symbolic self-completion on personal homepages. In: Proceedings of the 6th international conference on Work With Display Units (WWDU 2002), ERGONOMIC Institut für Arbeits- und Sozialforschung, Berlin, pp 270–271

Desmet PMA, Hekkert P (2002) The basis of product emotions. In: Green W, Jordan P (eds) Pleasure with products: beyond usability. Taylor & Francis, London

Draper SW (1999) Analysing fun as a candidate software requirement. Pers Technol 3(1):1–6

Gaver WW, Martin H (2000) Alternatives. Exploring information appliances through conceptual design proposals. In: Proceedings of the CHI 2000 conference on human factors in computing, ACM, Addison-Wesley, New York, pp 209–216

Glass B. (1997) Swept away in a sea of evolution: new challenges and opportunities for usability professionals. In: Liskowsky R, Velichkovsky BM, Wünschmann W (eds) Software-Ergonomie'97. Usability Engineering: Integration von Mensch-Computer-Interaktion und Software-Entwicklung, B.G. Teubner, Stuttgart, pp 17–26

Hassenzahl M (2002a) The effect of perceived hedonic quality on product appealingness. Int J Hum Comput Interact 13(4):479–497

Hassenzahl M (2002b). Character grid: a simple repertory grid technique for web site analysis and evaluation. In: Ratner J (ed) Human factors and web development, Lawrence Erlbaum, Mahwah, NJ

Hassenzahl M, Kekez R, Burmester M (2002). The importance of a software's pragmatic quality depends on usage modes. In: Proceedings of the 6th international conference on Work With Display Units (WWDU 2002), ERGONOMIC Institut für Arbeits- und Sozialforschung, Berlin, pp 275–276

Hassenzahl M, Platz A, Burmester M, Lehner K (2000). Hedonic and ergonomic quality aspects determine a software's appeal. In: Proceedings of the CHI 2000 conference on human factors in computing, ACM, Addison-Wesley, New York, pp 201–208

Hassenzahl M, Trautmann T (2001) Analysis of web sites with the repertory grid technique. In: Proceedings of the CHI 2001 conference on human factors in computing. Extended abstracts, ACM Press, Addison-Wesley, New York, pp 167–168

ISO (1998) ISO 9241: ergonomic requirements for office work with visual display terminals (VDTs)—Part 11: guidance on usability, International Organization for Standardization, Geneva

Janlert L-E, Stolterman E (1997) The character of things. Des Stud 18:297–314

Jordan P (2000) Designing pleasurable products. An introduction to the new human factors. Taylor & Francis, London, New York

Kunze E.-N. (2001) How to get rid of boredom in waiting-time-gaps of terminal-systems. In: Proceedings of the international conference on affective human factors design, Asean Academic Press, London

Logan RJ (1994) Behavioral and emotional usability: Thomson consumer electronics. In: Wiklund M (ed) Usability in practice. Academic Press, Cambridge, MA

Mackenzie C (1997) Where are the motives? A problem with evidence in the work of Richard Thaler. J Econ Psychol 18:123–135

McGrenere J (2000) "Bloat": the objective and subjective dimensions. In: Proceedings of the CHI 2000 conference on human factors in computing. Extended abstracts, ACM Press, Addison-Wesley, New York, pp. 337–338

Monk AF, Frohlich D (1999) Computers and fun. Pers Technol 3(1):91

Monö RW (1997) Design for product understanding: the aesthetics of design from a semiotic approach. Liber AB, Stockholm

Ortony A, Clore GL, Collins A (1988) The cognitive structure of emotions. Cambridge University Press, Cambridge, MA

Prentice DA (1987) Psychological correspondence of possessions, attitudes, and values. J Pers Soc Psychol 53(6):993–1003

Sandweg N, Hassenzahl M, Kuhn K (2000) Designing a telephone-based interface for a home automation system. Int J Hum Comput Interact 12(3&4):401–414

Schmitt BH (1999) Experiential marketing. Free Press, New York

Schwartz SH, Bilsky W (1987) Toward a universal psychological structure of human values. J Pers Soc Psychol 53(3):550–562

Seifert K, Baumgarten T, Kuhnt T, Hassenzahl M (2001) Multimodale Mensch-Computer-Interaktion: Tool oder Gimmick? In: Timpe K.-P, Marzi R, Karavezyris V, Erbe H.-H, Timpe K.-P (eds) Bedienen und Verstehen. 4. Berliner Werkstatt Mensch-Maschine Systeme, VDI-Verlag, Düsseldorf, pp 275–291

Chapter 20
Making Sense of Experience

Peter Wright, John McCarthy and Lisa Meekison

Authors' Note Funology 2, Chaps. 4 and 7

When Andrew and Mark first approached us about the 2nd edition of Funology with the offer of revising our two chapters or leaving them as they were, we opted for the latter. We felt there was no single dimension along which we could revise the chapters in any helpful way such as adding a case study or revising the theory. But when Andrew suggested we should write an endnote about what had happened since 2003, we thought it might be a good idea to try and pull together a few threads of experience, thinking, and publication that might be helpful to readers approaching these chapters for the first time or even people revisiting them and wondering what happened next.

Back in 2003 when the first edition came out, the idea that people's experience of technology might be worth studying as part of the process of designing and evaluating technology was unusual. Indeed around that time, the link between HCI and design was relatively under-developed, found mainly in books like Donald Norman's *The Design of Everyday Things* (Norman 2013) and *Emotional Design* (2005). Not unlike, Norman's journey through Cognitive Science to Interaction Design, we came to Interaction Design through the study of human factors in complex safety critical settings such as air traffic control centres and aircraft cockpits—the antithesis of "fun" settings. The transition from human factors to experience-centred design reflected a need we had identified to get some kind of hold on *what it was like* to be in control of dispatching emergency ambulances or directing aircraft on approach to land, or dealing with cockpit emergencies, and not just how the information was processed and the tasks performed.

P. Wright (✉)
Open Lab, Newcastle University, Newcastle upon Tyne, UK
e-mail: p.c.wright@newcastle.ac.uk

J. McCarthy
School of Applied Psychology, University College Cork, Cork, Ireland

L. Meekison
http://lisameekison.com/

© Springer International Publishing AG, part of Springer Nature 2018
M. Blythe and A. Monk (eds.), *Funology 2*,
Human–Computer Interaction Series,
https://doi.org/10.1007/978-3-319-68213-6_20

One of our chapters in Funology, *Making Sense of Experience* (Wright et al. 2004), was our first attempt to open up a space for thinking about what it might mean to study human experience in HCI, what scholars of human experience thought the phenomenon was, and how best to talk and think about it. We needed a chapter like this because nobody seemed to be writing or talking about it, and there wasn't a language of experience available within the human factors community. This search for an understanding of human experience that would be helpful for HCI and Interaction Design took us into disciplinary fields quite distant from our own training in psychology, disciplines like film making and literary theory as well as the pragmatic philosophy of Dewey (1934).

The didactic approach we took in *Making Sense of Experience* was to offer a 'framework', to give designers a language and a set of constructs with which to talk about experience. Because of the holistic approach we had adopted, the framework came with suitable caveats about avoiding reductive abstractions. Understandably, people began to ask two questions at conferences and in talks: "I don't get it, can you explain these concepts more clearly?" and "I get it, but how do I use it?" Our response, that the framework wasn't meant to be an operationalisable definition of experience needed clarification, something we tried to do in a number of publications that followed. The clarification was that the framework was meant to de-centre information processing as the best way HCI researchers could understand 'users' and instead put people's lived experience, their feelings, values, stories and ways of making sense of their interactions with the world, at the centre of HCI enquiry. For this insight, we owe a great debt to the inspirational Phil Agre (1997) not just for what he said about human experience and computing but for what he said about how to move things from the margins of academic discourses to their centres.

We put de-centring into practice in our chapter, *The Value of the Novel in Designing for Experience* (2004) by offering a re-reading of traditional HCI techniques for representing 'the user' (such as scenarios and personae) through the dialogical lens of our experience framework. But we wanted to go deeper into the philosophy and psychology of human experience in order to make a more substantial step towards placing human experience centrestage in HCI. So in our monograph *Technology as Experience* (2004), we took a deep dive into human experience. Residues of the framework are still there and there are case studies to illustrate how an experiential way of seeing the world plays out in situations like ambulance despatch as well as on-line shopping. But the main work of the monograph was to deepen the repertoire of concepts and genres to support discourse about experience and technology. Concepts such as *life as lived and felt*, Dewey's distinction between *an experience* and *the cumulative experience* that is a person's life can be found there, but in addition, we developed the notion of *dialogue* in a way which brings the relationship between *self-and other*, *openness* (*unfinalizability*), and *multi-voicedness* to the fore.

Although we felt *Technology as Experience* offered a more substantial foundation for the ideas in our *Funology* chapters, we had a lingering concern about how it could be used in design and how young researchers could be taught it. What does it really mean to design for openness or becoming? How does a relational or dialogical perspective on experience change how we design for experience? What should a young HCI researcher or interaction designer do? Difficult questions indeed.

The search for answers took us on another interdisciplinary journey beyond Dewey and Bakhtin to, Jay's *Songs of Experience* (2005), to narrative psychology (Bruner 1987, 1990) and to storied lives and experiences (Bruner and Turner 1986). It also took us to feminist theory and methodology (Belenky et al. 1986), and beyond. This journey is described in *Experience-Centred Design* (2010), in which we put narrative or more precisely *stories*, centrestage as one of the most fundamental ways for designers to understand human experience and we explored what it might mean for researchers to work with stories. Also, through concepts such as *active listening*, we tried to capture how stories are co-constructed through dialogue (the idea of anticipation, surprise, openness, and learning), and we re-examined the concept of *knowing the user* through a consideration of what it means to *understand* rather than to *know*. This led us to explore *empathy* both as a form of perspective taking and as an emotional-volitional orientation to 'the other' through relations of care, concern and hope (see also Wright and McCarthy 2008). Around this time, we started working with Jayne Wallace who through her making and her conversations showed us what it means to put this dialogical understanding of experience into design practice (Wright et al. 2008). Our early work together deepened our understanding of the varieties of experience of using and making which in subsequent papers gave depth and texture to ideas of enchantment and beauty not as a quality or attribute of a product but as a process, something we had tentatively discussed, but without the benefit of a designer's sensibilities, in our second Funology paper, *The Enchantments of Technology* (McCarthy and Wright 2004).

Seeing the relationship between designer and user as empathic, creative and dialogical critically shifts the agenda for design methods, and also shifts previously peripheral methods and tools, such as cultural probes and experience prototyping centrestage. The role of design methods is no longer confined to *representing the user* in such a way that *design implications* can be drawn. The suggestion that cultural probe returns aren't used to infer design implications but rather to allow designers to immerse themselves and respond directly or begin a conversation with the user makes perfect sense within a dialogical framing that places empathy and responsiveness at its centre. For designers and users in dialogue, the creative activity that they engage in is designed to allow individuals who have different perspectives on a problem, issue, setting or context, to respond to it from their own *centre of value*, informed by the encounter with the others' perspectives. Dialogical engagement does not presume the user becomes the designer or the designer becomes the user, quite the opposite because the designer would have nothing to offer if his or her experience were identical to that of the user. But what does it mean in practice to do this kind of design research?

In 2010, Pete moved to *Open Lab* (then called *Culture Lab*, see Wright and Olivier 2012). At Open Lab, we have had the opportunity to work with researchers living the experience of doing experience-centred design. We began by trying to understand the qualities of dialogical design encounters and what it means to frame participatory design methods as dialogue. We worked with researchers whose backgrounds ranged from computing through design to social sciences and arts and beyond. With them we explored how researchers designed together with communities over extended periods of time and built up relationships with them. We worked in sites which stretched traditional ways of doing HCI, such as a centre for Black Asian, Minority Ethnic, and Refugee (BAMER) women (Clarke et al. 2013), a medium-secure forensic prison (Thieme et al. 2016), hospitals (Bowen et al. 2011), live events (Hook et al. 2013), and rural communities (Taylor et al. 2013). These usually involved long-term embedded engagements in which a single researcher or small team spend 3 or more years working with a single community.

In *Taking [A]Part* (McCarthy and Wright 2015), we offer an account of what we learned working in these complex social settings. We take inspiration from participatory arts and political philosophy and offer an approach that tries to eschew the twin tyrannies of tokenism and exploitation. We explore what it means to have a *voice* rather than *a say*, how individuals participate at the intersection of multiple subjective positionings, and how participation can be a heavily textured and variable experience for all parties. From a design perspective, we have begun to appreciate that participation itself has to be designed, and the way it is designed has material effects on not only the experiences of participants but also the qualities of outcomes. The way in which individuals (not only those formerly called 'users', but all participants) are subjectively positioned by the process can open up or close down possibilities for voice, negotiation and self-organization. In light of this, we argue that the measure of participation should not be consensus but *dissensus*. Inspired by Rancière's analysis of *the sensible* (e.g. 2006), we argue that dissensus is the engine of creativity in participatory design projects, and that the best participatory projects seek to reconfigure what is sensible (that is to say what is hearable, visible, and even thinkable) to participants in order to create space for making previously unimaginable possibilities thinkable (Wright and McCarthy 2015).

So, the two chapters in Funology from all those years ago, set us off on a journey, to explore what it might mean to take human experience seriously in HCI. And for us, the bottom line is that experience-centred design must configure those people who participate as co-producers—skillful, knowledgeable, resourceful, and situated agents. Done well, experience-centred projects can produce mutual learning and transformative outcomes for all participants. Done badly, they can serve to exploit labour, invalidate local knowledge, neutralize voices, and silence dissent (McCarthy and Wright 2015).

1 Introduction

Ehn and Löwgren (1997) use the term *quality-in-use* to refer to a range of aesthetic, ethic and functional qualities that need to be considered in design. Qualities such as enjoyment, fulfilment and fun are not properties of technology. They are better thought of as outcomes of certain kinds of experience with—or through—technology. So if we are to understand what might make a particular product or design more pleasing or enjoyable to use, it would seem sensible to begin by trying to analyse experience of use.

There has of course already been much interest in experience. For example, Turkle (1995) explored the social meaning of computers, the cultures of computing and the impact of the Internet on our sense of self. The concept of *user experience* has also come to dominate in consumer arenas such as branding and electronic commerce (see Pu and Faltings 2000; Lee et al. 2000, for example). There is however an uneasy silence as to what actually constitutes experience. Questions such as how to set boundaries distinguishing a specific user experience from a general flow of experience, how to account for subjectivity, and whether it is possible to design experience, have remained conspicuously unanswered. In short, despite a growing acceptance of the need to focus on experience, the concept of user experience is not well developed conceptually. Without conceptual development, there is a danger that user experience and related concepts such as trust, loyalty, identity, and engagement will not be fully realized in studies of people and technology.

Elsewhere in this volume, Hassenzahl adopts a psychological approach to modelling experience in order to answer some of these questions. He attempts to identify the components of experience and to map the external physical properties of the world of artefacts onto their psychological effects, in the context of the user's purposes. His work is a step towards an explanatory model of experience in the traditional scientific sense. Our project here is slightly different. Our aim is to explore an approach which is holistic, constructionist and pragmatic in Dewey's sense of the term. Our theorising is thus of different quality but nevertheless aimed at being practically useful. Rather than isolate the elements of experience we seek to understand their interaction and how they mutually constitute each other. We also seek a stronger account of sense-making as the central process of experiencing.

2 Conceptualising Experience

The turn to experience that we have outlined above reflects an attempt to engage with the 'felt life' of people participating in activities with—and through—technology. Worthy though this aspiration is, there is also a risk. 'Experience' is an elusive concept that resists specification and finalisation. In recent attempts to introduce 'experience' into consideration of relations between people and technology, it has been confused with subjective feelings, behaviour, activity, social practice, and knowledge. Thus the first major task in developing an experiential theory is to provide a basis for understanding experience that is not confused with any of these.

Dewey's philosophy of experience was geared towards a clarification that would end the tendency to reduce. For example, he pointed out that the most debilitating reduction to private, subjective experience is a feature of modern times.

> To the Greeks, experience was the outcome of accumulation of practical acts, sufferings and perception gradually built up into ... skill ... There was nothing merely personal or subjective about it. (Dewey 1934, p. 198)

He argued that experience is the irreducible totality of people acting, sensing, thinking, feeling, and meaning-making in a setting, including their perception and sensation of their own actions. He offered a useful definition along these lines. Experience, he wrote,

> ... includes *what* men do and suffer, *what* they strive for, love, believe and endure, and also *how* men act and are acted upon, the ways in which they do and suffer, desire and enjoy, see, believe, imagine - in short, processes of experiencing ... It is "double barreled" in that it recognises in its primary integrity no division between act and material, subject and object, but contains them both in an unanalyzed totality. (Dewey 1925, pp. 10/11)

According to Dewey experience is constituted by the relationship between self and object, where the self is always already engaged and comes to every situation with personal interests and ideologies. People and setting are also changed by experience, and the unity of any experience is itself a moving, fragile, fleeting event. He also made a distinction between experience and *an* experience, where we have *an* experience only when the material experienced runs its course to fulfilment and has its own individualising quality and self-sufficiency, an emotional unity that gives experience aesthetic quality.

In a move that is particularly useful for our purposes, Dewey also described the conditions of aesthetic form as an analytic tool to be used when we encounter problems with the unfolding of an experience—what went wrong, why we lost interest, why others got restless, etc. His conditions include: continuity, cumulation, conservation, tension, anticipation, and fulfilment. As the framework we describe in the next section draws heavily on Dewey's conditions, we will not dwell on them here.

Bakhtin (1986, 1993), a philosopher with a more literary bent than Dewey, provided a complementary account of the relationship between experience and meaning-making, through which we have read Dewey's conditions. Bakhtin's work is most useful in our consideration of the personal qualities of experience such as trust, identification, loyalty, and commitment. In terms of ongoing experiences of technology, Bakhtin's work shifts our focus from the immediate quality of an experience (absorbing, captivating, irritating) to the sense we make of an experience in terms of our experience of our selves, our culture, and our lives. Specifically, I might find a particular web site absorbing during my first few visits but if it is not integrated into 'my life'—if it does not fit with my sense of my self—it is likely to become less absorbing as time goes on.

For Bakhtin, the unity of an experience in action and an account of the meaning made of it is never available a priori but must always be accomplished. To understand how it is accomplished we must understand Bakhtin's central contribution in this area, the idea of dialogicality. Dialogicality refers to the presumption that in human activity there are always at least two consciousnesses involved. The activity of an individual sitting reading a book involves dialogues between the consciousnesses of the reader and writer and it is possible to argue, the consciousnesses of the characters in the book. We argue that web sites and other technology touchpoints can also be seen as abstractions from dialogues and that attention to the dialogical may yield fresh insights. The site of dialogical knowledge is never unitary. It is always relationally based. That is to say that knowledge of the self always emerges as an expression of self-other relations and similarly knowledge of an object always emerges as an expression of subject-object relations.

What dialogism means for this paper is that any discursive account of an experience, including a person's own account to themselves of an experience of buying through the Internet, is incurably social, plural, and perspectival. In Bakhtin's terms, it is *interanimated* with the discourses of others. For example, my sense of myself as someone who supports small local bookshops is interanimated by discourses on the values of global capitalism, the importance of choice provided by small specialist booksellers, and the centrality of a personal relationship in choosing which books to buy. These discourses however might be accommodating of an Internet bookseller who appears to try to develop a buyer-seller relationship with me based on an understanding of my reading preferences, provides specialist choices, and seems to support small specialist booksellers. If my book buying activity moves from the small local bookshop to an Internet seller who present themselves as engaging meaningfully with some of these discourses and also has other qualities of interest to me—e.g. speedy fulfilment of an order—then my sense of my self is subtly changed through dialogue with that bookseller.

This relational subjectivity is also reflected in the analyses of reflective practitioners. For example, writers and film-makers practice their craft under the influence of a practical understanding of experience and how they might influence or help create experience. Elsewhere in this volume, we appeal to Boorstin's (1995) analysis of Hollywood movie making as inspiration for an account of the enchantments of technology.

Our aim in drawing together the writings of philosophers such as Dewey and Bakhtin and reflective practitioners such as Boorstin is to try to understand experience well enough to understand how Internet shopping designers and brand managers deploy technical knowledge to help their user-consumers create a fulfilling interactive experience. To do this we created a framework, based on the above work, that pulls together a set of concepts that can be used as tools to analyse user experience with emerging technologies.

3   The Framework

We will describe the framework in two parts, the first is concerned with describing experience from four points of view which we refer to as the four threads of experience. The second part is concerned with how we *make sense* in experience.

Before describing these it is important to make some observations which we hope will avoid some possible misinterpretations of what is written here.

There is no simple mapping between the framework and the concepts of Dewey and Bakhtin. Dewey's concern for a holistic and interactionist approach is manifest in our characterisation of the four threads of experience, some of Boorstin's perspectivalism is reflected in this too. In terms of sense making, Dewey's concepts of anticipation, reflection and the pre-linguistic sense of meaning are echoed in our account. Bakhtin's concern for unity as an active accomplishment is reflected in our characterisation of appropriating, reflecting and recounting. But more subtly, what we have tried to do is understand Dewey through Bakhtin's dialogical lens. Thus self-other relations as continually constructed, permeates our account throughout.

Dewey's pragmatic account of experience emphasised that experience cannot be reduced to fundamental elements but only exists as relations. Experience is essentially holistic, situated and constructed. Thus it would be mistaken to approach experience in a way a classical physicist might approach the study of matter—to identify a substance, define it in terms of molecules, atoms and sub-atomic particles. We share Dewey's view on the holistic nature of experience, yet we see the need to be able talk about and describe experience in ways that can be understood. The analytical prose we use in this report to talk about the framework might suggest we are attempting such a reductive approach. This is not our intention. Rather we intend to connote a space within which things can be juxtaposed, related, separated, coalesced but never isolated.

3.1 The Four Threads of Experience

First of all we begin by identifying four threads of experience. We have found it helpful to think of these as four aspects as four inter-twined threads making up a braid.

3.1.1 The Compositional Thread

The compositional structure of an experience is that aspect which is concerned with the part-whole structure of an experience. In an unfolding interaction involving self and other this could be thought of as narrative structure, action possibility, plausibility, consequences and explanations of actions. In an experience of an artwork, a poster or brand image it can be thought of as the compositional elements of the image their relations and implied agency. If you are asking questions like; "what is this about?", "what has happened?", "what will happen next?", "does this make sense?", "I wonder what would happen if?" then you are thinking about the compositional structure of experience.

3.1.2   The Sensual Thread

The 'look and feel' of a physical artefact or a web page are part of what we refer to as the sensual thread of experience. More generally, the sensual thread of experience is concerned with our sensory engagement with a situation. The sensations in an experience which we variously term thrill, fear, excitement are sensual, as are feelings such as walking into a room and finding it welcoming, a sense of belonging, a slight sense of unease or awkwardness in a conversation. Sometimes the sensual defies precise description but can affect our willingness to become involved. The look and feel of a mobile phone may be as important a determinant of our decision to become (or not to become) a mobile phone user as the functional possibilities it offers.

3.1.3   The Emotional Thread

The emotional thread of experience includes anger, joy, disappointment, frustration, desperation and so on. These are stark examples, but other more subtle things that are included here are fulfilment, satisfaction, fun and so on. We can reflect on the emotional thread of our own experience or we can through empathy relate to the emotional thread of others' experiences. Relating to a character, in a movie is an obvious example, but we might also empathise with the designer or retailer of an e-shopping site even though they are not materially present.

We need to distinguish between the sensual and emotional threads since we can engender emotions associated with achievement through the exercise of control over sensations such as fear or anxiety. This is this case for example when a rock climber climbs a dangerous peak. For Csikszentimihalyi (1990), flow states result from precisely this kind of balancing. We can also engender emotional states such as bliss through the deliberate neutralisation of sensual states such as excitement and anxiety as is the case with meditation.

Emotions are not just passive responses to a situation. Our actions and our understandings may be motivated by emotional aspects just as surely as they may be motivated by our intellectual or rational understandings of action possibilities and consequences. We may act through compassion or morality just as surely as we may act through a rational assessment of actions and their likely consequences or some other utilitarian process.

3.1.4   The Spatio-Temporal Thread

All experience has a spatio-temporal thread. Actions and events unfold in a particular time and place. When we are rushed we may feel frustrated and perceive space as confined, closeting. In addition, emotional engagement can make our sense of time change, hours can fly by in minutes. Pace may increase or decrease and our sense of space may open up or close down. Both space and time may become connected or disconnected as an experience unfolds. We might also distinguish between public and private space, we may recognise comfort zones and boundaries between self and other, or present and future. Such constructions affect experiential outcomes such as willingness to linger or to re-visit places or our willingness to engage in an exchange of information, services or goods.

3.2 Making Sense in Experience

People do not simply engage in experiences as ready-made, they actively construct them through a process of sense making. This process of sense making is reflexive and recursive. It is reflexive in the sense that we are always viewing experience through a person. Whether that is the first person or the third person or whether it is by recounting an experience to oneself or for others. This is not to be understood in some scientific way as an unfortunate consequence of our means of measurement. Rather it is central to what it means for something to be an experience. Without self and other, or subject and object interacting reflexively, there can be no experience. It is recursive in the sense that we are always engaged in sense making. Even when we reflect on experience as a completed object, we are having an experience.

Before describing sense making in detail however, we should note that the different sense making processes we describe are not linearly related in cause and effect terms. For example, in anticipation of some future planned action we may reflect on the consequences of that action, which may engender a certain sensual response. How we recount our experience to others may change how we reflect on it and so on.

3.2.1   Anticipating

When we think of the form of experience we need to extend our account beyond the beginning and the end of an episode. When experiencing on-line for the first time a well-known off-line brand we do not come unprejudiced to that on-line experience. We bring with us all sorts of expectations, possibilities and ways of making sense of an episode. In anticipation we may have a sensation of apprehension or possibly excitement. We may expect the experience to offer certain possibilities for action or outcome and may raise questions to be resolved. We will also anticipate the temporal and spatial character of the experience. We may come to the experience with a desire for fulfilling certain needs or we may be looking for inspiration. It is natural to think of anticipation as something that is prior to what ever it is an anticipation of, and this is true. But anticipation is not just prior. The sensual and emotional aspects of anticipation and our expectation of the compositional structure and spatio-temporal fabric of what follows, shapes later parts of the same experience, it is the relation between our continually revised anticipation and actuality that creates the space of experience.

3.2.2 Connecting

When a situation first impacts our senses before even giving meaning to it, the material components impact on us to generate some response, pre-linguistically. In the spatio-temporal aspect this may be an apprehension of speed or confusing movement or openness and stillness for example. An immediate impression of one frequently visited web-site is of redness and flesh tones which immediately gives an impression of sleaziness, yet on closer inspection it is a quite respectable e-commerce site. For the sensual aspect materially connecting may engender an immediate sense of tension or perhaps a thrill of novelty. For the emotional and causal aspects connecting may engender nothing more than a sense of relief or anticipation at something happening.

3.2.3 Interpreting

Giving meaning to an unfolding experience implies for the compositional and emotional threads, discerning the narrative structure, the agents and the action possibilities, what has happened, what is likely to happen and how this relates to our desires, hopes and fears and our previous experiences. We may sense the thrill of excitement or the anxiety of not knowing how to proceed or what will happen or where we are. At an emotional level, on the basis of our anticipation we may feel frustration or disappointment at thwarted expectations or we may regret being in this situation and have a desire to remove ourselves from it. On the basis of our interpretation falling short of our anticipation, we may reflect on our expectations and alter them to be more in line with the new situation.

3.2.4 Reflecting

At the same time as interpreting we may also make judgements about the experience as it unfolds and place a value on it. In reflecting on causal aspects, can we make any sense of things? Are we satisfied with a sense of progress or movement towards completion? From an emotional aspect, do we feel we are getting any sense of fulfilment or achievement? How does the experience tally with our anticipation and how do we feel about being in this situation at this time? From a sensual perspective are we anxious, bored, or excited. In addition to reflecting *in* an experience, we also reflect *on* an experience after it has run its course to completion. This often takes the form of an inner dialogue with oneself or with others. It is a kind of inner recounting. It serves to help us relate the experience to others in an evaluative way in support of appropriation and recounting which in their turn help us reflect.

3.2.5 Appropriating

A key part of sense making is relating an experience to previous and future experiences. In appropriating an experience we make it our own. We relate it to our sense of self, our personal history and our hoped for future. We may change our sense of self as a consequence of the experience, or we may simply see this experience as 'just another one of those'. The degree to which an experience changes our sense of self may also be the extent to which we see it as something we identify with and want to experience again. In relating experience to our future and past we also may look afresh at the experience or the setting engendering the experience. Sensual aspects of an experience may become just another "white knuckle ride" or they may become unique moments such as the unforgettable sense of immersion in pure translucent colour when as a first-time scuba diver, we descend into "the blue". The emotional aspects of flow and engagement during a session of the computer game "unreal tournament" may be quintessentially cathartic, providing a means of escape from the mundanities of everyday life. The compositional aspects of an experience may relate positively to our sense of self or not. Do we feel it is morally right or socially acceptable to go shopping at a virtual supermarket? How do we reconcile shopping at amazon.com with our commitment to 'the small bookshop' and to the concept of personal service? Is the experience of using a mobile phone one on which new possibilities for action in our everyday life become apparent or is it yet another concession to an undesirable technological future?

3.2.6 Recounting

Like reflecting and appropriating, recounting, takes us beyond the immediate experience to consider it in the context of other experiences. It is through a process of internal recounting that we reflect and appropriate experiences, but having appropriated an experience it is also natural to recount it to others. In this way we savour it again, find new possibilities and new meanings in it and this often leads us to want to repeat an experience—to go shopping again, to buy another book or to take another holiday. Experience often takes on different meanings or is giving different value when recounted in a different place at a different time—Marco Polo is reputed to have once said that adventures are hardships and sufferings had in the re-telling. Through recounting to others we draw out an evaluative response from others which changes our own valuation of it. We might for example relate our experience of mobile phones or e-shopping as a zealot but through dialogue with others become something of an apologist.

4   The Framework in Use

We have used the framework as a starting point for data collection in a study of brand and online experience with the consultants Siegelgale UK. We presented the framework to the company as a written report and an audio-visual presentation. The presentations indicated how the framework could be used to assess the qualities of an on-line shopping site. The company, with our assistance then went on to design its own evaluation exercise using the framework. The context of use we chose involved a variety of experiences of a particular highly branded organisation, Virgin. In a small qualitative study, seven participants were asked to use Virgin services in a number of channels including 'bricks and mortar', web, and WAP. Each of these involved different activities like buying a CD, getting advice on savings for retirement and so on.

They introduced their participants to the ideas of the framework and provided them with a notebook—come—checklist which they called a diary. Each notebook page was divided into sections corresponding to the sense making processes (e.g. anticipating, connecting, interpreting, etc.) and was accompanied by a checklist of concepts and guidewords from the framework. Participants were then asked to go off and have their Virgin experiences. In addition to their diary, they provided an oral account during a one-on-one debriefing afterwards. Both the contemporaneous notes and the debriefing were informed by a series of relatively open-ended questions. These were designed to facilitate the construction of a narrative of the experience that would engage with the concepts of the framework.

The study is reported elsewhere (McCarthy et al. 2002), but to give a flavour of the results, we present here an example of a user's experience of buying a CD at Virgin Megastores on-line. He wanted to find out the cost of a Nelly Furtado album and when it will be out and also the price of the Coldplay album.

Anticipating

Expecting it to do all the things as in the store and more ... find out information about artists, albums, gig information ... download free or pay audio and visuals, magazines books ... order on line, reserve on line.

Connecting/interpreting/reflecting

Very busy site- lots of flashing and whizzing going on [this in relation to virgin.com portal page] scanning to find search or section. I selected the Megastore link, I've gone some-where but not sure where as nothing is orienting me. In fact, what is the relationship between the tab and the Megastore link on the left hand side?... Is this an American site of a British site? Not much use to me- I'll find a review of coldplay and buy it from a music store instead. Pity though there are no star ratings for the album. Nothing on Nelly.

Recounting

Not that useful to me as it is very US oriented. Not as rewarding an experience as I had hoped for, it was more of an information site as opposed to an interactive site. Cannot download audio or video cannot buy online or reserve online. Disappointed and unlikely to recommend the site to anyone.

Siegelgale also recounted to us their own reflections about their experiences with the framework. One of the most important features for them was the importance of being able to question expectations and anticipations. They also concluded that aesthetics (by which they meant the sensual and emotional aspects of experience) is the key to being able to articulate user experience. Their view was that although they could understand the idea of sense-making and its components (connecting, interpreting, reflection and so on) these were difficult to work with and often enough it was sufficient to talk about three of the four threads of experience (sensual appearance, compositional structure, emotional unity). They also suggested that a concept of physicality or embodiment seemed to be lacking in the framework. While we might consider such issues within the spatial-temporal thread, our participants wanted something more direct to capture the similarities and differences between on-line and offline, between actually physically handling objects and reading about their descriptions.

Although some of the difficulties experienced by Siegelgale suggest to us that the framework approach requires some fine-tuning and some clarification of concepts if it is to be used as a tool for analysing experience, nevertheless these results are promising. They suggest that the underlying concepts are manageable and usable by practitioners as a way of thinking about user experience.

5   Conclusions

It is common these days to hear the term experience in connection with theme parks, Hollywood blockbusters, pre-packaged adventure holidays and so on. Experience is used to sell. We are guaranteed an 'experience of a life time' all we need to do is show up with the right amount of money. But Dewey and Bakhtin show us that experience is as much a product of what the user brings to the situation as it is about the artefacts that participate in the experience. What this position implies is that we cannot design an experience. But with a sensitive and skilled way of understanding our users, we can design *for* experience.

Design for experience requires the designer to have ways of seeing experience, to talk about it, to analyse the relations between its parts and to understand how technology does or could participate to make that experience satisfying. The framework we have presented here is not a method for analysing experience, rather it is a set of conceptual tools or a language for thinking and talking about experience. It is intended to help make visible what we consider to be the essential characteristics of experience. Characteristics that differentiate from behaviour, practice, knowledge and other more familiar psychological categories. The case study, informal though it was, suggests that the framework can provide practitioners with an understanding of the concept of experience that would help them design for experience.

References

Agre P (1997) Computation and human experience. Cambridge University Press, Cambridge, UK

Bakhtin M (1986) Speech genres and other late essays. University of Texas Press, Austin, TX

Bakhtin M (1993) Toward a philosophy of the act. University of Texas Press, Austin, TX

Belenky MF, McVicar Clinchy B, Goldberger NR, Tarkle M (1986) Women's ways of knowing: the development of self, voice and mind. Basic Books, New York

Boorstin J (1995) Making movies work: thinking like a filmmaker. Salaman James Press, Beverley Hills, CA

Bowen S, Dearden D, Wolstenholme D, Cobb M (2011) Different views: including others in participatory health service innovation. In: Proceedings of the second participatory innovation conference, University of Southern Denmark, pp 230–236

Bruner J (1987) Life as narrative. Soc Res 54:1–17

Bruner J (1990) Acts of Meaning. Harvard University Press, Cambridge, Mass

Bruner EM, Turner V (eds) (1986) The anthropology of experience. University of Illinois Press, Urbana

Clarke R, Wright P, Balaam M, McCarthy J (2013) Digital portraits: photo-sharing after domestic violence. In: Proceedings of the Chi'2013. ACM, New York, pp 2517–2526

Csikszentimihalyi M (1990) Flow: the psychology of optimal experience. Harper and Row, NY

Dewey J (1925) Experience and nature. Open Court, LaSalle, IL

Dewey J (1934) Art as experience. Perigree, New York

Ehn P, Löwgren J (1997) Design for quality-in-use: human-computer interaction meets information systems development. In: Helander M, Landauer TK, Prabhu P (eds) Handbook of human-computer interaction, 2nd edn. Elsevier Science, Amsterdam, NL, pp 299–313

Hook J, McCarthy J, Wright P, Olivier P (2013) Waves: exploring idiographic design for live performance. In: Proceedings of the Chi'2013. ACM, New York, pp 2969–2978

Jay M (2005) Songs of experience. University of California Press, Berkeley

Lee J, Kim J, Moon JY (2000) What makes internet users visit cyber stores again? Key design factors for customer loyalty. In: Proceedings of CHI'2000, The Hague, Amsterdam. ACM Press, New York, pp 305–312

McCarthy J, Wright P (2003) The enchantments of technology. In: Blythe M, Monk A, Overbeeke C, Wright PC (eds) Funology: from usability to user enjoyment. Kluwer, Dordrecht, pp 81–90

McCarthy J, Wright P (2004) Technology as experience. MIT Press, Cambridge, MA

McCarthy JC, Wright PC (2015) Taking [A]part: the politics and aesthetics of participation in experience-centered design. MIT Press, Cambridge, MA

McCarthy J, Wright P, Meekison L (2002) Characteristics of user experience of brand and e-shopping. Paper presented at the international symposium of cultural research and activity theory, ISCRAT 2002, Amsterdam, The Netherlands

Norman DA (2013) The design of everyday things. MIT Press, Cambridge, USA

Norman DA (2005) Emotional design: why we love or hate everyday things. Basic Books, New York

Pu P, Faltings B (2000) Enriching buyers' experiences: the SmartClient approach. In: Proceedings of CHI'2000, The Hague, Amsterdam. ACM Press, New York, pp 289–296

Rancière J (2006) The politics of aesthetics: the distribution of the sensible. Continuum International Publishing

Taylor N, Cheverst K, Wright P, Olivier P (2013) Leaving the wild: lessons from community technology handovers. In: Proceedings of ACM CHI'2013 conference on human factors in computing systems. ACM Press, pp 1549–1558

Thieme A, McCarthy J, Johnson P, Phillips S, Wallace J, Lindley S, Ladha K, Jackson D, Nowacka D, Rafiev A, Ladha C, Nappey T, Wright P, Meyer TD, Olivier P (2016) Challenges in designing new technology for health and wellbeing in a complex mental healthcare context. In: Proceedings of CHI'2016, pp 2136–2149

Turkle S (1995) Life on the screen: identity in the age of the internet. Phoenix, London

Wright P, Olivier P (2012) Digital interaction culture lab. Interactions Magazine 19:1

Wright PC, McCarthy JC, Meekison L (2004) Making sense of experience. In: Blythe M, Monk A, Overbeeke C, Wright PC (eds) Funology: From usability to user enjoyment. Kluwer, Dordrecht, pp 43–53

Wright P, McCarthy J (2004) The value of the novel in designing for experience. In: Pirhonen A, Roast C, Saariluoma P, Isom H (eds) Future interaction design. Springer, Berlin

Wright P, McCarthy J (2008) Experience and empathy in HCI. In: Proceedings of Chi'2008. ACM Press, pp 637–646

Wright PC, McCarthy JC (2010) Experience-centered design: designers, users, and communities in dialogue. Morgan and Claypool

Wright P, McCarthy J (2015) The politics and aesthetics of participatory HCI. Interactions Magazine 22:6. ACM Press, New York, pp 26–31

Wright P, Wallace J, McCarthy J (2008) Aesthetics and experience-centred design. ACM Transactions on Computer-Human Interaction (TOCHI) 15:4. ACM Press, New York, pp 1–21

Chapter 21
Enjoyment: Lessons from Karasek

Petter Bae Brandtzæg, Asbjørn Følstad and Jan Heim

1 Introduction

What makes some experiences enjoyable, and other experiences not? How can we understand enjoyment in human factors design; what components should we consider when we are designing for enjoyment? This chapter explores a theoretical model for understanding the components and nature of enjoyment, and how HCI (Human Computer Interaction) professionals can use the model to predict and evaluate enjoyment. The model is a modified version of Robert Karasek's well-known *demand-control-support model* used in work and organisational psychology (Karsek and Theorell 1990).

Enjoyment is a subjective experience that may be understood in relation to theories of motivation. Two distinct types of motivation for engaging in an activity may be distinguished. Extrinsic motivation depends on the reinforcement value of the outcome of the activity, and parallels the idea of 'technology as tool'; in a traditional usability perspective, whether or not the technology functions as a means to complete well-defined tasks, particularly work related tasks. Intrinsically motivated action is perceived as rewarding in it self, and is a parallel to the idea of 'technology as a toy'. In their study of motivation and computers in the workplace, Davis et al. (1992) conclude that both intrinsic and extrinsic motivation explain workers' intentions when using computers.

It is central to the understanding of user behaviour that a complex pattern of behaviour may consist of both extrinsically motivated tasks and intrinsically motivated activity. As an example, search behaviour on the WWW will often not follow a strict goal oriented pattern. Rather the user may soon be lured away from her search task by an interesting piece of information that diverts her attention in a joyous short-lived oasis of distraction before she returns to her original search

P. B. Brandtzæg (✉) · A. Følstad · J. Heim
SINTEF Digital, Oslo, Norway
e-mail: Petter.B.Brandtzag@sintef.no

© Springer International Publishing AG, part of Springer Nature 2018
M. Blythe and A. Monk (eds.), *Funology 2*,
Human–Computer Interaction Series,
https://doi.org/10.1007/978-3-319-68213-6_21

chore; probably only to be lured away a second time. Even the most boring tasks may include refuges of intrinsically motivated activity, in the same way, joyous activities may involve extrinsically motivated tasks. These two kinds of behaviour may be close in time and space; but they still involve dissimilar sets of human factors issues and different theoretical assumptions.

1.1 Karaseks Demand-Control-Support Model

Karasek's paradigm describes a simple theoretical framework of good and healthy work. The model postulates that job satisfaction and well-being result not from a single aspect of the work environment, but from the joint effects of the experienced demands of the work situation, the decision latitude available to the worker, and finally the degree of social support from co-workers and management. Job satisfaction and well-being occur when job demands, job decision latitude and social support, are high (Karasek and Theorell 1990).

The goal structure of a good working life in Karasek's terms is similar to the goal structure of enjoyable activities. A good working life is not seen as the engagement in a series of well-defined tasks, achieving well-defined goals. Rather it consists of an interwoven complexity of activities in dynamic environments with several actors and conflicting interests. Nielsen (1996) has proposed that in business or work it is becoming common to cater to subjective whims and satisfaction. It is also interesting that work has been referred to as "hard fun" (Jensen 1999). Traditional usability assessment does not address this aspect of work life at all, but is mainly focused on optimising task performance for the lonely worker in a static work environment. However, it should be noted that Shneiderman (1987) refers to "subjective satisfaction" as one of several usability goals.

We contend that the demand-control-support model will be useful in the investigation and understanding of the enjoyable experience, because the model includes components that seem to be universal in the understanding of activities associated with well-being. Csikszentmihalyi (1992) states: "It would be a mistake to assume that only art and leisure provide optimal experience" (p. 52). It should then be reasonable to expect that an adjusted version of the Karasekian model may help to understand and predict fun and enjoyment in human factors design. Literature on the theory of fun and enjoyment in support of the proposed model will be reviewed below.

2 Demands and Enjoyment: Challenge and Variation

Challenges and variation of procedures and tasks are regarded as important aspects of the demands of a work situation. Variation reflects the degree of an active experience, which is essential to the design of good jobs (Karasek 1979; Karasek and Theorell 1990). The concept of demands is understood as the degree of variation and challenge experienced by users of technology, and is also an important aspect of enjoyment. Products and services for enjoyment do not necessarily provide a given task to be performed, but rather a notion of "a good experience". The central element in the optimal experience is that the activity is a goal in itself (Csikszentmihalyi 1992).

The concept of challenging demands is not unknown in the human factors literature. According to Skelly (1995) variation is a well-known means of exploiting the element of curiosity or surprise. It has similarly been argued that a certain degree of unpredictability is important for the experience of fun (Davenport et al. 1998).

Thackara (2000) points out that future human factors design should take into consideration the fact that people enjoy being stimulated.

2.1 Challenges

> A lot of pieces that you deal with are very straightforward … and you don't find anything exciting about them… but there are other pieces that have some sort of challenge … those are the pieces that stay in your mind, that are the most interesting. Csikszentmihalyi (1992, p. 51).

Many people use their leisure and spare times to solve hard puzzles or seek out difficult challenges. As an example, computer games are often experienced as fun when they have a certain level of difficulty. This can be explained by the notion that a dynamic environment is associated with challenges that invites activity and involvement. Those users meet challenges that are stimulating and encourage creativity. The users get the opportunity to test their own skills (Holmquist 1997).

Csikszentmihalyi (1975) refers to other examples of challenging demands: surgeons performing difficult operations or rock climbers struggling to scale an unclimbed mountain peak. Situations like these may be intensely demanding, but at the same time they may elevate the individual to a level of optimal experience or a "flow" experience: An experience that takes the individual to a state of absorbing engagement. In literature, "demands" in terms of challenges and variations are connected to the opportunity and motivation to learn (Karasek and Theorell 1990). The opportunity to acquire mastery may promote a feeling of self-confidence and an intrinsic motivation to use a particular technology.

Correspondingly, Springel (1999) argues that the computer is becoming a device for stimulation. Springel suggests that the growing interest in games, Web-entertainment, online chat etc. foreshadows a new attitude towards media. In turn this may shape a different form of media, which directly engages and challenges users and allows them to take active roles in co-creating the new media experience. A variant of this is the concept of reality TV, such as "Big Brother"

Fig. 1 Big Brother in Norway—one of the most watched and successful TV programmes ever shown on Norwegian Television

(see Fig. 1). As a viewer you can vote for the participants you like or dislike. You can also choose to follow the program in different media channels such as the Internet and mobile phones and you have the opportunity to e-mail and chat with the participants in the show. The same tendency is reflected in other media channels such as MUDs (Multi User Domains). This trend is moving users away from being passive consumers to becoming active collaborators.

2.2 Variation—And the Surpassing of Users' Expectations

In the face of routine and repetitiveness, people easily get bored. Variation may be seen as hinged on assuming a universal human interest in novelty and fascination for surprises, spontancity, freshness and a certain degree of unpredictability. The importance of variation is congruent with the meaning of demands stated in the Karasek-model.

From a human factors perspective, the concept of variation involves the prediction that products or services with a static design, and at the same time no novelties or change, will lose users' interest. Today we see an increasing rate in the turnover of news, trends, systems, applications and products to meet the demands for variation and novelty. Karasek and Theorell claim that "new challenges must constantly be confronted - and offering them will be a significant challenge for work designers" (Karasek and Theorell 1990, p. 173). The same challenges will probably be significant for the human factors designer.

Unpredictability, or the element of surprise, is an important facet of variation. There are of course different levels of unpredictability. Regarding human factors design, it is unpredictability just on the edge of security, involving no more than minor risks or possible penalties, which will be of central relevance to fun (Davenport et al. 1998). Psychoanalytic theory also emphasises the importance of risk to jokes and fun (Freud 1960). Hassenzhal et al. (2000) state that products should have elements added in order to make them interesting, novel, or surprising. Designing for fun may involve the inclusion of something unexpected, an element of surprise and unrelated or opposing events. Surprise and unpredictability are also well-known approaches in marketing research to gauge consumer experience. Consumer experience is often understood in terms of the discrepancies between *ex ante* expectations of a product and the products *ex post* performance. The best predictors of a good experience are when the product actually exceeds the users' expectations (Oliver 1981).

Unpredictability and challenges may partially be in conflict with principles of traditional usability (Hassenzhal et al. 2000). Making something as simple as possible may make it boring. In the Karasek model the importance of the joint effects of demands and decision latitude is addressed. Karasek (1979) stresses the importance of matching challenges with individual skills and control. Csikszentmihalyi (1975) states that "flow" requires a subtle balance of not being too simple and not being too challenging. Demands without the experience of control will result in a stressful and frustrating experience, rather than the experiences of joy. Decision latitude or control without any demands will probably imply a passive and probably boring interaction. An enjoyable experience is dependent on the balance between demands and control.

3 Decision Latitude and Enjoyment: Skill Discretion and Decision Authority

User control is regarded as an important aspect of an enjoyable experience and is addressed through the concept of decision latitude. In consonance with the Karasek model, the concept is defined as including: the ability to use and develop skills, and the availability of decision-making authority or freedom of action. Decision latitude can also be seen in the light of engagement, a concept discussed by Brenda Laurel (1991). Engagement refers to the user's feeling of being in control of the interaction. Laurel writes about computer fiction, games, etc. and addresses the subject "I" who interacts in a virtual world. There should be nothing to mediate the communication between the user and the system. "I do, what I myself want and feel involved in what I am doing." (p. 116). Laurel suggests that the frequency of interaction, the range of possible alternatives available for selection at a given time and the effectiveness of the inputs influence engagement. The interface should enable the user to see the effects of her or his actions, to give a sense of agency or personal power.

3.1 Skill Discretion: The Opportunity to Use and Develop Skills

Getting to use one's own skills in the fullest range possible helps to make activities enjoyable. Everyone knows the truly intrinsic joy attached to engagement in activities that invite the utilisation of acquired skills. The ability to use and develop skills may be seen as related to the term "self-efficacy". Within social cognitive theory self-efficacy is the belief "in one's capabilities to organise and execute the courses of action required to produce given attainments" (Bandura 1997, p. 3). Self-efficacy involves monitoring and evaluating one's own actions, and influences the individual's decision about what activities to engage in, whether to proceed with the activity when faced with obstacles, and the mastery of that particular activity. Self-efficacy is not a measure of skill, but is closely related to the likelihood of the individual to reach the ability to using a given technology at a level of contentment. Factors such as unwanted complexity, unrealistic tolls on the knowledge of potential users, and other comfort issues faced by the users, may be construed as self-efficacy deficits. Low self-efficacy is likely to mean that the use of technology is not perceived as rewarding, following which it is less likely that the technology will be used in the future (Eastin and LaRose 2000). Conversely, high self-efficacy will be correlated with the tendency of increased technology use, and also the development of skill and mastery.

The development of skills may be seen as a path winding from the first faint attempts of the novice, to the full-blown repertoire of skills of the expert. Dreyfus et al. (1986) describe progress from novice to expert as going through qualitatively different stages; from early efforts of learning rules and repetitive training, to the non-reflective mastery of the expert. As skills develop, the engaged-in activities will gradually become intrinsically satisfying.

3.2   The Decision-Making Authority of the User

According to Karasek, elevating the level of the decision-making authority of the worker improves wellbeing at work. Similarly, providing the user with extended powers of decision-making may enhance the fun experienced in a user-technology interaction. The level of decision-making authority of the user is a consequence of constraints inherent in the relation between the user and the technology. Unlike constraints on decision-making in a workplace, rules, strict routines, organisational hierarchy and repetitiveness, the constraints of the user-technology relation are a consequence of the user not seeing or understanding the possibilities represented by the technology. Choi et al. (1999) address this point in their analysis of computer-game design factors. They suggest that fun computer games are characterised by the gamers freedom and leadership in the progression of the game's story. In order to make possible the leadership of the gamer, computer-games must give hints in a manner the gamers can perceive and understand. This, of course, without giving away the complete solution to the game.

Another example of user control is the opportunity of "fun design" through *personalization* of the technology. Nokia have served their users with products they can personalize with tremendous success. In the case of, for example, the Nokia 3350, users can personalize their phones by using a ringing tone as an SMS alert or compose their own SMS alerts for originality. A picture editor allows users to create and personalize picture messages for all occasions. The phone's rhythmic backlight alert accompanying a ringing tone also makes it fun to receive a call. Users can also opt for a fully personalized look by downloading profile names, logos and ringing tones. Other personalization options include an exclusive range of changeable Xpress-on covers in new and exciting colours.

These opportunities to personalize the technology give the user decision-making authority over the technology. It enables the user to influence and to create their own experience in a dialogue with the technology. The role of personalization may also be seen in relation to another phenomena, social cohesion or social identity; that you give the user a feeling of being part of a group, which will be discussed further in Sect. 4.2.

4 Social Support as Enjoyment: Co-activity and Social Cohesion

Many leisure activities may be characterised as socialising. Thackara (2000) states that providing the user with a sort of community is important in designing for a good user experience. Also an awareness of the conditions that support enjoyable social interaction is important in the design of systems (Monk 2000). The parallel in Karasek is social support. Karasek and Theorell (1990) define social support as "the overall levels of helpful social interaction available on the job from both co-workers and supervisors" (p. 69). Two element of this may be useful in explaining enjoyment: co-activity and social cohesion.

4.1 Co-activity

The concept of co-activity implies collective action; a user's social behaviour, not just human-computer interaction as a single user of the system. User studies have pinpointed the fun of doing things together (Mäkelä and Battarbee 1999; Mäkelä et al. 2000; Battarbee et al. 2000), and technology that promotes an opportunity for social interaction probably supports an enjoyable user experience. This is illustrated in Fig. 2, and the success of the multi-player options in Game Boy Advanced.

Communication and interaction are identified as the most popular activities on the Internet (December 1996). Likewise, the spread and use of mobile telephones can be seen as an example of a social need. Studies indicate that young people and adults use mobile telephones differently, where young people engage in expressive rather than informative use (Ling 1999). A Swedish field study that mapped the use of mobile telephones among youngsters, found that both the mobile phone and the information on it is often shared among users and made public in various ways. Young people use mobile phones for doing things together in collaborative, social action, rather than task-related communication (Weilenmann and Larsson 2000). The sharing of experiences, feelings and information is considered to be rewarding, pleasant and enjoyable. However, there may also be an element of competition or

Fig. 2 Game Boy Advanced:
with multi-player
opportunities

contest, which by nature is seen in relation to fun. Studies show that young people are likely to use the computer for playing games together, rather than playing in isolation (Wartella et al. 2000). These findings may be explained by the *social facilitation effect*; it is easier, and more rewarding and motivating to do things in the presence of others, because mere presence of others is arousing (Zajonc 1965). Children play more enthusiastically if a playmate is near by, even if only engaged in parallel play

4.2   Social Cohesion

Social cohesion is related to a social expression of being part of or attracted to a community. Social psychologists define cohesion as the attraction towards the group and motivation to participate in the activities of a group (Cartwright and Zander 1960). Being part of groups with high levels of social cohesiveness is positively related to individual wellbeing (Sonnentag 1996). The term *affiliation* is related to social cohesion. Affiliation occurs because social contact is rewarding. The rewarding aspect of affiliation includes emotional happiness and cognitive stimulation, opportunity for self-confirmation through the attention of others, opportunity for relevant self-knowledge through social comparison, and the opportunity of emotional support and sympathy (Hogg and Abrams 1993). Individually experienced rewards associated with social cohesion may be important to understand fun in human factors design. Mäkelä et al. (2000) found that children use digital images for joking, storytelling and sharing art. One of the purposes of sending images seemed to be to maintain attraction between group members. The social nature of technology use is also reflected in the boom of communities of young media users who create their own Web pages (Wartella et al. 2000). Such personal online publishing offers a fun way for young people to connect with their peers and others interested in the same topic. Jordan (1997) considers 'pleasure with products' to be characterised by social relations and communication enabled by the product. These products bring people together and provide topics of discussion or conversation.

Interactive technology is not, and should not be, socially isolating. On the contrary, it should be used for important social activity. High levels of social support are, according to Karasek and Theorell (1990), important in providing favourable effects in the interaction between demands and decision latitude. Human factors design should focus on the development of design, which provides more social opportunities, to facilitate enjoyable experiences.

5 Conclusion

Karasek's demands-control-support model, which predicts wellbeing and motivation in the context of work, is useful for understanding enjoyment in human factors design. When designing for enjoyment, professionals should consider demands, but at the same time allow a high degree of decision latitude and socially rewarding activity. The factors of demands and decision latitude have been treated separately, but it is important to address the crucial interaction effects between these. It has been explained that an enjoyable user-technology interaction depends on the interaction effects of challenge and use and the development of skills, as well as variation and the enabling of the decision-making authority of the user. How these joint effects enable an enjoyable experience will of course depend on the context. In addition, the effects of co-activity and design in support of social cohesion have been discussed. On this basis, three implications for the design of enjoyable technology may be formulated:

- *User control and participation, with appropriate challenges*: to enjoy technology the user should be enabled to carry out challenging activities. These activities should attract the user's attention and test his or her skills. Besides being challenging the design should allow the user to feel in control of the interaction. The user also needs to see the effects of her actions in order to give a sense of agency or personal power. To give the users an experience of active participation is central to an enjoyable experience.

- *Variation and multiple opportunities*: the user should be provided with a high level of variation by offering multiple possibilities and services. There should be an opportunity to personalise the product. This should be under direct user control, where the user explicitly selects between certain options. A key point is to give the user more than they actually expect.
- *Social opportunities in terms of co-activity and social cohesion*: the technology should give the user a feeling of being part of a group. The technology should also enable the users to do things together in social activities. A socially rewarding environment is necessary and essential for all humans, also when it comes to enjoyment.

Future research should focus on the development of reliable and valid measures of the factors and aspects that have been introduced. In addition the question of integrating knowledge and the evaluation of fun in the design process will be a major challenge in the future.

Acknowledgements We would like to thank our colleague Anne Lund.

References

Bandura A (1997) Self-efficacy: the exercise of control. W.H. Freeman, New York

Battarbee K, Mattelmäki T, Mäkelä A (2000) Design for user experience, method lessons form a design student workshop. In: Proceedings of the 1st Nordic CHI, Stockholm

Cartwright D, Zander A (1960) Group dynamics. Research and theory, 2nd edn. Row, Peterson and Company, Evanstone

Choi D, Kim H, Kim J (1999) Toward the construction of fun computer games: differences in the views of developers and players. Pers Technol 3:92–104

Csikszentmihalyi M (1975) Beyond boredom and anxiety. Jossey-Bass Publisher, San Francisco

Csikszentmihalyi M (1992) Flow. The psychology of happiness. Rider, London

Davenport G, Holmquist LE, Thomas M (1998) Fun: a condition of creative research. IEEE Multimedia 5(3):10–15

Davis FD, Bagozzi RP, Warshaw PR (1992) Extrinsic and intrinsic motivation to use computers in the workplace. J Appl Soc Psychol 22:1111–1132

December J (1996) Units of analysis for Internet communication. J Commun 46(1):14–38

Dreyfus HL, Dreyfus SE, Athanasiou (1986) Mind over machine: the power of human intuition and expertise in the era of the computer. Basil Blackwell, Oxford

Eastin MS, LaRose R (2000) Internet self-efficacy and the psychology of the digital divide. In: JCMC, 6,1. Retrieved April 16, 2001 from the World Wide Web: http://www.ascusc.org/jcmc/vol6/issue1/eastin.html

Freud S (1960) Jokes and their relations to the unconscious. Routledge & Kegan Paul, London

Hassenzhal M, Platz A, Burmester M, Lehner K (2000) Hedonic and ergonomic quality aspects determine a software's appeal. In: Proceedings of the CHI 2000 conference on human factors in computing systems. Apr 1–6, 2000, The Hague, Netherlands, pp 201–208

Hogg M, Abrams D (1993) Towards a single-process uncertainty-reduction model of social motivation in groups. In: Hogg M, Abrams D (eds) Group motivation. Social psychological perspectives. Harvester Wheatsheaf, New York

Holmquist LE (1997) The right kind of challenge. In: Braa K, Monteiro E (eds) Proceedings of the 20th informations systems research seminar in Scandinavia. IRIS 20. Department of Informatics, University of Oslo

Jensen R (1999) Dream society. The coming shift from information to imagination. McGraw-Hill Book Company, London

Jordan PW (1997) The four pleasures—taking human factors beyond usability. In: Proceedings of the 13th triennial congress of the International Ergonomics Association, vol 2. Finnish Institute for Occupational Health, Helsinki, pp 364–365

Karasek R (1979) Job demands, job decision latitude, and mental strain: implications for job redesign. Adm Sci Q 24:258–307

Karasek R, Theorell T (1990) Healthy work: stress, productivity, and the reconstruction of working life. Basic Books, New York

Laurel B (1991) Computer as theatre. Addison-Wesley, Reading

Ling R (1999) We release them little by little. Maturation and gender identity as seen in the use of mobile telephony, Telenor R&D Report 5/99

Mäkelä A, Battarbee K (1999) It's fun to do things together: two cases of explorative user studies. Pers Technol 3:137–140

Mäkelä A, Giller V, Tscheligi V, Sefelin R (2000) Joking, storytelling, artsharing, expressing affection: a field trial of how children and their social network communicate with digital images in leisure time. In: Proceedings of the CHI 2000 conference on human factors in computing systems. April 1–6, 2000. The Hague, Netherlands, pp 548–555

Monk AF (2000) User-centred design: the home use challenge. In: Sloane A, van Rijn F (eds) Home informatics and telematics: information technology and society. Kluwer Academic Publishers, Boston, pp 181–190

Nielsen J (1996) Seductive user interface. Retrieved April 12, 2001 from the World Wide Web: http://www.useit.com/papers/seductiveui.html

Oliver RL (1981) Measurement and evaluation of satisfaction in retail settings. J Retail 57 (Fall):25–48

Shneiderman B (1987) Designing the user interface: strategies for effective human-computer interaction. Addison-Wesley Publishing Co., Reading

Skelly T (1995) Seductive interfaces—engaging, not enraging the user. In: Microsoft interactive media conference. Retrieved April 10, 2001 from the World Wide Web: http://www.designhappy.com/sedint/TheMaze.htm

Springel S (1999) The New media paradigm: users as creators of content. Pers Technol 3:153–159

Sonnentag S (1996) Work group factors and individual well-being. In: West MA (ed) Handbook of work group psychology. Wiley, Chichester, pp 345–367

Thackara J (2000) The design challenge of pervasive computing. In: CHI, 2000. Retrieved April 1, 2001 from the World Wide Web: http://www.doorsofperception.com/projects/chi/

Wartella E, O'Keefe B, Scantlin R (2000) Children and interactive media. A compendium of current research and directions for the future. A report to the Markle Foundation. Markle Foundation. Retrieved February 2, 2001 from the World Wide Web: http://www.markle.org/news/digital_kids.pdf

Weilenmann A, Larsson C (2000) Collaborative use of mobile telephones: a field study of Swedish teenagers. In: Proceedings of the 1st Nordic CHI. Stockholm

Zajonc RB (1965) Social facilitation. Science 149:269–274

Chapter 22
Fun on the Phone: The Situated Experience of Recreational Telephone Conferences

Darren J. Reed

Author's Note, Funology 2
The original paper was written as a Research Fellow in the human–computer interaction group at York University. The research was centred around the study of naturalistic settings and the design of technology for older users.

Darren Reed, now a Senior Lecturer in Sociology, has gone on to study other areas of technology mediated communication from a conversation analytic perspective. These have included social interaction with robots, bus information systems, and the music sharing site Soundcloud. He currently looks at the relationships between 'digital creativity' and environmental monitoring.

2003 Chapter

1 Introduction

How do we know when we are having fun? A psychological approach can tell you what it feels like or how it is perceived; a sociological approach, on the other hand, can show how, as active participants in structures of social relevance, members of society *have fun together*.

We take the view that 'having fun' is a *situated and interactional experience*. By which we mean that identifiable social arrangements encourage, allow for, or engender a collective sense of enjoyable engagement. That they are identifiable means they are available for empirical observation. We have therefore a specific notion of fun in mind that might contrast with others (Blythe and Hassenzahl, this volume).

Goffman's notion of 'situated experience' is enlisted as a theoretical and conceptual basis, and conversation analysis (henceforth CA) as a methodology in our

D. J. Reed (✉)
Department of Sociology, University of York, York, UK
e-mail: darren.reed@york.ac.uk

© Springer International Publishing AG, part of Springer Nature 2018
M. Blythe and A. Monk (eds.), *Funology 2*,
Human–Computer Interaction Series,
https://doi.org/10.1007/978-3-319-68213-6_22

investigation of technologically mediated interaction. CA is a form of detailed 'naturalistic observation' that seeks to understand structures of meaning generated sequentially through interaction between individuals—typically in 'talk-in-action'. The combination of CA and Goffman—what we might call directed or applied CA—is seen in Hopper's (1992) analysis of play on the phone. Our contribution builds on this example in the situation of recreational telephone conferences, and aims to provide a viable approach for product development. Our efforts are preliminary, and offer only exemplars of analytic findings, but start to build a case for an empirical sociological account of technologically mediated experience.

2 Fun as Situated Experience

Erving Goffman is credited with bringing the individual into sociology and developing, according to Williams (1998), 'a distinctly sociological account of the person'. He does this by moving attention away from the individual's 'inner' life and toward 'externally observable forms of conduct' (ibid). He builds his notions of fun on the related area of play.

In sociology, play has historically been contrasted with work (Slater 1998) or conceived as meeting a range of social functions: from the socialisation of children to the large-scale development of culture (Bruner 1976; Huizinga 1949). By contrast, for Schwartzman (1978) play is a 'context of activity rather than a structure' (Sutton-Smith 1988: xi). Play is the product of action, creates its own context, and is freed from specified space and time.

Sociology's 'linguistic turn' in the 1960s (Lemert and Branaman 1997) brought an emphasis on the individual and an acceptance of communications-based notions of play. Stephenson (1967) sees media consumption as a form of play in which potential 'communication pleasure' involves complete and effortless engrossment. Bateson (1972) conceives of 'metacommunicative' cues that 'frame' behaviour beyond what is actually said or done; an example being the cue 'this is play', seen in the play fighting of animals. In what might be seen as a combination of these ideas, Goffman (1961) conceives of fun in terms of mutual engrossment in a social 'encounter' or 'focused gathering'. He says, 'When an individual becomes engaged in an activity … it is possible for him to become caught up by it, carried away by it, engrossed in it - to be, as we say, spontaneously involved in it' (Goffman 1961: 38). Fun is the sense of euphoria possible when there is 'spontaneous co-involvement', when all are engrossed in a commonly understood encounter.

Goffman therefore looks to identify the social propensities of experience: not as cognitive processes, but as *situated experience* within sociological arrangements. He does this by developing Bateson's idea of frames to cover all social meaning. Frames become 'principles of organization which govern events – at least social ones – and our subjective involvement in them' (Goffman 1974: 10–11). Goffman defines frame analysis as an examination of 'the organization of experience' (ibid: 11). Individual experiences in socially organized frameworks of meaning are more than mental emotions:

> frameworks are not merely a matter of mind but correspond …to the way in which an aspect of the activity itself is organized…Organizational premises are involved, and these are something cognition somehow arrives at, not something cognition creates or generates. Given their understanding of what it is that is going on, individuals fit their actions to this understanding and ordinarily find that the ongoing world supports this fitting. These organizational premises – sustained both in the mind and in activity – I call the frame of the activity.

> …activity interpreted by the application of particular rules and inducing fitting actions from the interpreter, activity, in short, that organizes matter for the interpreter, itself is located in a physical, biological, and social world. (Goffman 1974: 247)

To Goffman frames of meaningful experience are cognitive ('a matter of mind'), social (organized activity) *and* material (physical, spatial, temporal). What's more they are contextual and temporal: worked out in ongoing social activity.

2.1 Schemata of Frame Analysis

All social experience is made meaningful by frames, the most fundamental being 'primary frameworks'. There are two kinds of primary framework: *natural* and *social*. The first set of frameworks define situations in terms of physical contingencies that are not controlled by humans, such as the weather (Goffman 1974: 22); the second make sense of situations in terms of human intervention, activities, motives and the like. A barbecue, in which friends talk chat and eat, can be understood in terms of a social frame. Success or failure of the event might similarly be decided through a social frame of interpretation: people might not get on with each other, for example. On the other hand, a sudden downpour that sends guests scurrying inside without eating could be explained in terms of a natural frame, beyond the control of human actors (Fig. 1).

These primary frameworks however can be transformed into a new meaning. For example, the host of the barbecue might be held responsible for organising the event on the particular day if he or she was a professional weather person. The failure of the event might then be a matter of poor planning. Goffman says the meaningful experience can be 're-keyed'.

A key is,

> the set of conventions by which a given activity, one already meaningful in terms of some primary framework, is transformed into something patterned on this activity but seen by the participants to be something quite else. The process of transcription [or transformation] can be called keying. (Goffman 1974: 43–44)

Goffman identifies five basic keyed frames: make believe, contests, ceremonials, technical redoings, and regroundings (Goffman 1974: 48). The first of these the *make-believe key* playfully transforms a serious frame into a non-serious one and is

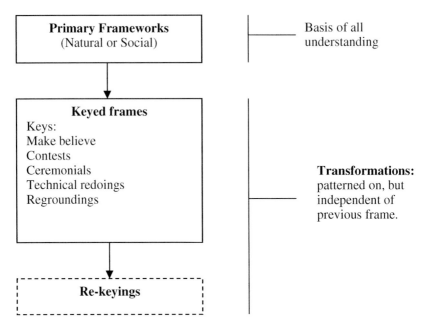

Fig. 1 A schemata of frame analysis

normally for the benefit of an audience. Daydreaming is an example of when the person employing the key and the person understanding the meaning are the same. Radio, television and theatre are examples of make believe keys because they involve 'dramatic scripting' according to Goffman. According to Manning (1992: 214) 'Make-believe [keys are] only sustained by a considerable collective engrossment in the transformed frame, and as a result, spontaneous make-believe keys are likely to be short-lived'.

An important point to be made about keys is they can also be transformed into new meaning—they can be *re*-keyed. Each re-keying is a transformation, a new frame patterned on the one preceding it. A situation may have many layers of meaning, or 'laminations',

> … the outermost lamination of a frame of a [theatrical] play tells us that what is happening is make-believe, even though an inner lamination tells us, perhaps, that Romeo and Juliet are very much in love. (Manning 1992: 126)

The potential confusion possible with multiple layers of meaning is avoided through clear frame 'anchors'; one example being 'brackets', which denote when a frame starts and ends. Like mathematical formula, brackets can be 'internal' or 'external': external brackets are not formally part of the activity to be framed, e.g. the raising of the theatrical curtain; internal brackets mark off 'strips' of the ongoing frame as having a separate meaning.

2.2   *Frames as Interactional Achievements*

An element that is underdeveloped in Goffman's frame analysis, is the *interactional achievement* of frames, specifically through the negotiation of bracketed meaning. When talking about playfulness, for example, Glenn and Knapp (1987) see 'a series of framing signals *mutually negotiated* by the participants' (emphasis added, p. 52). Frames of meaning are produced in real time interactive behaviour.

One way to extend Goffman's conceptual apparatus is to apply conversation analysis (CA), which is interested in the sense people make in sequences of action, as a temporal and emergent feature of people's interaction.

2.2.1   Thick External Bracket

Jason Rutter analyses comedy compéres' introduction talk in the 'Framing of Response' of the audience, which engenders 'alertness' and 'involvement' (Rutter 2000: 471). In this way, the comedy routine is framed as a comedic strip of experience,

> The introduction sequence is invariably a feature of stand-up openings and holds a position as forerunner to the entrance of comedians and their first joke. It provides a foundation for the performance and prepares the audience by establishing stand-up conventions, expectations and situation for the comedy to take place. (Rutter 2000: 482)

He makes a point about the keyed nature of jokes, in relation to the broader bracketed comedy routine when he comments,

> Given this organization, jokes performed by stand-up comedians cannot be seen as isolated texts. They cannot be seen as being hermetically separated from the ongoing performance, as they are located within, and part of, the developing interaction of stand-up. (Rutter 2000: 481)

While compéres' introductions have a conventional nature, Rutter notes an introduction sequence by Johnny Vegas—himself a comedian—that completely undermines these conventions. In a Goffman sense, the introduction-keyed frame is itself re-keyed.

Compéres' introductions are sequences of interaction (between compéres and audience) that we might think of as 'thick' brackets. Goffman details such a thick bracket in a description of a radio broadcaster introducing a live concert (Goffman 1974: 263). The framing work is achieved over a period of time, and may itself include internal brackets. The jokes are themselves transformed frames within the comedic episode based upon the compére's external bracketing.

2.2.2 Internal Brackets as an Interactional Achievement

In an explicit combination of conversation analysis (CA), Hopper (1992) uses the ideas of meta-communication and frames to understand the *interactive* nature of play on the phone. He underlines interactional achievement when he says 'Parties must … work out the course of play-in-progress' (Hopper 1992: 176). He notes that after ten years of looking for the 'elusive beginning bracket carrying the message 'this is play" he has instead come to the conclusion that it is misleading to think in terms of a single 'keying message signal' (p. 175). Instead, he says, telephone play takes the following form,

CAROL:	have you ate t'day?
(0.5)	
CAROL:	Eaten?=
Rick:	=eh heh heh=
CAROL:	=day hay huh huh eated. .hhhh=
Rick:	=No I-I-I- already eated heh heh heh

Transcript 1. *Telephone play, Hopper (1992:175–6)*

Rather than a specific cue, the extract is shot through with 'play-relevant keying' in the form of a *speech error* in the first line, *laughter* after the error is corrected and *repetition* in the last two lines. Hopper concludes 'The play frame is created and sustained through each of these interactive details. There is no single front bracket for play in this episode, but rather play's interactive management occurs across this entire fragment' (Hopper 1992: 176).

In short, 'Meta-communicative framing comes about not through individual message units that accomplish bracketing, but by interactive displays across speakers' turns. Each such indication of play's possible relevance may be confirmed, denied, ignored, or transformed by what happens next' (Hopper 1992: 177).

The example of CA and frame analysis allows for an appreciation of a number of specific interactional details such as episodic bracketing which may involve 'thick brackets' of interaction, and the play relevant keying of activity in a series of interactional turns. In each instance, successful framing is a transformation of the underlying frame.

3 Analysing Fun on the Phone

Goffman's concepts of fun and frame analysis and their application to real instances of activity through conversation analytic method inform our analysis of fun on the phone. This analysis is part of the ethnographic stage of work carried out at York

Fig. 2 Mutual elaboration of
conception and analysis

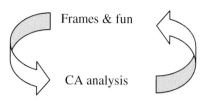

Frames & fun

CA analysis

University in an initiative investigating the 'good recreational experience' in technologically mediated interactions funded by PACCIT.[1]

The friendship links scheme[2] provides telephone conferences for isolated older people in London. Groups of four to eight people come together for an informal half hour chat each week and a trained coordinator or 'facilitator' encourages a friendly and lively atmosphere. With the consent of participants, several calls were recorded, transcribed[3] and analysed. This analysis involved combing the concepts in frame analysis and 'directed' conversation analysis. Rather than a one-way methodological process (i.e. concept informing observation), there developed a 'mutual elaboration' of concept and analysis (Reed 2002), wherein the concept of frames motivated looking at the data in a particular (directed) way and at other times CA provided the lead for conceptual development. Indeed 'analysis' and 'conceptual basis' developed together in a continual circular manner (Fig. 2).

For purposes of narrative integrity the following section contains exemplar 'findings' from this process that necessarily simplifies this relationship.

3.1 *From Formal to Fluid Interaction*

General observation of the telephone conferences revealed an interesting feature: facilitated calls started out stilted and formal, but at some point became a relaxed interaction between the whole group. The move from one to the other was highly relevant to us; the question being how it occurred. What became apparent was that we could understand this move in terms of particular interaction that framed the behaviour in different ways. We began to talk about this in terms of the move from 'formality' to 'fluidity'.

---

[1]People at the Centre of Communication and Information Technology.

[2]We would like to thank and acknowledge community network, the organisers of this scheme for their invaluable help.

[3]An integral part of the conversation analytic method is the detailed transcription of talk. Rather than a 'record' of the talk, transcripts and their generation are viewed as an essential part of the analysis process. However at all stages the original recording is regarded as the primary data and is re-turned to when reading the transcript.

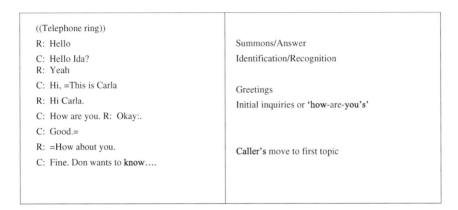

Fig. 3 Canonical opening of telephone call. Adapted from Schegloff (1986)

3.1.1 The Primary Frame of the Telephone

One way in which CA helped the conceptual development was in understanding the primary frame of the telephone conference. There has been a great deal of work on telephone conversations in CA that recognises the limitations of the audio channel and the consequences for behaviour. Schegloff (1986), for example, shows that there is a ritualised sequence of interaction—what he calls the 'canonical telephone opening'—at the beginning of a two person telephone call, to deal with questions about who is speaking to whom, about what, and who gets to speak next (Fig. 3).

The need to deal with such issues as 'access', 'recognition' and 'turn ordering' are emphasised within telephone conferences when there are more people involved in the call. In face-to-face conversation, the more people in an interaction the more difficult a conversation is. Atkinson (1982: 97) explains that ordered conversation is based upon 'solutions being found to what can be characterized as the problem of achieving and sustaining the shared attentiveness of co-present parties to a single sequence of actions'. The greater the number of people in a group the more difficult shared attentiveness is to generate, and the greater the likelihood of formal methods of coordination (ibid.). So, for example successful classroom interaction often depends upon rules about who gets to speak when. Wayne Beach suggests that facilitated group discussions lie toward the formal end of a continuum of formal and informal talk, what he calls a 'casual-institutional continuum' (Beach 1990: 201). All of which goes towards explaining why business meetings and telephone conferences have formal agendas.

Recreational telephone conferences have a natural frame of meaning based upon the physical (visual) limitations of the technology. From this *natural* frame, there needs to develop a *social* frame of socialising. This is achieved through particular social rituals.

3.1.2 Framing the Call—Introductions as Thick External Brackets

Given the primary frame of the telephone conference, several issues have to be addressed immediately upon the arrival of each new participant. Rituals have developed between switchboard operator, in group conversation facilitator, and new participant to do this.

In the calls we examined, an operator gains access to the ongoing call and introduces each new participant by directly addressing the facilitator. This 'direct introduction' leads to a 'directed welcome' segment between new participant and facilitator, which follows many of the ritualistic elements of a two-person call (greetings, how-are-you's, move to first topic). In the following example the call has been going for some time,

```
1     O:    [Excuse me lisa ]=
2     L:=y ↑Yes ↓Op[erator
3     O:           [It's the operator I've got ida joining you ↑now=
4     L: =ok R ROse had to go because she's not very well this
5     L: [morning
6     D: [oh poor thing
7     L: alright?=
8     D:=mmm=
9     L:=He↑llo Ida
10    I: Hello dear
11    L: hwa=
12    I: =i've got a dreadful coawd
13    L: ↑O↑h dear
14    I: Yeah I started off with hay>Fever Last< (.) week an I
15    I: finished up with larynGitis
```

Transcript 2. *Friendship links*

In the above example the Facilitator Lisa gives the merest hint of a how-are-you 'hwa' (line 11) and Ida immediately replies with news about her health (denoted by the latching convention '='). What follows is a small conversational sequence between the operator and Ida, which ignores the remaining group already on the line. This conversation introduces a particular conversation topic—that of Ida's health. In the next example, it is the facilitator who formulates the current topic at the end of the 'access/directed introduction/directed welcome' sequence with 'we were just saying it's a lovely morning…' (line 12),

```
1    O:   [ek (.) scyoos me again lisa Dorothy joining you?
2    L:   thank you very much
3    R:   >I dint have to put an extra blanket on<
4    L:   aahh [hhh
5    D:        [hhh
6    L:   morning dorothy?
7    D:   good morning lisa how are you
8    L:   I'm fine an an you alright?
9    D:   a:n so an so
10   L:   oh [oh]
11   D:      [ah] hh hh hh hh
12   L:   w we were just saying it's a lovely morning and its not so
13   L:   hot
14   D:   yeah dats right dis the humid weder doesn't agree with me hh
15   D:   hh hh hh
```

Transcript 3. *Friendship links*

In both cases—and indeed on every occasion of new participant introduction in this tape—the access/directed introduction/directed welcome sequence led to a period of two person (facilitator/new participant) conversation. At times these periods were extensive, leaving the remaining participants to just listen. The access/directed introduction/directed welcome sequence also impacts the ongoing conversation topic as mediated by the facilitator.

3.1.3   Two Forms of Internal Brackets—Formal and Fluid

Within the calls we identified two forms of framing activity, both aimed toward multiparty interaction. The first is a coordination effort on the part of the facilitator; the second form of 'frame breaking' initiated by participants.

Part of the facilitator role—engendered, as we noted, by the opening sequences of new participants—is the coordination of ongoing talk. A rather nice example of this coordination is a form of 'next turn allocation' we have called 'participation in the round'. Figure 4 shows some examples, the 'in-the-round' character being most apparent in the first example.

Evidently the facilitator's wish is to encourage talk from all; by allocating turns to each person their participation is ensured.[4] With each new participant comes a period of two-person interaction. So, while geared toward equality of activity, in the first instance it does not encourage multi-person interaction. This form of framing through internal brackets leaves the interaction with a formal structure.

---

[4]Facilitator professionals characterise such turn allocation as a form of 'ice-breaker' that gets the participants talking.

Example 1

```
L: so what did you do over the weekend, if we start off with you
L: renie.(.) did you do anything excitin(0.2)

((16 lines of interaction between Renie and Facilitator in which the
topic moves to the tennis, followed by …))

L:[ .hh what about] (.) dorothy do you watch the did you watch the
tennis

((23 lines of interaction between Dorothy and Facilitator))
```

Example 2

```
((prior conversation about school experiences))
L: (0.2) dorthy when you were at school obviously it was a different
process to the one we had=

((42 line of interaction between Dorothy and Facilitator))
```

Example 3

```
L: .h y you were saying um d d dorothy do do do you read a lot.

((9 lines of interaction between Dorothy and Facilitator, followed
by….))

L: WHAT ABOUT YOU RENIE you read novels?

((29 lines of interaction between Renie and Facilitator))
between Dorothy and Facilitator))
```

Fig. 4 Participation in the round

In some ways, these strips of formalised two-person interaction give way to more relaxed interaction. One way to understand how this occurs is to say that the current formal frame must be 'broken'. An example of frame breaking is through the 'flooding out' of laughter (see Goffman 1974: 350–9 for an account of 'flooding out').

Laughter is an accountable matter if placed wrongly; even one turn distance brings questions such as 'what's funny?' (Sacks 1995 VI: 746) and there is an imperative to 'get the laughter in' immediately upon its relevance. However, laughter is one activity that is available to people that, while 'tied' to the previous utterance, does not have to respect the general conversation rule of 'one person [speaks] at a time' (Sacks 1995: V1: 745).

```
1    L: w there you go (.) there's was I going to talk about the tennis
2    L: an I've been
3    L: squashed .hh [hh
4    D:          [hh hh hh hh hh hh hh hh [hh hh hh hh hh] hh
5    I:          [hh hh hh hh hh hh hh hh [hh hh hh hh hh] hh ahh
6    R:                        [well never mind dea:r]
7    I: wait a minute I gotta geta hanki
8    L: ok
9    R: (.) No well I think if it hadn't been for the raining I think
10   R: he would have won
```

Transcript 4. *Friendship links*

In the above extract, multi-person overlapping laughter (denoted by laughter tokens 'hh' in square brackets) appears to free up the interaction. This is one of the few occasions early in the interaction, for example, in which a participant jumps into the conversation—or 'self-selects'—(Renie at line 9) when others have been speaking in a period of two party interaction.

In the above example the Facilitator Lisa takes an in breath at line 3 ('.hh') that signals the onset of laughter. This is a minimal form of 'invitation to laughter'. A clearer example can be seen in the following example,

```
1    R: my grandchildren wont get m(hh)arried [hh hh hh
2    D:                         [ah hi h h h >h h h<
3    L: [they wont get ↑married]
4    R: yeah my grandson keeps saying (.)you gotta wait gran
5    R: she'll ave you'll ave we're ave great grandson (.) .hh you
6    R: gotta ang on till then [so i said ↑ow long do you think i'm=
7    D:                [Yeah
8    R: =gonna w(hh)ait [so he said i've got you on for the first=
9    D:                [exactly
10   R: =babys(hh)iter
11   D: [huh huh huh huh huh huh huh
12   L: [hh hh hh hh hh
13   R: i woont care i'm ninete:
14   L: [hh hh hh hh h
15   D: [oh: nn a:n
16   L: oh well you got a way to go ye:t
```

Transcript 5. *Friendship links*

Renie signals the humorous nature of her comment about her grandchildren not getting married with an inserted laughter token in line 1 with 'm(hh)arried', in what might be called a 'laughter voice' ('hh' denotes out breath). Renie goes on to explain the relevance of the comment in lines 4–10 and again signals with an inserted laughter tokens in 'w(hh)ait' and ''baby(hh)iter'. The 'punchline' comes at line 13 'i wont care i'm ninety:'. Instances such as these provide for the interactional achievement of playful keying according to Hopper,

> The first laugh raises the question 'is this play' and the second laugh ratifies play as a live possibility. The play frame is keyed not by just the first laugh, but by the shared laugher. (Hopper 1992: 180)

Interactionally achieved play frames are fragile and can easily be undermined with a serious comment (Drew 1987). In the above extract Renie's comment about her age is turned into a comment about longevity by Lisa the facilitator. Alternatively, initial invitation laughter tokens may not be 'picked up' in the first place. Further to this, Hopper says that the play frame must 'periodically be resustained' (Hopper 1992: 180) with further laughter.

At times in the interaction the tension between the need for coordination and fluidity is marked by the first countering the second. In the following example, the 'serious' talk that signals the end of the play frame is a turn allocation of the facilitator (lines 9–10). Another participant has been introduced into the group but is not feeling very well, she leaves immediately. To which Renie comments,

```
1    R: sounds if she's got dehidratid
2    L: (0.2) year dear that's (.) a shame ↑is'nt ↑it
3    R: yeah
4    D: ( )
5    L: right well there's just the three of us then girls
6    R: ah [he he he he ]he he he
7    D:    [he he he he]
8    R: ge:rwls loveley
9    L: so what did you do over the weekend, if we start
10   L: off with you rose.(.)
```

Transcript 6. *Friendship links*

The description 'girls' (line 5) is keyed as funny by Renie's 'invitation to laughter' at line 6 ('ah') and ratified by Dorothy at line 7. Lisa's 'so what did you do…' re-keys the interaction as a formal matter of coordination.

We see throughout the conference calls small segments of flooding out of the two party interaction at moments of shared laughter. At those moments it is possible for the formalized two-party interaction to be undermined and there occurs a momentary 'free-for-all', wherein any person can potentially talk. Segments of shared laughter and play-framing in combination with other elements can precipitate longer segments of fluidity in which the generic two party structure is turned to a multi-party interaction with much overlapping speech. These instances of spontaneous co-involvement are based upon these earlier moments of framing.

Summing up, these telephone conferences have a primary frame of seriousness due to the need for coordinated activity with the constraints of the telephone: opening sequences of telephone conferences along with coordination efforts, characterize them as formalized two-party frames of interaction. The ideal state for a recreational telephone conference is spontaneous co-involvement, which is worked toward through frame breaking activities such as shared and invited laughter.

4 Conclusion

Fun in telephone conferences is defined in this analysis in relation to what it is not: having fun is a matter of transforming the (necessarily) formal structured basis of activity into moments of triviality and playfulness. And as such it complements early sociological appreciation of play as not work. However, by understanding fun as the consequence of particular interpretive transformations in ongoing interaction, our sociology of fun becomes a dynamic conceptualisation: fun is always interactionally achieved by active social actors.

Our investigation of telephone conferences through applied or directed CA allows an appreciation of the interactionally situated experience of fun. A question might be how these insights benefit the future design effort. One way is that they can be recruited to inform experimental interventions. For example, we might ask how changes in the opening routines affect interaction patterns and see if fluidity can be reached more quickly.

On a more general note, we might ask how we can make telephone conferences more 'fun'. Initial answers appear counter-intuitive: fun is tied to engendering structure, and allowing for its re-framing; to have fun, we must have seriousness first.

References

Atkinson JM (1982) Understanding formality: notes on the categorisation and production of 'formal' interaction. Br J Sociol 33:86–117

Bateson G (1972) Steps to an ecology of mind. Ballantine Books, New York

Beach WA (1990) Language as and in technology: facilitating topic organization in Videotex focus group meeting. In: Medhurst MJ, Gonzalez A, Peterson TR (eds) Communication and the culture of technology. Washington State University Press, Pullman, pp 197–219

Bruner JS (1976) Nature and uses of immaturity. In: Bruner JS, Jolly A, Sylva K (eds) Play. Its role in development and evolution. Penguin Books, New York, pp 28–64

Drew P (1987) Po-faced receipts of teases. Linguistics 25:219–253

Glenn PJ, Knapp ML (1987) The interactive framing of play in adult conversation. Commun Q 35:48–66

Goffman E (1974) Frame analysis. An essay on the organization of experience. North Eastern University Press, Boston

Goffman E (1961) Fun in games. In: Goffman E (ed) Encounters: two studies in the sociology of interaction. Bob Merril, Indianapolis, pp 15–81

Hopper R (1992) Telephone conversation. Indiana University Press, Bloomington; Indianapolis

Huizinga J (1949) Homo Ludens. Beacon Press, Boston

Lemert C, Branaman A (1997) The Goffman reader. Blackwell Publishers Ltd, Oxford

Manning P (1992) Erving Goffman and modern sociology. Polity Press, Cambridge

Reed DJ (2002) Observing and quoting newsgroup messages: method and phenomenon in the hermeneutic spiral. Unpublished doctoral thesis

Rutter J (2000) The stand-up introduction sequence: comparing comedy compéres. J Pragmat 32:463–483

Sacks H (1995) Lectures on conversation, vols. 1&2. Blackwell, Oxford UK; Cambridge USA

Schegloff EA (1986) The routine as achievement. Hum Stud 9(2–3):111–151

Schwartzman H (1978) Transformations: the anthropology of children's play. Plenum, New York

Slater D (1998) Work/Leisure. In: Jenks C (ed) Core sociological dichotomies. Sage Publications Ltd, London, pp 391–404

Stephenson W (1967) The play theory of mass communication. Transaction Publishers, New Brunswick

Sutton-Smith B (1988) Introduction to the transaction edition. In: Stephenson W (ed) The play theory of mass communication. Transaction Publishers, New Brunswick

Williams R (1998) Erving Goffman. In: Stones R (ed) Key sociological thinkers. Macmillan Press Ltd, London, pp 151–162

Chapter 23
The Enchantments of Technology

John McCarthy and Peter Wright

Authors' Note Funology 2, Chaps. 4 and 7

When Andrew and Mark first approached us about the 2nd edition of Funology with the offer of revising our two chapters or leaving them as they were, we opted for the latter. We felt there was no single dimension along which we could revise the chapters in any helpful way such as adding a case study or revising the theory. But when Andrew suggested we should write an endnote about what had happened since 2003, we thought it might be a good idea to try and pull together a few threads of experience, thinking, and publication that might be helpful to readers approaching these chapters for the first time or even people revisiting them and wondering what happened next.

Back in 2003 when the first edition came out, the idea that people's experience of technology might be worth studying as part of the process of designing and evaluating technology was unusual. Indeed around that time, the link between HCI and design was relatively under-developed, found mainly in books like Donald Norman's *The Design of Everyday Things* (Norman 2013) and *Emotional Design* (2005). Not unlike, Norman's journey through Cognitive Science to Interaction Design, we came to Interaction Design through the study of human factors in complex safety critical settings such as air traffic control centres and aircraft cockpits—the antithesis of "fun" settings. The transition from human factors to experience-centred design reflected a need we had identified to get some kind of hold on *what it was like* to be in control of dispatching emergency ambulances or directing aircraft on approach to land, or dealing with cockpit emergencies, and not just how the information was processed and the tasks performed.

One of our chapters in Funology, *Making Sense of Experience* (Wright et al. 2004), was our first attempt to open up a space for thinking about what it might mean to study human experience in HCI, what scholars of human experience

J. McCarthy
School of Applied Psychology, University College Cork, Cork, Ireland

P. Wright (✉)
Open Lab, Newcastle University, Newcastle upon Tyne, UK
e-mail: p.c.wright@newcastle.ac.uk

© Springer International Publishing AG, part of Springer Nature 2018
M. Blythe and A. Monk (eds.), *Funology 2*,
Human–Computer Interaction Series,
https://doi.org/10.1007/978-3-319-68213-6_23

thought the phenomenon was, and how best to talk and think about it. We needed a chapter like this because nobody seemed to be writing or talking about it, and there wasn't a language of experience available within the human factors community. This search for an understanding of human experience that would be helpful for HCI and Interaction Design took us into disciplinary fields quite distant from our own training in psychology, disciplines like film making and literary theory as well as the pragmatic philosophy of Dewey (1934).

The didactic approach we took in *Making Sense of Experience* was to offer a 'framework', to give designers a language and a set of constructs with which to talk about experience. Because of the holistic approach we had adopted, the framework came with suitable caveats about avoiding reductive abstractions. Understandably, people began to ask two questions at conferences and in talks: "I don't get it, can you explain these concepts more clearly?" and "I get it, but how do I use it?" Our response, that the framework wasn't meant to be an operationalisable definition of experience needed clarification, something we tried to do in a number of publications that followed. The clarification was that the framework was meant to de-centre information processing as the best way HCI researchers could understand 'users' and instead put people's lived experience, their feelings, values, stories and ways of making sense of their interactions with the world, at the centre of HCI enquiry. For this insight, we owe a great debt to the inspirational Phil Agre (1997) not just for what he said about human experience and computing but for what he said about how to move things from the margins of academic discourses to their centres.

We put de-centring into practice in our chapter, *The Value of the Novel in Designing for Experience* (2004) by offering a re-reading of traditional HCI techniques for representing 'the user' (such as scenarios and personae) through the dialogical lens of our experience framework. But we wanted to go deeper into the philosophy and psychology of human experience in order to make a more substantial step towards placing human experience centrestage in HCI. So in our monograph *Technology as Experience* (2004), we took a deep dive into human experience. Residues of the framework are still there and there are case studies to illustrate how an experiential way of seeing the world plays out in situations like ambulance despatch as well as on-line shopping. But the main work of the monograph was to deepen the repertoire of concepts and genres to support discourse about experience and technology. Concepts such as *life as lived and felt*, Dewey's distinction between *an experience* and *the cumulative experience* that is a person's life can be found there, but in addition, we developed the notion of *dialogue* in a way which brings the relationship between *self-and other*, *openness* (*unfinalizability*), and *multi-voicedness* to the fore.

Although we felt *Technology as Experience* offered a more substantial foundation for the ideas in our *Funology* chapters, we had a lingering concern about how it could be used in design and how young researchers could be taught it. What does it really mean to design for openness or becoming? How does a relational or dialogical perspective on experience change how we design for experience? What should a young HCI researcher or interaction designer do? Difficult questions indeed.

The search for answers took us on another interdisciplinary journey beyond Dewey and Bakhtin to, Jay's *Songs of Experience* (2005), to narrative psychology (Bruner 1987, 1990) and to storied lives and experiences (Bruner and Turner 1986). It also took us to feminist theory and methodology (Belenky et al. 1986), and beyond. This journey is described in *Experience-Centred Design* (2010), in which we put narrative or more precisely *stories,* centrestage as one of the most fundamental ways for designers to understand human experience and we explored what it might mean for researchers to work with stories. Also, through concepts such as *active listening*, we tried to capture how stories are co-constructed through dialogue (the idea of anticipation, surprise, openness, and learning), and we re-examined the concept of *knowing the user* through a consideration of what it means to *understand* rather than to *know*. This led us to explore *empathy* both as a form of perspective taking and as an emotional-volitional orientation to 'the other' through relations of care, concern and hope (see also Wright and McCarthy 2008). Around this time, we started working with Jayne Wallace who through her making and her conversations showed us what it means to put this dialogical understanding of experience into design practice (Wright et al. 2008). Our early work together deepened our understanding of the varieties of experience of using and making which in subsequent papers gave depth and texture to ideas of enchantment and beauty not as a quality or attribute of a product but as a process, something we had tentatively discussed, but without the benefit of a designer's sensibilities, in our second Funology paper, *The Enchantments of Technology* (McCarthy and Wright 2004).

Seeing the relationship between designer and user as empathic, creative and dialogical critically shifts the agenda for design methods, and also shifts previously peripheral methods and tools, such as cultural probes and experience prototyping centrestage. The role of design methods is no longer confined to *representing the user* in such a way that *design implications* can be drawn. The suggestion that cultural probe returns aren't used to infer design implications but rather to allow designers to immerse themselves and respond directly or begin a conversation with the user makes perfect sense within a dialogical framing that places empathy and responsiveness at its centre. For designers and users in dialogue, the creative activity that they engage in is designed to allow individuals who have different perspectives on a problem, issue, setting or context, to respond to it from their own *centre of value*, informed by the encounter with the others' perspectives. Dialogical engagement does not presume the user becomes the designer or the designer becomes the user, quite the opposite because the designer would have nothing to offer if his or her experience were identical to that of the user. But what does it mean in practice to do this kind of design research?

In 2010, Pete moved to *Open Lab* (then called *Culture Lab*, see Wright and Olivier 2012). At Open Lab, we have had the opportunity to work with researchers living the experience of doing experience-centred design. We began by trying to understand the qualities of dialogical design encounters and what it means to frame participatory design methods as dialogue. We worked with researchers whose backgrounds ranged from computing through design to social sciences and arts and beyond. With them we explored how researchers designed together with communities over extended periods of time and built up relationships with them. We worked in sites which stretched traditional ways of doing HCI, such as a centre for Black Asian, Minority Ethnic, and Refugee (BAMER) women (Clarke et al. 2013), a medium-secure forensic prison (Thieme et al. 2016), hospitals (Bowen et al. 2011), live events (Hook et al. 2013), and rural communities (Taylor et al. 2013). These usually involved long-term embedded engagements in which a single researcher or small team spend 3 or more years working with a single community.

In *Taking [A]Part* (McCarthy and Wright, 2015), we offer an account of what we learned working in these complex social settings. We take inspiration from participatory arts and political philosophy and offer an approach that tries to eschew the twin tyrannies of tokenism and exploitation. We explore what it means to have a *voice* rather than *a say*, how individuals participate at the intersection of multiple subjective positionings, and how participation can be a heavily textured and variable experience for all parties. From a design perspective, we have begun to appreciate that participation itself has to be designed, and the way it is designed has material effects on not only the experiences of participants but also the qualities of outcomes. The way in which individuals (not only those formerly called 'users', but all participants) are subjectively positioned by the process can open up or close down possibilities for voice, negotiation and self-organization. In light of this, we argue that the measure of participation should not be consensus but *dissensus*. Inspired by Rancière's analysis of *the sensible* (e.g. 2006), we argue that dissensus is the engine of creativity in participatory design projects, and that the best participatory projects seek to reconfigure what is sensible (that is to say what is hearable, visible, and even thinkable) to participants in order to create space for making previously unimaginable possibilities thinkable (Wright and McCarthy 2015).

So, the two chapters in Funology from all those years ago, set us off on a journey, to explore what it might mean to take human experience seriously in HCI. And for us, the bottom line is that experience-centred design must configure those people who participate as co-producers—skillful, knowledgeable, resourceful, and situated agents. Done well, experience-centred projects can produce mutual learning and transformative outcomes for all participants. Done badly, they can serve to exploit labour, invalidate local knowledge, neutralize voices, and silence dissent (McCarthy and Wright 2015).

1 Introduction

Ask people about their enchantments and they will talk about being absorbed in a film or painting, glimpsing another world in a story, or being totally blown away by the engine of a motorbike. An artist might talk about the intense, sensuous experience of a single line in a painting. Having made the line, the artist is taken outside of herself for a moment, into the clarity of the line. A mechanic might talk about the pleasure of working with a particular motorbike. Mesmerised by the beauty of the engine, absorbed in each and every challenge, aware of the pleasure in a skill. Others may simply be enchanted by the stones on a beach, the smile on a baby's face, the view from the top of a hill.

There is no telling what holds the power to enchant any one of us. Whereas the motorbike engine would leave me cold, I can become absorbed in a painting or a person's smile. The things that have the power to enchant me, that grab my attention and shape my desires, seem written on me by the history of my experience in my culture. I trace my attachment to Apple computers to my enchantment with the Apple II, a computer that didn't seem like the ponderous computers we used at work. It was lively, engaging, different, the rave of another youth subculture. I loved and became hooked on the surface of it and the pleasure of transacting in what seemed to be a common language with something intelligent. DOS never did that for me. I found the requirement to communicate using the machine's language disenchanting.

Although our enchantments are our own, it is possible to clarify some aspects of the enchantments of technology. It is also worthwhile to do that. The dominance of function in the design of computer systems frequently produces disenchantment tales: stories of workers resisting new technology and students bored or cynical about it; stories of very powerful computational devices separated from the sensuousness of interaction and the meaningfulness of action. As a counterpoint to these tales of disenchantment, we explore the potential for people to be enchanted with technology by: clarifying what we mean by the experience we call enchantment; exploring sources of enchantment in film as an example of the experience of enchantment with a contemporary cultural form; and teasing out implications of enchantment for feelings and imaginative activity in relationships between people and technology.

2   What Do We Mean by Enchantment?

Rather than examine the relationship between people and technology through the study of form or function, we focus on experience. According to Dewey (1934), experience is constituted by the relationship between self and object, the concerned, feeling person acting and the materials and tools they use. It includes what people do and what is done to them, what they strive for and desire, and how they feel, fear, believe, hope, enjoy, and imagine. Experience registers life as lived and felt. In a companion chapter in this volume, (Wright, McCarthy, and Meekison), we present a framework for analysing experience. In the current chapter we focus on the power of technology to enchant.

Our particular concern with the power of technology to enchant is motivated by the capacity of enchantment to evoke both the transformative openness and unfinalisability of experience and the capacious potential of imagination to power holistic engagement by bringing past or future meanings into present action, making the mundane creative. Rendering people's experience with technology as transformative, imaginative, and creatively meaningful is the proper corrective against pervasive disenchantment tales. Technology may indeed be used to oppress, suck meaning from activity, alienate people from communities, mechanise, monitor, and rationalise but it also enchants, pleases, energises and clarifies.

Enchantment is akin to Dewey's 'holistic engagement' between the sensing person and their environment, which Dewey sees as necessary for peoples' growth and development. At the heart of this holistic engagement is an integrated sensual experience, where the senses unite to reveal the qualitative immediacy of the situation. Dewey uses the example of a mechanic who, when wholly engaged in his job, sees, hears, smells, and touches the engine. He senses it, and through his engagement with it, senses what is wrong. He is completely attentive, engrossed, intensely concentrated, and immersed or lost in an activity. In contrast, Dewey argues that when the senses are compartmentalised, "we undergo sensations as mechanical stimuli or as irritated stimulations, without having a sense of the reality that is in them and behind them" (Dewey 1934, p. 21). As stimuli, sensations are robbed of their power to enchant.

Holistic engagement captures some of what it means to be caught up but not what it means to be carried away. According to Bennett (2001), when we are enchanted we are "both caught up and carried away" (p. 5). When enchanted, although we are momentarily spellbound, our senses seem heightened. We notice lines, colours, and sounds that we have not previously noticed. And even as our senses are sharpened and intensify, as we see with heightened clarity, we are transfixed. Mesmerised by the sharpness with which we see the beauty of a scene, we are carried outside of ourselves. Caught up in wonder at the object and carried away by our senses.

According to Gell (1992) we are carried away by the power behind the enchanting object. For him, the enchantment of technology is in its becoming rather than its being. By this he means that the power of technology to enchant resides in our sense of wonder at the skill of the maker of the technology. For him, the Trobrianders' use of the prow-board of the Kula canoe is a case in point. The board is a visually intricate display or surface designed to dazzle anybody looking at it and put them off their stroke for a moment. Gell attributes the power of the canoe-board not to the visual appearance per se but to the fact that mild disturbances caused by the captivating visual effects of the board are interpreted by the viewer as evidence of magical power emanating from the board. They can't imagine how it came into being so it must be magic. The magical power is in the idea one has of the board coming into being.

Gell's analysis of the enchantment of technology describes a serious emotion, something like the awe one feels in a religious setting or in response to a mystery. Those seeing the prow-boards are in awe of the powers of those who had the power to make them. This disorients and frightens them. Without a sense of charm, delight, and pleasure, this description of enchantment is incomplete. Enchantment also describes a sense of pleasure borne of the experience of novelty. There is pleasure in the enchantment of seeing something new in a stone on the beach or a line in a painting. Perhaps it is the pleasure of experiencing becoming, the openness of everything. For us, enchantment is not only in the skill of the maker but also in the becoming that is all round us, the openness or unfinalisability of the world.

Imagine a world in which every-thing is open and unfinalised. In this world, we take no-thing for granted, for every-thing is always becoming, always on the way to being. We can't just assume that things are the way we see them. The stone on the beach, the computer on which I am writing, my body, our definition of personhood, the physical world, and time—all always becoming. Bakhtin, a philosopher and literary scholar, described a world in which all is always becoming, and reflected on how we can know and understand in such a world (see Bakhtin 1981, 1984, for example). For Bakhtin, the site of knowledge is never unitary, rather knowing is better seen in terms of dialogue. Dialogically, everything is perceived from a unique position in space and time. In a world in which knowing is decentred, whatever is observed is shaped by the position from which it is observed. Bakhtin goes so far as to refer to perception and observation as authorship, a constructive act of meaning making by an author, an "I" with no referent other than a person who is always changing and different. The 'thingness' of everything is worked at and is a provisional finalisation in a relational process between observer and observed. There is every possibility that, at a different time and from a different perspective, every-thing might be different. For Bakhtin, the world is an open place full of potentiality, freedom, newness, and surprise. In this world, the potential for enchantment rooted in the experience of novelty is everywhere.

The important point for us about Bakhtin's work is that he was describing what we seem to miss in the ordinary, everyday, world because we have already finalised it in our minds. We have closed our minds off to the potentiality of the physical, biological and social world, having already decided what everything is instead of looking closely. Because of this we fail to notice the essential creativity of our relationship with every-thing that is ordinary. In a world that is always becoming, we are compelled to create the thing-ness and event-ness of the world. And it is so ordinary that we miss it. For example, most of us already know what a body or a person is. We also know that the physical world is fairly solid. Instead of settling for this given knowledge, Bakhtin celebrates the human body in the act of becoming, never finished, continually built, swallowing and being swallowed by the world, and Prigogine and Stengers (1984) describe the physical world "as seething and bubbling with change, disorder, and process" (p. xv).

In the prosaic world in which we live, all encounters contain the possibility of something unexpected. Enchantment begins when engagement with the unexpected creates a 'moment of pure presence' (Fisher 1998). According to Fisher, the object with the power to enchant "does not remind us of anything we know and we find ourselves delaying in its presence for a time in which the mind does not move on by association to something else". In a moment of enchantment, one of the millions of stones on a beach stands out as unique. We notice something about its shape, colour, or smoothness, the clarifying sensuous experience of which holds us in that time and place for a moment, totally engaged with and fascinated by the stone.

According to Bennett (2001), the surprise of the encounter contains two feelings at the same time: the pleasurable feeling of being charmed by something that is new and singular, and not yet processed and categorised; and the disorientating feeling of being taken out of one's habitual sense-making dispositions. Bennett suggests that the effect of enchantment is a mood of fullness and liveliness, a sense of heightened perception and concentration, and a feeling of excitement about life.

In the account of enchantment that we have presented so far, sites of enchantment are everywhere to be found as long as we are prepared to encounter the novelty in things and events. The experience of enchantment is constituted by the relationship between a person who is fascinated by the singularity of objects and events and those objects and events. In terms of relations between people and technology, we might want to think of enchantment as a state of interactive fascination between the person enchanted and the source of enchantment. Our task in the next section then is to say something about how this interactive fascination can be evoked in interactive technologies.

3   Enchantments of Technology

Some contemporary technologies are readily associated with the experience of enchantment. The settings, characters, and activities of computer games create worlds in which children become caught up and to which they are carried away. Handheld computer games, such as Gameboys, seem to absorb children and teenagers for hours on end. Teenagers are not just satisfied with their mobile phones. They are bewitched to the extent that the primitive input and output devices matter very little to them. In the magical world of text messaging, where new communication media and the cachet of the mobile are dazzling, enchantment overwhelms function. Adults also have moments of enchantment with mobiles such as when a father's experience of time, space, and presence are transformed by speaking to his young daughter who is on the London Eye when he is at his desk in York.

How do we design for the potentiality that makes something as ordinary as a phone enchanting? Elsewhere in this volume, Sengers argues that we do this by focusing on the user's ability to engage in complex interpretation using cultural background knowledge. In another chapter in this volume, Dix's deconstruction of people's experience of traditional Christmas crackers, which enables him to re-mediate the experience, can be seen in a similar light. Both seem to suggest that, if we understand the richness and complexity of users' responses, simple artefacts can be designed to facilitate rich experiences. In a similar spirit, we examine the richness and complexity of people's responses to film, not to re-mediate the film experience (many games manufacturers have already tried that) but to make visible aspects of experience with contemporary technologies at play that would otherwise remain unseen.

Boorstin (1990), a writer and producer of Hollwood films, describes the rich and complex response of filmgoers' to film that enables filmmakers to enchant with even the simplest filmic experience. He suggests that the key is to understand that we experience or watch movies in three ways. Each way has a distinct pleasure and magic associated with it. Boorstin refers to his three ways of seeing as the voyeuristic eye, the vicarious eye, and the visceral eye.

The *voyeuristic eye* is a way of experiencing film in terms of the simple joy of seeing the new and the wonderful. It refers to a way of looking that gets up close to things and really looking at them but becoming bored as soon as the experience of seeing the newness of the thing has run its course. As the mind's eye it can be quite sceptical, and it requires a high level of plausibility and credibility. When it experiences events that seem implausible, it is inclined to disengage. What appears on screen must contain surprise and plausibility to seduce and enchant the voyeuristic eye. There is no magic without a new look at things, but a new look that makes sense in the world being experienced even if it is a fictional or fantasy world.

In the early days of cinema the voyeuristic eye was seduced by the magic of the projected moving image. As the projected image became passé, more was required: talk to go with the action, more and more precise synchronisation of talk and action, high precision editing, adherence to a more sophisticated grammar of cinematic action that took account of how the audience would fill in the gaps, colour, action and images that could not be accomplished other than with film. As we viewers become more experienced with a medium, it takes more to enchant the voyeuristic eye. However designers should bear in mind that less can be more enchanting as long as it gives us something new and wonderful. For example, Boorstin describes films by Charles Eames as miniature masterpieces because they made us see anew simple objects like spinning tops. These short films did this without story, characters, or even a point. They stood or fell in terms of voyeuristic pleasure on the back of the visual logic threading through the images.

The *vicarious eye* is attentive to the emotional substrate of action rather than to its internal logic and plausibility. It may be an even more powerful factor in our experience than internal logic because we can make allowances for what seems illogical if we are made aware of an emotional truth underlying it. While the magic of film can be threatened voyeuristically if a viewer feels 'that could not happen', it is threatened vicariously if a viewer feels 'he wouldn't do that'. No matter how implausible the action, if we are won over by the character, we may be convinced that 'yes, *he* would do that'. But it can never be a character separate from the world created in the film. For the vicarious eye, the basic unit is not the beat of the story but the moment of the character. In great moments, story time stands still and the pleasure of the new and wonderful is irrelevant. But a film cannot be enchanting if it is made up of great moments alone. The editor has to create a rhythm and movement between the voyeuristic and vicarious to keep us engaged—intellectually, emotionally, and valuationally.

The *visceral eye* is attuned to first-hand experience of thrill, joy, fear and abandonment. Here the character is a conduit for the viewer's feelings rather than the other way around. Unlike the vicarious eye, the visceral eye is not interested in characters in the empathic sense, it is interested in having tokens for our sense of thrill or fear. As we feel the thrill and fear of people on a roller coaster ride we are not empathising with them, rather we are having our visceral experience through their activity. However it is a thrill or fear cosseted by the knowledge that it is not actually you that is being attacked by an Alien or free-falling from 10,000 ft. As character is not empathic in the visceral experience, it is closer to montage than story, narrative, or connectedness. It consists of moments of gut reaction and as viewers we are carried along by those moments as if on a roller-coaster.

Of course the magical experience can have visceral, vicarious, and voyeuristic elements. In film, visceral alone can't be enough. We build up resistance to every thrill the director creates for us so that, as Boorstin puts it, when evaluating the visceral aspect of the magic of film: twice as much comes off half as effective. The visceral impact of films like Psycho and Alien depend as much on the characters we identify with as the thrills of action. The visceral shock of the shower scene in Psycho depends on Hitchcock's manipulation of the story to that point such that a tale of love and embezzlement becomes a tale of murder. He builds up to the visceral moment by playing with our voyeuristic and vicarious pleasures.

Boorstin's (1990) analysis of the magic of movies suggests that, in a media—savvy world, a combination of wonder at the new, sensuous experience, and emotional response to characters is required to create an enchanting experience. In the context of our analysis of enchantment, we read this analysis as replacing a relatively undifferentiated, cognitive approach to seeing, such as Gell's, with an active, differentiated, aesthetic approach. In contrast with Gell's monological perspective on enchantment, Boorstin's analysis develops from an understanding of the play of appearance as dialogical, with multiple perspectives on novelty, emotional tone, and sensuousness in constant interaction with each other against a shifting magic standard.

This analysis of how people experience film may inform the design of technologies in a number of ways. Following Dix, we could treat this analysis as a deconstruction of experience with one of the most significant mediums of the twentieth century, with a view to informing design of experience with new media such as computer games, virtual reality, and the cyberspace of MUDs and MOOs. The re-mediation of film experience in new media (see Bolter and Grusin 2001, for example). Alternatively, the analysis might be used to complement the application of concepts from experience of narrative in design with a set of concepts from experience with film. Or, following Sengers, we could use it to understand the abilities of people to interact with new technologies and media and to reveal the cultural resources we use to make these interactions personally meaningful. Film plays with our cultural knowledge of genre, storytelling conventions, visual logic, and the language of film. It creates a mediated inter-subjective experience by addressing itself to the ways in which we experience, see, and make sense and by assuming that we understand its modes of expression. Over time, new media will also have to nurture similar relationships and might benefit from an analysis of existing popular media such as film. Each of these uses of the analysis would take another chapter to develop. Our aim in presenting an analysis of enchantment with film here has been to take a first step by enhancing our sensibility to experience, especially enchanting experience with mediating technologies. One of the main ways in which we have done this is by illustrating the potential complexity of the user's response to and experience of these technologies, making visible the enchantment of novelty, emotional identification, and visceral thrill, which might otherwise remain unseen.

4   Enchantment in the Space of Public Appearance

It is easy to dismiss enchantment as trivial as having to do with play or entertainment and not the important things in life. Or even to treat enchantment as a modern day opiate of the masses: enchantment as a stylistic marketing device to cover up deficiencies of function and woo the unsuspecting consumer. Our final move in this chapter is a brief defence of enchantment in the context of people's interactions with technology. Our defence argues against the disenchantment tale of enchanting technology inevitably mesmerising—in the sense of controlling—the person using it. Elsewhere in this volume, Blythe and Hassenzahl have usefully critiqued the political analysis that associates mass media enchantment with passivity and cultural duping, suggesting a psychological corrective to enquire into people's enjoyment of TV and other media of entertainment. We want to raise similar concerns with readings of enchantment that downplay the activity of subtle participation.

Internet stores are interested in customer loyalty and commitment and try to enchant each customer by making transactions personal. They engage with the identity of the shopper inferred from a history of transactions. With respect to product design and branding, it is not just the functionality but also the style of a mobile phone that matters to the owner. We can dismiss this as a matter of 'style over substance' or we can inquire into the substance of people's experience of enchantment with the style of a product. What vitality does a sense of style bring to people's relationships with their mobile phones? Attending to style in this way does not stop us being critical but it does stop us being dismissive.

Apple have made style central to the relationships people have with their computers. The colour and transparency of the IMac and the titanium casing, slim body, and lightness of the G4 are examples of computers marketed as something more than what we traditionally think of as computers. Indeed this has been Apple's stock in trade since the Apple II. For Apple, computers are not just computational devices they are objects to be with. We are encouraged to see them as sensuous objects that, in our presence, become sensing, sensual, sense-making subjects. Referring back to Boorstin's analysis, we can see in the Apple Mac style sensitivity to our desire for novelty with emotional integrity. The design becomes a meditation on the computer and computation itself, as Eames' films meditate on the spinning top and in the process on the medium of film. A computer dialogically interanimated with voices of adventure and mesmerising technology from the space age (titanium and very light, almost defying gravity) and of contemporary consumer product aesthetic. But do we passively consume this message or complete the experience ourselves? Do we, like the enemies of the Trobrianders, dazzled by the canoe prow board, take leave of our senses and give into the power behind the technology? Or do we retain our sense while enjoying the vivifying pleasure of something new and wonderful?

DeCerteau (1984) argued that people who are dismissed as passive consumers often resist definition of themselves by others through the use they make of what is given. They can be enchanted by the technology they buy and still make their mark by giving the technology a personal meaning, for example, creating sub-cultural text messaging languages or using electronic bookshops as handy bibliographic databases. The response of organisations such as the Billboard Liberation Front in editing Apple's (and many other company's) billboard advertisements suggests resistance to the message and raises questions about who owns the power to enchant. The Billboard Liberation Front and other culture jammers (Klein 2000), who parody advertisements and billboards to radically alter their message, draw the advertisements, the brand, and the products into a moral-political discourse. Moreover they do it by trying to make better use of enchantment than the advertisers. The parodied advertisements evoke pleasure at the experience of novelty by creating images that are charming as well as damning. We are caught up and carried away by the metamorphing of the image and the movement between cultural worlds entailed in that morphing: is it an advertisement or not, is it the original or has it been changed?

Our approach to questions of the mindfulness or mindlessness of enchantment with technology turns on our understanding of the power of technological mediation of experience to press into the gap between feelings and expression or imaginative activity. This understanding offers up a range of possibilities that can take shape only as hypotheses at this stage. Table 1 provides a simple space in which to question about issues of interpassivity and interactivity, which we briefly address below in response to Zizek's critique of the enchantment of cyberspace as inevitably promoting passivity.

Zizek's (1999) critical analysis of interpassivity in cyberspace points to the potential for unhelpful passivity in relations with technology. He draws attention to people allowing themselves to be caught up in activity in order to avoid feelings. Zizek would argue that our enchantment with electronic pets and cyber friends is a form of *one-sided interpassivity* that enables us to engage in the activities of caring without having the responsibility and feelings of mutually relating. We have the experience of caring without the complexity of a relationship with another person, who may care in return or may be indifferent to being cared for.

However, if the electronic pet is not fed, it dies. It can be argued that this context of caring with limited responsibility provides the kinds of experiences that enable children to learn about relationships. In contrast, Zizek sees it as promoting interpassivity, with the cyber and the virtual as agents of mediation sustaining the subject's desire while acting as agents of prohibition of its full expression and gratification. He sees *one-sided interactivity* as acting through another agent, so that

1 An analysis of the possibilities of interactivity and interpassivity with technology playing in the gap between feelings and expression

<table>
<tr><td colspan="2" align="center">Interactivity</td><td></td></tr>
<tr>
<td rowspan="2">Mutual</td>
<td>Dialogue.
Culture jamming
Christmas crackers
(Dix this volume)
The Influencing
Machine (Sengers,
this volume)</td>
<td>Acting through an
agent
Electronic pet?
Cyber friend?</td>
<td rowspan="2">One-sided</td>
</tr>
<tr>
<td>Lovers being quiet
together.
Texting to just be in
each other's presence</td>
<td>Greek chorus doing
the feeling for us.
Throwing oneself into
rituals of activity in
order not to
experience a feeling.
Electronic pet?
Cyber friend?</td>
</tr>
<tr><td></td><td colspan="2" align="center">Interpassivity</td></tr>
</table>

my job is done while I remain passive. As we have seen in brief reflections on culture jamming and DeCerteau's treatment of the strategic action of consumers, these apparently one sided relationships may in fact involve a subtle, expressive response. One such response occurs when the technology mediating experience facilitates *mutual interactivity* as is the case in episodes of culture jamming, creative use of the technology given, and dialogue. The child who uses her mobile phone to call her father who is 200 miles away acts from feelings to imaginative use of technology. Another possibility, *mutual interpassivity*, points to the expressiveness of being quiet together. The clearest model is lovers being quiet together.

A jazz riff, a film, or a piece of technology might play me. I may be caught up and carried away by any of these things. However this does not render me passive or a victim to their powers of enchantment. Far from it, it may be that in recognising the creativity of my relationship with them, I begin to understand my emotions and the communication between me and others that makes me what I am and what they are—always becoming. Technology that enchants: a computer that allows me to question what it is to be a computer; textual communication, the limitation of which, requires me to be creative and expressive; objects or installations that are sensitive to the moment and to my sense of wonder and emotional integrity. Technology that enables me to change.

References

Agre P (1997) Computation and human experience. Cambridge University Press, Cambridge, UK

Bakhtin MM (1981). In: Holquist M, The dialogic imagination: four essays. University of Texas Press, Austin, TX

Bakhtin MM (1984). In: Emerson C, Problems of Dostoevsky's poetics. University of Minnesota Press, Minneapolis

Belenky MF, McVicar Clinchy B, Goldberger NR, Tarkle M (1986) Women's ways of knowing: the development of self, voice and mind. Basic Books, New York

Bennett J (2001) The enchantment of modern life: attachments, crossings, and ethics. Princeton University Press, Princeton

Bolter JD, Grusin R (2001) Remediation: understanding new media. MIT Press, Cambridge, Mass

Boorstin J (1990) Making movies work: thinking like a filmmaker. Silman-James Press, Beverley Hills

Bowen S, Dearden D, Wolstenholme D, Cobb M (2011) Different views: including others in participatory health service innovation. In: Proceedings of the second participatory innovation conference, University of Southern Denmark, pp 230–236

Bruner J (1987) Life as narrative. Soc Res 54:1–17

Bruner J (1990) Acts of Meaning. Harvard University Press, Cambridge, Mass

Bruner EM, Turner V (eds) (1986) The anthropology of experience. University of Illinois Press, Urbana

Clarke R, Wright P, Balaam M, McCarthy J (2013) Digital portraits: photo-sharing after domestic violence. In: Proceedings of the Chi'2013. ACM, New York, pp 2517–2526

DeCerteau M (1984) The practice of everyday life. University of California Press, Berkeley, California

Dewey J (1934) Art as experience. Perigree, New York

Fisher P (1998) Wonder, the rainbow, and the aesthetics of rare experiences. Harvard University Press, Boston Mass.

Gell A (1992) The technology of enchantment and the enchantment of technology. In: Coote J, Shelton A (eds) Anthropology, art, and aesthetics. Clarendon Press, Oxford, pp 40–63

Hook J, McCarthy J, Wright P, Olivier P (2013) Waves: exploring idiographic design for live performance. In: Proceedings of the Chi'2013. ACM, New York, pp 2969–2978

Jay M (2005) Songs of experience. University of California Press, Berkeley

Klein M (2000) No logo. Flamingo, London

McCarthy J, Wright P (2004) Technology as experience. MIT Press, Cambridge, MA

McCarthy JC, Wright PC (2015) Taking [A]part: the politics and aesthetics of participation in experience-centered design. MIT Press, Cambridge, MA

Norman DA (2005) Emotional design: why we love or hate everyday things. Basic Books, New York

Norman DA (2013) The design of everyday things. MIT Press, Cambridge, USA

Prigogine I, Stengers I (1984) Order out of chaos. Flamingo, London

Rancière J (2006) The politics of aesthetics: the distribution of the sensible. Continuum International Publishing

Taylor N, Cheverst K, Wright P, Olivier P (2013) Leaving the wild: lessons from community technology handovers. In: Proceedings of ACM CHI'2013 conference on human factors in computing systems. ACM Press, pp 1549–1558

Thieme A, McCarthy J, Johnson P, Phillips S, Wallace J, Lindley S, Ladha K, Jackson D, Nowacka D, Rafiev A, Ladha C, Nappey T, Wright P, Meyer TD, Olivier P (2016) Challenges in designing new technology for health and wellbeing in a complex mental healthcare context. In: Proceedings of CHI'2016, pp 2136–2149

Wright P, Olivier P (2012) Digital interaction culture lab. Interactions Magazine 19:1

Wright PC, McCarthy JC, Meekison L (2004) Making sense of experience. In: Blythe M, Monk A, Overbeeke C, Wright PC (eds) Funology: From usability to user enjoyment. Kluwer, Dordrecht, pp 43–53

Wright P, McCarthy J (2004) The value of the novel in designing for experience. In: Pirhonen A, Roast C, Saariluoma P, Isom H (eds) Future interaction design. Springer, Berlin

Wright P, McCarthy J (2008) Experience and empathy in HCI. In: Proceedings of Chi'2008. ACM Press, pp 637–646

Wright PC, McCarthy JC (2010) Experience-centered design: designers, users, and communities in dialogue. Morgan and Claypool

Wright P, McCarthy J (2015) The politics and aesthetics of participatory HCI. Interactions Magazine 22:6. ACM Press, New York, pp 26–31

Wright P, Wallace J, McCarthy J (2008) Aesthetics and experience-centred design. ACM Transactions on Computer-Human Interaction (TOCHI) 15:4. ACM Press, New York, pp 1–21

Zizek S (1999) The fantasy in cyberspace. In: Wright E, Wright E (eds) The Zizek reader. Blackwell, Oxford

Chapter 24
The Semantics of Fun: Differentiating Enjoyable Experiences

Mark Blythe and Marc Hassenzahl

Authors' Note, Funology 2

Fun is not always fun; being made fun of, for example, is not always an unequivocal delight. The English idiom "taking the piss" means making fun of someone in a critical or even aggressive manner. Comedians take the piss out of politicians in this way and occasionally this can make the audience feel as if they are involved in some form of political action. The British satirist Peter Cook once remarked that this type of activism was best exemplified in the wonderful cabarets of the 1930s "which did so much to prevent the outbreak of the Second World War". Nevertheless, enjoyment is political.

In the 1960s, Debord described a society where people were so dazzled by the productions of the cultural industries that they became much easier to manipulate and control. This vision still resonated in 2003 when the first of edition of Funology was published. It is easy to forget that this was before Facebook (2004), before YouTube (2005) before Twitter (2006) before the iPhone (2007) before Pornhub (2007) and before Angry Birds (2009). Yet even then, in what younger people may think of as some pre-digital dark age, we were worried about the ways technology was making us into the passive spectators of our own lives.

These days, the crowds on the metros of the world are at once present and absent—their ears are stopped with headphones and their eyes directed at screens depicting any world but the one they are in, we are approaching a state of continual distraction. Distraction is a very particular and (in this chapter) a somewhat worrying form of

M. Blythe (✉)
Northumbria University, Newcastle upon Tyne, England, UK
e-mail: mark.blythe@northumbria.ac.uk

M. Hassenzahl
Ubiquitous Design/Experience and Interaction, University of Siegen, Siegen, Germany

enjoyment. Why are we using our phones to move colourful pieces of candy into alignment with one another rather than read the Dostoevsky novel which might also be stored on the same device? When the Italian radical Antonio Gramsci was put on trial the prosecutor said "we must stop this brain working for twenty years". To do this in the twentieth century meant putting him in prison, today the same effect might be achieved by giving him an iPad.

For Slavoj Zizek enjoyment is a "disturbed pleasure" which can include pain. At a trivial level this is obviously true: even games must frustrate and baffle us if we are to enjoy eventually mastering them. But there are painful aspects of enjoyment that are not immediately apparent to us. If we are unhappy it may, on some level, be because we enjoy being unhappy. This is counter-intuitive but think, for example, of the troubled satisfaction we might take in cutting dead some bastard who has betrayed us. (Not us of course, we would never blank anyone, if Mark fails to say hello it's because he is very short sighted). But theoretically it's possible isn't it, painful, nasty fun? The point is that enjoyment can be mean, it can be part of a struggle.

The word fun, as the chapter points out, entered the language when the division between work and free time became more rigid. The idea of leisure and fun in the sense of frivolous distraction from the serious business of life, is a response to enclosure, the mechanisation of time and industrialisation. An ethnography by Paul Willis of motor car workers in the 1970s identified two forms of conflicted fun. The first was "taking the piss"—we already know this one—, the second was "flymping". Here shop floor workers would work extremely hard and fast for a short period of time in order to take longer and unofficial breaks. Today the divisions between work and leisure time are beginning to break down. When the site of labour was a shop floor, workers had to find secluded places to take their breaks. When the site of labour is a computer screen digital flymping is easy—from the earliest days of computing in the workplace there have been solitaire screens behind spreadsheets. At the same time, though, work screens are now behind the games we play in our living rooms. Capital has invaded aspects of our lives that previously were beyond its reach. Activities once only associated with free time or space have been commoditised: giving someone a lift in your car, putting someone up for the night in your spare room, doing somebody a favour—and technology plays a key role in making this happen. In this chapter we speculate, like many other researchers at the time, about ways in which the tedious might be made more fun. Besides computer games, enjoyment, fun, pleasure was absent when it came to technology. But that was long ago and now enjoyment is imperative. A poster campaign around one UK university campus extolls students to "Enjoy every minute".

A story by Zizek nicely illustrates the Lacanian notion that the hectoring superego's injunction is always—enjoy!

> A father works hard to organise a Sunday excursion, which has to be postponed again and again. When it finally takes place, he is fed up with the whole idea and shouts at his children: 'Now you'd better enjoy it!'. (Zizek 1999)

In a recent novel by a buzz feed journalist called "Start Up" the protagonist receives an email from a colleague copied to all of her co-workers inviting them to a pole dancing class. She reflects:

> Now you were expected to engage in forced organised fun with people you worked with. And it seemed to her that the definition of fun had been majorly stretched. When she was in her twenties people were way too jaded to think that something like a pole dancing class with colleagues was even remotely cool. She also felt like there was something slightly more insidious going on about how you were supposed to feel like your work was your everything - where you got your paycheck yes but also where you got fed and where you found your social circle. Everything had started to bleed into everything else. (Shafrir 2017).

If work is fun then fun is also work.
2003 Chapter

1   Introduction

Over the last 20 years repeated attempts have been made in HCI to put enjoyment into focus. However, it is only recently that the importance of enjoyment, even in serious applications, has been widely recognised by the HCI community.

Typical of a relatively new area of investigation is the lack of an agreed set of terms: enjoyment, pleasure, fun and attraction are often used interchangeably. But do they really refer to the same experiences? Of course, in common speech pleasure, enjoyment and fun are almost synonymous and this is not an attempt to fix the language. None of these terms are reducible to single definitions but for the purposes of this chapter we will propose a difference between pleasure and fun in an attempt to delineate distinct forms of enjoyment.

The chapter begins with a consideration of the psychological account of peak experiences and how this might relate to less intense activities. After exploring the semantic and cultural connotations of the word fun the chapter goes on to consider the historical and political construction of leisure in the West. The final sections outlines distinctions between "fun" and "pleasure". It is argued that pleasure is closely related to degrees of absorption while fun can be usefully thought of in terms of distraction. The distinction has important implications for design. It is argued that repetitive and routine work can be made fun through design while non-routine and creative work must absorb rather than distract if they are to be enjoyable.

2 Pleasure from a Psychological Perspective: Flow

Csikszentmihalyi's (1975) study of "flow" is one of the few psychological accounts of pleasure. After studying diverse groups, such as rock climbers, chess players and dancers, who were engaged in self motivating activities, Csikszentmihalyi discovered a common characteristic of their experiences. "Flow" was a term used by the participants themselves to describe a peak experience of total absorption in an activity. Csikszentmihalyi identified the conditions for flow as: a close match between skill and challenge, clear goals and constant feedback on performance. It was characterized by a decrease in self-consciousness and time distortion in that an hour might seem like a minute (Csikszentmihalyi 1975). Flow experiences may be experienced in non-leisure and serious contexts.

The term "micro-flow" was coined in order to catalogue small periods of activities which are not necessary, yet are engaged in routinely, for example, chatting, doodling and stretching. These activities are intrinsically satisfying, although they do not induce the deep and intense experience of flow. Csikszentmihalyi (1975) suggested that these apparently unnecessary activities are in fact vital to our well-being. Doodling, for example, may aid concentration in a dull meeting. However "micro-flow" is a less well defined concept than flow and does not adequately account for less intense experiences.

Flow addresses a "deep" kind of enjoyment which may be only rarely achieved (and actually called for). To experience flow, we have to go beyond our own limits. This, however desirable from a humanistic view, is not the type of enjoyment most people choose. Most of the time, more superficial, shallow, short-term and volatile "pleasures" are in the fore. Or as Seligman and Csikszentmihalyi (2000) put it:

> Why do we choose to watch television over reading a challenging book, even when we know that our usual hedonic state during television is mild dysphoria while the book will produce flow? (Seligman and Csikszentmihalyi 2000)

The answer to this question may be, in part, political. The next section considers the history of the word fun and offers an account of the leisure industry and mass media in relation to their development in the West.

3 The Politics of Fun

An examination of the changing uses of the word "fun" as illustrated in the Oxford English Dictionary demonstrates that fun, meaning—diversion, amusement, jocularity—appears relatively late in the language. (The following citations are all taken from the OED http://dictionary.oed.com/). In the earliest records, its meaning is—to fool, to cheat or hoax: "She had fun'd him of his Coin" (1685). Although this usage continued it was superseded in the eighteenth century "Tho he talked much of virtue, his head always run upon something or other he found better fun" (1727). In the mid eighteenth century Samuel Johnson described it as a "low cant word", its disreputable aspect continued into the nineteenth century "His wit and humour delightful, when it does not degenerate into 'fun'" (1845). The use of the word in the phrase "to make fun of—" also appears in the eighteenth century: "I can't help making fun of myself" (1737). Similarly, fun as in exciting goings on appears relatively late: "The engineers officers who are engaged in carrying out some of the Sirdar's plans get much more than their fair share of 'the fun'" (1897).

It was, then, at the turn of the eighteenth century that the language required and developed the word fun in something like its current form. It is not fanciful to relate this semantic development to the industrial revolution. When British society was industrialised and class relations came to be organised around production and labour rather than feudal ties, a "low cant word" appeared which signified the absence of seriousness, work, labour. When production is mechanised, when labour processes are rationalised, when time is ossified to demarcate work and leisure, the word fun appears as its correlative. As Thompson (1963) pointed out, the working class was there at the moment of its own making. The word fun then has a political dimension. It still retains its "low" associations. Fun remains a form of resistance in the workplace, the fun of "the laff", the piss-take (Willis 2000). Fun can be seen both as a resistance to the rigid demarcation between work and leisure and also as a means of reproducing that dichotomy.

The rigid division between work and leisure and the rise of the cultural industries are relatively recent phenomena. Writing on the cultural industries of the nineteen fifties, Adorno and Horkheimer (1986) pointed to the similarities between the ways in which leisure and work time were structured and monitored. For these authors, the cultural industries exacerbated the artificial division between enjoyment and the rest of life: "Amusement under late capitalism is the prolongation of work." (Ibid: 137). Although amusement is sought as an escape from mechanized work, mechanization determines the production of "amusement goods" with the result that leisure experiences are "inevitably after-images of the work process itself" (Ibid). For Adorno and Horkheimer, the cultural industries then encouraged passivity, operating as a hegemonic device and a means of mass deception.

These members of the Frankfurt school and other Marxist writers pointed out that leisure was structured to meet the demands of capitalist production and working days of alienated labour (Roijec 1985). The Situationists of the nineteen sixties argued that the entertainment industry and mass media had formed a "society of the spectacle" which enchants, distracts and numbs us, transforming us into the passive spectators of our own lives (Debord 1995). Fun is something we buy, something we consume, something that ultimately reproduces the situations of alienated labour that we are seeking to escape. This somewhat bleak view of fun can be related to the work of the cartoonist Bill Griffiths (see Fig. 1).

Bill Griffith's character *Zippy* wanders through consumer landscapes asking hopefully "Are we having fun yet?" There is something tragic about the look of these cartoons and about the question itself. The question suggests at once a promise and a betrayal. Like Seligman and Csikszentmihalyi's dysphoric TV viewers Zippy is probably not having fun even when he is told that he is.

Marxist analyses of the cultural industries and leisure are, of course, deeply unfashionable and have been criticized for their pessimism and elitism. Empirical studies on the actual uses of cultural products have show than consumption is not passive: private and individual meanings are invested in leisure activities despite hegemonic intent (Willis 1990). We do not watch TV solely because we have become the numbed spectators of our own lives, passively and joylessly consuming spectacles as "cultural dupes". Dysphoria is not the only result of watching TV. The experience may not be the intense peak that Csikszentmihalyi's chess players would call flow or Adorno might approve of but it is nevertheless in some sense rewarding. We believe that Csikszentmihalyi's humanistic and Adorno's pessimistic views can neglect the psychological reality of individuals—their need to be *absorbed* sometimes and to be *distracted* at others.

4   Context Dependency

It is important to consider enjoyment as a context dependent and relational phenomena. Enjoyment is never guaranteed. Think of activities associated with enjoyment: sex, dancing, riding, swimming, taking drugs, playing a game, talking, joking, flirting, writing, listening to music, looking at a painting, reading, watching a play, movie, or other entertainment. Each of these activities is enjoyable or not depending on the situation that the activity is embedded in. Each situation is a unique constellation of a person's current goals, previous knowledge and experiences, the behaviour domain, and applicable social norms. A ride on a roller coaster can be enjoyable, but maybe not after an enormous dinner. Activities or objects normally appreciated by a person do not necessarily or deterministically lead to

Fig. 1 Zippy The Pinhead

enjoyment. What may be enjoyable in one context (watching a soap opera with friends) might be utterly dull in another (watching a soap opera alone). A game we enjoyed playing yesterday might completely bore us today. Activities associated with enjoyment offer potentials for enjoyment rather than enjoyment itself (see Hassenzahl elsewhere in this book).

Enjoyment is, in the widest sense, context specific. Indeed the American philosopher John Dewey argued that all emotions are grounded in particular contexts of experience:

> There is no such thing as the emotion of fear, hate, love ... The unique character of experienced events and situations impregnates the emotion that is evoked. (Dewey cited in Jackson 1998, p. 11)

In this sense enjoyment doesn't exist in and of itself. It's a relationship between ongoing activities and states of mind.

Is it then impossible to define or categorise different forms of enjoyment? Can there be a body of knowledge about enjoyment, a "pleasure-based human factors," (Jordan 2000), a "funology" (Monk et al. 2002)? In Matt Groening's *Futurama* cartoon show there are theme parks on the moon designed by "fungineers". The idea is hilarious. How could fun be engineered? Taking enjoyment seriously is a paradox, which on the face of it, seems pretentious or simply silly. There are as many kinds of enjoyment as there are people in the world. In the novel *My Idea of Fun* Will Self assumes the character of a man who finds murdering tramps enjoyable (Self 1994). It may or may not be the case that psychopaths experience violence as enjoyment and we are in no more a position of authority in this matter than the grandiloquent author. But the existence of theme parks, and indeed all popular culture, suggests that there is a degree of common ground in our ideas of enjoyment, culturally specific though they may be.

5   The Experience of Fun and Pleasure

There are connotational and experiential differences between fun and pleasure. Fun has quite specific and differential everyday meanings. Pleasure as a term is more problematic. It is, like enjoyment, a superordinate term. In the following sections we discuss pleasure as a specific type of enjoyment rather than as a superordinate category. This distinct use of the word can be related to Aristotle's view of pleasure as sense stimulation through action. Commentators have argued that Aristotle saw pleasure as "the perfect actualisation of a sentient being's natural capacities, operating on their proper objects". This notion of pleasure as self-actualisation is echoed in Csikszentmihalyi's work and his emphasis on the importance of appropriate levels of challenge as a condition for flow. In the remaining sections then pleasure is thought of as distinct from fun in terms of intensity and its relation to action. More specifically we argue, that fun and pleasure can be thought of as experiences that generally differ in terms of distraction and absorption (see Table 1 for an overview of specific differences). This is not to suggest a polar dichotomy and it must be stressed that these experiences are fluid.

During the fleeting and amorphous experience of fun, we are distracted from the self. Our self-definition, our concerns, our problems are no longer the focus. We distract ourselves from the constant clamour of the internal dialogue. This is not meant to imply that fun is unimportant or by any means "bad". Its ability to distract with short-lividness and superficiality satisfies an important underlying psychological need.

In contrast, pleasure is a deeper form of enjoyment. The main difference between pleasure and fun is its focus on an activity and a deep feeling of absorption.

Pleasure, in this sense, is not short-lived. It may not even be spontaneous. It happens when people are devoted to an object or activity. It happens when people try to make sense of themselves—explore and nourish their identities. The objects or activities an individual is absorbed by make a connection to his or her self. They become important, relevant.

It has been argued that the dichotomy between work and pleasure originates in the protestant work ethic (Willis 2000). Clearly it is a false dichotomy: work can be a pleasure, it can be absorbing. But is it fun? The workplace can be the site of fun but it is generally in the context of a break from work. Fun cannot be serious and if it is then it ceases, in this sense, to be fun.

It is likely then that repetitive and routine work based tasks and technologies might be made fun through design but non-routine and creative work must absorb rather than distract if they are to be enjoyable. The infamous winking paperclip in word is clearly intended to be fun but most people find it annoying. It distracts rather than aiding concentration or absorption. A cute graphics approach may be appropriate to making repetitive or mundane tasks more enjoyable and Hohl et al. describe a good example of this in their chapter for this book. But such an approach can be hazardous if the experience that is being designed for should be pleasurable rather than fun.

In the following sections we discuss differences between fun and pleasure in more detail.

Table 1 Experiential and cultural connotations of fun and pleasure	Fun/distraction	Pleasure/absorption
	Triviality	Relevance
	Repetition	Progression
	Spectacle	Aesthetics
	Transgression	Commitment

5.1 Triviality and Relevance

The word fun in English carries cultural connotations of frivolity and triviality. Fun is an antonym of serious. In this sense science and art are not fun. Where there is an association with these endeavours and fun, it is with education. Occasionally pedagogues attempt to "make" science and art fun. The implication of this is, of course, that they are not already intrinsically fun themselves. Thus early educational software incorporated games to make the learning less serious, less unpleasant. But there is something uncomfortable about the yoking together of fun and serious applications. The fun elements in educational software can appear as bribes when they are not totally integrated (Laurel 1993, p. 74). They are confidence tricks; they are the spoonful of sugar that helps the bad medicine go down.

It may be that where learning and high art are enjoyable it is when they are totally absorbing in and for themselves. Opera, ballet, classical music, poetry, do not carry cultural connotations of fun but of pleasure. "High" art is not a distraction, indeed if our powers of concentration are not up to it they may actively bore us and cause anxiety. Art demands absorption and we are not necessarily prepared to commit that much of our attention to it. Fun may be banal and in some respects morally suspect. It can be malicious—I was just having a bit of fun. Game shows, quiz shows, reality shows are increasingly absurd and surreal and those that decry a "dumbed down" mass media are accused of elitism. In this sense, fun can function as a moral imperative—Western hedonistic culture frequently tells itself to—lighten up, live a little, get a life, have some fun.

Jordan distinguishes between needs pleasures, which move a person from discontentment to contentment, drinking a glass of water for example, and appreciation pleasures, where something is pleasurable no matter what the current level of contentment, drinking wine, for instance (Jordan 2000: 14).

> The important thing to note, then, is that pleasure can be thought of both as the elimination of, or absence of, pain and also as the provision of positive, joyful feelings. (Ibid: 15)

Fun is not necessarily the absence of pain or even the provision of a joy it is the absence of seriousness. An activity or object that is fun is trivial in the sense that it does not make a strong connection to the self. It is not necessarily personally relevant and meaningful. Distraction from the self requires this. A roller-coaster ride is fun, it dazzles the senses, but it is not revealing. After a roller-coaster ride you might realise that you have a weak stomach, but you are unlikely to uncover a hidden aspect of your personality. (However, if you take the roller-coaster ride in order to overcome strong personal fears then you will rather experience pleasure. This is also an example of the relational nature of experience.) Activities or objects that are absorbing, are personally meaningful. They become a part of one's self-definition. They are long-lived, i.e., people tend to stick to these objects and activities.

But how does relevance come about? One source of relevance has already been mentioned: opportunities for personal growth. Activities (and sometimes objects) can be self-revealing. For example, playing a part in a play may be a pleasure, because of the insights one gains while trying to relate to the figure in the play. Questions like 'How do I feel about the figure? Would I act the same or differently? How does it feel to give up my own personality for a while?' have the power to change ways of thinking about oneself. This is very different to the fun we get out of watching a second rate Sci-Fi movie such as *Barbarella*. Here distraction from the self is at the fore. It is important to note, that relevance does not depend on the activity or object per se. What seems to be a silly movie to us can be very relevant to others. A second source for relevance is memory. Every object or activity can have personally relevant meanings attached to it that go beyond the obvious. This can be a source of pleasure. Imagine a couple listening to *their* song—the song that reminds them of their first rendezvous. Besides the actual enjoyment of merely listening to the song, pleasant memories are triggered. These memories will add to the pleasure. This again, differs very much from listening to a radio playing in the background while doing the daily household chores. The former requires focus and absorption; the latter is a welcome distraction from an otherwise boring task. A third source of relevance is anticipation. Here fantasies about activities or objects that are about to happen are a source for pleasure. Both memory and anticipation require a high commitment to and focus on the activities and objects involved.

5.2 Repetition and Progression

Popular culture is based on repetition. Although there is repetition in "classical" music the repetition is focussed towards progression: the gradual change and development of themes and movements; pop music as a form, is based on repetition that does not necessarily progress: the alternation between verse and chorus and the relentless emphasis of a regular beat. The mainstays of popular entertainment are largely formulaic. Soap operas, sitcoms and game shows are all based on the repetition of particular themes. When sitcoms break the formulae—Niles finally getting together with his unrequited love in *Frasier* for instance, the show is rarely as popular. All popular sporting events endlessly repeat the same scenarios. Within all of these forms there must be infinite possible combinations which produce new events: the new pop song, the new episode of *Friends*, the next game of *Who Wants To Be A Millionaire*, the next world cup and so on. High culture may also depend on certain kinds of repetition, genre for instance, but it is not concerned with creating formulae. There could be no *Hamlet II*. High art is concerned with complete experiences. Popular culture is concerned with cycles of sameness, endless variation within self-replication. Games, whether physical or virtual, also depend on variable repetition. Consider the number of physical games that involve bouncing a ball, or the act of bouncing a ball itself. There is a comfort and a joy in the act. In computer games there is not only the physical repetition of hitting buttons on a keyboard or a joy pad but also the repetition of virtual action on the screen: running, jumping, hitting, shooting, dying.

Pleasure can be thought of in terms of progression rather than repetition Progression stimulates, it makes us think, it surprises. Surprise marks the central difference between satisfaction and pleasure. Satisfaction is the emotional consequence of *confirmed* expectations, whereas pleasure is the consequence of *deviations* from expectations. For example, a meeting with an important client that went better than expected or an unexpected pay rise. Here, the source of pleasure is not the actual outcome of the meeting or the size of pay rise—it is its unexpectedness. The notion that surprise may lead to pleasure, has an important implication, which can be circumscribed by the metaphor of a "hedonic treadmill" (Brickman and Campbell 1971, cited in Kahneman 1999). A novel object may be pleasurable but reactions to novel objects are not stable. The individual will adapt and the likelihood of pleasure derived from the novelty of a certain object will decrease. As Aristotle noted, pleasure decreases because the mind becomes less active, less stimulated as it becomes familiar with the novel object or experience. Instead of having fun by repeating familiar patterns, the pleasure-seeker will constantly explore new regions and domains in her pursuit of pleasure. Csikszentmihalyi's flow also depends on progression in this sense. It requires a close match between ability and challenge. Progression seems to be a necessary precondition for challenge; a challenge can only be set up, when there are things to do and it is clear what hasn't been done yet, pleasure involves the setting of plans and actions to meet these goals. Without the possibility of generating new and challenging goals pleasure, in this sense, is unthinkable.

5.3 Spectacle and Aesthetics

During fun the senses must be engaged, there must be spectacle. The bright and luminous colours of children's toys, the gaudy kitsch sets of the popular game show, the explosions of light and sound in popular film are instances of the spectacle of fun. Attention is "grabbed", we demand increasingly violent distraction; the leisure society is also the society of the spectacle. Spectacle and wild colour signal and signify fun. Subdued pastels do not. If there is an aesthetic of fun then it is gaudy, and fleeting, it bursts at the eye like a firework.

Aesthetic pleasures are more abstract and orderly (Duncker 1941, cited in Rozin 1999). The Gestalt of objects and activities, their regularity, symmetry, shapeliness, solidness reassures us. There is a danger of confusing aesthetics with *tastes*. It is now, more or less accepted in the field of aesthetics that judgements of taste are not universal or timeless but historically, culturally and socially specific (Devereaux 2001). However, within given cultures some aesthetic values can endure for a very long time as examples of "classic" architecture, sculpture and painting indicate. Thus, aesthetic values are something people share.

To return to the distinction between pleasure and fun, the fun of the spectacle is a result of the *intensity* of perceptual stimulation, whereas aesthetic value is concerned with the *quality* of perception.

5.4   Transgression and Commitment

What, is the "fun" of the practical joke, the wind up, the "piss take" the unexpected appropriation of a situation? The fun of the "laff" in the workplace involves a transgression, albeit temporary and playful, of accepted forms of work behaviour. Goffman (1972, p. 59) describes this as the "flooding out" of one social frame to another. Perhaps then, transgression can be thought of as an element of fun, if only in a temporary deviation from seriousness. The mechanics of the joke are reduced by some writers on comedy, to category mistakes or the coming together of independent frames of reference creating a conflict or tension which is relieved in laughter; the essential basis of comic devices then, is conflict. Bergson considered satire to be "a social sanction against inflexible behaviour" (cited in Skynner and Cleese 1993) The transgressions of "fun" like those of satire are "bites that are not bites" (Bateson 1972). They are safe transgressions within particular contextual boundaries.

Again, in relation to fun and the distinction we are trying to outline, transgression can be fun but commitment may be pleasurable. Being absorbed in an activity requires—first of all—a general acceptance of the activity, a commitment to the basic assumptions and rules underlying this activity. Imagine two people playing a game. For the first the game is appealing. She figured out strategies to win in the context of the game. She accepts the game. The activity of playing, understanding and using the rules absorbs her. She will experience pleasure. The other person finds the game boring, but wants to oblige the first person. In order to distract herself from the boredom she finds a way to cheat, to bend the rules. By doing this, she ridicules the game but she may now have fun playing it. Both players enjoy themselves but their experiences will significantly differ in quality.

6   Conclusion

To summarise, this chapter has argued that although words like fun and pleasure are closely related and may each function as a superordinate category for the other, there are experiential and cultural differences between them. Fun has been considered in terms of distraction and pleasure in terms of absorption. This is not to suggest that pleasure is a more worthy pursuit than fun, it is rather an attempt to delineate different but equally important aspects of enjoyment. It is possible to appreciate Shakespeare and still acknowledge that *The Simpsons* is the greatest achievement of Western civilisation. Both offer rich and fulfilling experiences but they are very different kinds of pleasures. As Peter Wright and John McCarthy argue elsewhere in this book, it is not possible to design an experience, only to design *for* an experience; but in order to do this it is necessary to have an understanding of that experience as it relates to and differs from others.

References

Adorno T, Horkheimer M (1986) Dialectic of englightenment. Verso, London

Bateson G (ed) (1972) Steps to an ecology of mind. Ballantine Books, New York

Csikszentmihalyi M (1975) Beyond boredom and anxiety: the experience of work and play in games. Jossey Bass Publishers, San Fancisco

Debord G (1995) The society of the spectacle. Zone Books, New York

Devereaux M (2001) The philosophical status of aesthetics. Available: http://aesthetics-online.org/ideas/devereaux.html

Goffman E (1972) Encounters: two studies in the sociology of interaction. Penguin, Harmondsworth

Jackson P (1998) John Dewey and the lessons of art. Yale University Press

Jordan P (2000) Designing pleasurable products: an introduction to the new human factors. Taylor and Francis

Kahneman D (1999) Objective happiness. In: Kahneman D, Diener E, Schwarz N (eds) Well-being: the foundations of hedonic quality. Sage, New York, pp 3–25

Laurel B (1993) Computer as theatre. Addison-Wesley, Reading, MA

Monk AF, Hassenzahl M, Blythe M, Reed D (2002) Funology: designing enjoyment. In: Paper presented at the CHI. pp 924–925

Oxford English Dictionary. Available at: http://dictionary.oed.com/entrance.dtl

Roijec C (1985) Capitalism and leisure theory. Tavistock, London

Rozin P (1999) Preadaption and the puzzles and properties of pleasure. In: Kahneman D, Diener E, Schwarz N (eds) Well-being: the foundations of hedonic psychology. Russell Sage Foundation, New York, pp 109–133

Self W (1994) My idea of fun. Penguin books, London

Seligman MEP, Csikszentmihalyi M (2000) Positive psychology: an introduction. Am Psychol 55:5–14

Shafrir D (2017) Startup! A novel. Little Brown and Company

Skynner R, Cleese J (1993) Life and how to survive it. Methuen, London

Thompson EP (1963) The making of the english working class. Gollancz, London

Willis P (1990) Common culture. Open University Press

Willis P (2000) The ethnographic imagination. Polity Press, Cambridge

Zizek S (1999) You may! The postmodern superego. In: London review of books, vol 21, no 6

Part VII
"Methods and Techniques"

Chapter 25
Measuring Emotion: Development and Application of an Instrument to Measure Emotional Responses to Products

Pieter Desmet

Author's Note, Funology 2

This chapter described the initial version of PrEmo (PrEmo1) that was introduced in 2002. Since its introduction, PrEmo1 was used to measure emotions evoked by a wide variety of products and other designed stimuli, such as wheelchairs (Desmet and Dijkhuis 2003), automotive design (Desmet and Hekkert 1998), mobile phones (Desmet et al. 2007), airplane meals, and functional fragrances (Desmet and Schifferstein 2012), serving both as a means for generating insights for new product conceptualization and as a means for evaluating the emotional impact of new design concepts. In 2013, a fully revised version was launched: PrEmo2, see Fig. 1.

The new version, introduced by Laurans and Desmet (2017), was based on insights gained in ten years of experiences with using PrEmo1. Three main improvements were made. The first is the character design. The PrEmo2 character was designed to be is less 'cute' than the PrEmo1 version. In addition, by adding detail to the facial expressions, the new character gives more reliable information about the expressed emotion. As a consequence, the recognition rate has increased significantly. The second improvement is in the set of emotions. PrEmo2 measures 14 emotions that are categorised in four domains: General well-being emotions (joy, sadness, hope, and fear), expectation-based emotions (satisfaction and dissatisfaction), social context emotions (admiration, contempt, shame, and pride), and material context emotions (attraction, aversion, fascination, and boredom). Whereas the PrEmo1 set was optimised for product appearance, the new set has broader application possibilities. The third improvement is a more extensive validation. The animations were validated in 8 studies across 4 countries (total $N = 826$), including China,

P. Desmet (✉)
Technical University of Delft, Delft, The Netherlands
e-mail: P.M.A.Desmet@tudelft.nl

© Springer International Publishing AG, part of Springer Nature 2018
M. Blythe and A. Monk (eds.), *Funology 2*,
Human–Computer Interaction Series,
https://doi.org/10.1007/978-3-319-68213-6_25

Fig. 1 Stills from the PrEmo2 "contempt" animation

the Netherlands, United Kingdom, and the United States (see Laurans and Desmet 2017). More information about PrEmo2 can be found on the website of the Delft Institute of Positive Design: www.diopd.org

2003 Paper

1 Introduction

Emotions enrich virtually all our waking moments with either a pleasant or unpleasant quality. Cacioppo and his colleagues wrote,

> emotions guide, enrich and ennoble life; they provide meaning to everyday existence; they render the valuation placed on life and property. (Cacioppo et al. 2001, p. 173)

These words illustrate that our relationship with the physical world is an emotional one. Clearly, the 'fun of use,' i.e. the fun one experiences from owning or using a product, also belongs to the affective rather than rational domain. The difficulty in studying affective concepts as 'enjoyment of use' and 'fun of use' is that they seem to be as intangible as they are appealing. Even more, rather than being an emotion as such, 'having fun' is probably the outcome of a wide range of possible emotional responses. Imagine, for example, the fun one has when watching a movie. This person will experience all kinds of emotions, such as fear, amusement, anger, relief, disappointment, and hope. Instead of one isolated emotion, it is the *combination* of these emotions that contributes to the experience of fun. It is not implausible that the same applies to other instances of fun, whether it is sharing a joke, using a product, or interacting with a computer.

So far, little is known about how people respond emotionally to products and what aspects of design or interaction trigger emotional responses. In order to support the study of these responses, a measurement instrument was developed that is capable to measure combinations of simultaneously experienced emotions: the Product Emotion Measurement Instrument (PrEmo). This chapter discusses the development of PrEmo in the context of existing instruments. In addition, an illustrative cross-culture study is reported, in which emotions evoked by car models have been measured in Japan and in the Netherlands.

2 Approaches to Measure Emotion

Before one can measure emotions, one must be able to characterise emotions and distinguish them from other states. Unfortunately, although the concept of emotion appears to be generally understood, it is surprisingly difficult to come up with a solid definition. In the last 100 years, psychologists have offered a variety of definitions, each focussing on different manifestations or components of the emotion. As there seems to be no empirical solution to the debate on which component is sufficient or necessary to define emotions, at present the most favoured solution is to say that emotions are best treated as multifaceted phenomena consisting of the following components: behavioural reactions (e.g. retreating), expressive reactions (e.g. smiling), physiological reactions (e.g. heart pounding), and subjective feelings (e.g. feeling amused). Each instrument that is claimed to measure emotions in fact measures one of these components. As a consequence, both the number of reported instruments and the diversity in approaches to measure emotions is abundant. In this chapter, the basic distinction is made between non-verbal (objective) instruments and verbal (subjective) instruments.

2.1 Non-verbal Instruments to Measure Emotions

This category comprises instruments that measure either the expressive or the physiological component of emotion. An expressive reaction (e.g. smiling or frowning) is the facial, vocal, and postural expression that accompanies the emotion. Each emotion is associated with a particular pattern of expression (Ekman 1994): for example, anger comes with a fixed stare, contracted eyebrows, compressed lips, vigorous and brisk movements and, usually, a raised voice, almost shouting (Ekman and Friesen 1975). Instruments that measure this component of emotion fall into two major categories: those measuring facial and those measuring vocal expressions. Facial expression instruments are based on theories that link expression features to distinct emotions. Examples of such theories are the Facial Action Coding System (FACS; Ekman and Friesen 1978), and the Maximally Discriminative Facial Moving Coding System (MAX; Izard 1979). Generally, visible expressions captured on stills or short video sequences are analysed. An example is the Facial Expression Analysis Tool (FEAT; Kaiser and Wehrle 2001), which automatically codes videotaped facial actions in terms of FACS. Like the facial expression instruments, vocal instruments are based on theories that link patterns of vocal cues to emotions (e.g. Johnstone and Scherer 2001). These instruments measure the effects of emotion in multiple vocal cues such as average pitch, pitch changes, intensity colour, speaking rate, voice quality, and articulation.

A physiological reaction (e.g. increases in heart rate) is the change in activity in the autonomic nervous system (ANS) that accompanies emotions. Emotions show a variety of physiological manifestations that can be measured with a diverse array of techniques. Examples are instruments that measure blood pressure responses, skin responses, pupillary responses, brain waves, and heart responses. Researchers in the field of affective computing are most active in developing ANS instruments, such as IBM's emotion mouse (Ark et al. 1999) and a variety of wearable sensors designed by the Affective Computing Group at MIT (e.g. Picard 2000). With these instruments, computers can gather multiple physiological signals while a person is experiencing an emotion, and learn which pattern is most indicative of which emotion.

The major advantage of non-verbal instruments is that, as they are language-independent, they can be used in different cultures. A second advantage is that they are unobtrusive because they do not disturb participants during the measurement. In addition, these instruments are often claimed to be less subjective than self-report instruments because they do not rely on the participants' own assessment of the emotional experience. For the current application however, this class of instruments has several limitations. First, these instruments can only reliably assess a limited set of 'basic' emotions (such as anger, fear, and surprise). Reported studies find a recognition accuracy of around 60–80% for six to eight basic emotions (see Cacioppo et al. 2001). Moreover, these instruments cannot assess combinations of simultaneously experienced emotions. Given these limitations, it was decided not to use this approach for measuring emotions evoked by products.

2.2   Verbal Instruments to Measure Emotions

The limitations of non-verbal instruments as discussed above are overcome by verbal self-report instruments, which typically assess the subjective feeling component of emotions. A subjective feeling (e.g. feeling happy or feeling inspired) is the conscious awareness of the emotional state one is in, i.e. the subjective emotional experience. Subjective feelings can only be measured through self-report. The most often used self-report instruments require respondents to report their emotions with the use of a set of rating scales or verbal protocols.

The two major advantages of the verbal instruments is that rating scales can be assembled to represent any set of emotions, and can be used to measure combinations of emotions. The main disadvantage is that they are difficult to apply between cultures. In emotion research, translating emotion words is known to be difficult because for many emotion words a one-to-one, 'straight' translation is not available. Between-culture comparisons are therefore notoriously problematic. To overcome this problem, a handful of non-verbal self-report instruments have recently been developed that use pictograms instead of words to represent emotional responses. An example is the Self-Assessment Manikin (SAM; Lang 1985). With SAM, respondents point out the puppets that in their opinion best portray their emotion. Although applicable in between-culture studies, these non-verbal scales also have an important limitation, which is that they do not measure distinct emotions but only generalised emotional states (in terms of underlying dimensions such as pleasantness and arousal). Consequently, a new instrument for measuring the emotions evoked by products was developed. This instrument combines the advantages of existing non-verbal and verbal self-report instruments: it measures distinct emotions and combinations of emotions but does not require the participants to verbalise their emotions.

3 The Product Emotion Measurement Instrument

Does my question annoy him? Is she amused by my story? In the face-to-face encounters of everyday life we constantly monitor and interpret the emotions of others (see Ettcoff and Magee 1992). This interpretation skill was the starting point for the development of PrEmo. PrEmo is a non-verbal self-report instrument that measures 14 emotions that are often elicited by product design. Of these 14 emotions, seven are pleasant (i.e. desire, pleasant surprise, inspiration, amusement, admiration, satisfaction, fascination), and seven are unpleasant (i.e. indignation, contempt, disgust, unpleasant surprise, dissatisfaction, disappointment, and boredom). Instead of relying on the use of words, respondents can report their emotions with the use of expressive cartoon animations. In the instrument, each of the 14 measured emotions is portrayed by an animation by means of dynamic facial, bodily, and vocal expressions. Figure 2 shows the measurement interface.

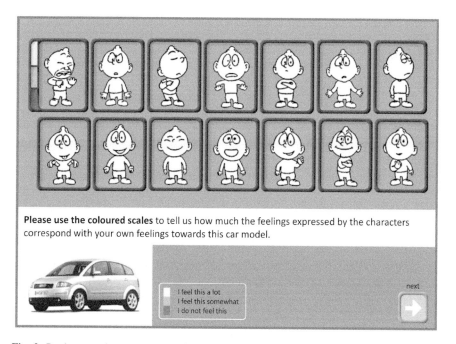

Fig. 2 Product emotion measurement instrument interface

The procedure of a PrEmo experiment is self-running. The computer screen displays instructions that guide respondents through the procedure, which includes an explanation of the experiment and an exercise. The program's heart is the measurement interface, which was designed to be simple and intuitive in use. The top section of this interface depicts stills of the 14 animations. Each still is accompanied by a (hidden) three-point scale. These scales represent the following ratings: "I do feel the emotion," "to some extent I feel the emotion," and "I do not feel the emotion expressed by this animation." The rating scales are 'hidden behind' the animation frames. A scale appears on the side of the animation frame only after the animation is activated by clicking on the particular still. The lower section of the interface displays a picture of the stimulus and an operation button. During an experiment, the respondents are first shown a (picture of a) product and subsequently instructed to use the animations to report their emotion(s) evoked by the product. While they view an animation, they must ask themselves the following question: "does this animation express what I feel?" Subsequently, they use the three-point scale to answer this question. Visual feedback of the scorings is provided by the background colour of the animation frame.

3.1   Emotions Measured by PrEmo

The 14 measured emotions were selected to represent a manageable cross-section of all emotions that can be elicited by consumer products. For this selection, a multistage method was used. First, a set of emotions was assembled that is sufficiently extensive to represent a general overview of the full repertoire of human emotions. This set of 347 emotions was compiled by merging and translating reported lists of emotions. In the first study, participants ($N = 20$) rated these emotions on the dimensions 'pleasantness' and 'arousal,' which represent the dimensions of the 'Circumplex of Affect' developed by Russell (1980). Both dimensions were rated on a three-point scale: pleasant-neutral-unpleasant, and calm- moderate-excited respectively. In addition, participants marked emotion words with which they were not familiar. On the basis of these ratings, the set emotions was divided in eight categories (see Table 1). Note that one combination, i.e. moderate- neutral is not included. It is left out because it is not considered to be an emotional category in the

Table 1 Emotion categories

Category	Amount of included emotions	Category	Amount of included emotions
Excited pleasant	30	Calm unpleasant	34
Moderate pleasant	53	Moderate unpleasant	61
Calm pleasant	24	Excited unpleasant	46
Calm neutral	14	Excited neutral	20

Circumplex model. Emotions that were ambiguous or marked as unfamiliar were omitted from the set.

In order to further reduce the set, the second study was designed to select those emotions that are most often elicited by products. In this study, participants ($N = 22$) used a rating procedure to indicate which emotions they often, and which they do not often experience in response to product design. They were instructed to do this for each of the eight emotion sets. On the basis of the sum scores, 69 emotions were selected that are evoked regularly by product design (the sum scores of these emotions were significantly higher than the average score).

Subsequently, in the third step, the set was further reduced by eliminating those emotions that are approximately similar to others in the set. Participants ($N = 40$) rated the similarity of the emotions in pairs. With the use of a hierarchical cluster analysis, the set of 69 emotions was reduced to a set of 41 emotions. In a final study, participants ($N = 23$) rated all 41 emotions on a five-point scale (from 'very relevant to product experience' to 'not relevant to product experience'). On the basis of the mean scores, the final set of 14 emotions was selected. Although, evidently, products can elicit more than these 14 emotions, these are the ones that can be considered to occur most frequently. Moreover, PrEmo requires a set that can be surveyable. The set of 14 is regarded as a workable balance between comprehensive and surveyable. Note that a detailed report of the selection procedure can be found in Desmet (2002).

3.2   Dynamic Cartoon Animations

The idea to use expressive portrayals of the 14 emotions was based on the assumption that emotional expressions can be recognized reliably. Ekman (1994) found that facial expressions of basic emotions (e.g. fear and joy) are not only recognised reliably, but also univocally across cultures. As the emotions measured by PrEmo are subtler than the basic emotions, more information than merely the facial expression is needed to portray them reliably.

Our approach to this problem was to incorporate total body expression, movement, and vocal expression. It was decided to use a cartoon character because these are often particularly efficient in portraying emotions. This efficiency is achieved with abstracting which reduces the emotional expression to its essence. Abstracted portrayals can make the task of recognizing emotional expressions easier because the amount of irrelevant information is reduced (Bernson and Perrett 1991). Moreover, with cartoon characters it is possible to amplify (or exaggerate) the expressive cues that differ between emotional expressions (see Calder et al. 1997).

A professional animator designed the character and created the animated expressions. A vocal actor synchronized the vocal expressions. To enable the animator to create clear portrayals, a study with actors was conducted. In this study, four professional actors (two males, two females) were instructed to portray each of the 14 emotions as expressive and precise as they could. These portrayals were

inspiration

disgust

Fig. 3 Two animation sequences

recorded on videotape and analysed by the author and the animator. On the basis of this analysis, the animator created the animations. By ways of example, Fig. 3 shows the animation sequences of *inspiration* and *disgust*.

3.3 Validity and Reliability

The validity of PrEmo, i.e. the degree to which it accurately measures the emotions it was designed to measure, was assessed in a two-step procedure. The first step was to examine the validity of the animations. An important requirement was that PrEmo should be applicable in different cultures or language areas. Therefore, the study included participants from four different countries ($N = 120$; 29 Japanese, 29 United State citizens, 33 Finnish, and 29 Dutch participants). Participants were shown three animations and asked which of these three best portrayed a given emotion. Of the three animations shown, one was designed to portray the given emotion, and the other two portrayed other emotions (yet similar in terms of pleasantness and arousal). The animation that was supposed to portray the given emotion was considered valid when it was selected more often than could be expected by chance. A strict significance level (i.e. $p < 0.001$) was applied because it was important to identify also slightly inaccurate animations. On the basis of the results, it was concluded that in order to be valid, the animations portraying *desire* and *disappointment* needed further development. These two animations were found to be invalid in Japan and therefore adjusted on the basis of a study with four Japanese actors.

The validity of the instrument was examined in a second study ($N = 30$). In this study, both PrEmo and a verbal scale were used to measure emotions evoked by six chairs. The level of association between the results obtained with PrEmo and those obtained with the verbal scales was analysed. The correlations between emotion scores measured with the two methods were high (r varied from 0.72 to 0.99) and all but one (i.e. *amusement*) were significant ($p < 0.05$). For each emotion a repeated measures MANOVA was performed to examine interaction effects between chair model and instrument (i.e. either verbal scale or PrEmo). None of the analyses found a significant interaction effect between chair and instrument.

In agreement with the high correlation, these findings indicate that the participants did not respond differently to each of the chairs as a result of the measurement instrument applied. Based on these results, it was concluded that PrEmo is satisfactory with respect to its convergent validity. Moreover, participants reported in a questionnaire that they preferred using the animations to using words for reporting their emotional responses. The animations were found to be more intuitive in use and, importantly, much more enjoyable.

4   Cross-Cultural Application

The application possibilities of PrEmo have been explored with a between-culture study in which emotions evoked by six car models (see Fig. 4) were measured both in Japan ($n = 32$) and in the Netherlands ($n = 36$). It was decided to use cars because in previous studies we found that car models that vary in appearance can elicit strongly different emotions (see e.g. Desmet et al. 2000). Participants were matched on gender and age (20–60 years old). In a written introduction, it was explained that the purpose of the experiment was to assess emotional responses to the car designs. After the introduction, participants were shown a thumbnail display that gave an overview of all the models. Subsequently, photos of the six car models were presented in random order. After looking at a photo, participants reported their response with the 14 PrEmo animations.

In order to obtain a graphical representation of the results a correspondence analysis was performed with two factors: emotion (14 levels) and car combined with culture (12 levels). Correspondence analysis is a technique for describing the relationship between nominal variables, while simultaneously describing the relationship between the categories of each variable. It is an exploratory technique, primarily intended to facilitate the interpretation of the data. Figure 5 shows the

A (Audi A2)   C (Mazda Demio)   E (Toyota bB)

B (Fiat Multipla)   D (Opel Zafira)   F (Toyota Funcargo)

Fig. 4 Stimuli used in the application study

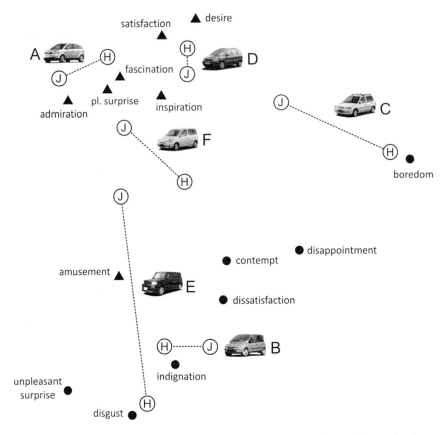

Fig. 5 'Product and emotion space' of Dutch ('H') and Japanese ('J') participants for six car models

two-dimensional solution of the analysis, which explains 90.3% of the total variance: the 'product and emotion space.'

This product and emotion space visualises the associations between the car models and the reported emotional responses. Pleasant emotions are indicated with a triangle and unpleasant with a circle. The results of the Japanese participants are indicated with a 'J,' and those of the Dutch with an 'H.' The distances between the car models reflect the relationships between them (with similar models plotted close to each other). Similarly, the distances between the car models and the emotions reflect the relationship between them. This means that car models that are plotted close to each other evoked similar emotions, whereas those plotted at a distance from each other evoked different emotions. Cars A and D, for example, evoked similar emotions, whereas Cars A and B evoked noticeably different emotions.

In the product and emotion space some effects catch the eye. Clearly, the degree to which car models differ from each other also varies. The difference between Cars A and D, for example, is smaller than the difference between Cars A and B.

Moreover, some car models appear to have elicited mainly pleasant emotions (e.g. Car D), some mainly unpleasant (e.g. Car B), and some both pleasant and unpleasant (e.g. Car F). In addition, two between-culture effects can be observed. First, the degree to which the emotional responses of the cultures differ depends on the car model. The space indicates that cultural differences are greatest for Cars E and C. Cars, A, B, and D, on the other hand, appear to have elicited similar emotions in Japan and in the Netherlands. Secondly, the product and emotion space indicates that the Japanese experienced generally higher ratings on pleasant emotions than the Dutch. The three car models that showed the largest cultural differences elicited more pleasant emotions in Japan than in the Netherlands.

4.1 Between-Culture Differences

The correspondence analysis is an exploratory technique, primarily intended to facilitate the interpretation of the data. Because it is not appropriate to draw conclusions, the observed between-culture effects have been examined in more depth, with an analysis of variance. For each emotion a two-way repeated measures MANOVA was performed with car (six levels) as within-participants factor, culture (two levels) as between-participant factor, and the emotion as dependent variable. Some interesting culture effects have been found. For three emotions, cultural differences independent of car model were found. Japanese participants showed higher mean scores on the following emotions: *admiration*, *satisfaction*, and *fascination* ($p < 0.01$). This may point to a cultural difference in how car models are experienced: apparently Japanese people are generally more admiring of, satisfied, and fascinated by car models than the Dutch. Some car x culture interaction effects indicated that there are also cultural differences in responses with respect to the particular car models used in the study. Interaction effects were found for *disgust*, *unpleasant surprise*, *dissatisfaction*, *amusement*, *admiration*, and *satisfaction*. For example, the Dutch participants were not amused by the same car models as the Japanese.

A notable finding was that, contrary to expectations, cultural differences cannot be explained by product-familiarity. For instance, for Car B (Fiat Multipla) no significant cultural differences were found with respect to the emotions it elicited. This was not expected, because the Dutch participants were familiar with this model, and the Japanese were not. These findings confirm the idea that in product development, cultural differences must be recognized, and that these differences are both difficult to predict and to explain. Companies involved in 'global marketing' should be aware of these differences and should perhaps develop various design strategies for different cultures, instead of attempting to market identical products in different countries.

5 Discussion

The unique strength of PrEmo is that it combines two qualities: it measures distinct emotions and it can be used cross-culturally because it does not ask respondents to verbalise their emotions. In addition, it can be used to measure mixed emotions, that is, more than one emotion experienced simultaneously, and the operation requires neither expensive equipment nor technical expertise. And, also important, respondents reported that the measurement task with PrEmo is pleasant or even enjoyable. A limitation for the application in human computer interaction is that the 14 measured emotions represent a cross-section of emotions experienced towards static product design. It is not said that this set also represents emotions that are experienced towards dynamic human product interaction. Some emotions may be over-represented, whereas others may be missing. Before PrEmo is applied for the measurement of emotions evoked by interacting with a computer (or any other product) it must be determined if the 14 emotions are adequate and, if not, the set animations should be adjusted.

What is the point of measuring emotions evoked by products or computer programs? More interesting than discovering *which* particular emotions are evoked by a set of stimuli, is to understand *why* those stimuli evoke these particular emotions. This information can be used in the development of new products, to elicit pre-defined emotion profiles. Hence, the interpretation of PrEmo results requires theoretical propositions about how product emotions are related to the product's appearance and interaction, and the characteristics of the person who experiences the emotions. In cognitive emotion psychology, emotions are regarded as outcomes of appraisal processes. According to Frijda (1986), emotions are elicited when a subject appraises a stimulus as important for the gain of some personal concern. A concern can be any goal, standard, attitude, or motive one has in life, e.g., achieving status, feeling safe, or respecting the environment. In following Arnold (1960), Frijda argues that when we appraise a stimulus as beneficial to our concerns, we will experience positive emotions and try to approach this particular stimulus. Likewise, when we appraise a stimulus as conflicting with our concerns, we will experience negative emotions and try to avoid it. As concerns are personal, different subjects have different concerns. As a result, individual subjects will appraise a given product differently. As different types of emotions are evoked by different kinds of appraisals, appraisals can be used to differentiate emotions (e.g., Ortony et al. 1988). For the 14 emotions measured by PrEmo, Desmet (2002) described the specific appraisal patterns underlying each emotion. Understanding these patterns could guide designers in controlling the emotional responses to their designs.

A second application possibility of PrEmo is to use it as a means to communicate emotional responses to products. The emotional aspects of a design can be difficult to discuss because they are often based on intuition. The 'product & emotion space' that results from a PrEmo experiment makes the intangible emotional responses tangible. In various design workshops, the space has proven to be a valuable support to discuss emotional aspects of design in a design team. In addition, designers found it to be effective when used as a means to communicate, argue, and defend their ideas to non-designers who are also involved in the product development (e.g. marketing, engineering, etcetera).

The decision to design an instrument that measures both pleasant *and* unpleasant emotions was based on the notion that unpleasant responses are as interesting as the pleasant. What are the characteristics that make one product more enjoyable or attractive than another? Some of us find riding a roller coaster fun, whereas others would not want to be found dead in one. Some consider the fear experienced when thrown from a bridge with elastic tied to one's ankles to be fun whereas others prefer to play a game of bridge. Whatever the interpersonal differences in what we find to be fun, it would clearly be incorrect to assume that that fun is related only to pleasant emotions. Frijda and Schram (1994) stated that art often elicits paradoxical emotions, that is, positive and negative emotions simultaneously, and that it is precisely these paradoxical emotions that we seek and enjoy. In the words of Frijda (p. 2) "we enjoy watching tragic miseries, and we pay fair amounts of money to suffer threat and suspense." It may be interesting for designers and design researchers to investigate the possibilities of designing such paradoxical emotions. Eventually, these efforts may result in products that are unique, innovative, rich in their interaction, interesting, and fun to use.

Acknowledgements This research was funded by Mitsubishi Motor R&D, Europe GmbH, Trebur, Germany. Paul Hekkert (Delft University), Jan Jacobs, and Kees Overbeeke are acknowledged for their contribution to this research. Animated characters were drawn by Peter Wassink.

References

Ark W, Dryer DC, Lu DJ (1999) The emotion mouse. In: Proceedings of HCI international'99, Munich Germany, August 1999

Arnold MB (1960) Emotion and personality: psychological aspects, vol 1. Colombia University Press, New York

Bernson PJ, Perrett DI (1991) Perception and recognition of photographic quality facial caricatures: implications for the recognition of natural images. Eur J Cogn Psychol 3:105–135

Cacioppo JT, Berntson GG, Larsen JT, Poehlmann KM, Ito TA (2001) The psychophysiology of emotion. In: Lewis M, Haviland-Jones JM (eds) Handbook of emotions, 2nd edn. The Guilford Press, New York, pp 173–191

Calder AJ, Young AW, Rowland D, Perrett DI (1997) Micro-expressive facial actions as a function of affective stimuli: replication and extension. Pers Soc Psychol Bull 18:515–526

Desmet PMA (2002) Designing emotions. Unpublished doctoral dissertation

Desmet PMA, Dijkhuis EA (2003) Wheelchairs can be fun: a case of emotion-driven design. In: Proceedings of the international conference on designing pleasurable products and interfaces, June 23–26 2003, Pittsburgh, Pennsylvania, USA. ACM publishing, New York

Desmet PMA, Hekkert P (1998) Emotional reactions elicited by car design: a measurement tool for designers. In: Roller D (ed) Automotive mechatronics design and engineering. ISATA, Düsseldorf, Germany, pp 237–244

Desmet PMA, Schifferstein NJH (2012) Emotion research as input for product design. In: Beckley J, Paredes D, Lopetcharat K (eds) Product innovation toolbox: a field guide to consumer understanding and research. Wiley, Hoboken, NJ, pp 149–175

Desmet PMA, Hekkert P, Jacobs JJ (2000) When a car makes you smile: development and application of an instrument to measure product emotions. In: Hoch SJ, Meyer RJ (eds) Advances in consumer research, vol 27. Association for Consumer Research, Provo, UT, pp 111–117

Desmet PMA, Porcelijn R, Van Dijk M (2007) Emotional design: application of a research based design approach. J Knowl Technol Policy 20(3):141–155

Ekman P (1994) Strong evidence for universals in facial expressions: a reply to Russell's mistaken critique. Psychol Bull 115(2):268–287

Ekman P, Friesen WV (1975) Unmasking the face: a guide to recognizing emotions from facial cues. Prentice-Hall, Englewood Cliffs, NJ

Ekman P, Friesen WV (1978) Facial action coding system: a technique for the measurement of facial movement. Consulting Psychologists Press, Palo Alto, CA

Ettcoff NL, Magee JJ (1992) Categorical perception of facial expressions. Cognition 44:227–240

Frijda NH (1986) The emotions. Cambridge University Press, Cambridge

Frijda NH, Schram D (1994) Introduction to the special issue on emotions and cultural products. Poetics 23(1–2):1–6

Izard CE (1979) The maximally discriminative facial movement coding system (MAX). Instructional Recourses Centre, University of Delaware, Newark

Johnstone T, Scherer KR (2001) Vocal communication of emotion. In: Haviland-Jones MLJM (ed) Handbook of emotions, 2nd edn. The Guilford Press, New York, pp 220–235

Kaiser S, Wehrle T (2001) Facial expressions as indicator of appraisal processes. In: Scherer K, Schorr A, Johnstone T (eds) Appraisal processes in emotion. Oxford University Press, Oxford, pp 285–300

Lang PJ (1985) The cognitive psychophysiology of emotion: anxiety and the anxiety disorders. Lawrence Erlbaum, Hillsdale, NJ

Laurans G, Desmet PMA (2017) Developing a non-verbal emotion self-report tool for categorical emotions. J Des Res (in print)

Ortony A, Clore GL, Collins A (1988) The cognitive structure of emotions. Cambridge University Press, Cambridge

Picard RW (2000) Towards computer that recognize and respond to user emotion. IBM Syst J 39 (3/4)

Russell JA (1980) A circumplex model of affect. J Pers Soc Psychol 39:1161–1178

Chapter 26
That's Entertainment!

John Karat and Clare-Marie Karat

John Karat (1949–2015).
John was a longtime member of the SIGCHI leadership, inspiring and supporting the community to take a worldwide view of HCI and its membership. This global view extended to his chairing the International Federation for the Information Processing Technical Committee on HCI (IFIP TC13). In addition, he was Editor-in-Chief of the Kluwer Academic Publishers that published the 2003 Edition. He served on the ACM SIGCHI Executive Committee for decades and was actively involved in ACM CHI and DIS, and the IFIP INTERACT conferences in the HCI field. When John was not meeting with HCI colleagues from around the world, he was a Research Staff Member at the IBM TJ Watson Research Center. John conducted HCI research and published numerous papers and patents on a variety of topics including privacy, personalization, conversational technologies, and information management. Over his career with IBM development he researched and advised on design collaboration, researched and developed speech-based systems, including the design of IBM's first large-vocabulary desktop speech recognition system and early electronic medical record systems. To honor John's many contributions, he was awarded the ACM SIGCHI Lifetime Service Award and was named a Distinguished Scientist by ACM. John's legacy will live on in the SIGCHI and IFIP communities. In honor of John's contributions to the usable privacy and security community, and his dedication to mentoring students and community service, the USENIX sponsored SOUPS conference (Symposium on Usable Privacy and Security) has instituted the annual John Karat Usable Privacy and Security Student Research Award for top graduate students following in his

J. Karat · C.-M. Karat (✉)
IBM TJ Watson Research Center, Hawthorne, NY, USA
e-mail: cmkarat@gmail.com

© Springer International Publishing AG, part of Springer Nature 2018
M. Blythe and A. Monk (eds.), *Funology 2*,
Human–Computer Interaction Series,
https://doi.org/10.1007/978-3-319-68213-6_26

405

footsteps with excellent scientific research, mentoring, and community service. It was awarded for the first time in 2016.

2003 Chapter

1 Introduction

1.1 *Motivation for the Project*

What kind of entertainment do people want from a Web site on art and culture? And what is the appropriate context of use for people to enjoy entertainment on the Web? Is this an individual activity or one that that would be most engaging if it occurred in a social context? And would people like to do this in multiple locations, for example, at home, work and on the road? At this time, there really is no Web experience similar to the most common entertainment activity, namely, watching TV (Vogel 1998). In this chapter, we present our experience in developing an entertainment W site for art and culture where the user-centred design (UCD) process led us to the design of TV-like, streaming, multimedia experiences delivered over the Web and similar to TV documentaries, but enriched by hotlinks enabling user control of the experience and access to extra content. The "less clicking, more watching" design approach that emerged through the research is in contrast to the prevailing notion that entertainment on the Web must be highly interactive and participatory as in the model for video games and chat rooms. Although almost one-half of Internet users spent some time with other members of their household every week (Cole 2000), there are few online entertainment opportunities appropriate for such group experiences. We explored this possibility by testing individual as well as small group use of the Web entertainment prototypes. The chapter begins with a brief discussion of the entertainment concept, followed by a description of the UCD process through which the design emerged for the initial prototype, discussion of the usability testing of the enriched prototypes, and a discussion of the lessons learned about the research topic and the methods employed to address it.

1.2 *Entertainment on the Web*

Many traditional forms of entertainment such as talking, reading, listening to music, watching movies and TV, playing sports and games, shopping, cooking, gardening, eating, drinking, visiting museums, attending cultural events have their counterparts on the web. Talking and gossiping have a forum in electronic chat rooms; reading news on the is becoming increasingly popular; the previously solitary video-game experience has found new meaning in the networked game era; and shopping has gigantic proportions on the Web, newly augmented by the thrills of on-line auction.

As stated previously, our research examines possible Web counterparts for a TV-like experience, i.e., Web-based "watchable" entertainment experiences provided on the screen of a desktop or laptop computer. Currently, few Web sites have experienced success in this arena, and those that have are of limited scope (The Economist 2000). The best examples are sites featuring animated cartoons, often based on parody, such as Joe Cartoon (www.joecartoon.com); sites that show short films, previews, and commercials such as Atom Films (www.atomfilms.com); and the "Web cam" phenomenon.

The three most common explanations for this shortage of options are the lack of bandwidth for video; the inadequacy of the desktop sitting position; and the need of interactivity in Web entertainment (The Economist 2000). However, networked video games have shown that the first two problems are not enough to deter entertainment: pre-downloading and local computer graphics rendering can deal with bandwidth problems, and people seem to sit forever in front of video-games.

So, if interactivity is the defining component of Web experiences, then the concept of a "watchable", TV-like Web experience is a contradiction in terms. In fact, throughout the development of this project, Web designers repeatedly told us that people are entertained by computers only when actively interacting with the content (Skelly et al. 1994, Webster and Ho 1997, and Murray 1997). This belief is strengthened by the repetitive failures of the traditional entertainment industry to create Web entertainment. The first cycle, fuelled by the success of the *"The Spot"* (www.spot.com) and by the MIT Media Lab advocating interactive TV, failed spectacularly in 1997 both for Microsoft and AOL (see Geirland and Sonesh-Kedar 1999). The dot.com phenomenon of 1999/2000 spurred a new wave of projects that also ended mostly in failure, particularly in the case of Steven Spielberg's www.pop.com, the Digital Entertainment Network, and Pseudo (www.pseudo.com) (Red Herring 2000; The Economist 2000). The opposite model, making TV into a Web device, has also mostly failed, notably in the case of WebTV (The Economist 2000).

Does that mean "… *the Internet will not be the main vehicle for electronic entertainment* …" (The Economist 2000, p. 32)? Although we do not have a definitive answer to this question, our work in the e-culture project, described in the remainder of this chapter, suggests that people not only want and like to watch TV-like Web experiences, but also that those experiences may be significantly different from both traditional TV viewing and Web-surfing.

2 Method

2.1 Overview of User Centred Design Approach

What kind of entertainment do people want from a Web site on art and culture? To answer this question, we conducted a variety of UCD activities including interviews with curators and cultural programmers, focus groups with a range of participants in different cities in the United States, interviews with visitors to museums in New York City, online surveys of museum Web sites, and usability walkthroughs and test sessions with prototypes of the design concept for the art and culture site. The detailed description of all of these UCD activities and results is beyond the scope of this chapter. Please see another publication (Vergo et al. 2001) for details regarding the curator and visitor interviews, surveys, and focus groups. In this chapter, we will describe the group usability walkthroughs and the individual and small group usability test sessions with target users that informed the design of the prototypes of the Web site.

2.2   Users

Based on existing research about Internet users, information from the cultural institutions with which we were partnering to develop the Web site, and from IBM, we defined our typical user as a person at least 9 years old who spends an average of 10 or more hours a week on a computer, and of that time, five or more hours are spent on the Internet. Our target users attended at least one cultural event in the last 12 months.

2.3   Iteration 1: Usability Design Walkthroughs

Based on the results of the interview and survey data from visitors to museums and online museum sites, the team developed five early design concepts to walkthrough with target users in group settings. The usability design walkthroughs were run in 12 sessions with a total of 70 participants ranging from 9 to 72 years in age who were screened for cultural interest and experience with the W. Participants were first shown "best of breed" excerpts of existing Web sites related to culture, and then they were presented mock-ups of new design ideas. Qualitative and quantitative data were collected. The mock-ups of design ideas shown in the second part of the usability walkthroughs encompassed five different design approaches for exploring cultural content:

1. A filtering system based on direct manipulation of large databases with visual feedback (such as in Alberg and Shneiderman 1994). Current uses of this visualization design for database information are the display of chronological events in patient medical records, judicial records for juveniles, and inventories of movies.
2. A set of lenses (tools) to manipulate the way content could be viewed (such as in Stone et al. 1994). This approach allows the user to select a lens for "history" or "music" and place it on a visual display of artwork in order to learn about events in history or the music at the time the artwork was created.

3. A chat system where people could talk about a particular art work (such as Viegas and Donath 1999). This approach resembles a virtual art exhibit: art objects are distributed around a "room" and the user is represented by a bubble on the screen and can drag their bubble near others in order to chat with them and join their conversation.
4. A notebook system where the user collects and comments on artistic content, and later publishes the notebook for public/private viewing. This approach enables the user to create and keep a personal "scrapbook" of their experience in viewing or hearing artistic content.

5. A multimedia system where the user watches guided multimedia tours, interacting whenever interested in related information. This approach resembles a guided tour in a museum, but is augmented by the ability of the user to take control of course of the tour and see related information or detailed information on the cultural content of the tour.

A major finding of the usability design walkthroughs was that most participants viewed unfavourably Web sites involving active interaction with content or other visitors they did not know. They viewed these more interactive design concepts as work-like experiences, not entertainment. They saw little value in interacting with other people who were not acknowledged experts in the cultural area being presented in the tour. The guided tour format was clearly the best received among the design ideas. Participants strongly suggested the replacement of text by audio. We summarized these findings by hypothesizing that in this domain of entertaining Web experiences, users wanted *less clicking, more watching*. Users were comfortable with the idea of a streaming experience that leads them through artistic and cultural artefacts where, unlike television, the stream can be paused, replayed, or interrupted for further exploration. Users have a strong desire for the availability of related information through hypermedia links and in-depth analysis of the works of art. Users were adamant about having a "human voice" behind the multimedia experience, that is, a personal viewpoint and narrative in the exposition of the content.

2.4 Iteration 2: The Design and Evaluation of the Interactive Prototypes

Based on the results of the usability design walkthroughs, we developed a design concept for the cultural Web site based on the idea of providing users multimedia tours guided by experts, artists, or celebrities. In our design, a tour presents information to the user continuously, from beginning to end, unless the user chooses to explore related material or to exercise control. To cope with the requirement for a minimum of 56Kbps bandwidth, we decided to explore multimedia experiences primarily based on still pictures and sound with minimal use of video. At 56Kbps, a continuous video stream is of insufficient quality, but at that speed it is possible to download combined audio and images that have reasonable quality. The primary use of still photographs also reduced production costs since shooting video is more expensive than using still pictures accompanied by recorded audio. As a note, because of copyright issues, the W site is for IBM internal use only at this point in time, and is not accessible via the World Wide Web.

In our design, the main multimedia experience, or main tour, is composed of multiple scenes connected linearly that play continuously to tell a story from the tour guide's perspective. The tours resemble a short TV documentary and play within a

Web browser window. The main tour is enriched by the addition of user controls such as pause/resume, a navigation map to enable scene changes, and by the inclusion of hot spots for two different kinds of related content that we labelled side tours and branches. A side tour is a self-contained multimedia segment focusing in depth on some aspect of the tour. A branch is a static Web page with text, pictures, and links to related information on a specific subject. Since side tours were more costly to produce than branches, we produced side tours only for highly desirable related information.

Figure 1 shows a snapshot of a tour with key features enlarged and presented to the right of the main screen. The majority of the screen area is filled with tour content (pictures, text, occasionally very short segments of video). On the bottom left-hand side, a pictorial navigation map gives the user an idea of their position in the tour, the duration of different scenes in the tour, and the proportion of the tour remaining.

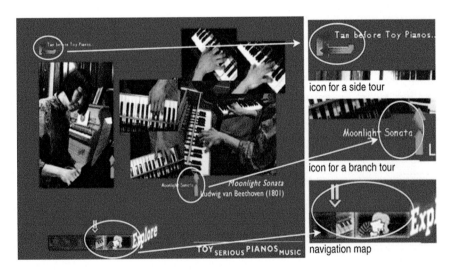

Fig. 1 Typical scene of a tour with its navigation map, links to a side tour, and two branches. Expansions of three user interface components shown on right

Rolling the mouse over the map presents textual information about each scene, while clicking on the picture of a scene interrupts the current scene and immediately starts the scene corresponding to the clicked image. As the tour progresses, hot spots indicating the availability of side tours and branches appear on the screen. These hot spots remain for a minimum duration of 10 s and then fade away. The hot spots appear when the content relates to them and fade away after the related part of the tour has finished. When a side tour is selected, the main tour is interrupted and the side tour is played. When a side tour finishes, the main tour resumes from the point where it was left. A click on a branch pauses the tour and opens a new window on the browser, displaying the Web page associated with the branch. To resume the main tour, the user must click on the pause/resume icon above the map.

All the tour content including the scenes from the main tour, side tours, and branches is available from the Explore Page at the end of the tour. Figure 2 depicts the Explore Page for the tour shown in Fig. 1. Clicking on the tour map restarts the tour from the beginning of the scene that is clicked. Similarly, clicking on side tours and branches immediately starts them. The user can access the Explore Page at any time during a tour by clicking on the corresponding hot spot on the right of the map.

We developed two prototypes. The first tour featured the work of a toy pianist, Margaret Leng Tan. In the tour, the pianist talks about her involvement with toy pianos, how music is arranged for a toy piano, and her connections to Schröeder, the famous cartoon character created by Charles Schulz. Two side tours describe the history and mechanics of toy pianos and the work of Margaret Leng Tan before

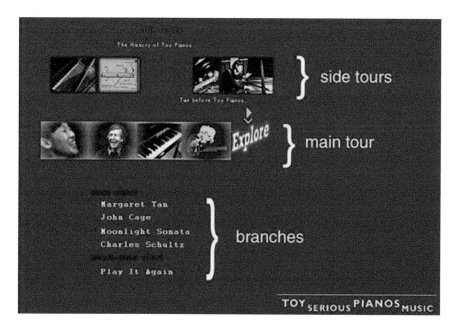

Fig. 2 Exploration page that allows access to the main tour, side tours, and branches

becoming a toy pianist. The main tour lasts 4:15 min and the side tours take 1:18 min and 0:50 min, respectively. The tour also includes five branches.

The second tour focused on Ludwig van Beethoven and his Ninth Symphony. Three side tours are provided; one about Beethoven's deafness, and two side tours that enable the user to explore Beethoven's scores and his Heiligenstadt Testament. Beethoven's main tour lasts 10:10 min and the first side tour is 2:00 min in length. The other side tours, since they incorporate interactive elements, have no fixed duration, although their exploration typically took 1:00 min each. There are also five branches available for user exploration.

The usability test of the two tours focused on answering the following questions:

1. Can a Web tour experience with "less clicking, more watching" be entertaining and engaging?
2. Are users satisfied with the level of interactivity designed in the tours?
3. Do users who report higher subject matter appeal interact more and/or spend more time on the tour?
4. Do users look for related information during the streaming portion of the tour or from the Explore Page?
5. Does social context (singles or pairs) have an effect on reported levels of our subjective measures?

2.4.1   Usability Test Procedure

Participants in the study were assigned to one of two conditions: they experienced the Web sites alone (Singles) or in pairs (Pairs). The experimental procedures were very similar for the two groups. Participants in our study completed three Web experiences, based on our two pilot tours. We used the first two experiences as learning trials and only analyzed data from the third. Participants first experienced both the low-interactivity and high-interactivity versions of one tour and then experienced the high-interactivity version of the other tour. Low-interactivity tours had limited play control (only pause and resume) and no side tours or branches. During the third tour, the users were provided a fully interactive tour and were free to interact with it however they desired.

The participants were recruited based on the profile defined in the "Target Users" section above. The age range of the subjects was from 21 to 55 years old. Eight participants were randomly assigned to the singles condition, and eight groups of two participants each were randomly assigned to the pairs condition. Participants in the pairs condition all knew each other before the session. Participants sat facing a 17-in. personal computer monitor placed on a table with a keyboard (which was not used) and a mouse. The sites that the participants evaluated were presented in a full screen Netscape 4.7 browser window. After participants had filled out a pre-session questionnaire, the experimenter then set the browser to the first site, briefly introduced it, and then left the room to observe the session from the control room. After each tour, participants filled out a post-session questionnaire (PosSQ) describing their experience. Subjects were instructed to spend as much time on each tour as desired and to tell the experimenter when they were done. After all three tours were completed, the experimenter interviewed the subjects using the debrief question-naire (DQ).

Immediately before the first experience with a high-interactivity tour, partici-pants were directed to take at least one branch and one side tour during the main tour, and also told that they could explore as much of the information as they wanted on the Explore Page. Participants in the pairs condition were instructed to take turns controlling the mouse and to make sure that each of them take at least one side tour or branch selection. Before the final (high-interactivity) experience, par-ticipants were told to interact with the tour as much or as little as they wanted. The main part of the PosSQ was a set of four questions asking the users to rate the level of engagement, entertainment, satisfaction with the level of interactivity and subject matter appeal of each of the tours they experienced, using a seven point Lickert scale.

We analysed the videotapes and logged the user's mouse activity as follows. We counted the number of times the participants moved the mouse pointer so that it was located on an object that could be selected (rollovers), and the number of times an object was actually selected. The objects could have been branches, side tours, or navigation map scenes including the Explore Page. Since the length of the two tours differed, we calculated a "time in exploration" as the total time the participants spent on the tour, minus the base time of the main tour itself (if no branches, pauses or side tours were taken): Time (explore) = Time (total) − Time (base) where, Time (total) was the total time a user spent on the tour, from the moment they started until they announced they were done. Total time included the time spent with the tour, pauses, plus the time spent on all branches and side tours.

3   Results

The data presented are from the PostSQ, the DQ, and user mouse activity that were analyzed for the third, high-interactivity experience of each participant or pair of participants.

3.1   User Ratings of Tours

The means for user ratings of the four aspects of the tours (how engaging, enter-taining, satisfied with the level of interactivity, and appealing the subject matter was) were all above neutral (4.0), ranging from 4.63 to 5.56 (see Table 1). The Beethoven tour was slightly more positively rated than the Tan tour on all four measures, however, there were no statistically significant differences. Entertainment, engagement, satisfaction with interactivity, and subject matter appeal were all positively inter-correlated, with engagement and entertainment the most highly so. The range of the inter-correlations was 0.59 to 0.89.

We explored the relationship between explore time spent on the tours and the user's subjective ratings of the tours. Participants spent an average of 3:52 min of explore time on Beethoven and 6:12 min on Tan ($p > 0.12$). There were no statistically significant differences on the amount of time spent on the tours by singles and pairs. Also, there were no significant correlations between the four subjective measures and the explore time spent on the tours. Thus the amount of time participants spent in exploring related information was not a factor in their subjective ratings of the tours.

We analysed the videotapes of user mouse activity and the summary data are reported in Table 2. The results show that users interacted an average of 16.2 times during each of the tours. Given that the participants spent an average total time of 10:30–14 min on the tours, this meant that a participant was clicking about once a minute during the experience. The number of interactions before reaching the Explore Page was similar to the number of interactions after reaching the Explore Page, thus our assumption that participants would interact both during and after the tour was supported. There were no statistically significant differences in the types of user activity.

We next analysed the relationship between user mouse activity and the user's four subjective ratings of the tours. Results show that user mouse activity was negatively correlated with engagement and entertainment both before and after the Explore Page (see Table 3). Subject matter appeal was negatively correlated with mouse activity after the Explore Page. This means that participants who watched the tours more, and interacted less, were more engaged and entertained, and found

Table 1 Means for singles, pairs and all users on engagement, entertainment, satisfaction with interactivity, and subject matter appeal (7-point Likert scale, 1 = most negative, 7 = most positive)

Means	Singles	Pairs	Total
Engagement	5.56	5.50	5.52
Entertainment	5.00	5.38	5.25
Satisfaction with Interactivity	5.37	4.63	4.87
Subject matter appeal	5.12	5.19	5.17

Table 2 Summary of user mouse activity during the final tour

Type of user Mouse Activity	Mean actions Before Explore Page	Mean actions after Explore Page
Branches taken	1.91	2.33
Side tours taken	0.83	1.58
Scene changes made	1.75	1.25
Rollovers made	3.79	2.75
Total actions taken	8.29	7.91

Table 3 Correlations of user mouse activity with engagement, entertainment, satisfaction with interactivity, and subject matter appeal

User mouse activity	Engagement	Entertainment	Satisfaction with interaction	Subject matter appeal
Total mouse activity	−0.44*	−0.48*	−0.44*	−0.45*
Mouse acts before Explore Page	−0.41*	−0.35	−0.29	−0.01
Mouse acts after Explore Page	−0.46*	−0.44*	−0.19	−0.43*

*Correlations significant at the $p < 0.05$ level

the material more appealing. Conversely, those participants who were less entertained and engaged were more likely to interact with the tour. From the DQ, results showed that 18 out of 24 participants said they would like to have the multimedia experiences similar to this again.

3.2 A Comparison of Singles and Pairs

We analysed our data for differences between single participants and pairs of participants. There were no significant differences between the means for singles and pairs on any of the four subjective measures or on mouse activity of any type. In the debrief questionnaire, 10 out of 16 pair participants reported that they thought the experience was more fun as a pair than it would have been had they experienced it alone. Of the 18 out of 24 total participants who said they would return to the site, 12 said they would want to do it with family and friends rather than alone.

3.3 User Debrief Data

At the end of the sessions, all 24 participants reflected on their experiences and answered a set of open-ended questions from the experimenter. Participants thought that the W experiences flowed well and were entertaining. They stated that the navigation map worked as expected, was valuable, but would be even better if continuous control over the flow of the experiences were provided through rewind, fast forward, and jump capabilities to position the tours anywhere within a scene. When asked about the side tours and branches, participants responded that they were also valuable, worked as expected, and would be even better if they were more in-depth and contained more content. They were adamant that the Web experiences should be just one level deep in terms of access to related information. Participants said they liked the integration of the various forms of media with the presentation of the story, and the user control and interactivity choices they had were the best parts of the multimedia experience. Participants said they would like to have these types of experiences at work, home and on the road. They thought it would be a valuable and refreshing break to be able to take 5–10 min at work and enjoy a cultural experience like these in the study. They also said they would really enjoy being able to explore the tours with friends and family and said they might preview these cultural experiences alone at first, and then engage others at home and in various mobile computing situations in the communities in which they live to experience the tours with them.

4   Discussion and Conclusions

Is the Web an interactive medium? Most certainly yes, but not exclusively, as this research shows that users indicated satisfaction with watchable experiences. We found an interesting design dilemma. Users want to have continuous control over cultural Web experiences, however, those who report the highest satisfaction, entertainment, and engagement with the experiences use the controls the least. Both the literature (Laurel 1993; Murray 1997) and our informal experience with Web-designers during the project suggest a strong disbelief in TV-like experiences on the Web. Although our research (and the participants' views) was heavily influenced by the pervasiveness of TV as the primary entertaining experience for people, we do not regard our results as an endorsement of TV as the ultimate entertainment experience. Instead, the results indicate that a major factor in entertainment is who we are entertained by and not the level of audience control over the entertainment experience. It is important to note that our design was defined as much by the idea of *"human voice"* as by the *"less clicking, more watching"* paradigm. In other words, perhaps people have both a remarkable interest in the flow and experience of listening and watching stories, and they are engaged by a storyteller as a respected person with a point of view. In this light, TV can be considered a highly developed and engineered storytelling medium, while the Web is still trying to discover how to tell good stories.

   Another key point is that this research topic is a complex and multifaceted one. This study reports significant correlations between user behaviour and self-reported entertainment and engagement in the range of 0.43 to 0.48. This means that through an examination of the variables in this study, we are able to explain about 20% of the variance in user behaviour. This topic is likely one where a user's behaviour is influenced by many factors. The results regarding social and physical context support the notion of designing entertaining experiences for singles and groups in various locations. Participants thought the design of the tours made them wonderful as "refreshing breaks" at work and that they would watch them multiple times with friends and family and explore the links in various ways. The lack of statistical difference in the ratings by singles and pairs confirms the value of the experiences in different social contexts.

Certainly the tour content has an impact on how entertaining users perceive the tours to be. We expected that people who were more interested in a topic would spend more time on a tour by exploring more related information. The results did not support this idea. We found that subject matter appeal was not related to the duration of the experience, but was negatively correlated with the level of inter-activity by the users. These surprising results warrant further investigation.

As a methodological note, we found utilizing both small group design walk-throughs and individual evaluation sessions to be highly beneficial. In the small group sessions, we were able to collect data from a large number of people about a large number of design alternatives to filter the data and arrive at a design for individual testing. The individual sessions were time and labor intensive for the team to carry out, however we were rewarded with rich and detailed design information from the participants. We feel that the two methods are appropriate for different stages in the design process, and that they provide complementary information.

The Web enables a wide range of entertaining experiences for users, and there is much more to learn about the potential for entertainment on the Web. Future research can help to build a framework for understanding this topic through an in-depth investigation of user reactions to Web experiences of varying duration, content, with various types of interaction possible, social context, and physical settings.

References

Alberg C, Shneiderman B (1994) Visual information seeking: tight coupling of dynamic query filters with starfield displays. In: Proceedings of CHI'94, Boston MA, April 1994. ACM Press, pp 313–317

Cole JI (2000) Surveying the digital future. UCLA Center for Communication Policy, Los Angeles, California

Geirland J, Sonesh-Kedar E (1999) Digital Babylon: how the geeks, the suits, and the ponytails tried to bring hollywood to the internet. Arcade Publishing, New York, New York

Laurel B (1993) Computers as theatre. Addison-Wesley, Reading, MA

Murray JH (1997) Hamlet on the holodeck: the future of narrative in cyberspace. The Free Press, New York, NY

Red Herring (2000) The sorry state of digital hollywood. Available: http://www.redherring.com/mag/issue85/mag-sorry-85.html

Skelly TC, Fries K, Linnett B, Nass C, Reeves B (1994) Seductive interfaces: satisfying a mass audience. In: CHI'94 conference companion, Boston MA, April 1994. ACM Press, pp 359–360

Stone M, Fishkin K, Bier E (1994) The movable filter as a user interface tool. In: Proceedings of CHI'94, Boston MA, April 1994. ACM Press, pp 306–312

The Economist. (2000). A survey of E-entertainment, 7 Oct 2000

Vergo J, Karat C, Karat J, Pinhanez C, Arora R, Cofino T, Riecken D, Podlaseck M (2001) "Less clicking, more watching": results from the user-centered design of a multi-institutional site for art and culture. In: Bearman D, Trant J (eds) Museum and the 2001: selected papers from an international conference. Archives & Museum Informatics, pp 23–32

Viegas A, Donath C (1999) Chat circles. In: Proceedings of CHI'99, ACM Press, pp 306–312
Vogel HL (1998) Entertainment industry economics, 4th ed. Cambridge University Press, Cambridge, UK
Webster J, Ho H (1997) Audience engagement in multimedia presentations. Data Base Adv Inf Syst 28(2):63–77

Chapter 27
Designing for Fun: User-Testing Case Studies

**Randy J. Pagulayan, Keith R. Steury, Bill Fulton
and Ramon L. Romero**

Author's Note, Funology 2

When we originally wrote this chapter, one of my goals was to establish some awareness (and possibly even some credibility) in two different directions. As psychology researchers born out of academia and the Human-computer Interaction/Human Factors fields, very few of our research colleagues (if any) were talking about applying or adapting existing HCI/HF methodologies into the world of video games. Common conversations at that time revolved around silly debates for the appropriate sample size for a usability test as opposed to challenges that arise when faced with creating an experience that is 'appropriately difficult'. This makes sense though, because why would an established field like HCI focus on unique challenges seen in video games when games weren't seen as a legit form of computing? The idea of someone from a video games studio, publisher, or console maker showing up at a CHI conference was unheard of. At the same time, the video games world was largely unfamiliar with the techniques or approaches to user-centered design thinking that have matured from the 1980s up to that point. The closest anyone would come to user research was "playtesting", which was still a term in the video games industry referring to bringing in your best friends to play your unfinished game in return for some pizza. In addition, anything that involved an actual customer was lovingly called a 'focus test' driven from marketing. Thus, we, at Microsoft, and a handful of others, straddled an interesting existential misery of not really fitting on either side of this equation. We were strangers in a foreign land.

R. J. Pagulayan (✉)
Microsoft, Redmond, USA
e-mail: Randy.Pagulayan@microsoft.com

K. R. Steury
Microsoft, Redmond, USA

B. Fulton
Amazon, Seattle, USA

R. L. Romero
Sony Interactive Entertainment America, San Diego, USA

© Springer International Publishing AG, part of Springer Nature 2018
M. Blythe and A. Monk (eds.), *Funology 2*,
Human–Computer Interaction Series,
https://doi.org/10.1007/978-3-319-68213-6_27

Fast forward to the present (or let's just say 15–20 years later), and the world has changed. The video games industry has embraced the notion that applied HCI techniques can be a difference maker in creating awesome gaming experiences. Small studios to big game publishers have invested heavily in applied psychology research and it has evolved to the point of being a legitimate discipline that sits right next to production, design, art, quality assurance, localization, and so on. No longer are the days when you show up at a meeting and the game teams politely (and sometimes impolitely) ask, "Who the hell are you and why the hell are you here?". And, academia and the HCI industry has recognized the innovations and contributions that the ongoing pressures of video gaming requires. Universities across the world have courses in user research in games, there are professors of 'games user research', and any given HCI-related/Psychology conference attendee expects the presence of someone from a video game company. We've come quite a long way in the games user research discipline, and I'm proud to say that the timing was right for us to see the evolution from the start. And with that, I hope you enjoy reading this piece, a snapshot in time that still brings some relevance to the games user research practitioner to this day.

2003 Chapter

1 Introduction

The goal of this chapter is to demonstrate that extending current usability methods and applying good research design based on psychological methods can result in improved entertainment experiences. This chapter will present several case studies where user-centered design methods were implemented on PC and Xbox games at Microsoft Game Studios. The examples were taken from several series of larger studies on Combat Flight Simulator (PC), MechWarrior 4: Vengeance (PC), Halo: Combat Evolved (Xbox), and RalliSport Challenge (Xbox). These examples were chosen to illustrate a variety of user-centered methods and to demonstrate the impact user-centered design principles can have on an entertainment product. Furthermore, these examples are presented in a way that illustrates a progression from addressing usability issues similar to those found in productivity applications, to extending usability methods to address more unique aspects of game design, to using survey methods to address issues related to fun for which standard usability methods do not suffice. For more detailed descriptions of Microsoft Game Studios user-testing methods and laboratory facilities, see Pagulayan et al. (2003).

1.1 Methods and Games

The many similarities between productivity applications and games suggest that traditional discount usability methods would be suitable in the entertainment domain. Games have selection screens and menus just like other software applications. Task persistence, performance, ease of use, learnability, and all the potential obstacles to efficiency and productivity are found in games as well. However, it is possible to conceptualize usability in games as including other areas of game design, such as the comprehension of rules and objectives, control of characters, and manipulation of camera (view), to name a few. A game designer must script an experience within a game, so an extension of usability techniques from productivity applications to games becomes clearer. Usability testing in games (and productivity applications for that matter) can be viewed as an experiment designed to assess whether users will interact with a given product in the way the designer intended.

Video games also differ from productivity applications in a variety of ways. For example, productivity applications represent tools as a means to an end, whereas games are designed to be pleasurable for the duration of gameplay. This distinction means that we must ensure that the user has an appropriate level of challenge and engagement while playing a game, rather than focusing on how efficiently they can achieve their goals. Goals are often defined externally in productivity applications, whereas games define their own goals. This implies that goals and objectives in games must not only be clear to the user at all times, but that they must be interesting as well. Another difference between games and productivity applications lies in the number of choices available. While there are relatively few productivity applications designed for a specific purpose, there are many games. Games must compete with other game titles for consumer attention, but they must also compete with different forms of entertainment, such as watching television or reading a book (Pagulayan et al. 2003). The implication is that games that are not immediately enjoyable will be dismissed quite easily in favor of other forms of entertainment.

It is our intent to show that the HCI field has access to methods drawn from current usability methodologies and experimental psychology that can be adapted to address issues in the entertainment industry. Extending current usability methods and utilizing good research design can result in improved entertainment experiences.

2 Case Studies

Below are four examples of user-centered design methods that were used on video and computer games at Microsoft Game Studios. The first case study begins with more traditional usability methods. Subsequent case studies extend the use of current methodologies to situations that demonstrate how one can address some of the unique issues encountered with games.

2.1 Combat Flight Simulator

This case study is an example of a usability issue identified using limited-interactive prototypes. This method is not substantially different from standard usability methods. The focus of the usability test in this case study was on the game shell screens for the PC game, Combat Flight Simulator (CFS). Game shell screens consist of all menus and screens (e.g., Main Menu, Options, etc.) which are used to set the particular desired gaming experience, but not encountered during actual gameplay. For example, in the game shell a user can often change the difficulty level of the game and set other gameplay parameters.

A usability test was performed early in the development cycle of CFS and used a limited-interactive prototype of the game shell screens (i.e., functional screen widgets including radio buttons, check boxes, and toggles). Techniques utilized included user and task observations, scenarios, and thinking aloud protocols (e.g., Nielsen 1993). Some of the tasks presented to users were exploratory and not performance based. In general, the users' tasks were to explore a particular game shell screens while thinking aloud.

In CFS, parameters users can modify using the game shell include aircraft selection, start location, time of day, weather, number of enemies, and difficulty of enemies. During the exploratory task for the Options screen, a usability problem was detected related to terminology. "AI" is an acronym for artificial intelligence commonly used in the video and computer games industry to refer to the behavior of non-human entities. For example, the cars players drive against could be referred to as "AI" as well as the marines that fight along side of players, providing that in both situations the cars or marines are not actively controlled by a human player. In CFS, "AI" referred to the skill level of computer-controlled enemy pilots, which confused participants in several different ways. Figure 1 represents a portion of an Options screen presented to participants.

The term "AI Level" was not well understood by most—in fact, just two of seven understood the intended meaning of the term. In addition, the options presented for AI Level (Low, Med, High) failed to provide helpful context cues. The main problem was with the term "AI". While it is a jargon term for game developers, it is unfamiliar to many gamers. It is worth noting that all participants had experience with computer games, and the majority of them also had experience with flight games. A poor choice of fonts further compounded the problem. The font used was sans serif resulting in participants confusing the uppercase "I" with a lowercase "l", or the numeral "1". This introduced incorrect cues that some participants used as a basis for guessing (e.g., "is it ALtitude"?). These quotes from participants were representative of the problems they encountered.

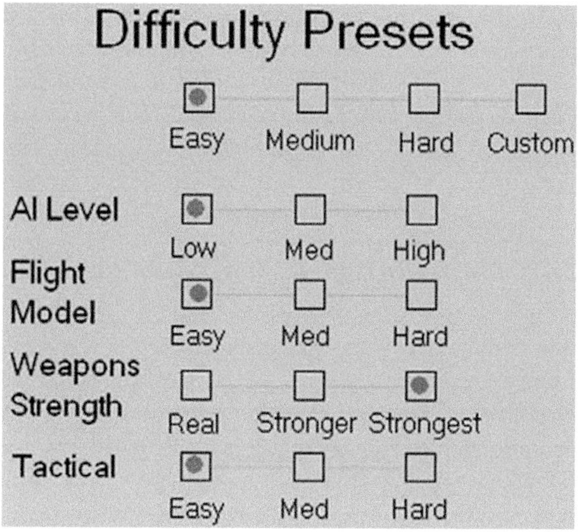

Fig. 1 Portion of Options
screen presented to
participants

Difficulty Presets

	Easy	Medium	Hard	Custom
	◉	☐	☐	☐

AI Level	Low ◉	Med ☐	High ☐
Flight Model	Easy ◉	Med ☐	Hard ☐
Weapons Strength	Real ☐	Stronger ☐	Strongest ◉
Tactical	Easy ◉	Med ☐	Hard ☐

A-One-level? I don't know what that means. I have no idea...

A. I. uh..this uh...AL level? A.I. level? I don't understand this terminology.

A-One-level, I'm not sure what that means, or AL level.

Figure 2 represents the final solution. The term "AI Level" was replaced with
"Enemy Level". The options that were originally "Low," "Medium," and "High"
were replaced with "Rookie," "Veteran," and "Ace". Subsequent testing detected
no difficulties with these terms, and the options are relatively clear in the context of
a WWII fighter pilot game. This terminology has remained consistent through
subsequent releases of the Combat Flight Simulator series.

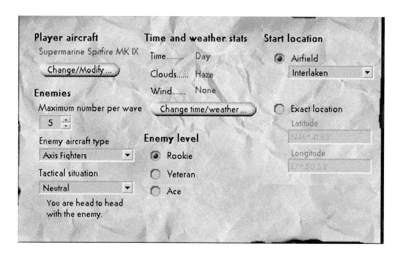

Fig. 2 Final interface for Options screen

In this example, information obtained from users through current usability practices was used to improve upon the usability of the game shell. This is a case where the user goals for the options screen did not substantially differ from user goals in productivity applications; the tools were a means to an end. Users should be able to set their preferences with maximum learnability and minimum errors.

2.2   MechWarrior 4: Vengeance

The popular MechWarrior series is based in a science fiction future where wars are fought by elite warriors who pilot giant destructive walking tanks called Battlemechs. One of the defining features in MechWarrior games is that they are action-oriented. However, they are also designed as simulations, which are games that attempt to model the complexity of the real world. Mastering the complexity is part of the fun for those who enjoy this type of game. For example, in MechWarrior, all weapons were designed with limitations that could exist in the real world, such as limited ammunition, differential firing rates for different weapons, and the potential for the Battlemech to overheat and shut itself down with the overuse of some weapons. This level of complexity can be quite intimidating for novice users. A simplistic solution to this problem would be to make the game simpler. However, because some of the game's fun stemmed from its complexity, the goal was to help novice users sufficiently master the intricacies of the game in an enjoyable way. In the following case study, user-testing methods helped identify areas of assistance for novices who experienced difficulties controlling their Battlemechs when playing MechWarrior 4: Vengeance (MW4), without sacrificing the complexity that helps define the MechWarrior series.

The development team established a goal of making MW4 as approachable as possible for novice users. In particular, novices needed to be able to use the basic movement and weapon controls with minimal difficulty. The controls in MW4 are complex because Battlemechs are capable of a movement called torso-twisting, where the top part of the Battlemech can rotate to look in a direction different from its movement trajectory, similar to a gun turret on a modern tank. Weapons are placed in the upper part of the torso, so rotating the upper torso is fundamental to aiming weapons while playing the game. To help users coordinate these movements, on-screen aids were included as part of the standard interface.

Figure 3 is a screenshot of the MW4 interface during gameplay. There are two items on the screen that were created to aid users. The green cone-shaped item within the circular radar placed at the bottom-center of the screen represents the user's line of sight, or field-of-view. It is marked Field of Vision. Figure 4 is an enlarged version. In this situation, the torso is rotated slightly to the left relative to the bottom portion of the Battlemech. If the cone funneled out toward the right, the line of sight would be to the right of the Battlemech.

The second item is a green horizontal scale placed in the center of the screen beneath the targeting reticle where a horizontal bar represents the amount of torso

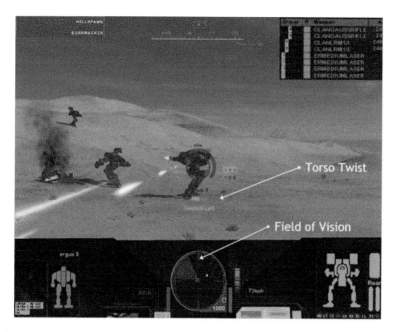

Fig. 3 User interface for MechWarrior 4 during gameplay

Fig. 4 Field of Vision
indicator and Radar

rotation (see Fig. 3). This is marked as Torso Twist. This provided users with a graphical cue for torso rotation on the x-axis relative to the bottom part of the Battlemech. For example, if a Battlemech is facing straight forward (i.e., the torso is perfectly aligned with the lower portion of the Battlemech) there is no horizontal bar beneath the reticle. The length of the horizontal green bar is a function of the distance the torso moves away from the center in either direction. In other words, the length of the bar grows as the Battlemech rotates further from perfect alignment of the upper (torso) and lower portions of the Battlemech. In Fig. 5, the green bar

Fig. 5 Torso Twist indicator.
a Slightly turned to the left,
b severely turned to the left

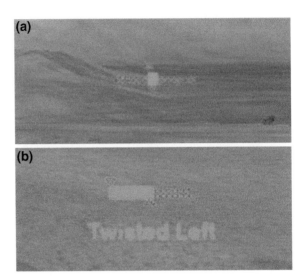

represents the Battlemech rotated slightly to the left in panel a. In panel b, the Battlemech is rotated much further away from the center toward the left.

Usability testing of these interface items included common pass/fail and error counting techniques. During this test it was shown that novice users struggled with the basic controls of the game. In particular, they struggled to control the Battlemech's torso, frequently confusing the direction of the upper torso for the direction in which the Battlemech was moving. The fundamental problem was that few people were using the on-screen cues created to avoid this exact problem. Four of seven participants did not notice or use the field of view reference (green cone), and no participants noticed or used the green rotational bar. Only two participants demonstrated a skillful use of torso twisting, but all demonstrated difficulties steering their Battlemech.

It became apparent that torso-twisting with Battlemechs is a skill with a gradual learning curve. Based on these results and other convergent evidence, the development team chose to create a training mission with the primary goal of educating users about the visual cues in the interface while giving them a relatively safe environment to practice the use of torso-twisting. Follow-up testing showed that the training mission decreased the torso-twist problem, but did not eliminate it. In a follow-up study, all participants knew how to use the on-screen visual cues to correct or avoid torso-twisting problems, though they still struggled to control their Battlemechs. The critical difference between the game prior to the addition of the training mission and after is that although users still experienced difficulties learning the controls, they were empowered to fix the problems themselves after completing the training mission.

Educating novice players through training makes the game more accessible to a larger population without alienating the current population of MechWarrior gamers.

The training mode gives more time to novice users to adapt to the game's complexity, and for more experienced users, the training mode is optional.

2.3 *Halo: Combat Evolved*

A skilled game designer can create an experience that people consistently enjoy. However, as with any product, a user may not use it in the way the designer intended. In the case of video games, this may lead to an experience that is less fun for the user. This case study presents an example from the development of Halo: Combat Evolved (Xbox). It presents one of the biggest challenges in game design: making the user feel like they are making interesting decisions while ensuring that they are playing the game in a way that leads to the most fun. This example illustrates how discount usability methods enabled the designers to see how users approached their game, to understand barriers that were preventing users from experiencing the game as intended, and to refine the game so that it drew the user into the optimal experience.

One of the greatest difficulties involved in game design is perfecting the balance between scripting an experience intended by the designer without making the user feel restricted in the choices they can make. For example, a designer may want the user to move the character to a particular location in the virtual world, where an exciting battle awaits them. This could be accomplished by creating a narrow pathway, or by placing barriers in the virtual world restricting the user to only one path, with no other choices. However, such restrictions may be unsatisfying to users because they feel like they have no control over their decision-making. Alternatively, the designer could implicitly control the user's behavior by giving him or her free range of movement, but placing something attractive (e.g., a new weapon) at intermediate points between the user's current location, and the designer's desired location (i.e., the location of the exciting battle). This may draw the user down a certain path, while making him or her feel in control. While tactics such as this are common, there are a number of different variables encountered in game design that interact with one another, making it very difficult to predict exactly how users will play the game.

Different mechanisms affect a user's experience of combat in games. Because Halo is a game about combat, it was very important to the team that combat be compelling. The development team spent a lot of time refining the aiming controls, enemy behavior, weapon variety, and the layout of the environments to make combat as enjoyable as possible. In addition, the designers created a wide variety of combat situations and areas where users could engage in different tactics to succeed. Certain situations were created where a particular strategy would yield the most rewarding and fun experience. The assumption was that if users approached the combat situations as the designers intended, users would have as much fun as the designers did.

The Halo team wanted to see whether the typical user would have fun playing the game, and if not, what issues were blocking them from doing so. To provide this information, eight participants who liked to play console action games were recruited to play the game in a usability test format.

Participants were presented with a mission and asked to play the game as they would at home. For this test, traditional error counts and pass/fail criteria were not used. Instead, the focus was on the tactics that participants used to complete the level in comparison with the 'ideal' tactics created by the designers. Designers were present at all sessions. This was critical because they best understood how they wanted players to respond to their game, and they could also see, first hand, what cues users were not seeing.

The designers had created large, outdoor environments that gave players more freedom to approach situations in a variety of ways. However, the large environments afforded players the ability to see enemies from a long distance. Testing revealed that participants were trying to fight enemies from much further away then the designers intended. During usability sessions, novice players began shooting at the enemy as soon as they could see them, but the designers had wanted the players to first move closer to the enemies, and then fight. While all users were able to successfully complete the missions using these tactics, the game was designed to be most enjoyable when combat was engaged at a much closer distance. At greater distances, the experience was less satisfying. For example, the weapons were not designed to be effective at these longer distances resulting in participants complaining that their weapons were not accurate enough.

From observing the usability sessions, designers quickly identified several solutions. First, they revised subtle cues that they had placed in the aiming reticle. The aiming reticle refers to a graphical indicator in the center of the screen that represents where the user is aiming a weapon. In the original design, at any distance from the enemy, the reticle would turn red when placed over an enemy, informing the user that their weapon was aimed properly at the enemy. The designers modified the reticle behavior so that it turned red when the enemy was targeted, but only at a combat distance intended by the designers. Therefore, when the player aimed at the enemy from a long distance, the reticle would not change color, providing a subtle cue that they were not at an optimal distance, thus encouraging users to move closer. The designers also increased the diameter of the reticle to emphasize its inaccuracy at greater distances. Second, the team modified the enemy behavior. When the user shot at the enemy from a much greater distance, the enemy now would not return fire, but would dodge the shot or move behind a rock or tree, requiring the user to move closer. When the user reached the optimal engagement point, the enemy would step out of cover and engage in battle with the player. At other times, the enemy would approach the player to maintain the intended combat distance.

Subsequent usability testing showed that these changes encouraged the majority of users to engage in combat from the intended distance. Users were able to see more of the interesting components of the game, were more satisfied with the performance of their weapons, and as predicted, enjoyed the game more.

In standard usability practices, the goal of the test is to ensure that the design of product effectively supports what the user is trying to accomplish. In this case study, the perspective of usability test goals has been shifted away from the end user, and toward the designer. In the Halo example, the goal was to ensure that the user experience matched designer intention. These types of situations often require less structure than more common usability methods.

Once we are confident that users are playing the game in the way that designers intended, we can begin to get feedback about subjective preferences on the game's design. Although this may not require direct observation, a larger number of users must be utilized to ensure reliability of participant response, as the next case study illustrates.

2.4 RalliSport Challenge

RalliSport Challenge (RSC) is a racing game on the Xbox console. As in many popular racing titles, the Career Mode is one of the core pieces of this game. In order to understand how user-centered design methods were able to influence the design of this game, one must first understand how the Career Mode functions in RSC.

The Career Mode starts with a limited set of cars available to the user, and a limited set of driving events they can participate in. Depending on the users' performance, new cars and more challenging events are made available to users as they progress through their careers.

Each event consists of several stages (or races). After each stage, users gain points based on a number of factors; top speed, what place they finished (e.g., 1st, 2nd, etc.), lap times, and amount of car damage. Once an event is over, all of the points gained from each stage are added up to obtain the user's total points. The user's goal is to accrue a specified number of points to gain access to the next set of cars and events that are currently unavailable to them. Reaching the specified point total is the only way to progress through the game. This is often referred to as "unlocking" cars and events.

This design is not unique in racing games. It is common to specify a certain performance criterion in order to progress through the game. However, keeping in mind that playing games is a choice (i.e., users are not forced to play), the manner in which this design is implemented could cause frustration for many users. In RSC, the issue revolved around the users' ability to re-race a stage within an event to earn more points.

To better illustrate this issue, consider a scenario where a user has three racing events that are currently available to them, each comprising several stages: Event A (stages A-1 through A-4), Event B (stages B-1 through B-5), and Event C (stages C-1 through C-6). To unlock the next set of events and cars, the user must accumulate a minimum set of points by racing through most (and sometimes all) of the stages in Events A, B, and C. If a user races all Events, but does not earn enough points to unlock new cars and events, what options are they left with? According to the developer's design, their only option would be to re-race an entire Event, including all stages within that Event. If the user selected Event C, all of the previous points earned from each stage (C-1 through C-6) are erased, and the user must then re-race stages C-1 through C-6.

The developers preferred this model because they felt the game would be too easy otherwise, and also because this system is much closer to the actual sport of rally racing. This is a valid argument. Games should not be too easy and must ensure the appropriate level of challenge and success. On the other hand, this strategy potentially puts users in the situation where they lose points on several stages they are already happy with. If the user performed very well on C-1, C-2, C-3, C-5, and C-6, but performed very poorly on C-4, they would lose points on all stages, which could result in extreme frustration.

To address this issue, a series of tests was designed to assess how frustrating this situation may be to users. The goal in designing these tests was to simulate as close as possible a situation where users would actually play through a series of events, but not accrue enough points. Simply asking users about this situation would not suffice; users had to experience it. In addition, data collected from a traditional usability study sample size would not be sufficiently reliable to measure a subjective preference. Therefore, a survey methodology that would allow for a larger sample size and not require direct observation was used.

Working with designers, a mock career mode was created that consisted of three Events, each with a minimum of four stages. Secondly, a profile that described the population of users was created. In this case, nearly thirty users who frequently play racing console games were brought in. To prevent contamination of the data, standard experimental research protocols were implemented. The laboratory was set up so that responses from each user could be collected independently, with no interaction between users.

Participants were instructed to play through the mock career mode as if they were at home. Once they finished the events, they were informed that they did not earn enough points to unlock the next set of cars and events. At this time participants were given the following options, and asked which option they would use for unlocking the next set of events:

(a) Re-race any of the previous 3 EVENTS to improve your total point score; However, you must race ALL stages within the EVENT(S) you choose;
(b) Re-race CERTAIN STAGES in any EVENT of your own choosing to improve your total score;

Then participants were asked how they felt about the following methods for unlocking the next event.

(a) Re-racing an entire event (including all tracks within the event);
(b) Re-racing any stage of my own choosing within an event;

Lastly, users were asked the same question, but presented with a situation where the Events consisted of nine stages (which was representative of some Events in the actual game, although they did not experience this).

The data gathered provided fairly convincing evidence against forcing users to re-race the entire Event. The majority of participants reported a preference for being able to re-race a single stage as opposed to an entire event and this preference became even stronger when presented with the nine stage event. A smaller percentage of users preferred to be able to re-race only entire events, suggesting that they agreed with the developers, who felt that allowing users to re-race stages made the game too easy.

Based on these survey data, the recommendation was to implement multiple difficulty levels, one of which allowed the re-race option (Normal difficulty), but the more advanced levels that did not (Difficult or Advanced difficulty). This solution met the needs of both user preferences, those who want to be able to re-race any stage at any time, and "hard-core" users who prefers more of a challenge and penalty when playing through their career. This was presented to the developer, who in turn implemented the functionality into the design of the final retail product—Normal (no re-race) and Easy (re-race). The developer felt the no re-race option was truer to the sport of rally racing, which is why it was termed Normal. More importantly, the decision to implement multiple difficulty levels was also validated by some reviews from popular gaming magazines and gaming sites after the game was released.

> ...punishingly difficult and repetitive, depending on which of the two available skill levels you choose when you start the game....With no "re-do" button available in Normal mode to simply restart a race if you screw up, some gamers will find themselves angered and frustrated with having to start over from the beginning of the series, while others will embrace the degree of challenge with open arms. Mahood 2002

> We were also ready to slap the game down for being too hard, until we tried it on the easy difficulty setting. Normal difficulty, as well as giving you some stiff times to beat, doesn't allow you to retry a stage – if you mess up, you have to restart the whole event. Even racing veterans would be well advised to play Easy mode to get to grips with the tracks at first, as this allows you to retry. Smith 2002

In this example, the question being addressed (to allow re-racing or not) did not have a clear answer before data were collected, and did not have a specified task-based performance criterion. Finding a solution for this design question was not necessarily related to the learnability of the product, or the efficiency of use, or the number of errors a user would make. The solution was based on what was most fun for users.

3   Conclusion

There are a variety of approaches that can be taken from existing HCI-related methods to address the unique aspects of user-centered design issues found in games. Although the case studies discussed above focus on specific issues, they should also be viewed as a broad representation of some methods currently being utilized by the User-Testing Group at Microsoft Game Studios. Each of these examples is part of a larger series of tests performed throughout the development cycle of each of the games. It is not sufficient to run a single test for one game. As in any HCI-related field, to be maximally beneficial, user-testing must be integrated into the development process from the very beginning.

Each of the methods presented in the case studies has limitations, and should only be used when appropriate. A good user-testing engineer must understand the points in the development cycle at which certain methods should be used to address particular problems, and convince development teams of the general value of data collected through user testing.

In general, standard usability methods are not appropriate when assessing subjective preferences, due to the small sample size that accompany typical usability studies. With fewer participants, reliability of the results comes into question. However, usability methods used in conjunction with more open-ended tasks provide rich information about how a user interacts with a product which can provide a designer with greater understanding of how a user approaches and plays their game.

Larger sample methods are good for assessing subjective preferences, but they are limited because they rely on self-report. Usability studies, which typically focus on a user's actions, are better for obtaining in-depth, behavioral information. Just as in traditional usability testing, it is also impossible to look to the user to tell you how to design a game. Users can only evaluate aspects of the game ("this is fun", "that is not"). From there, it is up to the user-testing engineer and designer to work together to define the root causes of issues and create compelling solutions.

Unfortunately, there is no set formula for how the pieces fit together. The variability in the process through which games are developed must be reflected in the methods used for addressing user-testing issues.

Furthermore, there are many challenges with obtaining feedback on all areas of the user experience when playing games. For example, the survey methods that we utilized in testing adults proved to be problematic when trying to obtain feedback from children on child-oriented games. Additionally, testing multiplayer portions of a game (when two or more people play the same game together) presents further difficulties. Interactions between participants do not allow us to assume that the observations are independent, which makes it difficult to know the extent to which the results (i.e., subjective ratings) are influenced by the players' social interactions within the game, or pre-existing social relationships between them.

Finally, most of our research focuses on the initial experience of a game. Information based on this stage of the user's experience is critical because designers and developers are so immersed in the game that they often forget what the initial experience is like for someone who has never played their game. While many development teams would also like to obtain detailed feedback about gameplay that occurs beyond the initial experience, doing so with current development practices has many practical constraints. As such, these issues need to be explored further to provide cost-efficient user-testing methods to obtain feedback on all these areas of the user experience.

For more in-depth case studies, see Pagulayan et al. (2003) (MechCommander 2, Oddworld: Munch's Oddysee, and Blood Wake) and Medlock et al. (2002) (Age of Empires II).

Acknowledgements Special thanks to the development teams on CFS, MW4, Halo, and RSC and Digital Illusions. In particular, we'd like to express our gratitude to Christina Chen, TJ Wagner, David Luehmann, Jaime Griesemer, John Howard, Hamilton Chu, and Peter Wong for permitting us to use the actual games in our examples. We'd also like to thank the Microsoft Game Studios User-Testing Group, in particular, Lance Davis, Kyle Drexel, Kevin Goebel, David Quiroz, Bruce Phillips, John Davis, and Kevin Keeker.

References

Mahood A (2002) RalliSport challenge. Off Xbox Mag, April 2002. Imagine Media, Inc, Brisbane

Medlock MA, Wixon D, Terrano M, Romero RL, Fulton B (2002) Using the RITE method to improve products; a definition and a case study. In: Humanizing Design, Usability Professional's Association 2002 annual conference; 2002 July 8–July 12; Orlando, Florida. Usability Professionals' Association, Chicago, IL

Nielsen J (1993) Usability engineering. Morgan Kaufmann, New York

Pagulayan RJ, Keeker K, Wixon D, Romero RL, Fuller T (2003) User-centered design in games. In: Jacko J, Sears A (eds) The human-computer interaction handbook: fundamentals, evolving technologies and emerging applications. Lawrence Erlbaum Associates, Mahwah, pp 883–906

Smith M (2002) Reviews: RalliSport challenge. British Telecommunications plc

Chapter 28
Playing Games in the Emotional Space

Kristina Andersen, Margot Jacobs and Laura Polazzi

Author's Note, Funology 2

The FARAWAY project started at an interesting time and place, late autumn 2001, during the first year of the Interaction Design Institute in Ivrea. In, what felt like the beginning of entirely new time, we found ourselves thrown together in a small Italian town, ready to build new things, but also quietly wondering what we were doing there: Where were we? How had it come to this?

At this time, the institute employed a small number of researchers, who were free to pursue their research interests in small group collaborations, in a similar way to other research labs at the time. In retrospect, this was a unique opportunity; we were not Ph.D. students under pressure to develop individual research statements, and we were not post docs hired to work on already defined and funded projects. Instead, we were essentially given time and opportunity to describe and build, what we thought was the most important thing right now. This does not mean that we did not face challenges and resistance, we absolutely did, every presentation was a fight to secure continuous support and to insist on the unofficial slogan of the FARAWAY project: This is Not a Product.

The collaboration was in some sense born from this resistance, at the time it was not very common to consider emotional aspects of technological designs, but we were reading art and game theory and books like The Gift by Mauss and Decartes Error by Damasio. Also, we were temporarily displaced ourselves, and in retrospect there was no other project we could have done.

Looking back at the text now, it is striking how much of these concerns still seem unresolved. The framing of enjoyment and positive emotional engagement is maybe a little naive now, in the age of phishing and doxing, and there is clearly an

K. Andersen (✉)
TU/e, Eindhoven University of Technology, Eindhoven, The Netherlands
e-mail: kristina@tinything.com

M. Jacobs
Mia Leher and Associates, Los Angeles, USA

L. Polazzi
Experientia, Milan, Italy

© Springer International Publishing AG, part of Springer Nature 2018
M. Blythe and A. Monk (eds.), *Funology 2*,
Human–Computer Interaction Series,
https://doi.org/10.1007/978-3-319-68213-6_28

entire underside to the work that we did not consider at the time: How do we design for the ending of relationships, for privacy, for protest and protection? We can still proceed in our considerations of symbolic emotional value of technology, but now we must do so in the realisation that hope, love and longing must come with an acknowledgement of dread, loneliness and grief. Doing so, will not make our work any less humane and hopeful, but it will make it more inclusive and considered.

2003 Chapter

1   Introduction

The nature of user studies is changing, reflecting changing visions of both *what* is relevant to study and *how* to study it in order to understand the user. The focus has progressively moved from the task to the activity including an analysis of not only people's actions but the social, cultural, and economic factors that influence their behaviour. Now, with the so-called ubiquitous technologies moving out of the office and pervading people's personal sphere, designers must examine different facets of the user; in order for a product to be good (and successful), it should be both satisfying and pleasurable in the private life of the individuals. From a methodological point of view this raises new challenges: How do we ensure that an application is pleasurable for a given audience? How do we collect data pertaining to the personal sphere and useful in understanding the experiences of people and their emotions? How do we interpret this data in order to inform the design process?

The work of Gaver et al. (1999) suggests that methodologies inspired by artistic or creative techniques can be successfully applied to user studies. Using tools that are both provocative and aesthetically pleasant to involve people in engaging experiences is a possible way of accessing their intimate sphere and collecting revealing data. We believe that these kinds of surveys are particularly useful as preliminary studies in the design of enjoyable applications. Although the results they produce are not scientific in a traditional sense, they can be applied in a rigorous manner. While designing for pleasure, the information we receive may not always be measurable and must be interpreted using alternate methods.

This paper presents the approach we used during the FARAWAY project. The aim of the project was to explore how new technologies might support remote communication between people in affectionate relationships, decreasing the perception of distance; an additional objective became investigating the means with which to access and study peoples' experiences inside this domain. As a result of our investigations, we propose considering fun and pleasure as qualities of both the products we design and the methodology we adopt to involve users in the design process. Creating enjoyable experiences for the users we want to understand can provide revealing insights for the development of enjoyable applications.

Our approach is based on the concept of game play. Fun, enjoyment, and emotions are spontaneous aspects of human life, difficult to isolate, observe or measure. However, they can be provoked. Like narrative and poetry, games are valuable instruments that can trigger sincere emotional response in artificial situations. Designing games also means designing a framework for experiences, which is the final objective of designing applications for fun or pleasure. And designing games could prove to be a way to understand and influence the design of interactions. Our study shows that using games results in a high level of spontaneous participation from users, increasing both their degree of involvement in the design process as well as the value of their contribution. Hence, evaluating the level of participation and enjoyment for a game can be a way to test the validity of a solution.

2 FARAWAY

The FARAWAY project starts from the sense of privation experienced by people that are emotionally close but physically distant. In order to mitigate this sense of loss, loved ones seek alternative means of communicating. Linguistic studies demonstrate that for people in an intimate relationship the very objective of mediated communication is to feel each other's presence. Telecommunication is used primarily by loved ones 'to express a wish to be together' (Channel 1997: 144) as opposed to any actual exchange of verbal content.

Existing media only partially supports these kinds of interactions. Real-time communication creates a sense of each other's presence by virtue of the simultaneity of the exchange, yet current media offer a narrow channel for people who are communicating in order to do affective work. One disadvantage in distance communication with respect to face-to-face interaction concerns the lack of sensorial richness in current technological artefacts. A series of design and research projects like 'LumiTouch' (Chang et al. 2001), 'The Bed' (Dodge 1997), 'Kiss communicator' (Buchenau and Fulton 2000) and 'Feather, Scent and Shaker' (Strong and Gaver 1996) have already experimented with the use of non-verbal languages and sensorial modalities to exchange emotional content over distance; in these projects interaction modalities like blowing, touching and squeezing have been successfully incorporated into concepts and prototypes for sending and receiving messages remotely.

However sensorial weakness is not the only limitation of existing artefacts. Most of the effort in the telecommunication industry goes towards increasing the bandwidth of the channels, the miniaturisation of the technology, and the ubiquity of the services. Yet, as Taylor and Harper (2002) assert in their study of mobile phone usage amongst teenagers, designing communication technology requires a more articulated and profound understanding of the user's practices and their capacity to build and share meaning. In the context of FARAWAY and from an interaction design perspective, we explored new ways of conveying presence and emotions over distance focusing more on people's experiences and desires than on the technology itself. Understanding how this kind of communication works was, for us, a starting point to imagine how interaction design might support and enrich it.

One of our main routes of investigation involves symbolic objects and their power to embed presence and affective meaning. Thanks to specific, yet diverse rules, human beings are able to transform items into virtual 'traces' and tokens of affection for something or someone that is not present. In other words, these objects become symbolic surrogates of presence. Religious objects like the Eucharist and symbols of love like wedding rings or friendship bracelets are examples of this process. How does symbolic investment work in affectionate relationships? How can these mechanisms be utilised to design an artefact? These types of issues are dependent on cultural factors and even idiosyncratic values and attitudes. As designers, we felt the need to study the latter empirically, exploring and testing our ideas through a user survey. But the private nature of this type of communication is an serious constraint; how could we access the desires and behaviours of people inside their affective relationships? Instead of observing, we chose to create experiences that provoked response from the users. The core idea being to gradually shift from the existing to the new by creating, collecting and interpreting individual experiences within the defined design space.

In order to create meaningful experiences we had to create a meaningful context. Generally, people enter the emotional sphere by virtue of their relationship with a loved one. In our case we had to artificially trigger this process, simultaneously keeping the natural qualities of the emotional experience. The IF ONLY games are the method we developed for this purpose.

3 Another Reality

The idea of using games was influenced by game theory and techniques of arts movements. A game is a way to create another reality and allow people to enter that reality. According to Mataes (2001), games, like poetry and movies, demand what Coleridge calls a 'suspension of disbelief' (Coleridge 1817: Vol. I., Chap. 14); when participants are immersed in this kind of experiences, they are willing to accept the internal logic of the experience, even though this logic deviates from the logic of the real world. Murray (1998) goes further and suggests that creators of games need to not only suspend disbelief but 'actively create belief' by allowing players to manipulate objects and engage in enactment rather than processing descriptions.

The same kind of principles is evident in the artistic methods of the surrealist movement. Through playful procedures and methodologies of the fantastic, surrealist games (Gooding 1995) lead to out of the ordinary situations where the player is allowed to express her or himself in a more spontaneous and intuitive way. In other words, these methods stimulate the players to suspend their disbelief and access a 'surrealist' realm of creation. For example, in one of the more direct surrealist techniques, automatic writing, participants are encouraged to write whatever comes to mind, in a *stream of consciousness* style where nothing is corrected or rewritten. The unexpected material produced by this free associative writing game reveals unconscious thoughts and desires that may not otherwise be accessed.

The IF ONLY games refer to game theory and surrealism on two levels. On one level we created a 'surrealist' situation that allowed our players to produce and contribute emotionally truthful content and on another level we allowed them to experiment with non-conventional methods and use their creativity in order to investigate and express that content in new ways.

4 Designing the Games

When they are in an intimate relationship, people usually develop a universe of specific codes, languages and references that identify their private mode of communication. These elements become the constituents of an alternative space, or 'emotional space', that people enter while communicating with their loved ones. The IF ONLY games aim at re-producing this emotional space as a framework for the players to experience different kinds of emotional communication. Each game invites the players to experiment with a new and unusual way of communicating presence and emotions within the emotional space created by an affective relationship: by creating objects, by wearing particular clothes, by sharing physiological information, etc. In order to trigger expressive responses and allow players to provide real emotional content, the games have been designed using specific procedures, language and graphics.

The Interaction Design Institute Ivrea, where the project was initiated, is full of people from different countries who are away from home and who participate in long distance communication with loved ones on a daily basis. We chose students, researchers, and staff of the Institute as our players. The same group was maintained throughout the experience. The players were invited to play through 'invitation cards' (Fig. 1) containing instructions or recipes. The invitation cards were in some cases complemented with a 'comment card' or with a 'questionnaire' card containing specific questions about the experience created by the game. During one week the game cards 'appeared' on the desks of the players each morning. The players would then leave the results on their desks the following morning from where they would 'disappear,' thus making the exchange of cards and results a game in itself. At the end of the first week the 'mystery' was revealed during a collective meeting where we presented the project and lead a discussion about the overall experience and the specific games. The following games used the same modality, but with a smaller set of people.

an invitation to the game of
∽ THIS IS HOW I FEEL ∾

You find yourself in a small town. You did not use to live here.
You arrived from somewhere else. Let's call it
BiggerCity. You left someone behind. Let's call that person
DistantOne.

THERE IS A BOX ON YOUR TABLE

Unwrap the box. Open the Box.
Taste a Leone. Empty the rest onto the table.

Take the weekend. Ask yourself how you are feeling. You want to
send this feeling to DistantOne. What does this feeling look like?
Can you taste it? Can you smell it?

Take this feeling and put it in the box. What would you call this
feeling. Put the name of the feeling on the outside of the box.

Leave the box on your desk on Monday morning.

∽ IF ONLY ∾
game .01

Fig. 1 Game card of the game 'This is how I feel'

5   The IF ONLY Games

The IF ONLY games consist of 13 games played by up to 44 players. The games as a method are still being developed towards a point where they can be used to fully involve users' desires and dreams not only in an initial information gathering process, but for use during the entire span of a design process. In this sense the following games and the context they were played in can be seen as an initial proposal and a pilot study. In the following we take examples from three of the games and the responses of the players.

5.1   *This Is How I Feel*

This game initiates the IF ONLY series, inviting the player to participate. The overall objective is to introduce and establish an emotional connection between the individual players and IF ONLY, and to test the player's willingness to play as well as generate an index of representations of emotions. The focus is on the modalities of expression of emotional content and the related levels of earnestness and intimacy.

The card (Fig. 1) sets the context and mood and invites the players to the first game. DistantOne is introduced as the loved person with whom the player wants to communicate. The player is asked to express an experienced emotion within the three-dimensional space of a candy box. By collecting and analysing these communications instead of actually allowing them to be exchanged, the results are a series of one-way messages from the player.

The task takes a complex concept, an emotion, which is by definition impossible to express fully and suggests to represent it though an 'impossible' medium, the limited three dimensional space inside a given box. The underlying idea is to counter an impossibility with another impossibility. Within the scope of the games this logical contradiction is meant to liberate the player from the difficulty of the undertaking and encourage lateral thinking.

The game generated a large variety of responses. The players put written texts, colours, textures, images, objects, combinations of objects, etc. in their boxes. Many of them tried to communicate physical sensations like warmth and coldness, light or shining by using objects with particular sensorial properties (e.g. metal for cold); others (Fig. 2) created a suggestive experience for their DistantOnes by providing them with tools to interact with. Surprisingly, a task people found difficult was generating a verbal description for the represented emotion.

> I got a little frustrated trying to resolve the complexity of it in a single word. (C.)

> I found choosing a word to describe what I felt was very standard, I mean, it was impossible. (A.)

Fig. 2 'I can hear you' is an example of response where the player used an object to create an emotional experience for the loved person. The seashell contained in the candy box is an invitation for the receiver to listen to a sound that is emotionally charged for the sender

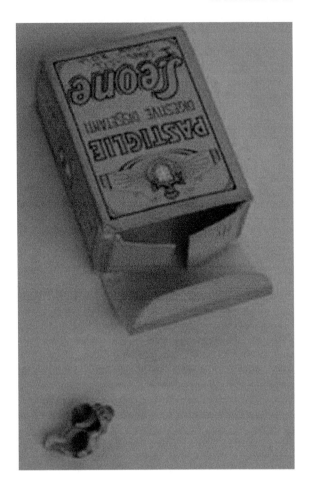

Words are normally considered as a natural and effective way to communicate to someone else how or what we feel. This game suggests how other media can be powerful in translating our inner states into a message.

5.2 Here I Am/Take Me with You

This game is played in two parts. The first part is an exercise in self portraits, asking the players to manifest their identity in a small object, 'LittleYou.' In this sense the task is to wilfully transfer ones own presence into an object. The objective of this game is to explore different modalities and styles in creating a symbolic self-representation. How do the players perceive their own presence and how do they translate this perception in something physical?

The second part is testing whether this transfer has been successful by letting someone else relate to the object and look for levels of identification, care and affection. The players are asked to exchange and take care of each other's self-representations, 'LittleOther' (Fig. 3). The couples of players exchanging objects are created randomly. At the end of the game each player is given a questionnaire to compile. The objective of this game is to test the ability of the 'LittleYou's' to convey the presence of their creators, looking for levels of personification in the objects, the transfer of emotional meaning from the creator to the care-taker and the degree of ownership with both players.

This game revealed the nature and qualities of the objects as conveyers of presence. The results from part 2 show a high level of personification in the way the players refer to the 'LittleOther'. Besides the use of 'her or him' (see Fig. 4), this attitude is generally reflected in the answers to the questionnaires and in the later debriefing discussion. The 'LittleOthers' are perceived as live entities and in many cases the players explicitly project the qualities of their creators onto them.

He's fragile so I feel very protective of him (K.)

[LittleOther is] shy like the creator but I like it for this reason. (S.)

Fig. 3 One player with her 'LittleOther'

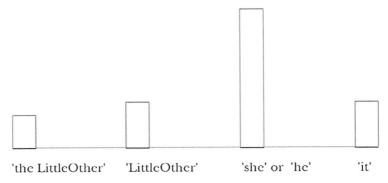

Fig. 4 How the players refer to their 'LittleOther' in the questionnaires

This process of projection strongly interacts with the relationship between the caretaker and the creator of the object. In the above case the 'power' of the 'LittleOther' even went beyond the purposes of the games, by mediating an interesting social interaction between the two players during both the day of the testing and at the later group discussion.

> How you made it? Very fast, I think. And like wrapped it from the heart and that's not the way you normally work. I would like to encourage you to do it more because I was immediately taken away by the way you did it. (M.)

Other interesting aspects concern the level of attachment to the 'LittleOthers' and its relationship with the emotional and personal effort employed by the creators in building the 'LittleYou'. According to the questionnaires, 13 players out of 17 did not want to part with their 'LittleOther'. In these cases, the other person's representation ('LittleOther') was much more important than their own ('LittleYou') and the separation at the end of the day was perceived as a loss.

> If it's possible to keep her that would be perfect (F.)

> I was more attached to the LittleOther than LittleMe. I just made my LittleMe and then spent the day with LittleOther. I actually forgot about LittleMe (M.)

In conclusion, the game confirmed the power of the objects as vessels for presence and pointed out a series of mechanisms in the processes of personification and emotional attachment. At a general level the objects created during part 1 were successful in provoking emotional reactions and interactions in part 2. The crucial role in the presence attribution seems to be played by the ritual of the exchange itself; it is this ritual that allows the object to acquire an emotional and symbolic meaning. As one player pointed out, the 'LittleOther' itself becomes live because 'it comes from a live person', once given its symbolic meaning the object continues to transmit the presence of the other.

5.3 *You Gave Me This*

This game further explores the potential of a ritual gift exchange. 'You gave me this' is entirely based on the concept of symbol; the purpose is to determine how the ritual of exchange and symbolic investment can relay the presence of another person. It is based on the results of the previous games but differs in that we have designed and produced a game piece, a 'bean', through which the game is conducted. This game also introduces a new element by letting the players connect and play with their actual 'DistantOne'.

The bean is a padded object sewn in white fabric (Fig. 5). It is activated and invested with symbolic meaning by the sender (player 1) and then given to the receiver (player 2) who will have it in a fixed state (on). As the input modality the first players are given the opportunity to put artefacts of their choice inside the bean. These secret artefacts are attributed to the bean in order to symbolically invest it with the presence of player 1. The beans are then sent to DistantOne (player 2), who takes care of the bean for one week. Although player 2 is unaware of what is inside of the bean, they do know that player 1 has placed a secret object inside.

The results are gathered through a questionnaire for the sender and a diary for the receiver. Results from this game are still being collected at the time of writing but it is possible to see some tentative directions already. An important tendency is that the senders are generally happy to express themselves through the piece, even if the mode of interaction and the media itself is strictly predefined.

How did you feel during this game?

Introspective and happy (C)

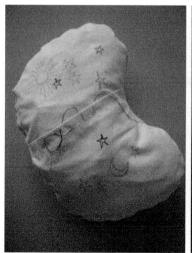

Fig. 5 A bean 'activated' by player 1. The player put an object of his choice into the pocket of the bean and sew the pocket. The bean is then sent to the player's DistantOne

What were you trying to express?

To share sensual experiences of subtle parts of life (L)

The biggest question in relation to the beans is if they are successful in conveying the presence of the distant loved one, decreasing the perception of distance. The diaries of the receivers provide us with important clues in this sense. On one side, the bean seemed to encourage a process of personification, with many receivers giving the bean a name.

> This bean should have a name P. Bean - P for short. P is a significant name between the sender and myself. (C)

> The bean is definitely a girl and its name is 'beany'. (M)

> I decided to call the bean 'Human bean'. (F)

Similarly, emotions are attributed to the bean itself along with actual character traits.

> I think it is quite a sad bean, maybe not sad but definitely suffers from a little melancholia now and then. And it seems lonely - lost. (L)

> She always makes me smile whenever I look at her - she is so cute and happy. (M)

However, the qualities of these emotional states seem to be reflections of the nature and state of the relationship between the sender and the receiver, with a love relationship having a higher tendency to attribute feelings of loneliness to the bean. Very often the correlation between the bean and the sender is explicitly mentioned in the diaries of the receivers and the symbolic object is recognized as a powerful substitute of the loved person.

> It makes me miss her less because I feel like we are together somehow. (M)

> Beany is my substitute for the time being and that kind of works out for me. (M)

The beans seem to inspire very different levels of care and responsibility; several players jokingly blame themselves for forgetting the bean for a short time while in few cases the bean is neglected for several days.

> I am a bad person. Today I forgot to bring 'Beany' with me to Copenhagen to watch football. (M)

> Several days have gone by and I have completely forgotten about the bean. (L)

However, the desire expressed by most of the players to keep the bean after the experiment can be considered as a sign of emotional attachment to the object. A more difficult element to decipher is the role played by the secret inside. The most common response is to not look inside, yet the behaviours vary from player to player. Someone identifies the content of the pocket with the heart of the bean, confirming its role as a vessel of presence, while, in the most extreme example, a player looses interest in the bean after deciphering its content.

In conclusion, the mechanism of investing the object with a symbolic act and the ritual of exchange were successful in triggering emotional response and in some cases affective attachment in the receiver. The diaries of several players also seemed to confirm a process of projection of the loved person's presence into the bean.

6 Conclusions

From the experiences conducted so far with the IF ONLY methodology it is possible to draw two kinds of conclusions. One kind concerns the method itself and its suitability for obtaining responses from the users. In this light, the first general observation to make is that the players genuinely enjoyed playing the games. They are very busy people and discovering that so many of them not only participated but also had a good time doing so, was a positive surprise for us. This also made it possible for us to collect a wide and diverse collection of content. The level of participation was especially high considering that we didn't make any agreement with the players beforehand or put any pressure on them to continue to play. We estimated that about 38 people of the 44 that we invited, played at least one of the games. It was evident that playing was fun just by observing people during the different experiences or talking with them informally. Yet, more explicit confirmations came from the comments expressed by the players in a collective meeting at the end of the first set of games and in the comment cards of the players. Examples of feedback from these two sources are:

> Everything was beautiful and cool! (M.)

> The first day I was a bit nervous, I was like 'oh no, nothing that takes any time, I'm so stressed' and then I was 'I like that it takes some time' because the time it took was as much as I gave it and I needed to give it so much time. (L.)

Another important element is the emotional participation that we obtained from the players. Coherently with our objectives, the games were successful in making people 'suspend their disbelief', 'enter the emotional space' and actively act and react within it. This emotional participation was high and progressively augmented throughout the games. Already in the first game many of the responses contained strong and revealing content. As time went by, the IF ONLY style became more and more familiar for the players and the immersion became more complete. Evidence of this process is in the way in which the language of the games is appropriated; for example in the 'Here I am' game, the words 'LittleYou' and 'LittleOther' are recurrently used both in the 'comment cards' and in the informal conversations between the players. 'LittleYou' and 'LittleOther' rapidly became recognised characters of a common imaginary world.

> The LittleOther is with me for dinner, We will be back soon. Don't worry. (M)

The second type of conclusion concerns the success of the methodology in generating knowledge valid to inform the design of technological applications in this field. The games provided useful insights about what people value in emotional communication, what might allow them to sense the presence of a distant loved person and how emotional content can be creatively expressed. The whole set of results suggested a wide range of directions that interaction design might take in order to enrich existing communication practices; some of them, particularly related to the examples described here, can be summarised as follows.

The first concerns the transmission of emotional content over distance. The games (see 'This is how I feel') indicate that one of the most powerful ways to express emotions is exchanging sensorial information like texture or colours, and creating an experience for the receiver. Both these modalities of expression might easily be supported by existing or upcoming technology and open interesting opportunities for interaction design. Simple sensorial information might be used to enrich existing objects (wearables, pieces of furniture, linens) with new function-alities or completely new products could be invented. More complex solutions might be created allowing users to actively modify their environment or design aural, visual and tactile experiences for the receiver.

Another source of inspiration is the observed process of symbolic investment. The results of the games suggest various ways in which technology might enhance this process; providing people with ways to augment objects with simple beha-viours or digital traces that would hold their presence for another user. A potential design direction is represented by the concepts of ritual and exchange. The rules incorporated into the games seem to have been successful in supporting the pro-cesses of coding and decoding symbolic meaning. It would be interesting to experiment with how those or similar rules could be incorporated into various systems of communication mediated by technology.

From our experience we learnt that, in order to be successful, these methods need to be applied iteratively. Although the games played so far have allowed us to investigate general ideas and test some low-tech artefacts, further iterations are required in order to move towards real technological solutions. The next step of the FARAWAY project is to design new games incorporating the use of working technological prototypes. Each prototype will be a tool to explore a concept from the first game series in more detail and to investigate which input and output modalities, rules of interaction and technological solutions might support the communication process. These games will be played with people in real long distance relationships using similar rules and the game structure of the previous phase.

The 'IF ONLY' method has already provided us with valuable information about sharing emotion over distance as well as ideas about presence and symbolic investment. We believe that further games will generate additional useful data that could be used in order to transform these interaction elements into viable products or services.

References

Buchenau M, Fulton J (2000) Experience prototyping. In: Proceedings of DIS '00. ACM Press, NY, pp 424–433

Chang A, Koerner B, Resner B, Wang X (2001) Lumitouch: an emotional communication device. In: Extended abstracts of CHI '01. ACM Press, NY, pp 313–314

Channel J (1997) 'I Just Called to Say I Love You': love and desire on the telephone. In: Harvey K, Shalom C (eds) Language and desire. Routledge, London

Coleridge ST (1817) Biographia Literaria; or biographical sketches of my literary life and opinions. Rest Fenner, London

Dodge C (1997) The bed: a medium for intimate communication. In: Extended abstracts of CHI '97. ACM Press, NY, pp. 371–372

Gaver B, Dunne T, Pacenti E (1999) Cultural probes. Interactions 6(1):21–29

Gooding M (ed) (1995) A book of surrealist games. Shambhala Redstone Editions, Boston, MA

Mateas M (2001) A preliminary poetics for interactive drama and games. In: Proceedings of SIGGRAPH '01, art gallery, art and culture papers, pp 51–58

Murray J (1998) Hamlet on the holodeck. MIT Press, Cambridge, MA

Strong R, Gaver B (1996) Feather, scent, shaker: supporting simple intimacy. In: Videos, demonstrations, and short papers of CSCW'96. ACM Press, NY, pp 29–30

Taylor A, Harper R (2002) Age-old practices in the 'New World': a study of gift-giving between teenage mobile phone users

Chapter 29
Deconstructing Experience: Pulling Crackers Apart

Alan Dix

Author's Note, Funology 2

It is 15 years since this chapter was first written and there have been developments both in virtual crackers and in design and creativity methodology during that time. The original conclusions mentioned that deconstruction–reconstruction was "*one part of a systematic armoury for the design and remediating of experience.*" This note will expand a little on this methodological message and on the changes since that time.

Unknown knowns—externalising tacit knowledge—A central part of the deconstruction–reconstruction process is identifying the core elements. On the whole, surface elements are not hard to identify: if you ask experts in any field they are usually able to name the key concrete nouns and verbs in their area. However, the experienced effects are more difficult: while concrete nouns are easy, often more abstract nouns are far harder to name, in particular qualities, concepts, criteria, and design dimensions. They are unknown knowns: tacit knowledge, things that you know, but don't know that you know.

Making this tacit knowledge explicit is a core goal of *externalisation*; elsewhere Layda Gongora and I have argued that this externalisation can lead to a step change in understanding, making it possible to reason and discuss about one's own knowledge (Dix and Gongora 2011). Indeed this is at the heart of the higher levels of Schön's (1984) reflective practice and precisely how the experience deconstruction–reconstruction gets its power.

Fun but not engaging—seeking critical transitions—The examples in the case study derived some of their concepts from analysis and some, apparently, from thin air. However, even the more systematic analysis does not in itself create the concepts, instead this is another largely tacit expert skill. Part of the 'armoury' are techniques to help this externalisation process.

A. Dix (✉)
University of Birmingham, Birmingham, UK
e-mail: alan@hcibook.com

© Springer International Publishing AG, part of Springer Nature 2018
M. Blythe and A. Monk (eds.), *Funology 2*,
Human–Computer Interaction Series,
https://doi.org/10.1007/978-3-319-68213-6_29

451

One approach is to try to find *critical transitions* (Sas and Dix 2009), pairs of concrete examples that are as similar as possible to one another and yet one of them has some hard to frame property and one does not. In articulating the difference, often criteria or design dimensions emerge. This may involve deliberately engineering scenarios where you have a visceral reaction, and then use this as a trigger for reflection and analysis.

One example of this was in work with Masitah Ghazali trying to better understand what makes an experience 'fun'. One question that emerged was whether all fun experiences are also engaging. It was easy to find engaging experiences that were not fun, but every 'fun' experience also seemed to be engaging, This vocabulary of experience is complex and tacit, we easily recognise 'fun', even the fact that others find something 'fun' that we do not, but find it hard to explain.

To explore this we started with an experience that was boring, neither fun nor engaging: waiting for a kettle to boil when you are desperate for a cup of tea. We then tried to modify it until it was fun, but still not engaging. One idea was a tweeting kettle, when the kettle boiled the pressure of the steam would pop up a small bird (plastic not live!), which would then tweet as the steam vented. However, this was itself slightly unsatisfying, possibly 'funny' rather than 'fun', or perhaps a moment of 'fun' ending an otherwise unfunny experience … but then how is that unlike a shaggy dog story?

The critical transition of the example was from 'no fun' to 'fun', but recognising the visceral sense of dissatisfaction had turned a question about 'fun' and 'engaging' to an apparently closer one between 'fun' and 'funny'. We then went on to explore linguistically, looking at the variety of Malay words that might translate as 'fun'. Whereas most languages 'cut up' the conceptual space of concrete objects in a similar way, the less delineated abstractions of perception and emotion are often dealt with differently; languages 'chop' the space in different directions, and in their cross-cutting expose finer distinctions.

(Re)coding dialectic—visceral reactions for theory development—Another technique that exploits this combination of critical transition and visceral reaction is *(re)coding dialectic* (Dix 2008). Perhaps you have performed a grounded theory (Glaser and Strauss 1967) analysis of interview transcripts, or perhaps you come to the transcripts with an existing theoretical construct such as actor-network theory (Latour 2005). You need to know whether your theory, indicative or theoretical, is adequate for the data. The grounded theory analysis relies on its process, and actor-network theory on its theoretical foundations, and previous utility in other domains, but have they worked here, now, on your data?

(Re)coding dialectic starts by coding the data with the vocabulary of the theory —in the case of inductive theories a recoding as they had already started with a theory-free coding.

Some parts of the transcript may be hard to code with the existing categories. This may be because it is irrelevant to the purpose at hand, but the gap may suggest broadening your remit, or expanding your vocabulary.

Perhaps most interesting are those parts that can be coded, but where the coding feels inadequate; for example, an account of pulling crackers, or watching Soviet-era protest theatre (agitprop) might equally be classed as "leisure activity". Often saying "just a" helps to make the issue obvious: "Brecht's play is *just a* leisure activity". You look for that visceral reaction, the sense of insufficiency, the feeling in your stomach that your vocabulary does not adequately represent the phenomena. Again this is sometimes fine, for the purposes of your analysis the expression is acceptable, even if it might be inadequate in other contexts. However, in examining the dialectic between term and thing, word and world, and attempting to explain it, "it is not sufficient because …", new insights, new criteria, new distinctions emerge—your analytic vocabulary and model of the world become richer.

Bad Ideas—breaking boundaries and mapping the domain—Perhaps the most successful, and certainly most fun, technique in uncovering criteria and dimensions is *Bad Ideas* (Dix et al. 2006). Normally, during brainstorming or similar ideation exercises, you try to think of good ideas, the emphasis is typically on non-judgemental idea generation, to encourage out-of-the-box thinking, and occasionally technology is used to allow anonymity. However, in practice it is hard not to generate small changes to existing ideas, and to feel the need to defend one's own idea.

Bad Ideas does the opposite, it asks you to think of bad or plain silly ways to tackle whatever issue or design problem is at hand; an example during one session was 'an inflatable dartboard'. Because the ideas are deliberately bad, you have increased freedom to be divergent to explore the far-flung reaches of the design space, however inhospitable. Furthermore, if it was deliberately a *bad* idea, then there is less of the emotional attachment that creates defensiveness; you can be free to critique it.

However, Bad Ideas does not stop with idea creation, but follows this with a series of questions to prompt reflection: "it is bad, but why is it bad?", "is there anything good about the bad idea?", "is there something that has the bad property, but is not bad?". This investigation can sometimes itself lead back to a good idea: the bad idea allowed exploration and the critique brought the bad idea back into 'good' territory, but in a new and unexpected place.

As important, the probing questions force one to articulate critical distinctions in the design space. Initially the very craziness of the idea suggests instant criteria, but the "sharing the property but not bad" question (or equivalent for good features) forces finer distinctions—critical transitions again.

Extensions by others—Other researchers have developed work based on some of these techniques, either working with the author or individually.

While virtual crackers were about creating virtual experiences from physical ones, often making things physical, or embodied can help externalise them. Paula de Silva (2012) worked with specialised forms of Bad Ideas in addressing ubiquitous computing design; this included participants creating models of proposed devices. Similarly, Layda Gongora's RePlay method (Gongora and Dix 2010) uses theatre improvisation techniques to explore both physical design questions, but also more abstract concepts, for example encouraging participants to act out computers and networks.

Finally Clare Hooper created a systematic method TAPT (Teasing Apart and Piecing Together) based on the deconstruction–reconstruction methods of this chapter (Hooper and Millard 2010). Her work was targeted at software engineers who are used to more systematic methods, and so she constructed a series of worksheets within a step-by-step method.

Virtual Crackers—aging, but soldiering on—Since this chapter was first written, virtual crackers had a brief foray into Facebook. This required a similar redesign process due to the nature of what was possible given the Facebook interface and API at the time—although still digital, effectively a different medium to plain web pages. In particular, there used to be a limit on the number of posts that an application could make on a user's behalf, and so the user had a finite 'box of crackers' which were used up during the day and replenished at night (another job for TorQil the cracker elf). Unfortunately, only a year or so after, Facebook changed its model for application developers, and so Facebook Christmases were once again bereft of crackers.

However, virtual crackers on the web are still going strong. They have had a few minor upgrades over the years, but with the same, now web-retro, look and feel. They no longer attract the same level of 'fan mail' as they did in their first days, but for various people they have themselves become part of the traditions of Christmas, which they initially emulated. Perhaps most interesting was the conversation with someone born in a non-crackers country, who told me how in her first Christmas in the UK, she had known what to expect because she had previously received virtual crackers—definite hyper-reality.

2003 Chapter

1   Words

> *the cursed animosity of inanimate objects* (Ruskin)

I was recently shown the above quotation. It was quoted in a book by Madeleine L'Engle (1980, p. 11). She does not just quote this, but says "What I remember from Ruskin is ...". It is not just a quote from Ruskin, but for her it is THE quote. The significance was not only personal for her, the reason it was shown me was because it made an impression on my wife and the reason I quote it here was because it also made an instant impression on me. What about you?

So why is it such a powerful phrase?

First it is something instantly recognisable with which we can all resonate. L'Engle talks about tangled coat hangers, but I am sure we all have stories about doors that won't lock or unlock, drawers that get stuck, cars that start every morning except the morning of that job interview.

But if it were just the sentiment L'Engle probably would have not remembered the exact words.

Let's look closer.

I think it is instantly obvious that the phrase turns on the two words "animosity" and "inanimate". Structurally in the sentence they sit opposite one another, but furthermore the two words have a similar look "...anim..." and sound[1]. Resonance in speech brings the words together in our minds—a frequent 'trick' of poets and orators.

But then the words tease us. They sound very similar, but one has the prefix "in". So the sentence appears to say: "the X of non-X". There is a dissonance, an apparent contradiction within the surface form of the utterance. Digging a little deeper, as soon as one thinks about the meaning of the individual words, this dissonance evaporates. The word "animosity" is about enmity whereas "animate" is about life. So at a semantic level there is no contradiction. However, think yet deeper and again we are struck by the dissonance of ideas—"animosity" presupposes intent and personality, attributes of the living not the inanimate. Dissonance resurfaces in pragmatics.

Yet the sentence, however paradoxical, is also familiar. Resonance and dissonance in form and meaning.

~ ~ ~

The idea of "deconstructing experience" can sound alien—somehow wanting to take apart something integral and personal. By understanding and rationalising experience don't we devalue it? However, the process of analysing and deconstructing[2] aesthetic experience is well established in literary, graphic and musical art.

This analysis and deconstruction is not just an academic exercise for the critic or interested observer. Instead the artist is aware and using this knowledge of form and technique to guide and support the creative process.[3]

Let's look again at some of the things we have learnt from Ruskin's quote:

(a) the use of similar sounds to bring words into contrast
(b) the use of sentence form to do the same
(c) the use of parallels between surface form and deep meaning
(d) the use of paradox (also seen in oxymoron)

---

[1]In fact the two words "animosity" and "inanimate" come from a group of related Latin words derived from "anima"—breath, soul or life and "animus"—spirit or mind.

[2]Note I am not using "deconstruction" here in the recent traditions of post-modern criticism, but in a broader looser sense of just taking apart, teasing out the strands that make something what it is ... and, in this context, especially those that make something 'work' as an experience or as a designed artefact.

[3]The vocabulary of literary and other artistic criticism is large and rich. For example, the Penguin Dictionary of Literary Terms and Literary Theory (Cudden 1998) contains over 4500 terms. Poets and artists are amongst those expanding and using this language. For example, Gerard Manley Hopkins coined the term "sprung rhythm" to describe a metrical form of his own verse, which was also found in far earlier writing, and in so doing both reinforced his own style and influenced later poets (Hopkins 1918).

Understanding these it is possible to start to use them oneself. Let's take the second and try to make something using them:

She fans the glowing embers while ice gathers on the sill

Not great poetry, but note by pivoting the sentence on the conjunction "while", the two words "embers" and "ice" are in some way brought together, and in their contrast focus the contrast of the two clauses.

And even in writing this section I've deliberately used the rest of the techniques. Notice the repeated use of the words "resonance" and "dissonance". The words sound similar (they rhyme!) and hence call themselves together, yet they are opposites. But furthermore as words they are opposites at two levels. We use them for ideas and concepts—hence we could say that the idea of enmity of a non-living thing is in some way dissonant and yet that the idea itself somehow resonates with our personal experiences. However, the words can also be used about sound—the surface form—and indeed that is their origin: things that sound good together and those that don't. So opposites have been brought together by sound and meaning—the surface reflects the deeper meaning.

This is the power of analytic deconstruction—it gives us tools for thought and the means for construction of something new.

2   Pictures

So we have seen how deconstructing a paradigmatic example can uncover techniques that can be used to construct new things. This deconstruction to understand is the very stuff of science and academic enquiry giving rise to theory, the language of generalisation. The application of this theory to guide the construction of new things is the essence of design.

However, when the topic of deconstruction is human experience and aesthetics, we do not expect a theory that, like physical theory, completely explains and allows us to predict the exact form of future things. Instead, these humane theories are potential pathways, more like worn tracks on grassland than signposts on roads.

In the example we found techniques that can be taken away and applied again and again. However, this process of deconstruction and reconstruction can be applied in a more situated and contextual fashion in order to understand a particular artefact and redesign it for a slightly different setting or for a different media. It is the latter—the changing of media—that will be the main focus of the rest of this chapter.

~ ~ ~

Graphic designers have faced a rapid change in their discipline over recent years. After a century or more of growing understanding of print media in magazines, books and posters, the computer has completely upturned patterns of work. For many the drawing board has all but been replaced with the workstation and tablet. Effects such as feathering of images, morphing and layering, that would have in the past required great expertise in draughtsmanship, painting and photographic manipulation, have become possible in a few clicks in Photoshop.

But as well as changing the tools to produce images on traditional media, they have increasingly been called upon to design for new electronic and often inter-active media—initially CD-ROM delivered content and increasingly the web.

It took cinema more than a generation to move from a filming of theatre to a creative discipline in its own right with its own vocabulary, reference works, and rich genres. Graphic design has been expected to make a greater transition largely within the last 10 years.

There are two major ways in which we, as humans, make old experience available to new situations. One we have already discussed—theory and abstraction. The other is perhaps more grounded—examples and analogy. The latter is probably the major way in which a lot of visual design works, allowing incremental progress through reapplication of the familiar. But analogy does not promote more fundamental leaps.

This entrapment by the incremental has been a problem especially with web pages, where designs that look good on paper or even on screen fail when transferred to the web. Sometimes the only way in which the design could be rendered was as a single large bitmap, or collection of bitmaps leading to slow download times and often strange alignment problems as formatting differed between web browsers.

The problem is that the web appears at first to be a medium just like a computer screen—after all that is where a web page appears. However, the internal structure of web pages and dynamics gives it different properties. To design for the web one needs to understand those properties.

the golden rule of design

understand your materials

One example of this are images rather like those in Fig. 1i. A strong frame (the box) with some element, usually a curve or angled line crossing the frame. Although this is drawn more iconically, this may be the design for the page as a whole with text and further graphics within and around the frame.

This sort of design is very common, but translates badly to the web environment. This is because of the crossing highlighted in Fig. 1ii. This requires either that the

(i)   (ii)

Fig. 1 Breaking boundaries

(i)   (ii)

Fig. 2 Gestalt flow

whole image is a bitmap, or that different parts are very precisely aligned. However, the slightly different formatting on different browsers means that attempts to fragment the image lead to unexpected spaces or poor alignment.

This has got better recently as more recent versions of browsers have allowed more precise positioning, but the difficulty of achieving this type of effect (and others) is one reason why designers often turn to Flash splash screens even for fairly static content.

However, if we dig more closely and ask *why* the image is the way it is more solutions become apparent. The use of a strong element breaking a boundary is used because it gives a sense of dynamism. We know that web pages can render rectangular frames very easily. We also know that precise positioning is possible, but we would like to convey the idea of the strong image crossing the boundary.

Look at Fig. 2i. Although the lines do not actually cross the boundary, our visual gestalt 'fills in' the gap and the lines still appear to cross. Note the use of several smaller lines rather than one big one, this means that precise alignment is not critical.

Note here we are deconstructing and reconstructing in two senses. First we are taking surface elements: the box, the angled lines, and re-placing in the new image. However, more important we are also looking at the underlying effects of those visual elements: the breaking of boundaries, the dynamism, and how to achieve these experienced effects in a different medium. Some of the precise visual features are lost in the redesign: the actual crossing, the single line is changed to several lines, but the underlying feelings are reproduced (Table 1).

Table 1 Deconstruction and reconstruction of the image

Original image (Fig. 1i)	New image (Fig. 2i)
Surface elements	
Strong box	Strong box
Single thick diagonal	Several thin diagonals
Actual crossing	Not present
Experienced effects	
Breaking boundaries	Gestalt feeling of boundary crossing
Dynamism by crossing	Dynamism by gestalt crossing plus multiple lines suggest movement

Of course this does not create a solution, these are just sketches, a particular design would require more detailed work, but the deconstruction and reconstruction opens up the design space. If the constraints were different we would need to look for different solutions. For example if the image were in fact a company logo that we needed to preserve in appearance we could not take the liberties we did in Fig. 2i. Instead we might use a toned down version as a background image, or perhaps use a small version with large elements that emphasise key visual features as in Fig. 2ii.

The general lesson is that as we move between media we need to deconstruct the effects that make the experienced image and reconstruct those, not the surface image. This will typically include preserving certain surface features, especially if these are themselves evocative, but we can move away from reproduction to reconstruction.

3   Crackers

Now those of you who do not come from Britain or Anglicised parts of the world probably do not know what a cracker is. Crackers are tubes of paper pinched in near each end to make a tubular 'package' in the middle (see below). Two people pull the cracker, one holding each end. Inside the cracker is a tiny amount of gunpowder so that when the cracker eventually pulls apart it also makes a loud bang. Then, from inside usually fall three things: a motto or joke (usually a very bad joke), a paper hat and a small plastic toy.

It was nearing Christmas 1999 and aQtive, a start-up company I was involved with, wanted something to send to friends and contacts … and perhaps spread a little the brand name! There were numerous electronic card sites, this was passé—couldn't a hi-tech company do better. Then, one day whilst driving on the motorway the idea came—why not an electronic cracker?

Of course, it is not as easy as that. Real greetings cards are flat, largely printed, arrive in the post. Although different in electronic form than on cardboard, there is not a great gulf. In contrast, real crackers are solid (well not flat), are used together with someone else, not just looked at, but pulled and things found inside. It is not clear that any electronic version could work—because the experience would be too different and too impoverished.

In fact, Virtual Christmas Crackers were a great success and many of those who received them from aQtive sent them on to others. Sadly their life time is short (about 3 weeks leading up to Christmas), so they are not a major year-round

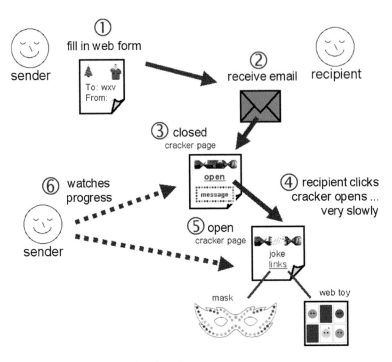

Fig. 3 The process of sending a virtual cracker

product, but each year since they have been equally successful and attract frequent 'fan mail' to TorQil the cracker elf.

Virtual Crackers were successful because they did not simply try to emulate real crackers, but in some way captured aspects of the essence of the experience—deconstructing the experience of real crackers made of paper and gunpowder and reconstructing it in the very different medium of the web.

We'll look briefly at how virtual crackers work for the sender and receiver and then examine more deeply this process of deconstruction and reconstruction.

The sender's interface starts off very much like any electronic greeting card. There is a web page where you fill in your email address and name, the recipient's email and name and a short message to be delivered with the cracker (Fig. 3, step ①). Again, rather like an electronic greeting card, you get to choose a general cracker theme (Christmas, Valentines, New Year) and a design for the outside of the cracker.

When the sender is satisfied the form is submitted and an email is sent to the recipient (step ②). The email contains a URL where the cracker can be found, again like most electronic greetings cards. However, clicking the email does not lead to the full cracker contents, but instead to a "closed cracker" page with the outside of the cracker and button to press (step ③). When this button is clicked the cracker pulls apart, but very very slowly—almost painfully so (step ④). When the cracker image has pulled apart the web page is replaced with an "open cracker" page and a 'bang' sound (step ⑤). Only then can the recipient see the joke and links to further pages with a 'web toy' (an animated GIF or applet game) and a mask. The mask is on a page of its own and is big enough that if you print it out you could cut out the mask and wear it.

The sender also has a URL both on a confirmation web page and in an email sent at the same time as the recipient's email. The sender's web page only shows the outside of the cracker until the recipient has opened it (step ⑥). So the sender can't peek ahead of the recipient!

4   Experience

The operation of the virtual crackers sounds a bit like a mixture of electronic greetings cards, a direct translation of some aspects of physical crackers, and some ad hoc additions. In fact, looking more closely we can see that the virtual crackers are a reconstruction of a deconstruction of the real cracker 'experience'. Virtual crackers succeed not because they replicate real crackers, but because they capture the essence of the experience: an experience that is interactive and multi-party.

We'll look at some of the facets of this deconstructed experience in turn (summary in Table 2). The table classifies these facets into surface features and experienced effects as in Table 1. However, this distinction is a little arbitrary; for example, it was not clear how to classify 'surprise' (due to bang). This is natural as the surface features of course give rise to the experienced effects.

> *I think your crackers are fantastic!!*
> *These are very cool! Well done!*
> cracker feedback

Design: Although there are expensive crackers for high class dinners, on the whole crackers are a cheap and cheerful part of Christmas celebrations—crepe paper, simple designs, plastic toys, looking good for a while and then torn apart. The web pages reflect this, simple bold graphics and page design. Furthermore, the cheap materials of crackers mean that sometimes the 'bang' doesn't work, etc. The virtual crackers use dynamic effects that tend to be flaky and

Table 2 The crackers experience

	Real cracker	Virtual cracker
Surface elements		
Design	Cheap and cheerful	Simple page/graphics
Play	Plastic toy and joke	Web toy and joke
Dressing up	Paper hat	Mask to cut out
Experienced effects		
Shared	Offered to another	Sent by email, message
Co-experience	Pulled together	Sender can't see content until opened by recipient
Excitement	Cultural connotations	Recruited expectation
Hiddenness	Contents inside	First page—no contents
Suspense	Pulling cracker	Slow … page change
Surprise	Bang (when it works)	WAV file (when it works)

browser dependent. Even with great care in construction dynamic web material tends to be less than perfect. This becomes 'forgivable' because it merely picks up existing qualities of the real crackers! If instead one wanted a virtual Fabergé egg things would be different. The experience would be one of opulence and would require meticulous design—quirky, unreliable, even minutely imperfect web pages would be unacceptable.

Play: Real crackers contain a joke (usually a very bad joke) and some sort of toy: plastic ring or figure, tiny game, etc. This was the easiest aspect to translate to the virtual experience except that the toy becomes an electronic toy: either an animated image or a small web game.

Dressing up: The other thing inside a real cracker is a paper hat. The first thought for this was to show a 'smilie' face with a hat on it. This would have been fun and pretty, but hardly captured the essence of a paper hat. A paper hat is something you can put on—dress up in. The next thought was to have a separate page with a hat that you could print out and cut out. However, this would need at least one glued joint and some quick measuring of head circumferences showed it would need to be in two parts. The solution eventually adopted was a cut out mask. This fits on an A4 or US letter paper and is a different way to 'dress up'. Often people do not put on the hats from the crackers, but they would be upset if the hat was not there. With virtual crackers we do not expect that many people actually print and cut out the masks, but the fact that you *could* leads to an apparent tangible experience.

Shared: With real crackers, each place setting typically has a cracker and the person will offer their cracker to pull together. It is a shared experience. Because the virtual crackers are offered, albeit by email and not in person, they also have aspects of this sharedness. The mail to the recipient and the cracker web pages all emphasise who the cracker has come from, so the sharedness is reinforced throughout.

> *your virtual crackers are the bomb!*
>
> *they are too cool to be kept to myself*
>
> cracker feedback

Co-experience: A harder aspect of the cracker experience is the physical pulling. This is clearly very tactile, and pressing a mouse button hardly compares! The fact that the sender cannot see the inside of the cracker until the recipient has opened it does add a little to this sense of co-experience, but it is perhaps one of the weaker aspects. If combined with an instant messaging technology, perhaps it would be stronger.

Excitement: Real crackers are pulled in the middle of a party or celebration meal. Although they are just made of paper and plastic there is a real excitement about pulling them. This is partly because of the situation, but partly because of the cultural connotations that go with them: childhood Christmases, family celebration. Virtual crackers are able to recruit some of this excitement, because people associate them with the real thing. Often feedback to TorQil has mentioned this sense of nostalgia. Although the focus is on the deeper aspects of experience, it is the surface visual characteristics that give the instant familiarity. Recall the Ruskin quote. It is often the nature of aesthetic experiences that they rely on a confluence of surface attributes and deeper meaning.

> *Thank you for putting a smile on my face and bringing back*
>
> *some funny memories! My mother is from England and I grew*
>
> *up pulling the "real" crackers during the holidays.*
>
> cracker feedback
>
> *This is such a great idea! As an ex-pat Brit' I have missed*
>
> *Christmas crackers all the years that I have lived in the USA*
>
> cracker feedback

Hiddenness: the contents of a real cracker are hidden until the cracker is pulled apart. Similarly with virtual crackers, the first page the recipient sees when the cracker URL is followed does not show the joke etc. Only when the cracker is 'opened' does the recipient (or sender) see inside.

Suspense: Although crackers are made out of paper they are surprisingly difficult to pull apart. There is a sense of growing suspense as you start to pull and pull. Sometimes even frustration when the paper never seems as if it is going to break. In fact, for children I've occasionally had to make a small tear in the paper to make it break for them at all. Virtual crackers are, of course, not physically pulled, but the slow (painfully slow) movement of the halves of the cracker when the 'pull' button is pressed and the long wait until the contents are revealed adds to the sense of suspense.

Surprise: The pulling of the real cracker ends in the explosion as the cracker bursts open with a bang! Well, usually with a bang, sometimes they just come apart and the bang never comes. The opened virtual cracker also produces a 'bang' albeit simply a WAV file, and just like real crackers this 'bang' sometimes fails depending on browser capabilities!

~ ~ ~

Before moving on, I guess I should note that this analysis of the deconstruction and reconstruction of the crackers experience is itself partly a rational reconstruction of the process we went through in producing the final design. Virtual crackers succeeded partly because when faced with problems we explicitly tried to look for the underlying issues and aspects of real crackers in order to be able to recreate a similar experience in virtual crackers. But also there were times when we did not do this explicitly, but looking back we can see that virtual crackers succeeded because we unconsciously or perhaps even accidentally reproduced aspects of the deeper essence of the experience.

The above analysis should be read therefore rather like the analysis of the Ruskin quote. It may be that Ruskin was explicitly aware of the techniques he was using, as he was clearly reflective on the nature of art. But he was also a very practised, skilled and inspired writer, so it may be that these techniques were unconscious and unplanned. Or it may even be an accident and the fact that this quote is remembered was because it just happened to embody the right features. Whichever is true about that quote it is certainly the case that, for those of us without Ruskin's genius, more structured methods and heuristics can help us achieve more robust and effective prose.

Similarly, we know that the virtual crackers in some way 'worked' and in unpacking this we can perhaps move towards 'designing in' that success.

5 Reflection

Rather than starting with a 'method' and then applying it to examples to demonstrate utility, this chapter has progressed by successive revelation as we examined increasingly more complex examples of deconstruction and reconstruction of experience.

As previously noted, the process of deconstruction lies at the heart of science and academic study. The main use is to allow us to unpack the generic issues that underlie a particular instance in order to understand related phenomena elsewhere. All the points (a)–(d) we uncovered in Sect. 1 are of this form. Generic properties or facets of the Ruskin quote that we could use in other literary works. This is the sort of thinking that is common in detailed low-level literary analysis.

In fact, several of these points can clearly be generalised across media. For example, point (c) parallels between surface form and deep meaning can be seen as a version of Louis Sullivan's "form follows function" (About 2002; Miller 2000). Also point (a) says that things with similar surface characteristics are somehow 'brought' together by that. This is also a principle of visual perception used frequently in information visualisation, graphic design and fine art.

In other chapters of this book we can also see this process at work, in particular throughout most of the first part of this book insight from various related areas is being translated into general design advice and understanding. For example, Wright et al's identification of their four aspects of experience, or Reed's use of the existing sociology of play and playfulness. In the normal course of the academic process, these all attempt to produce generic universal principles and heuristics that can be applied to new problems and situations.

However, the graphic design example and even more the deconstruction of the crackers experience point to a more situated use of deconstruction that enables the reconstruction of the *same* experience in a different medium. Of course, I am using the word 'same' here cautiously—it is by no means an identical experience either encountering virtual crackers after real ones, or even the variants of the simple line and rectangle graphic. However, the essence of the experience is in some way captured.

In the case of the graphic design there are also general lessons like those from the literary analysis. For example, the principle of breaking boundaries to give dynamism can be deliberately used where a sense of dynamic is required. This is the sort of generic heuristic that can be found in more analytic discussions about design. This synthesis of new designs from a 'bag' of heuristics and guidelines that have been distilled from previous experience, this construction of the new based on the deconstruction of the old, this is the heart of more systematic design and engineering.

However, the new graphics in Fig. 2i, ii are not synthesised from scratch but instead borrow the precise set of deep characteristics found in the original graphic (Fig. 1i) and do so by embodying it in features that follow as closely as possible the surface features of the original. So, Fig. 2i is not just a different graphic that expresses dynamism, it does so by using the more particular technique of breaking boundaries. Not only this, but it uses a rectangle and an angled line. So, the final graphic is in some way recognisably consonant with the original and recognisable as being 'the same' in a different way.

Similarly with virtual crackers, if we had dug to the deepest level of the experience and then *only* asked "can we reproduce these", then we might have produced a totally new (and possibly successful) 'fun' and 'party-like' artefact, but it would not have deserved the name 'virtual cracker'. Blindly recreating surface features (like the image of the hat) in a different medium may *not* recreate the same experienced emotions and effects. So reconstruction in a new medium is not reproduction in that medium. However, equally we try to stay close to the original surface form in order to be 'the same' as the original.

It is interesting to note that the excitement of the virtual crackers borrows from the cultural nuances of the original, which are themselves evoked by similarity in surface features.

Looking at the deconstructed crackers experience, we could go on to abstract these to find some general principles to aid the design of experience in other domains. However, the most important lesson from this is not the particular deconstructed facets, but the process of deconstruction and reconstruction itself.

6 Distillation

Deconstruction of instances and analysis to form abstractions is the essence of science. Construction of new artefacts by the synthesis of these abstractions in new contexts is the essence of design. These can be applied to experience as in other domains. Of course, as we are dealing with human emotions the abstractions, like those in literature and art, are guidance and heuristics, not hard rules.

However, in the successive examples in this chapter, leading to the rich crackers experience, we have seen a movement from general principles to a more situated use of deconstruction and reconstruction as a process of analysing a particular experience in order to translate it to a new medium.

In Janson's History of Art (Janson 1977, p. 14), he shows how Manet's famous painting *Le Déjeuner sur l'Herbe* reproduces aspects of a previous engraving after Raphael and that engraving itself is based on older Roman sculptures. This process of inspiration across media clearly occurs naturally over time. However, the rate of change of digital media exceeds any previous times when reconceptualisation occurred between media. A more systematic approach to dealing with this transition is not just an academic luxury, but essential if design is to keep up with technical change.

This chapter offers one part of a systematic armoury for the design and remediating of experience.

http://www.hcibook.com/alan/papers/deconstruct2003.

Acknowledgements Virtual crackers are an online product of vfridge limited and can be seen (and experienced!) at: http://www.vfridge.com/crackers/.

The first version of virtual crackers was produced by aQtive limited in conjunction with Birmingham University Telematics Centre. Thanks especially to Ben Stone who produced the first cracker implementation. Since then they have evolved through comments from numerous people.

An early version of the analysis in this chapter was presented at the 2001 Computers and Fun conference (Dix 2001), where I received many helpful comments.

This is part of a wider study of the nature of technological creativity and innovation (see http://www.hcibook.com/alan/topics/creativity/) and this has benefited from discussions and input from many people and especially recent support from the EPSRC funded EQUATOR and CASCO projects.

References

About.com (2002) Master architect—Louis Sullivan. http://architecture.about.com/library/bl-sullivan.htm

Cudden JA (1998) The Penguin dictionary of literary terms and literary theory, 4th edn. Penguin, London

Dix A (2001) Absolutely crackers. Paper presented at Computers and fun 4, York, UK. http://www.hcibook.com/alan/papers/crackers2001/

Dix A, Ormerod T, Twidale M, Sas C, Gomes da Silva P, McKnight L (2006) Why Bad Ideas are a good idea. In: Proceedings of HCIEd.2006-1 inventivity, Ballina/Killaloe, Ireland, 23–24 March 2006. http://alandix.com/academic/papers/HCIed2006-badideas/

Dix A (2008) Theoretical analysis and theory creation. Chapter 9 In: Cairns P, Cox A (eds) Research methods for human-computer interaction, Cambridge University Press, pp 175–195. http://alandix.com/academic/papers/theory-chapter-2008/

Dix A, Gongora L (2011) Externalisation and design. In: DESIRE 2011 the Second international conference on creativity and innovation in design, pp 31–42. http://alandix.com/academic/papers/desire2011-externalisation/

Glaser B, Strauss A (1967) The discovery of grounded theory: strategies for qualitative research. Aldine, Chicago

Gongora L, Dix A (2010) Brainstorming is a bowl of spaghetti: an in depth study of collaborative design process and creativity methods with experienced design. In: First international conference on design creativity, ICDC 2010, Kobe, Japan, 29 Nov–1 Dec 2010

L'Engle M (1980) Walking on water. Lion Publishing, Tring, UK. ISBN 0 86760 341 0

Hopkins GM (1918) Preface to poems. (quoted in [[C98]] pp 854–855)

Hooper C, Millard D (2010) Teasing apart and piecing together: towards understanding web-based interactions. In: Proceedings of the WebSci10: extending the frontiers of society on-line, Raleigh, NC, US, 26–27 April 2010

Janson HW (1977) A history of art: a survey of the visual arts from the dawn of history to the present day. Thames and Hudson, London

Latour B (2005) Reassembling the social: an introduction to actor-network-theory. Oxford University Press

Miller C (2000) Lieber-Meister—Louis Sullivan: the architect and his work. http://www.geocities.com/SoHo/1469/sullivan.html

Sas C, Dix A (2009) Understanding and supporting technical creativity. In: Proceedings of HCIEd'09. http://eprints.lancs.ac.uk/id/eprint/42423

Schön D (1984) The reflective practitioner. Basic Books, London

Silva P (2012) Designing user interfaces with the BadIdeas method: towards creativity and innovation. Lambert Academic Publishing

Chapter 30
Designing Engaging Experiences with Children and Artists

Richard Hull and Jo Reid

2003 Chapter

1 Introduction

Product (and service) designers have long been concerned with the user's direct experience of their offerings, with ease-of-use often the primary design goal. However, we believe that the indirect experiences evoked by a product are at least as important to many users. For example, a comfortable bicycle seat is valued by most cyclists but the fun of speeding through the open country with good friends is more likely to motivate the purchase of a bicycle in the first place.

The 4D experience project at Hewlett-Packard Laboratories is concerned with exploring experiences of this second kind, particularly in the context of ubiquitous and wearable computing. We ask two key questions:

- What constitutes a compelling consumer experience?
- How can we deliver such experiences through emerging computer technologies?

Our research methodology combines technology development, experimental prototype deployments, and user research. In this chapter, we will sketch three attempts to develop systems that evoke engaging experiences in their users, review those exercises in the light of an underlying model of experience, and discuss the positive involvement of users and artists in the design process.

R. Hull (✉) · J. Reid
Hewlett-Packard Laboratories, Bristol, UK
e-mail: richard@calvium.com

Present address:
R. Hull · J. Reid
Calvium Ltd., Bristol, UK

© Springer International Publishing AG, part of Springer Nature 2018
M. Blythe and A. Monk (eds.), *Funology 2*,
Human–Computer Interaction Series,
https://doi.org/10.1007/978-3-319-68213-6_30

2 Zap Scan

Our first exercise involved the development of an exhibit for the Explore@Bristol hands-on science museum (Explore 2002) intended to demonstrate that engaging, fun experiences can be made from everyday office technology. The resulting exhibit, Zap Scan, allows users to draw a picture with supplied paper and crayons, scan that picture (or anything else) through a one-button interface, and see the scanned image appear on digital picture frames on either side of a vertical screen. Separately, if they wish, they can move to a nearby print station, select their image on a touchscreen, enter their name, insert a pound coin, and produce a glossy greetings card with their image on the front and their name on the back. Figure 1 shows Zap Scan in place at Explore.

Zap Scan was targeted at children aged between 3 and 12, many of whom were recruited in a multi-stage, participatory design process as users, testers, and informants (Druin 2002; Winograd 1996). At the outset of the project, we visited Explore to understand the environment, observe visitors, and review operational matters with staff. Based on these early observations and our previous research in image sharing, we chose a concept and value proposition—electronic display of scanned drawings—that would appeal to children in this age range. Then at various stages of development, we solicited input from the potential users as follows (see Fig. 2):

- Pre-school children at a local play scheme helped to test the concept using a standard display, scanner and printer
- Older children at a local primary school explored user interaction, flow and timing issues with an integrated prototype
- Children at a local junior school provided feedback on the overall system behaviour and appeal.

The completed exhibit was deployed in April 2001. Despite (or because of) its deliberately simple functionality, Zap Scan turned out to be very popular.

Fig. 1 Zap Scan deployed at Explore@Bristol

Fig. 2 Testing the Zap Scan concept and prototypes

For example, nearly thirty thousand images were scanned between April and October 2001, and between 17 and 37 cards were printed each day of the summer holidays. Users often spend a considerable time working on their drawings. Parents sit beside their children watching them draw or joining in themselves. Friends sit together and share in the excitement of seeing each other's drawings on the digital picture frames. Young children nudge their friends and point excitedly when a picture appears. Children also gaze in anticipation into the printer and watch the cards appear bit by bit.

So, lots of people, usually children, had lots of fun using a computer-based exhibit that actually did very little. What they seemed to enjoy was the act of drawing, social interaction, the joy of seeing their creative work on public display, and the attractive cards. In the later discussion, we will begin to explore why this might be so.

3   A Walk in the Wired Woods

A Walk in the Wired Woods is an art installation in which an exhibition of woodland photographs is augmented by a digital soundscape. Equipped with headphones connected to a small shoulder bag, visitors typically spend around twenty minutes wandering around the exhibition, viewing the photographs and listening to audio pieces chosen to enhance the images (see Fig. 3). The particular sounds heard by a

Fig. 3 Images of A Walk in the Wired Woods

visitor at any point are determined automatically by a small computer system in the bag that monitors the visitor's location within the exhibition space (Hull and Reid 2002; Randell and Muller 2001). For example, when standing close to certain photographs, a visitor might hear atmospheric music fitting the scenes depicted. As she moves on to other images, the music might be replaced by natural woodland sounds, or by a spoken fragment of woodland mythology. The overall effect is of a situated soundscape that might be characterized as "what you hear is where you are".

The content for the installation was developed in parallel with the underlying wearable computing technology with the deliberate intention of allowing each to influence the other. We formed a multi-disciplinary design team with artist Liz Milner and musician Armin Elsaesser to explore both *what* might be done with this new technology and *how* it might be achieved. This resulted in a number of possible technology enhancements, some of which were incorporated into the installation. For example, as it became clear that different soundscapes would require different audio characteristics such as mixing, looping, and fading with distance, we developed a HTML-like mark up language for specifying the behaviour of the visitor's wearable client with respect to particular content (Hull and Reid 2002).

The completed installation, incorporating around thirty pieces of situated audio, was deployed in the atrium of the Hewlett-Packard Laboratories building in Bristol in the early part of 2002 (Mobile 2002). During it's residency it was visited by several hundred people from a variety of backgrounds whose responses were overwhelmingly positive. Of course, the high quality of the photographs and music contributed significantly to this outcome, However, most visitors reported that the extra dimension added by the contextual juxtaposition of the two media adds significant further value. A simple ranking exercise revealed that more visitors likened the exhibition to a Walk in the Woods (something that it attempts to evoke but really is not) than to a museum tour (something that it really is) (Hull and Reid 2002). This reinforces our belief that it is possible to create a convincing and compelling experience with the kind of mobile technology that we can expect to become pervasive over the next ten years. Moreover, our own experience of the design process confirms our belief in the power of collaboration with artists to drive innovation in both technology and content.

4 Soundscape Workshops

In the Zap Scan exercise, children played the roles of user, tester, and informant, while the Walk in the Wired Woods explored the use of creative practitioners as full partners in the design process. Recently, we put those two elements together in a pair of workshops in which children were invited to take on the lead design role in the creation of digital soundscapes. The participants were 11–12 year old children drawn from two local secondary schools.

The first workshop involved ten pupils from John Cabot Technology College in Bristol and ran over two consecutive days. Pupils were first introduced to the technology, the idea of the soundscapes, the authoring tools, and the process of production. Then they were divided into working pairs and let loose. Each pair was assigned an adult enabler who encouraged, helped and observed but did not interfere with the creative process. At the end of the workshop, the pairs had produced five diverse soundscapes:

- a trip to the beach that starts with sounds of the car journey and ends with the sounds of sea surf and happy play
- a tiger cub riddle where audio clues were distributed along a visually marked path
- music from different parts of the world
- a walk-around radio station with different kinds of music in different areas
- a game to match flags from different nations to the appropriate music and sounds of that country.

Though these did not have the depth of professional works, they were a lot of fun and showed creativity, technical proficiency and hard work. One interesting aspect was that all five pairs had to use the same physical space for their soundscapes, leading to negotiations over physical props or visual signs that were used to anchor pieces of audio (see Fig. 4). It also forced the technical team to provide mechanisms for navigating between multiple digital soundscapes occupying a given space.

The second workshop, involving pupils from St Gregory's school in Bath followed a similar structure except that the two days of the workshop were separated by a week to allow time to explore design ideas and prepare materials. In particular, the participants spent two afternoons in school working with an artist in residence.

Fig. 4 Pupils from John Cabot creating a soundscape

Fig. 5 Plan of the St Gregory's soundscape

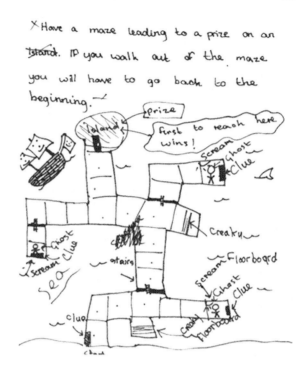

The result was a single design based on a haunted house that involved hand painted props, a tent and a video loop (see Fig. 5). The second day of the workshop was spent implementing as much of the design as possible in the time available. Again, the end result was imaginative and rich with content, and the workshop as a whole was considered a success by the participants and visiting staff.

The workshops confirmed that children are capable of adapting quickly to new technologies and creating novel and engaging applications. They are able to respond to the responsibility of the design partner role with ingenuity, creativity, and enthusiasm. On the other hand, both groups tended to produce soundscapes that had echoes of the examples that had been used to introduce them to the technology, though this may simply reflect the severe time pressure under which they worked. The pupils seemed very engaged in the design process and very satisfied with what they produced.

5 Discussion

The exercises described demonstrate that it is possible to deliver engaging experiences through systems with fairly simple functionality. We can begin to explain the positive responses of users to the resulting systems through the model of experience shown in Fig. 6.

The model is one of the outcomes of a study at the Explore@Bristol science museum in which we observed visitors interacting with the exhibits and explored their responses through discussion groups, interviews and questionnaires (Kidd 2003). It attempts to unpack the nature of engaging experiences by identifying three key dimensions that are likely to play a role in those experiences:

- Challenge, achievement and self-expression
- Social interaction, including bonding, sharing and competing with others
- Drama and sensation, including stimulating sights, smells and sounds, and other cues that trigger the imagination.

The model is inevitably partial and provisional, but we can begin to use it to interpret the success of our prototypes in terms of the elements that they provide. Our hypothesis is that experiences that contain some mix of these dimensions *will* be engaging. For example, Zap Scan scores highly on the social dimension as users congregate around the drawing area and display, and also provides an opportunity for self expression. In contrast, the Walk in the Woods emphasizes the Drama/Sensation dimension by stimulating the visitor's senses and imagination through music, sounds, images and narrative. Moreover, the exercises suggest that emphasizing a single dimension may be sufficient to engender a good experience, although some recent work based on a desert island soundscape suggests that

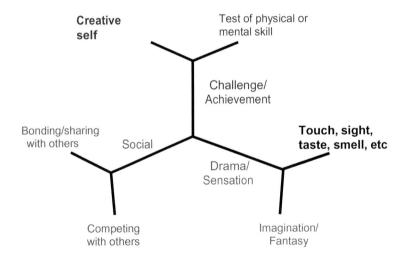

Fig. 6 A provisional model of consumer experience

adding a second dimension (challenge) to an experience with a strong single dimension (sensation) can further enhance that experience (Nethercott 2002).

The exercises also illustrate different types of user participation in the design process. Zap Scan is the most conventional example, with primary and pre-school age children playing the roles of user, tester and informant (Druin 2002). Their input helped to reinforce our belief in the value proposition of the system, and refine its presentation and operation. Overall, we consider that Zap Scan confirms the value of participatory design as espoused by many over the last two decades. Furthermore, one of the children was responsible for the exhibit's cool name, which was far better than any of the technical team's suggestions!

The Walk in the Wired Woods exercise was motivated by a desire to discover what artists would make of *and* with wearable computing technology. We work with artists because they tend to be imaginative, creative, demanding, meticulous, and extreme users of technology who are used to asking what-if questions. Given this perspective, we encouraged the artists to act as full design partners of both the experience *and* the underlying technology. Naturally, the artists tended to have more influence on the content, and the technical team on the technology, but both elements of the design clearly benefited from the collaboration. The participation of the artists ensured that the resulting installation was fascinating and that the technology evolved appropriately.

The children's workshops enrolled secondary school age children as full design partners, again to see what they would make of and with new technology. In this case, we provided prototype wearable computing technology, tools and training but left the experience design activity completely to the children. Naturally, the resulting experiences lacked some of the depth of more experienced practitioners, but the systems they managed to produce in a few hours were a great testimony to their creativity and application. The main result from our perspective is that it does seem possible that users will be able to create their own contexts for experiences using this technology much as people create their own websites today. This is crucial if the emerging technology is to be rapidly adopted and shaped towards its eventual meaning and value.

In conclusion, then, we believe that the exercises show that engaging consumer experiences can be evoked through applications of computing technology, that it is possible to model experiences in such a way as to inform their development, and that the involvement of users and creative practitioners in the design process greatly increases the likelihood of success.

References

Druin A (2002) The role of children in the design of new technology. Behav Inf Technol 21(1): 1–25

Explore (2002) Explore@Bristol. http://www.at-bristol.co.uk/explore

Hull R, Reid J (2002) Creating experiences with wearable computing. IEEE Pervasive Comput 1(4), Oct–Dec 2002

Kidd A (2003) Technology experiences: what makes them compelling? Hewlett-Packard Laboratories

Mobile (2002). In: Mobile Bristol conference. http://www.hpl.hp.com/hosted/mbristol

Nethercott J (2002) An experimental comparison of two soundscapes based on two theories of motivation: implications for the design of compelling experiences. M.Sc. dissertation, Department of Computer Science, University of Bath

Randell C, Muller H (2001) Low cost indoor positioning system. In: UbiComp 2001. ACM, Atlanta

Winograd T (1996) Bringing design to software. ACM Press, NY

Chapter 31
Building Narrative Experiences for Children Through Real Time Media Manipulation: POGO World

Antonio Rizzo, Patrizia Marti, Françoise Decortis, Job Rutgers
and Paul Thursfield

Author's Note, Funology 2

POGO World Fifteen Years Later

Antonio Rizzo, Francoise Decortis and Edith Ackermann

To return fifteen years later on POGO designed between 1997 and 2002, what meaning can it have? Answering this question calls for more questions: what is being done today in the high-tech development laboratories that think and design the class of the future? What educational philosophies are these developments based on? How are the educational environments designed? What are the resources for prototyping cyber-physical systems? Since the last century, activities that are considered important for children have changed radically (Yarosh et al. 2011). Yarosh and colleagues provide an overview of these dimensions and the values promoted by the international community of researchers and designers (e.g. "Interaction Design for Children", IDC), which have guided and led the design of technologies for children since the 2000s. They point to five orientations that have remained immutable throughout the last ten years of the IDC community: (1) a contribution to the social, intellectual and creative development of children; (2) a vision of children as active agents in the appropriation of technical devices, which requires involving children in the design and evaluation of these devices; (3) anchoring in the perspective of constructivism and constructionism; (4) a preference

A. Rizzo (✉) · P. Marti
University of Siena, Siena, Italy
e-mail: antonioriz@gmail.com

F. Decortis
Université de Vincennes - Paris 8, Paris, France

J. Rutgers · P. Thursfield
Philips Design, Eindhoven, Nederland

© Springer International Publishing AG, part of Springer Nature 2018
M. Blythe and A. Monk (eds.), *Funology 2*,
Human–Computer Interaction Series,
https://doi.org/10.1007/978-3-319-68213-6_31

for a new digital ecology in which traditional devices interact with new ones; and (5) the creation of gateways between physics and digital.

We suggest that the EU POGO project has been at the heart of the movement identified by Yarosh and his colleagues, and which, before his time, is an integrative (ecological), reflective (theoretical foundation), experiential (GUI to NUI), and methodological (empirical, grounded in field situations) example of this movement.

POGO has ventured on the path of augmented spaces, communicating tools, intuitive interfaces, seamless integration of the digital and the physical, making use of of the most advanced technologies, based on a vision of the future that integrates an aesthetic of innovative interaction modalities. POGO was a research project based on an educational philosophy rooted in the work of psychologists such as Bruner, Vygotsky, Piaget and Dewey, whose contribution has been to refocus education on building knowledge and know-how while they are engaged, with others, in projects that interest and involve them. From the perspective of these current dimensions, let us come to the eight reasons that underlie the relevance to revisit POGO fifteen years later.

1. The first reason why it is pertinent to return to POGO is above all theoretical and rooted in the very principle of the evolution of human cognition, both cultural and "cumulative". What distinguishes human culture from other species is its cumulative nature (Dean et al. 2012). This "cumulative" concept reflects our tendency as a species to re-use and to improve the artifacts that have been produced by our ancestors and by our peers. In designing POGO, **we focused on successful practices** in schools rather than those that presented problems, focusing on what works well rather than stopping at what does not work, as is often the case. This was also the key for integrating the digital with the physical resources (see below).

2. POGO, was built to be integrated, and to interact, with the most effective and common tools made available to children in their school environments. The idea was to conceive an environment that does not create a break with existing tools, but rather to make these new tools integrate into the classroom and be used concomitantly with the old ones, while offering new opportunities for interaction. The idea was not to substitute one tool for another but to **produce a "seamless" world between** existing and new tools, between so-called **"physical" tools and so-called "new" digital technologies**.

3. With POGO, we designed tools that would **merge artistic practices** such as drawing, painting, and sequential audiovisual art. We tried to exploit the work of Scott McCloud on the basic principles of sequential art by exploring the concept of the ellipse, the "two-box" sequence that creates a progression of time and movement. The camera, one of POGO's tools, was originally conceived and designed as a tool that allows children to capture a variety of viewpoints, not just the point of view of the person holding it in their hand.

4. POGO was built according to the principle of construction/deconstruction/ reconstruction: POGO constitute a family of objects (those they can bring, those they can conceive). Children and teachers can build by themselves, assigning

functions and meanings to them, playing and "pretending". In POGO, children had the opportunity to do, undo, and redo: they could try things, observe together constructed things, deconstruct, start over, transform, add pieces, remove them, all with great ease. The concept of **"Bricolage"** as presented by Levi-Strauss, which we can be translated as **"Tinkering"** (as developed by the Exploratorium Museum in San Francisco), refers to these relationships.

5. POGO was both a performing stage and a creation and editing workshop: a commonplace of creation and design, where each child, by creating and conceiving, brings a small stone to the building which sets itself up collectively. In designing POGO, we anticipated the importance of an environment in which children could speak about a **hundred languages** (cfr. Loris Malaguzzi), while appreciating the possibilities offered by each media: mixing performance and narration, use layer and level techniques, write, recite by adding voices, annotate an image or text, and so on.

6. Designing POGO has contributed to a vision of the learners as intelligent subjects that act to **develop and transform the world** in which they grow, a vision rooted in the work of Bruner, Piaget, and Vygotsky. In a synthetic way, let us retain here the dimensions of inter-subjectivity and the discursive and dialogic spirit. Piaget argued that role-taking and point-of-view skills, assembled in a cooperative spirit, not only make culture and language possible, but promote reasoning. Bruner (1996) further argues that, "we live most of our life in a world constructed according to the rules and devices of the narrative." This is why educational initiatives should promote the creation of a "metacognitive sensitivity" to give children an awareness of what the constructions and constraints of narrative impose on the reality they create.

7. Designing for the child in the perspective of long-term transformations is an essential issue. **Sustainability and affordability** are central here. We failed to create a sustainable environment. The possibility of observing the use of POGO in the long term was not possible, which we regretted. Our target was the school but we did not take into account the process of introducing POGO world in the school. An assessment made by McKinsey determined it was impracticable to create a commercial product.

8. POGO was designed by a multidisciplinary team of ergonomists, designers, psychologist, software and hardware engineers. The project was an opportunity for extensive production of mock-ups and prototypes. Yet the **prototyping tools** available at that time to embody in interactive artifacts the concepts and the envisioned scenarios were really inadequate. The hardest challenge was to merge the computational resources dedicated to interact with the physical world (e.g. microcontrollers, sensors and actuators) with the resources for producing complex elaboration and manipulation of the content (e.g. multimedia processing with full fledged computers). This brought some of us to engage in the design of new prototyping tools that would make physical computing simpler to afford and to use in interaction design projects (Rizzo et al. 2016).

2003 Chapter

1 Introduction

POGO world is an information technology environment to support the development of narrative competence in children. When we first started our user study for the design of POGO world we were conscious that the development of educational technologies calls for new interaction design approaches to overcome the limitations of current personal computer-based metaphors and paradigms. To help understand the limits and constraints of computer based technology for mediating educational activities and to have sound empirical evidence to share within our multidisciplinary design team, we carried out a longitudinal study in two European schools; one located in Siena, Italy the other in Brussels, Belgium. In this study, we observed and described more then 30 narrative activities, and we discovered that even though in one of the schools advanced digital technologies were available, and although there were several activities that included the use of computers, none of the narrative practices involving such equipment were perceived by the teachers as successful. Moreover, the teachers considered the introduction of computers into the activities that were successful a potentially disruptive factor that would prevent cooperation. The teachers supported their claim with evidence drawn from their own experience (UniSiena and UniLiegi 1999). Thus one of the main challenges we faced in designing a new system for interactive story building was to envision a new form of interaction that encourages creativity and cooperation and that did not jeopardise successful pedagogical activities currently used in the schools.

In the following we describe a general model of successful narrative activities that we developed from the field study in the two schools. We then present the POGO tools in some detail in order to briefly illustrate how the proposed tools embody the concept of *situated editing* as a metaphor to mediate interaction between children and the POGO world. Finally we present a summary of the tests carried out in the schools with the last version of the POGO world prototype. This provides some empirical evidence to show that POGO world does not jeopardize successful narrative practices, but empowers it.

2 Narrative Activity in Classroom

Through our observation of narrative activities in the classroom, we found that the cycle of creative imagination proposed by Vygotsky (1998) as a psychological process to account for the creation of knowledge occurring in the zone of proximal development, could be used to represent the different chronological and structural phases of a narrative activity. The cycle of creative imagination has four phases namely, *exploration, inspiration, production* and *sharing* and describes how the individual experiences the external world, elaborates the impressions received, assembles them in a novel way and shares this production with others. The narrative activities at school included a focus on all of these phases, often in linear sequences, sometimes with small loops or repetition, sometimes with a leap.

Exploration This consists of the interactions with the real world, which can be either direct or mediated by social relations. All narrative activities that we observed are rooted in the child's experience. The narratives represent things the child has seen, heard, touched or encountered in the museum, the forest, the seashore or even in the classroom but with the support of objects, instruments and people that create an event. This means that the teacher initially focuses on the sensory experiences of the child, which subsequently constitute the starting point for the theme and for the ideas. At this stage, the child uses instruments appropriate for exploration (e.g. dip net, shovel, microscope, points of view, etc.) and handles various materials (e.g. earth, shells, sand, photos, objects, etc.).

Inspiration This is a phase of reflection and analysis on the experience had during the exploration. The child is encouraged to think about the previous experience, discuss it and sort out the elements they gathered. The teacher supports the child in the analytic process and in the discussion of choices. Individual writing, drawing or group discussion usually supports this phase. In any case there is a moment when their ideas and thoughts are externalised in more or less lasting way.

Production This corresponds to the recombination of the elements dissociated and transformed during the previous phase. In other words, production is the moment where children, based on selections and choices of elements made previously, produce new content usually through a great variety of media. During this phase, the teacher's role is to supervise the organization of narrative content, as well as to ensure conformity with standard rules of story construction. The teacher also makes sure that the text is coherent and sufficiently rich. During this phase, the children mainly use their notebooks, pens and pencils for illustrations, cardboards, puppets, posters and bricolage sets.

Sharing This is the phase in which children's externalised productions start to exist in their social world. Children present the result of their production and verify the effects of this production on the others (e.g., children, teachers, parents). We observed that to conclude the activity, the teachers propose a moment of exchange and sharing of the narratives produced. In conventional activities, this is the phase when the teacher concentrates on the presentation of the final product created by the child. The most important document in this phase of the activity is the Story Notebook, which contains the final product: text and drawing. In other cases the sharing phase is a full-scale performance of groups of children or of the whole class.

The pedagogical activities observed in the schools were modelled as Narrative Activity Models (NAM) and were used in order to define the users' requirements and to relate them to the POGO concept and enabling technologies. A set of requirements were produced both to assess and refine the design concepts proposed by the industrial designer of Domus Academy and Philips Design and to foster further concepts to be expressed in mock-ups. One of the key concepts for the interaction design was the one of situated editing supported by "invisible computing" (Norman 1998), that is, to allow a seamless integration of the physical and virtual world through intuitive interaction modalities.

3 POGO World

From the NAM it was clear that in the school environment children use a number of tools to construct narratives. In referring to the phases that constitute the activity, we ascertained that the exploration phase is usually carried out in groups. Afterwards each child independently carries out the process of creating the narrative. The graphic illustration of the story is also done individually, during or after the verbal description. In other words, in conventional activities story making is mainly an individual undertaking. Nevertheless, we also observed several activities where all the children of the class or of sub-groups worked together to create a single story. The patterns of cooperation vary on a case by case basis, but generally the phases of group activity (choice of story subject and story line, etc.) alternate with individual creation (drawing, inventing dialogs, etc.), where each child makes his/her personal contribution to the construction of the story and the same tools are used to support both individual and group activities in a seamless way.

Most of the personal computer technology available in the school does not fit with this articulated way of carrying out narrative activities; the technology is just out of the loop.

3.1 The POGO Tools

The POGO environment can be thought of as a virtual story world, accessible through a number of interactive physical tools distributed in the environment. The active tools are the main interface to the narrative process. The functionality of the tools spans many areas from gestural (live performances), visual (manipulation of images and drawings) and aural (sounds and atmospheres), to manipulative (physical feedback, kinematics) and material (surface and texture, weight etc.). Although the system is computer-based, the standard computer interface of keyboard, screen and mouse has been replaced with a far more intuitive one. The interaction is very simple so that children can begin to play with no need for instruction.

The system has a number of tools that support the process we call situated editing. Raw non-digital media elements (e.g. drawings, sounds) can be converted into digital assets using tools for rich asset creation. These digital assets are stored on physical media carriers and can be used in tools that support story telling. With these tools assets come alive on a big projection screen, sound system, paper cardboard, paper sticks etc. The system provides tools to capture the creative end-results and share it with others using the internet in movies or in digital or paper

Fig. 1 The Beamer

based storyboards. POGO has been developed in a modular way allowing parts of the system to be re-used and combined for different purposes. The following tools compose the POGO world:

The *Beamer* is a threshold tool that connects the real and the virtual environment by allowing the passage of physical things into the virtual story world. The Beamer captures new story elements such as real world objects (including the children) or live video. It has a base unit integrating a horizontal LCD screen with a pressure sensitive touch panel, a video camera, a card reader, and a composition area (Fig. 1). Drawings and objects can be positioned in the composition area, and collages can be created there. The captured image can also be edited in the same composition area. The camera can be used to capture these elements and also as a simple, live video mode where the images are directly projected onto the walls.

The *Cards* are media for exchanging story elements such as sounds, pictures and video clips. They are a 'memory' for story elements that can be associated to real-world objects by physically attaching the card on drawings, clay models or toys. Cards contain a unique ID tag and are used as physical pointers to virtual story world elements. Whenever the card is activated (e.g. inserted in the silver mat) it displays the corresponding image or sound. When the children pop a story card in the slot on the table and press the record button, the pictures are stored on the card. If a child places a card in one of the pockets on the side of the silver mat, the pictures on the card are displayed as a background. If a child puts the story card into one of the Mumbos, whatever is on the card is shown on the mat in front of the background.

The *Mumbos* are tools to control foreground elements on the screen. Through the Mumbos, images can be animated (moved) and modified. For example, if the Mumbo is rolled, the image stored in the card contained in the Mumbo moves in the direction of the roll.

The **Camera** tool allows to record live video which can be stored in cards and displayed together with the other elements and characters. A controller allows the image size on the screen to be adjusted and photos to be taken which can then be inserted in the background (Fig. 2).

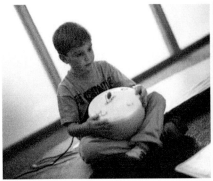

Fig. 2 Mumbo and camera tools

The *Settings*, comprise a silver mat surrounded by leather cushions and various tools. The mat is a screen on which to project images. However, it is also possible to project images anywhere in the physical environment, including onto the children's body (Fig. 3). The Background Composer is inserted into the setting. This allows up to three cards to be inserted to create a hierarchical background. Dropping cards into the Background Composer activates background images and/or related sounds in mixed media combinations. Background images can be created by the children (e.g., drawings, collages, composition of elements picked from the real world) or they can be selected from a database. The Background Composer provides a continuous output, so even if there is no card in it, a live video image is shown as background. In a sense, the live image allows children to "perform" a story in the real world on a virtual background.

The *Colour Wheel* is located in the setting and is used to set the background colour for the screen. It uses four joystick buttons for controlling the colour value, and the effect will be visible directly.

Fig. 3 Silver mat

Fig. 4 The Sound Twister

The **Sound Twister** allows activating sounds by inserting a sound card into the mat (Fig. 4). The Sound Twister Tool is used to playback sounds that are stored on cards. The tool consists of a number of pads. The sounds stored on a card are assigned to these pads and are played if a pad is pressed. Sounds generated in this way are mixed so that multiple sounds can be played at the same time.

The **Sound Mumbo** is used for sound playback using effects such as pitch shifting and echo. An effect can be selected by tilting the tool in the $X–Y$ plane. The sound source can be real-time using an embedded microphone for input, or can originate from sounds stored on asset cards.

The **Voice Tool** allows users to insert their voices into the story. A controller allows recording and modifying these voices. Children can speak in strange voices and can add echoes and noises.

The **POGO VCR** tool (Fig. 5) is used to capture everything that is played in the POGO system. It can record video and audio streams that are generated with the play PC and convert them into a movie file. The Recorder/Reader records and displays story scenes. When a card is inserted in one side of the Recorder/Reader, the scene is recorded in real time. When it is inserted in the opposite side, the recorded scene is displayed.

4   The Evaluation of POGO

Rather than performing formative and summative evaluation sessions, we decided to constantly assess the outputs of our design process with teachers, children and colleagues. Thus in a sense, we renounced formal evaluations in favour of a longer term qualitative assessment of the project outcomes. In particular, testing of the final prototype attempted to understand if and how the designed technology had

Fig. 5 POGO VCR

fulfilled the pedagogical goals, how the transition from current pedagogical praxis had been embodied into the praxis of the POGO design solution, and how the POGO environment could mediate the narrative cycle in various activity settings.

In the following, we report the evaluation activity performed at school on the final prototype with children aged from 6 to 8 years. All activities were designed, set up and co-ordinated by the teachers according to their pedagogical objectives. Different kinds of activities were proposed by the teachers ranging from free activities that were selected and coordinated by the children who were responsible for creating the narrative as they wished, to more structured activities proposed and coordinated by the teacher who decided timing, rules, content and dynamics.

All narrative sessions were videotaped and the dialogues transcribed. The resulting narrative productions were analysed and used as basis for debriefing sessions with teachers to analyse and interpret results. The results are described below.

4.1   Impact on the Narrative Activity Phases

Exploration POGO seems to integrate smoothly with the current practices of collecting story elements: children can bring personal objects, intimate memories, photographs etc. into the POGO environment and evolve them into elements of the narration. The POGO environment supports the transition from everyday life experiences to the fantastic world of narration by affording the collection of different media such physical objects, sound and noises and transforming them into virtual objects thus creating a rich repository of elements useful for the story. In particular the Beamer, which enables the user to import a virtual version of any sort of object, stimulates children to store an experience represented by the object itself. During the testing, the teachers encouraged the children to explore potential story elements by using the Beamer, projecting images of seashells gathered on the beach and mushrooms picked in the woods and so on. In traditional activities, they could only reproduce them in the form of a drawing, a kind of activity that is sometimes so time consuming as to prohibit a further elaboration of the drawings into story elements. With POGO, it is possible to immediately import an object, to draw and transform it in a virtual element of the story, so that the exploration phase is not conceptually separated from other phases of narrative activity. We consider this a clear added value over traditional ways of supporting the exploration phase.

Inspiration As happens with traditional activities, during the inspiration phase POGO can be used to encourage children to rethink an experience, to analyse its constituent parts and to express it orally or by drawing. The change with POGO is in the resources available to stimulate thought and decision-making. In comparison with traditional practices, the POGO tools seem to offer greater support to the children. The possibility of combining and recombining elements on the Beamer table, and of displaying the result on the screen in real time, facilitates experimentation and comparison of different solutions. In addition, screen displays have an amplifying effect that facilitates perception and information sharing. The tools support personal reflection, collective comparison and meaning negotiation. This was particularly evident in one activity "Mushroom Development", where the children spent nearly all the class in the inspiration phase. They placed themselves around the Beamer and the teacher encouraged them to use the material available to reconstruct the developmental phases of the mushroom. The Beamer acted as a support for handling the material, and for producing drawings. The teacher encouraged each child, in turn, to suggest ideas by modifying the material on the Beamer table. The screen enabled them to monitor their own production, as well as the productions of other children, providing a basis for further discussion. Intermediate products were stored in different cards recording individual as well as collective contributions.

Production Narrative activity in the school is very rich, rewarding and successful. In designing POGO, we learnt from the most successful existing practices and tried to amplify them in the learning process through the use of the POGO technology. The testing confirmed that we achieved this objective, and the production phase was one of the most surprising in terms of creative constructions made by the children. They made new connections among contents just by manipulating the tools. They explored the flexibility of the tools in representing and structuring the narration. Furthermore, the POGO tools allowed teacher and children to take clear roles, from guidance to content direction, technical direction, and performance. In general we assisted an interesting process of role diversification. We observed a division of labour during the creation of scenes between 'producers of content', responsible for arranging into acts and scenes, and 'technicians', responsible for producing elements of the story like backgrounds, characters, sounds, etc. (Fig. 6). The first group created the story, and the second was focused on realising them. The new role of 'technician' enabled shy children who generally do not participate enough in conventional activities, to be more involved, doing something that allowed them to join the negotiation process through their actions and not their words. The distribution of the tools and their location in space helped the diversification of roles: the

Fig. 6 Production with
POGO: distribution of roles

Beamer was the area to create contents (for technicians), other tools served to memorize (cards) or reproduce (screen) the results (the work of content producers).

During the test, we witnessed a process of scaffolding and fading that was perfectly realised through the tools. When the teacher provided less guidance (fading), the children produced their story with greater independence—as occurred in two activities—"Castle invaded by witches" and "Story of sound"—where the puppet show metaphor was employed. During these activities, the children spontaneously sat behind the Beamer, facing the settings. As soon as the activity began, the children started moving the silhouettes on the glass plate of the Beamer and kept track of the results projected on the settings. Their characters were animated against background scenery, like a Chinese shadow puppet show. The teacher then suggested improvised dialogues, giving voice to the characters, rather than describing the action, as they did in other activities.

The example reveals how role taking with POGO can be extremely varied and imaginative. Children can be actors during the performance, or spectators when looking at the screen.

Sharing The POGO tools can be used to amplify and enhance collective sharing of the children's production. Children can share both the creative process and the product of the narrative activity. This meta level of sharing stimulates meta-cognition and meaning construction and negotiation.

This effect is demonstrated by the fact that children insisted on 'redoing' the story many times, and presenting it to other children who had not participated in the production. This need for sharing can be explained by the fact that during the creative process the children concentrated mainly on creating a scene, but at the same time they were exposed to the global view of the narration. With POGO they learnt to change point of view and acquire a different perspective on the story, from local events to coherent plots. This effect is difficult to obtain with conventional tools like paper and pencil where all elements of the story are located on one or two

Fig. 7 Sharing with POGO: distributed activity in the space over different media tools

pages of a notebook. With POGO the contents are stored in different tools distributed in the space and represented by different media at the same time (Fig. 7).

5 Conclusions

POGO's challenge was to design innovative technologies for children that should be equally attractive, fun, long lasting and yet offering sound pedagogical learning opportunities to be seamlessly integrated in the current context of European schools. The results so far have been very encouraging. The POGO world does not replace any of the current tools that the teachers successfully use in their teaching practice, instead it empowers these tools and integrates them with new opportunities.

But the most important achievement of our research was the development of an educational tool that supports the entire cycle of creative imagination, letting it evolve as a never-ending creative process.

Indeed the POGO tools allow a rich sensorial interaction where physical and virtual elements of children's reality can be explored, analysed, decomposed, and recombined in new ways. The existing objects or the new one produced working with the different POGO tools can be captured by children and edited in real time. What a child builds or brings as a part of the personal experience can be combined with the products of other children in a continuous negotiation process where the evolution of transformations of the objects is recorded and the movement along this process of meaning construction can be used as a way to understand the other's points of view.

Moreover the physical objects that are produced in this iterative and combinatory activity remain live features of the process and can be used as the physical address for the articulated production of future creative activity.

The POGO project advocated a design that was focused on children and teachers' activities and grounded in thorough research into those activities. Its purpose was to provide a sustainable solution that can help children create and enjoy intellectually interesting activities.

POGO presents a new 'type' of system, an open system. It is a kind of 'personality', capable of intelligent responses. Depending on how it is used POGO it is a camera, a video recorder, a microphone, a display screen. POGO reacts to the user and adjusts its behaviour accordingly. It is open to change. And, while at the moment it supports children in building stories together, POGO offers many avenues of exploration. It proposes new ways of looking at interaction design, of handling knowledge management systems, of enhancing electronic learning for children of all ages. It points into new directions for collaborative working in the office and collaborative, creative activities in the home.

Acknowledgements This research was funded by the European Commission within the I3-Experimental School Environment Programme. We owe particular gratitude to the Hamaïde and Tozzi School's children and teachers for their enthusiastic and generous participation to the project. A special thanks goes to Laura Polazzi, Berthe Saudelli, Claudia Fusai, Gabriele Molari, and Barbara Castelli for the effort spent on the POGO project, in particular working with teachers and children. Our thanks go also to the peer reviewers and to the editors who give us support and very helpful comments.

References

Bruner JS (1996) The culture of education. Harvard University Press, Cambridge, MA

Dean LG, Kendal RL, Schapiro SJ, Thierry B, Laland KN (2012) Identification of the social and cognitive processes underlying human cumulative culture. Science 335(6072):1114–1118

Norman DA (1998) The invisible computer. MIT Press, Cambridge, MA

Rizzo A, Burresi G, Montefoschi F, Caporali M, Giorgi R (2016) Making IoT with UDOO. Interact Des Archit(s) 1(30):95–112

UniSiena, UniLiegi (1999) Narrative and learning: school studies. POGO deliverable n° 00001/v. 1

Vygotsky LS (1998) Imagination and creativity in childhood. In: Rieber RW (ed) The collected works of L.S. Vygotsky. Plenum, New York

Yarosh S, Radu I, Hunter S, Rosenbaum E (2011) Examining values: an analysis of nine years of IDC research. In: Proceedings of the 10th international conference on interaction design and children. ACM, pp 136–144

Part VIII
"Case Studies in Design"

Chapter 32
From Usable to Enjoyable Information Displays

Sara Ljungblad, Tobias Skog and Lars Erik Holmquist

Author's Note, Funology 2

Oh, enjoyable information displays! Where did it all go wrong?

We have chosen to leave this chapter unchanged, as it represents a capsule of future-oriented optimism in a subfield of interaction design that so far has failed to deliver any perceivable value whatsoever in the forms of commercial products or useful devices, despite over two decades of cumulative work in the research community. This chapter was written at a time when there was an intense wish to move human-computer interaction from the reigning desktop paradigm to something else—call it ubiquitous, ambient, tangible or calm computing, along with a myriad of other terms meaning basically the same thing. The overbearing sense was that it was unhealthy to spend too much time staring into a desktop screen that was, as Mark Weiser, the "Father of Ubiquitous Computing", said, "isolated and isolating from the overall situation". So along with literally thousands of other researchers, we built a parade of "calm" and "ambient" information displays that were supposed to sit in the background of human perception, seamlessly integrated into the real world, soothing and beautiful, informative and useful, yet at the same time aesthetically pleasing and enjoyable. Like a fine painting, only better.

Instead, what we got were tiny personal screens that captivate so much of our attention, and get in the way of so many social and professional situations, that the desktop seems like a model of restraint by comparison.

It is interesting that the main example we chose to work with in this article was the weather forecast. There is no doubt that this is valuable information; one of the

S. Ljungblad
University of Gothenburg, Gothenburg, Sweden

T. Skog
DIRECTV Latin America, Dallas, USA

L. E. Holmquist (✉)
Northumbria University, Newcastle upon Tyne, UK
e-mail: lars.erik.holmquist@gmail.com

© Springer International Publishing AG, part of Springer Nature 2018
M. Blythe and A. Monk (eds.), *Funology 2*,
Human–Computer Interaction Series,
https://doi.org/10.1007/978-3-319-68213-6_32

most popular software categories on smartphones is weather apps. But when you have the weather in your pocket, available any time you want it, why would you want to hang it on your wall? Certainly, neither we, nor anyone else I can recall, has so far managed to construct a weather display that successfully presents the information in a more subtle and pleasing way than a straight-up map or list of numbers and icons. (Not to mention that our own design was so obscure that almost nobody could understand it, and one test subject even mistook it for a diagram of the subway system!)

At least the obstacles are not necessarily technical or financial anymore; those exotic flat-screen displays we obtained at great cost in 2002 are now available to buy for almost nothing in the nearest supermarket. New illumination technologies like multi-colour light-emitting diodes (aka RGB LEDs) promise great flexibility and low power consumption, and could create new possibilities in interior design, such as instantly changing the colour and associated mood in a room. On the other hand, some of the display technologies we forecasted, like electronic ink and electroluminescent fabrics, have not made much notable progress, except in specialized applications such as e-book readers. This indicates that either there is still a lot of technical development to do, or, perhaps more likely, the call for wall-size ambient displays is simply much smaller than we imagined.

In any case, the research community has by no means given up. Over the years, there have been literally thousands of similar projects that have aimed to move information from off the desktop and into the environment. A quick Google Scholar search reveals that there were over 200 new papers using the keyword "ambient display" just in the last year (and almost 40 that specifically mention our particular sub-branch, "informative art")! Yet, we cannot think of a single commercially successful product in this space. One would think, with all the collective effort in imagining these future artefacts, by now you should be able to go out and buy a bus-schedule Mondrian or a weather-forecast globe in the nearest gadget store (and for a short while, thanks to MIT spin-off Ambient Devices, you actually could). But the reality seems to be that ambient information displays are simply not worth the money or effort. Perhaps, much like their spiritual co-traveler tangible interfaces, they represent a too simplistic and too limited approach to solving a vastly more complex problem. We simply are not prepared to add yet another physical device for every piece of information we need to access, when a single laptop or phone does the job so much better and more compactly.

However, there may be hope for these bold ideas yet. It is worth noting that despite the emphasis on enjoyment, much of the information that was actually pumped into these displays at the time were anything but—in the article, we are talking e-mails, stock quotes and local transport departures. Before social networks, the thinking was still very much office-bound and productivity-based, despite our honest efforts to do experiments out in the "real" world. Perhaps the error was not so much with the displays themselves, but with the information we chose to represent? In an age where every breakfast bun and every baby sneeze is circulated to our nine hundred or so closest friends, there should be an opportunity to make something more personal, less operational, in the area of enjoyable information displays. And in the constant presence of aggressively beeping devices, where we are forever near-sightedly prodding our small screens in search for the next like, the next comment, the next status update, the call for calm and reflection when it comes to designing information displays is more relevant than ever. Perhaps we should just not have tried to hang them on the wall.

(Lars Erik Holmquist, April 2017)

2003 Chapter

1   Introduction

When computer screens act as public information displays, they are usually designed to present information as efficiently as possible. This is appropriate considering the traditional view of usability, where you wish to achieve optimal readability. Think of timetables for buses and trains, lists of arrivals and departures at airports, parking meters, clocks, etc. They are all efficient in the sense that they successfully communicate the information they are supposed to, but they rarely feel exciting or aesthetically pleasing (see Fig. 1). At the same time, in the same places, you will find all kinds of adornments, placed there with the sole intent to entertain and stimulate the people spending time in these places. In a similar way, we decorate our homes and offices with posters, paintings and other decorative objects to create an environment that appeals to our senses.

If people like to surround themselves with decorative objects like posters and paintings and at the same time there is a need for information presentation in the human environment, why not incorporate the enjoyment factor into the design of information displays?

Fig. 1 Electronic timetables for public transportation are seldom attractive, designed to be usable rather than enjoyable

This need for beautification is an issue that has been identified by other people, and attempts are sometimes made to make existing information displays more enjoyable. For example, at the central train station in Göteborg, Sweden, a little wooden tower holds a computer screen listing current arrival and departure times (see Fig. 2). Despite the designer's effort to make the display more enjoyable by incorporating it in a decorative object, the dull appearance of what is presented on the screen makes it stick out rather than blend in. A more sensible way of designing a decorative information display might be to use people's aesthetic preferences as a starting point when developing information displays that are intended to be enjoyable rather than just usable. Imagine, for example, a weather forecast

Fig. 2 At the central station in Göteborg a computer screen is built into a tower to make the display more esthetically pleasing

presented in the style of a painting of a well-known artist such as Piet Mondrian, displayed on a flat panel screen on your living room wall.

This is a scenario that might be more feasible than one may think. Plasma and LCD screens are already advertised for hanging on the wall like paintings (T3, 2002). They are intended for viewing TV, video and DVD, but could also become part of spaces where we socialize, read and relax, e.g. displaying a decorative picture. In this case, the computer screen works fairly well as an adornment, but the possibilities that advanced technology offer are much greater. This is especially obvious if you consider that in the near future, things like curtains, walls, lamps and tables could be augmented to present real-time information. Computer technologies are becoming more affordable and unobtrusive every year and new technologies such as "electronic inks" and electro-luminescent materials may soon allow flexible materials such as paper and fabrics to become computer displays. If people are to have various objects in their everyday environments constantly presenting information, it is necessary to make the presentation blend into the surroundings.

Pictures are often used as decoration and considered enjoyable. A picture that presents complex information can be beautiful, and at the same time be a very effective way to describe, explore and summarize a set of numbers (Tufte 1984). Information visualization uses the possibilities given by computers to present overviews and manipulate large data sets or complex data. In this way, pictures make it possible to show information that would otherwise be hidden or hard to interpret (Card et al. 1999). Furthermore, advances in computer graphics are making it possible to dynamically transform pictures, e.g. with painterly rendering and color transformations, such as those found in Adobe® Photoshop® filters.

Informative Art is a playful combination of traditional wall decorations (such as posters and paintings) and dynamic computer displays (Redström et al. 2000). A piece of informative art looks like a piece of abstract art, but instead of providing a static image its visual appearance is continuously updated to reflect some dynamically changing information. The resulting visualization is then shown on a wall-mounted display to give the impression of an ordinary painting. Installations of informative art have previously been displayed in conference settings, such as SIGGRAPH Emerging Technologies (Skog et al. 2001).

1.1   An Example of Informative Art

To illustrate the concept of informative art, we will now describe the display seen in Fig. 3. The projected image is reminiscent of a Mondrian painting and provides a visual display of e-mail traffic for a group of people working in an office. The visualization has six colored fields, each of them reflecting the e-mail traffic one person has been involved in during the last 24 h. The more e-mail a person sends and receives the larger the field representing that person gets.

The colors of the fields indicate how much time has passed since a person last sent an e-mail. A field can be of any of the three primary colors Mondrian used for

Fig. 3 Informative art: a visualization of e-mail traffic inspired by the Dutch artist Piet Mondrian

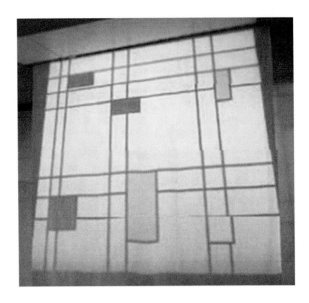

his compositions, i.e. red, yellow and blue. In our visualization, red indicates that a person is "hot", i.e. that she recently sent an e-mail. As time passes without this person sending an e-mail, the color "cools down" to yellow and finally, if the person has not sent an e-mail in a long time, the field turns blue.

The result is a "calm" display, inspired by Mark Weiser's idea of ubiquitous computing (Weiser and Seely Brown 1996) that is running in an office environment, constantly providing a group of people working there with updated information on their e-mail traffic.

2 A Case Study of Informative Art

In what sense would a piece of informative art be enjoyable? Would people actually be able to read it? To find out, we conducted a study of informative art in use. We wanted to get perceptions and opinions from as many people as possible, to explore if it would work in an everyday setting. The IT University in Göteborg, Sweden, where about 150 students are present everyday, was chosen as the setting for our study.

Before designing the piece, we conducted a pre-study involving 31 students, to develop ideas that would generate interesting data. Several suggestions were made, including timetables for public transportations, available classrooms, current news, etc. As one of the most common suggestions was a weather forecast, we chose to design a local weather forecast of Göteborg. How the information would be presented was not brought up in the pre-study, except that it would use graphical shapes rather than text. After the pre-study, the simple yet appealing structure of

Mondrian's compositions was chosen as inspiration for the piece, as it had previously been designed and was considered suitable for presenting information that was both readable and enjoyable in conference settings (Skog et al. 2001).

The resulting piece is similar to the Mondrian inspired visualization described earlier. This time, however, instead of a person's mailbox, each colored square on the display represents the weather for one day (see Fig. 4). The display is read western style, left-to-right, top-to-bottom. The first square (top-left) represents today's weather, the next top one tomorrow and so on. This gives a four-day weather forecast in the following way:

The size of each square reflects the temperature for that day. The warmer it is that day, the larger the square becomes.

The colors of a square show the weather condition of that day: yellow represents a sunny day, blue represents a rainy day and the remaining primary color, red, represents clouds.

The piece is implemented as a java application that retrieves the weather information from the "Yahoo!" online weather service (http://weather.yahoo.com). The application reflects the information on the Web dynamically and is updated every five minutes to mirror any changes. In this way, the resulting image reflects the current weather and a four-day weather forecast, while still being reminiscent of a painting in the style of Mondrian.

The visualization was shown for a week on a large flat-panel screen in an open public space at the University. We conducted two studies during this testing period, one of them being preceded by a brief explanation of the piece to a group of

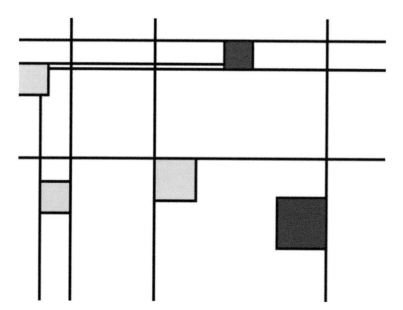

Fig. 4 A weather forecast of the Göteborg area, presented in the style of Mondrian

students. During the briefing about 30 students were told about the overall concept of informative art as well as how to read the piece. The two studies resulted in a total of 40 questionnaires, of which 15 came from students who had attended the briefing.

2.1 Comments from Students Who Attended the Briefing

When provided with a brief introduction, a majority of the students indicated that they could read the visualization, that they enjoyed it, and that it naturally blended into the background. Two students made the following comments:

> I think it is good because the information is transmitted in a simple way. I don't find the display annoying either, which is positive.

> Very good. Easy to learn and it immediately blended into the background and consciousness. Actually, I don't want it to be taken away.

However, it turned out that misinterpretations and misreadings did occur, even for students who had been to the briefing. One person made the following comment:

> An interesting and different way to show an uninteresting weather forecast. The question is how interesting it is to show a forecast with graphics. It would be more informative if it was the actual weather that was being presented.

In fact the piece was designed so that the interpretation of the coming four days temperature would be guided by the current weather, which together with its representation on the display, would serve as a frame of reference. Other people also seemed to have mistaken the display for showing a weather forecast for the coming five days, rather than today's weather and the coming four days.

One student suggested that blue could be associated with a blue sky rather than rain. Another subject found it hard to map the days to the position of the squares as they had different sizes and did not follow a line. This person also mentioned that she associated blue with cold and red with heat.

As we wanted a decorative image to become an information display, we suspected that some might find it hard to read. We also expected that people would be skeptical about this way of presenting information. Despite our fears, the comments were generally positive as the majority actually said that they liked it. One person appeared to be excited by the novelty and surprise introduced by the concept, commenting that it was:

> Fun and sensational!

2.2   Comments from Students Who Did not Attend the Briefing

Those who did not attend the briefing were usually not even aware that the piece was a weather display, let alone how to read it. As the piece only changed its visual appearance rarely, some students believed that it was a static image. One person made the following comment:

> ...this is fairly useless as a painting. The machine is meant for animated images, right?

This was not the only person who seemed to believe that the display showed desktop wallpaper or a temporary image rather than information:

> I don't know what it is but it looks like art.
>
> A bad paraphrase of a painting made by Piet Mondrian
>
> Some digital art

When we designed the piece we did not want it to be annoying or attention grabbing. Thus, instead of having gradual transitions or animations that would attract attention, the changes appeared instantly. Many who did not know that it visualized information commented that they had not noticed any changes. The fact that they did not perceive changes might have been affected by the fact that they neither expected it to reveal information, nor to change its visual appearance.

Some people clearly sensed that it probably was displaying some information, without knowing exactly what:

> It is hard to see if the pattern has changed. Whatever it shows is hard to read"
>
> I don't understand the content, but with some information about the context...

Without any given context, it seems very hard, if not impossible to know that informative art visualizes information, regardless of what it is.

Some students had their very own suggestion on what was visualized. For instance, one student believed that the piece was a map, showing cold and warm fronts. Another student had a more peculiar explanation (especially considering the University was located in Göteborg!):

> It is a network visualization, part of the subway in Stockholm.

Those who did not understand the information could still appreciate the piece as a decorative item, which some seemed to do, considering the following comments on the piece (Fig. 5):

> Inspiring!
>
> Pretty, however I think that it may be different (the attitude) depending on what kind of art you like.

Fig. 5 The local weather
forecast presented as
informative art at the IT
University in Göteborg

2.3 Discussion

Based on the results gathered from the students who had not attended the briefing,
we see that it is hard if not impossible to figure out what the information is without
an explanation. The fact that "we see things through the eyes, but we understand
things with the mind" (Solso 1999) is literally inevitable with informative art.
Whereas many information displays provide the context along with the information,
with informative art you need to know what to look for, in order to interpret it
correctly. Thus, you will only be able to read the information if you know how to
"decode" the visualization.

The majority of the students who had attended the briefing expressed in their
answers that they liked the piece; some indicated that the display was both
enjoyable and readable:

> It is a nice and easy way to get weather information

> I got an explanation right away, and then I understood what it visualized, I thought it was
> nice.

Perhaps an even more important result that this particular piece was enjoyable
and readable by some was that it suggested other applications. A student emphasized that not only did she find it inspiring, but also that the piece raised ideas on
how other information could be presented.

3 Conclusion

Informative art is not designed to display information in the most efficient way, but
rather in a fashion that appeals to people's sense of aesthetics. In general, it does not
provide the viewer with exact information, but instead gives an overview or

summary of some data, e.g. that the temperature will rise in the coming days, or an overview of the e-mail activity in a group.

It is possible, however, that people could eventually learn how to extract more detailed information from a piece of informative art so that in some cases, it could replace a numerically precise presentation. For instance, in the case of the Mondrian weather display, people might after some time learn to associate the size of a square with a certain temperature, rather than just get a sense of whether it will be colder or warmer in the next few days.

People who are not aware of the informative nature of a piece are likely to perceive informative art as pure decoration. Without an explanation it is extremely hard, perhaps even impossible, to know what a certain piece represents. This suggests the possibility of showing private information in a place where other people also spend time. For example, a piece of informative art could be placed in someone's office to give a daily update about her stock portfolio, or show the time elapsed in different projects. Visitors would look at the decoration, but not be able to read the information.

Is informative art the solution for getting rid of the boring information screens displaying arrivals and departures at airports, parking meters etc. in public places? Probably not; these information displays are designed with efficiency and readability as the most important design criteria. As they display information that has a need for exactness, this field of application is probably not the best one for something that has as a primary aim to be enjoyable. If the same information is presented for a group of people who read the timetables everyday, the time left until the next bus could indeed be presented with informative art.

In the future, informative art could give us a continuously updated overview of complex information and provide opportunities to expose and visualize information that is otherwise hidden or hard to interpret. For instance, context-related information, such as the amount of people in a building or the activity in a workplace could be visualized with graphical shapes and patterns. Such displays would be constantly running in the background in everyday environments, and could ultimately provide a form of natural or "calm" technology, combining an informative function with the aesthetic and visual appeal of traditional art.

References

Card S, Mackinlay J, Schneiderman B (1999) Readings in information visualization using vision to think. Morgan Kaufmann Publishers Inc, San Francisco, California

Redström J, Skog T, Hallnäs L (2000) Informative art: using amplified artworks as information displays. In: Proceedings of designing augmented reality environments (DARE). ACM Press

Skog T, Holmquist LE, Redström J, Hallnäs L (2001) Informative art. In: SIGGRAPH 2001 conference abstracts and applications (emerging technologies exhibition)

Solso R (1999) Cognition and the visual arts. The MIT Press, Massachusetts, Future Publishing LTD, London, UK (T3, Sept 2002)

Tufte ER (1984) The visual display of quantitative information. Graphics Press, Chesire Conneticut

Weiser M, Seely Brown J (1996) Designing calm technology. PowerGrid J 1(01). Available at: http://www.powergrid.com/1.01/calmtech.html

Yahoo weather service: http://weather.yahoo.com/

Chapter 33
Fun for All: Promoting Engagement and Participation in Community Programming Projects

Mary Beth Rosson and John M. Carroll

Author's Note, Funology 2

When we wrote this chapter in 2003, we were in the midst of a long-term engagement with residents of Blacksburg, Virginia, USA, as research participants in the Blacksburg Electronic Village project, the first Web-based community networking project in the U.S. (Carroll and Rosson 1996). During the ten years we lived in Blacksburg we continuously worked with a variety of community groups on various projects around the theme of integrating information and technology infrastructures into community life. Since that time, community informatics (the design and appropriation of computational systems in support of geo-located communities) has expanded and gained more prominence in HCI and CSCW research (Carroll and Rosson 2013). Through that same period, research on tools and methods for end-user programming and development has continued, though there has still been relatively little attention to community applications of novice programming (Paternò 2013).

Our current research in community informatics is situated in another small university town (State College, PA, USA). Our design palette has moved from more accessible desktop systems to 24 × 7 mobile computing devices and their apps as new infrastructures for community activity (Carroll et al. 2015). We decided not to revise this chapter, as it still reflects our enduring interest in motivating and elevating technology-mediated activities by residents to enhance participation, engagement and well-being of their communities. Instead we comment briefly on several themes in the chapter with illustrations from ongoing work.

With the emergence and pervasive adoption of smartphones, access to information and computation has become omnipresent for many individuals. At the same time, the footprint of the devices and the activities they can host has shrunk enormously—especially for younger community residents, there are strong expectations that all applications will be usable on a smartphone. This has shifted

M. B. Rosson (✉) · J. M. Carroll
The Pennsylvania State University, State College, PA, USA
e-mail: mrosson@ist.psu.edu

© Springer International Publishing AG, part of Springer Nature 2018
M. Blythe and A. Monk (eds.), *Funology 2*,
Human–Computer Interaction Series,
https://doi.org/10.1007/978-3-319-68213-6_33

the focus from activities that could involve significant "construction" effort (e.g., programming a visual simulation) to ones that leverage information that is collected automatically (e.g., from sensors and logs) or incidentally as a side effect of other activities (e.g., interactions via social media). As a result, many of the future opportunities for community applications will have a data-centric orientation. The concept of "supra-thresholding" (Carroll et al. 2015) is an example: it refers to computational methods for collecting and aggregating community information that on its own would be too sparse and distributed to grasp, but that can raise awareness and promote evidence-based decision making.

A central proposition in our original chapter is that by making programming fun we can attract and engage a more diverse constituency of community members in programming-related activities (e.g., senior citizens, youth). Fun continues to be an important user experience design goal for many situations and we always consider ways to enhance pleasure or enjoyment when we build community apps. But our approach to "fun" has become more nuanced and embedded in the particular apps we create. Examples include remembering and contributing to the history of a community location (Han et al. 2014a), contrasting and appreciating different views of community events (Han et al. 2014b), sharing a locale-specific image or thought that makes you happy in the moment (Carroll et al. 2015), and helping groups of elders coordinate everyday projects and events together (Wirth et al. 2016).

Another theme in the original chapter was the potential attractiveness of inter-generational activities, in particular collaboration across age groups who bring differing sources of motivation or expertise and thereby take on complementary roles. Intergenerational collaboration seems a highly significant community resource today in a context of shrinking and threatened social services, and a rapidly growing demographic of healthy aging people. We have investigated the concept of "developmental learning community": collectives comprised of indi-viduals with different knowledge and skill, and with a commitment to helping one another develop skills and knowledge (Rosson and Carroll 2006, 2013). In this broader view, including multiple generations in a new community activity is one step toward creating a fruitful context for shared learning and growth.

2003 Chapter

1 Introduction

1.1 *Programming as a Community Activity*

The increasing pervasiveness of community networks has opened new channels for community interaction (Carroll and Rosson 2001). Residents may email questions or suggestions to town officials or leaders of other organizations (Cohill and Kavanaugh 1997); parents may contact public school teachers online, and track their children's weekly activities through regular email bulletins; community elders may share their memories and wisdom with community youth (Carroll et al. 1999; Ellis and Bruckman 2001). However, such activities are discretionary: residents must first believe that the new opportunities will be rewarding, if they are to take the time to investigate and participate (Rosson et al. 2002a).

We are exploring the motivational characteristics of community-oriented collaboration in the CommunitySims project, where diverse members of our local community cooperatively design and build visual simulations that raise or illustrate community issues (Rosson et al. 2002a). Participants plan, share, and discuss their projects via a Web site [communitySims.cs.vt.edu]. Our initial studies have centred on interactions between middle school children and community elders. Prior work has shown that children of this age are able and motivated to work with visual simulations (Rader et al. 1997); elders may be less likely to become simulation programmers, but several studies have demonstrated their willingness and availability for youth-mentoring activities (Ellis and Bruckman 2001; Oneill and Gomez 1998; Wissman 2002).

Our earlier papers have described the problems experienced by students and elderly residents learning to use Stagecast Creator (Seals et al. 2002; Wissman 2002), the participatory design of community simulations (Rosson et al. 2002a), and the nature of cross-generational collaboration (Rosson et al. 2002b). In this brief case study, we focus more specifically on participants' subjective reactions to the community-oriented simulations and to the process of simulation programming.

2 The CommunitySims Project

2.1 The Stagecast Creator Environment

CommunitySims projects are constructed with Stagecast Creator, a visual programming environment designed to allow children and other nonprogrammers to build simulations by example (Smith and Cypher 1999). Users construct simulations by creating a "stage" (a rectangular grid) of animated characters. Each character is given one or more visual appearances, along with a set of rules enabling them to move, change appearance, create or delete other characters, and so on.

Figure 1 shows a CommunitySims project—a schoolyard fight. The students and the teacher are characters, as is the door. The visual *before-after rule* in the lower part of the figure illustrates the visual programming paradigm: if the "before" condition for a rule is met (the visual state of the world and the conditions specified on the left of the rule), the "after" actions are performed (in this case, each of the actions changes the character's appearance). The starting condition always specifies a visual context (here, the two boys next to each other, facing forward), though it may also specify values for variables defined globally or for each character. A key

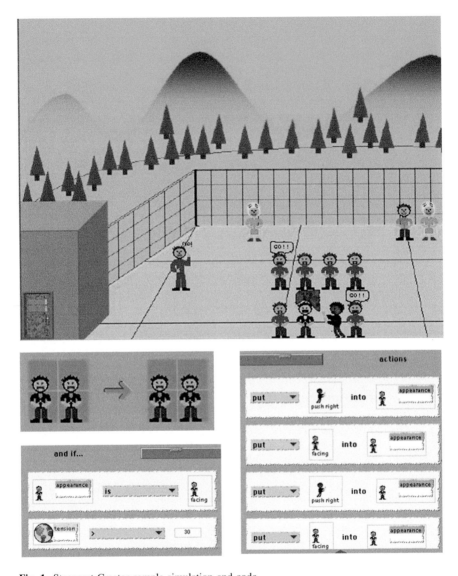

Fig. 1 Stagecast Creator sample simulation and code

challenge in Creator programming is to map simulation objects and behaviours onto
visual effects (Rosson et al. 2002a; Seals et al. 2002). For instance, in the
schoolyard fight, changes in "tension" cause the boys to begin pushing and hitting
each other.

2.2 *Cross-Generational Programming Workshops*

To study the collaborations between students and adult residents, we organized two community simulation programming workshops: three women and four boys came to the first workshop; one woman and three girls to the second.[1] Attendees were recruited through email messages or phone calls; each individual was offered a small stipend ($30) as a thank-you for participating in the one-day event.

We wanted the workshops to be a friendly and supportive environment in which middle school students and elderly women could meet and learn about one another, and collaborate on programming projects. Although most of the students knew each other in advance from school, and several of the women knew one another from other community activities, the students and women had never met; an important side goal of the workshop was to introduce them to each other.

Our research team was available to coach and answer questions as needed, so that participants did not feel that they were being "tested"; instead we encouraged them to try out the visual programming examples and tools, and to explore their own ideas for community simulations. Although we planned to characterize the workshop activities for research purposes, we also hoped to initiate and facilitate a small set of informed and motivated community members who could participate in future CommunitySims activities.

All of the participants had previously been introduced to the Creator programming. The women received their training as part of an experiment comparing the efficacy of two different tutorials (Wissman 2002); the students were trained during a study of collaborative learning with a minimalist tutorial (Seals et al. 2002). We limited the adult participation to women because our earlier work had suggested that elderly women have more intrinsic interest in the visual programming supported by Stagecast Creator than their male counterparts (Wissman 2002).

The students (ranging in age from 12 to 14) reported greater experience with computing than the women (all over 70 years of age). For example, students have had more years of computer use and describe a greater variety of computer-based activities, than the women. An important specific difference is the relative experience with "programming" activities, such as the creation of spreadsheets and Web pages. The students also reported experience with graphics or drawing tools; none of the women had used such tools.

Both workshops followed the same schedule and provided participants with the same materials and activities:

- Introduction to CommunitySims; brief statements of personal interests and background with computing.
- Use of CommunitySims Web site; logging on, opening, running, and commenting on the example simulations. The example simulations were:

---

[1]Two additional elderly women were scheduled to participate in the second workshop, but last-minute personal problems prevented them from attending.

- *Smoking Kids* (two kids smoke, get sick, collapse); *Schoolyard Fight* (kids argue, yell, fight until teacher arrives); *Flirting or Hurting* (cute guy harasses girls in hallway); *Noise Pollution* (noisy neighbourhood party); *Smart Road* (weather affects road conditions); *Cliques* (groups form on playground); and *Classroom Bully* (a boy beats up on other kids).

- Survey of subjective reactions to the example simulations, as well as on a larger set of hypothetical simulation features.
- Refresher tutorial on Creator; review of basic skills as well as more advanced techniques.
- Group formation, with each woman joining one or more students; due to absent participants, two girls were paired with researchers.
- Collaborative work extending 1–2 example simulations.
- Collaborative work generating and elaborating ideas for 1–2 new simulations.
- Collaborative work building a new simulation.
- Survey of general reflections and project goals.

Throughout the day, research team members assisted attendees and took notes. We used two digital recorders to capture the discussion among participants. In the following section, we discuss participants' general reactions to the workshop activities, along with more specific reactions to the example simulations and to a set of hypothetical simulation features. A more extensive analysis of the collaboration episodes between the students and women can be found in Rosson et al. (2002b).

3 Participant Reactions

3.1 General Reactions to Workshop Activities

At the end of each workshop, participants completed a survey that included questions about how easy it had been to extend or build new simulations, and what might be done to facilitate their shared programming with partners. The group was moderately positive about the overall collaboration experience (averaging 3.73 on a 5-point scale). During the workshop, we noted many cases in which students were advising one another, observing each other's progress, and in general promoting a sense of activity and excitement in the projects underway. However, several participants voiced concerns about the difference in ages:

W2: "I was overwhelmed and could not keep up with teenagers."

W3: "The young folks are so aggressive with the computer."

G1: "Just make sure that your partner is someone of around the same age so you will agree on more things."

These comments caused us to speculate that real-time collaborative programming may not be the most effective way to establish cross-generational interaction.

Our future research will focus on asynchronous collaboration where community elders suggest topics and guide students toward community issues; pair or small-group synchronous collaboration on programming projects will be limited to same-age participants.

Participants also rated their interest in future work with CommunitySims activities. Figure 2 graphs responses to four questions: the extent to which Creator simulations help to build community; whether participants want to build, or to refine simulations; and how well they know the Creator tool.[2] Whereas the students were moderately positive in these final ratings, the women's ratings suggest more uncertainty about future activities. Notably, the average student rating of Creator understanding was 4.0 whereas the women's average was 2.5. However, the women seem to have accepted our community education goals more than the students; the women's agreement that Creator simulations can help to build community was 3.25, compared to a rating of 2.71 for the students.

Participants' open-ended comments reinforced the patterns seen in the rating data. All 11 participants answered "yes" to a question asking if they wanted to continue in the CommunitySims project. But the nature of participants' future plans varied: three of the four boys tied their interest to game development (B1: "I'd like to make games out of existing sims"), whereas all four of the girls conveyed more general positive reactions (G3: "Yeah, I though it was really fun when we got to make our own world and that kinda stuff"). Although the women also answered affirmatively, each was careful to qualify her future involvement (W4: "Yes, but I need to have more knowledge about creating a CommunitySims project").

3.2 Reactions to the Example Simulations

A specific research goal of the CommunitySims workshops was to study the features of simulations that make them more or less appealing to the middle school students and the women. One source of relevant data comes from participants' use of and reactions to the seven example simulations. These simulations were explored during the initial use of the CommunitySims Web site, and were also used during the "refresher" training provided in the first few hours of the workshop.

During exploration of the example simulations, participants were encouraged to leave comments; across all seven examples, 22 comments were made by students, 4 by women. When we examined these comments, we found that the women commented on the community issue the simulation had been built to raise. For instance, W2 reacted to Noise Pollution: "I agree that courtesy demands speaking to the neighbours first before calling police. Also, where is a responsible adult?" In

---

[2]We provide average ratings and associated graphs as a way to point out interesting patterns, but have refrained from statistical tests or more conclusive inferences due to the very small sample size.

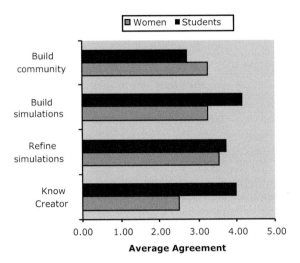

Fig. 2 Final ratings women ($N = 4$) and students ($N = 7$) regarding future work with CommunitySims

contrast, the students focused on simulation usability or realism problems (e.g., "OK...I don't see what is happening here. This one is too short to understand."). We speculate that the women took the topics of our example simulations more seriously, and that they were more motivated to initiate community-oriented discussion. This is consistent with their somewhat stronger agreement that Creator simulations can help to promote community discussion (recall Fig. 2).

After exploring the CommunitySims Web site, participants were asked to choose the example simulation that they thought was most *fun* to use, most *educational*, and *least useful*. There was considerable agreement about what was most educational (Smoking Kids, 7/11) and least useful (Smart Road, 8/11). There was less agreement about what was most fun, although 4/7 students chose Classroom Bully because "it was funny". In general, participants reported that they preferred simulations that had a clear message, or that "did" something. It is difficult to visualize the impact of a smart road (it measures changes in a car's movements) or a noisy neighbourhood party, whereas it is very obvious that a bully has hit someone, or that a kid has collapsed after smoking for a while.

3.3 Ratings of Hypothetical Simulation Characteristics

Also after exploring the sample simulations, participants completed 21 scales rating the extent to which a hypothetical simulation characteristic would make it more "fun". We created the list of possible characteristics by reviewing the simulations we had built or viewed, and by brainstorming characteristics we felt might be attractive. We included concepts we expected to be appealing to middle school students (e.g., cute, silly), but also "serious" characteristics that we thought might increase enjoyment by adults (e.g., educational, matching the real world).

Fig. 3 Ratings of hypothetical simulation characteristics, for boys ($N = 4$), girls ($N = 3$), and women ($N = 4$). The scales rated agreement (1 = Strongly Disagree; 5 = Strongly Agree) that the given feature would make a simulation more fun

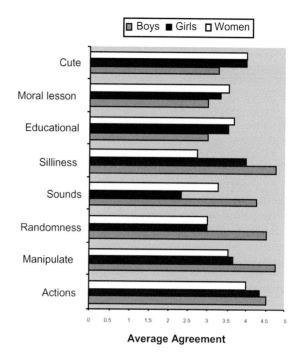

Many of the features produced fairly neutral responses, and did not evoke different responses from the boys, girls, and women. However, a few features produced more interesting patterns (Fig. 3). For example, the highest "fun" rating overall (4.27) was given to "The different actions and movements of the characters"; this rating also produced considerable agreement across the boys, girls, and women. This finding has a simple interpretation—the participants complained that several of the example simulations were boring or had too little action, so it is possible that frustration with these simulations caused action to be highly valued.

At the same time, Fig. 3 suggests that not all of the characteristics evoked the same ratings from the different subgroups. For instance, the boys were more positive than the women or girls about simulations that could be manipulated, or that included random behaviour or sounds. Although both the boys and girls gave "silliness" a relatively high rating, the boys tended to have less favourable reactions to educational themes, moral lessons, or "cuteness" in a simulation. Surprisingly, across all 21 ratings, the girls' reactions were more similar to those of the women ($r = 0.37$) than to the boys' ($r = 0.03$). This leads us to speculate that gender-related concerns or biases may better predict enjoyment of different simulations than age.

A simple interpretation of the boys' ratings is that they believe they will have more fun with simulations that appear and behave like computer games (randomness, manipulation, sounds). This interpretation is consistent with the general observation that all of the boys attempted to add game-like features to the example simulations (e.g., a flying bird that "bombs" other birds). In general, our results suggest a new requirement for the ongoing work of generating and refining community-related simulations—namely how to raise or provoke a real world community issue while also making the simulation interactive and game-like.

4 Discussion and Future Work

4.1 Summary of Workshops and Reactions

We conducted two workshops to investigate community residents' reactions to cross-generational programming, and to explore features of community-oriented simulations that might make them more or less intrinsically interesting to different user groups. Despite the small group size, we identified several interesting patterns in reactions to the community-oriented simulations.

All of the users agreed that the best simulations are those where the characters "do" something. However, the boys clearly viewed Creator programming as more of a computer game than did the girls or women. We observed that the boys spent considerable time making the example simulations more game-like. With respect to hypothetical features, the boys felt that features such as character manipulation, silliness, sounds, and randomness would make simulations more fun—the same features that would cause the simulations to be more like computer games. At the end of the day, several of the boys expressed an interest in converting CommunitySims projects into games.

4.2 Promoting Engagement and Participation in Community Programming

Our long-term goal is to promote the development and discussion of community simulations that are intrinsically interesting to all segments of the population. Given the findings of these workshops, we are beginning to explore techniques for making a simulation seem more like a game, but still express community issues. For example, the Noise Pollution simulation has been enhanced to include greater variety in the sounds it uses, to enable viewers to "summon" the police, and to send off the party-goers when the police arrive. Our design challenge is to find a way to make the programming fun without trivializing the underlying community issues.

As expected, the women seemed to take our vision of community interaction and education more seriously than the students. During their work with the students, they helped to ensure that projects contained community-specific content. They also contributed issue-oriented comments to the example simulations, and at the end of the day were more likely to agree that simulations could provoke community discussion. Student contributions tended to be more individualistic and game-oriented, emphasizing the importance of modelling by adult community members.

The differing expectations and reactions of our workshop participants has led to a more refined view of community participation in the CommunitySims project. We plan to recruit adult participants by emphasizing the importance of community discussion, pointing to the programming projects as a way of attracting youth to the topics. Where possible, we will join older residents with students willing to take an idea for a project and build it. At the same time, we will recruit students by trying to give the projects more of a game-like character, or perhaps challenging the students to make the projects game-like but still related to community concerns.

Community networks leverage and develop local resources through online collective endeavour. One of the most precious resources any community has is its elders. This has always been true, but today it may be more true. Our elders have been called the civic generation because of their lifelong commitment to community issues and institutions (Putnam 2000). CommunitySims is only a first step, but its goal is to leverage and develop this precious resource through mutually-engaging, cross-generational, collaborative learning.

Acknowledgements This work was supported by the National Science Foundation (NSF ITR EIA-0081102). We thank Cheryl Seals, Justin Gortner, Tracy Lewis, Jason Snook, and Erik Dooley for their help in planning and conducting the workshops.

References

Carroll JM, Rosson MB (1996) Developing the blacksburg electronic village. Communications of the ACM 39(12):69–74

Carroll JM, Rosson MB (2001) Better home shopping or new democracy? Evaluating community network outcomes. In: Proceedings of CHI 2001. ACM, New York, pp. 372–379

Carroll JM, Rosson MB (2013) Wild at home: the neighborhood as a living laboratory for HCI. ACM Trans Comput Hum Interact 20(3), Article 16

Carroll JM, Rosson MB, VanMetre CA, Kengeri R, Darshani M (1999) Blacksburg Nostalgia: a community history archive. In: Sasse MA, Johnson C (eds) Proceedings of INTERACT 99. IFIP, Amsterdam, pp 637–647

Carroll JM, Hoffman B, Han K, Rosson MB (2015) Reviving community networks: hyperlocality and suprathresholding in Web 2.0 designs. Pers Ubiquit Comput 19(2):477–491

Cohill AM, Kavanaugh AL (1997) Community networks: lessons from Blacksburg, Virginia. Artech House, Norwood

Ellis JB, Bruckman AS (2001) Designing Palaver Tree online: supporting social roles in a community of oral history. In: Proceedings of CHI'01. ACM, New York, pp 474–481

Han K, Shih PC, Rosson MB, Carroll JM (2014a, February) Enhancing community awareness of and participation in local heritage with a mobile application. In: Proceedings of the 17th ACM conference on computer supported cooperative work & social computing. ACM, New York, pp 1144–1155

Han K, Shih PC, Carroll JM (2014b) Local news chatter: augmenting community news by aggregating hyperlocal microblog content in a tag cloud. Int J Hum Comput Interact 30 (12):1003–1014

Oneill DK, Gomez LM (1998) Sustaining mentoring relationships online. In: Proceedings of CSCW'98. ACM, New York, pp 325–334

Paternò F (2013) End user development: Survey of an emerging field for empowering people. ISRN Software Engineering, 2013

Putnam R (2000) Bowling alone: the collapse and revival of American community. Simon & Schuster, New York

Rader C, Brand C, Lewis C (1997) Degree of comprehension: children understanding visual programming environment. In: Proceedings of CHI 97. ACM, New York, pp 351–358

Rosson MB, Carroll JM (2006) Developmental learning communities. J Community Inf 2(2)

Rosson MB, Carroll JM (2013) Developing an online community for women in computer and information sciences: a design rationale analysis. AIS Trans Hum Comput Interact 5(1):6–27

Rosson MB, Carroll JM, Seals CD, Lewis TL (2002a) Community design of community simulations. In: Proceedings of DIS 2002. ACM, New York, pp 74–83

Rosson MB, Seals CD, Carroll JM, Gortner J (2002b) Cross-generational learning and collaboration in simulation programming. Unpublished manuscript, Center for Human-Computer Interaction, Virginia Tech, Blacksburg, VA, 24061

Seals C, Rosson MB, Carroll JM, Lewis TL (2002) Fun learning Stagecast Creator: an exercise in minimalism and collaboration. In: Proceedings of HCC '02. IEEE, New York, pp 177–186

Smith DC, Cypher A (1999) Making programming easier for children. In: Druin A (ed) The design of children's technology. Morgan Kaufmann, San Francisco, pp 201–222

Wirth RJ, Yuan CW, Hanrahan BV, Carroll JM, Rosson MB, Bindá J (2016, November) Exploring interactive surface designs for eliciting social activity from elderly adults. In: Proceedings of the 2016 ACM on interactive surfaces and spaces. ACM, New York, pp 403–408

Wissman J (2002) Examining minimalism in training older adults in a software environment. Unpublished masters thesis, Department of Industrial and Systems Engineering. Virginia Tech, Blacksburg, VA

Chapter 34
Deconstructing Ghosts

Jonathan Sykes and Richard Wiseman

2003 Chapter

1 Fear Is Fun

Initially, you might not consider fear and fun as being emotionally related. However, by looking at some activities in which people participate, it is clear that fear can be fun. Rock climbing, bungee jumping, sky-diving, car racing, mountain biking, paint- balling, and white water rafting are all activities where fear is a significant factor in the enjoyment of the sport. It is the presence of adrenalin in the bloodstream, caused by fear, which attracts so many people to such activities.

If fear is to be used as the fun element upon which a video game is constructed, it is important that we choose the correct fear mechanism for exploitation. Gray (1971) assigns fears to one of two categories: learned or innate. Learned fears are those that we can attribute to a previous experience. A fear of hamsters, for example, may occur if bitten by a hamster during childhood. Because *learned* fears are directly linked to an individual's experience, they are not prevalent throughout society. In contrast, those fears which Gray argues to be 'innate' are found to affect a sizeable proportion of the population. Such phobias include fear of snakes, spiders, the dark, strangers, loud noises, falling, heights, death, imaginary creatures, etc. Because mass media, such as film and computer-games, attempt to appeal to the widest possible audience, it is 'innate' fears that are typically exploited.

J. Sykes (✉)
Brighton, UK
e-mail: drjonsyjes@wildux.com

R. Wiseman
University of Hertfordshire, Hatfield, UK

© Springer International Publishing AG, part of Springer Nature 2018   519
M. Blythe and A. Monk (eds.), *Funology 2*,
Human–Computer Interaction Series,
https://doi.org/10.1007/978-3-319-68213-6_34

The fear of ghosts and other supernatural entities is a widespread phenomenon. It is a fear already exploited as a means of having fun. Halloween is a celebration of all things that scare us. It is a time where people delight in both causing fear in others, and the experience of fear itself. Tales of the supernatural are common across all mediums of entertainment, including literature, cinema, theatre, television, and radio. This chapter looks at the possibility of enhancing the fun associated with 'horror' based computer-games by deconstructing paranormal experiences.

1.1 Deconstructing Ghosts

As part of the Edinburgh Science Festival 2001, Wiseman et al. (2002) investigated whether the reputably haunted rooms within rooms in the vaults (which do not have the same reputation for being haunted). In total, 220 volunteers spent approximately five minutes alone in one of ten vaults. They were then asked to record any unusual occurrences and rate how frightening the experience was. The participants were 'blind' to the reputation of the various vaults; i.e. they were not aware of which were thought to be haunted and which were not. Nonetheless, Wiseman et al.'s findings show a significantly strong correlation between rooms which were rated by participants as being 'more frightening', and rooms that have a reputation for being haunted.

Wiseman et al. comment on structural similarities between rooms that are reputably haunted. Room size was considered to be potential factors which may have led participants to feel unnerved. This is supported by Kaplan's evolutionary theory of environmental psychology, which argues that humans feel uncomfortable in environments where they feel exposed to the threat of attack—such as large open spaces, and dark, isolated environments (Kaplan 1987). If this is the case, it would seem plausible that in extreme circumstances the human brain could perceive illusory, paranormal-like experiences which convince the agent to flee to a safer place.

If stimuli which induce feelings of a supernatural presence in the real world can be introduced into the virtual world, computer game players will have the chance to enjoy the experience in a safe and controlled environment.

2   The Virtual Vaults

To determine whether the visual properties of a reputably 'haunted' environment might induce fear, a computer generated representation of Edinburgh's underground vaults has been constructed. This study is a repeat of Wiseman et al.'s (2002) investigation. However, instead of visiting the 'real' vaults, participants report how they feel in the computer simulated environment.

2.1 *Method*

As part of the Edinburgh Science Festival 2002, members of the public were invited to tour the virtual vaults. In total, thirty-nine people agreed to participate in the pilot study. Their ages ranged from 12 to 59. Because of the educative nature of the festival, many of the participants were at the younger end of the scale, with the mean age being 27. Children under the age of 12 were not allowed to participate, and children between the ages of 12 and 16 required guardian permission before they could take part.

Each session would begin with an initial briefing, informing participants of the previous study by Wiseman et al. (2002) and a short history of the hauntings associated with the vaults. The purpose of the briefing was to gain informed consent, and to increase anxiety in a way that mirrors the traditional ghost tour. To further increase anxiety, the tour of the virtual vaults was conducted on the stage of a local theatre, behind a heavy black curtain. Except for a narrow green spot light, the stage was in darkness. This maintained the 'creepy' context, and also enhanced the display image by increasing the contrast on the visual display unit.

When seated, participants were asked to wear a head-mounted display (HMD). Using the HMD the viewer can browse the simulated world in the same way you view the real world. As the head is moved, the screen updates the visual scene to represent the new perspective. This one-to-one mapping between head movement and screen display increases the level of presence, and can make the participant feel as though they are actually inside the virtual environment.

Movement around the virtual vaults was conducted by the experimenter. This was to allow the participant to concentrate on the visual experience, rather than on the control mechanism. The vault first experienced by participants alternated between participants, as did the direction of exposure. Therefore each trial would begin in a different vault to the last, and the tour would follow either a clockwise or anti-clockwise route through the environment. Participants were led to the centre of each vault where they were left to view their surroundings for approximately 30 s, at which point the experimenter would guide the participant through the virtual environment, into the next vault.

When viewing each vault the participant was asked to vocalise their experiences. They were asked to report whether they could see something, hear something, or feel something that might be considered unusual. The experiences were then recorded for later analysis.

2.2 *Results*

Of the thirty-nine participants who took part in the study, 64% reported an 'unusual' experience. Experiences ranged from apparitional sightings, a sudden chill on entering a virtual vault, feeling of another presence inside the vault, and the perception of breathing on the back of the neck, to the report of an itch, perceptions of levitation, feelings of discomfort and an increase in anxiety.

It may seem intuitive that younger participants would be more scared, and therefore more likely to report an experience. However, it can be seen in Table 1 that the ratio of experiences across age groups was largely consistent, with a slight drop (42%) for the 16–30 age group. There were no reported experiences by participants aged above fifty, but the number of participants within this age group was very low.

It is clear from Table 2 that the distribution of experiences in the virtual vaults, and the type of experience reported is not consistent across the vaults. Vault nine has a higher frequency of experiences, and they are predominantly visual in nature. This is in keeping with the paranormal phenomena reported in the real vaults, where vault nine is renowned for its high number of reported experiences. However, Spearman's analyses of correlation found no significant correlation between the distribution of experiences in the real vaults, and the distribution of experiences reported in the virtual vaults.

Table 1 Distribution of participants and experiences by age group

Age group	No. of participants	No. reporting experience	Percentage reporting at least one experience (%)
12 > 16	11	9	82
16 > 30	12	5	42
30 > 40	6	6	100
40 > 50	7	5	71
50>	3	0	0

Table 2 Distribution of experience across the real and the virtual vaults

Vault	Mean number of experiences		Number of experiences reported in virtual vaults		
Number	Real world	Virtual world	Visual	Auditory	Somatic
1	1.000	0.077	2	0	1
2	0.440	0.103	2	0	2
3	0.913	0.103	3	0	1
4	1.200	0.128	2	0	3
5	0.625	0.179	3	1	3
6	0.330	0.103	2	0	1
7	0.647	0.026	0	0	1
8	0.632	0.230	6	0	3
9	1.130	0.310	9	1	2
10	0.750	0.231	4	1	4

2.3 Discussion

If the experiences reported in Wiseman's initial experiment were evoked entirely by the visual scene we would expect the distribution of experiences reported in the virtual vaults to map onto the distribution of experiences found in the real vaults. However, this was not the case. We can therefore conclude that either the current simulation fails to deliver the same visual complexity necessary to produce such a response, or that the visual structure is not the exclusive factor in the formation of such experiences. The latter of the two explanations is perhaps more appropriate given that paranormal-like responses were recorded during the study, and that visual experiences were dominant, rather than olfactory (of which there were none) or auditory (of which there were 3). There are many sensory cues in the real environment that are difficult to simulate in a virtual environment. In the real vaults there is an unusual scent and a taste created by the dust, the floor is uneven under foot, there are air pockets which flow through some rooms, and there is the occasional muffled noise which can be heard from the street, three storeys above.

Had such stimuli been included in the simulation, the distribution of reports in the virtual world may have better matched the distribution of responses in the real world.

Although there was no mapping between the real and virtual worlds, the study proved to be particularly successful in producing paranormal-like experiences. Even though participants were aware that the scene they had witnessed was merely a simulation, and therefore presumably free of supernatural forces, they would consistently report experiences normally associated with haunted environments. An explanation of this phenomenon might lie in the participant's anxiety level, and the context provided during the briefing session.

During the briefing session participants were provided with reading material that might build on anxieties already present in the reader. Additionally, participants were isolated behind a curtain and asked to wear the HMD (a piece of technology which is still largely surrounded in mystery), thus raising anxiety levels even further. It is possible that participants were interpreting their anxiety with respect to the context surrounding the experiment, which in turn affected their perception of the virtual environment (Schachter and Singer 1962). If participants were aware of their own anxiety during the tour, through cognitive evaluation of their physical state they could conclude that they were feeling scared. What is particularly interesting is how the feeling of fear was so powerful that for many the result was a misinterpretation of the stimuli to the point of having paranormal experiences. The projection of anxiety onto ambiguous stimuli is supported by the finding that more visually based experiences occurred in room nine, a room that is visually complex, providing more opportunities to misinterpret an image.

It is clear that some environments can make us uneasy, and it is easy to see how we can frighten ourselves into seeing almost anything. The question is how this information might be used to build games that are scary and fun? If context is as important as suggested here, the building of plausible and involving narratives will help raise the player's level of anxiety and provide a framework to structure their experience. If a strong narrative is interwoven with an environment that builds an expectation of fear, and increases the player's anxiety, it seems that fear is a natural product.

Acknowledgements Mercat Tours (http://www.mercattours.com) for providing continued access to the underground vaults in Edinburgh. Ian Baker, Sarah Haywood, Rosie Pragnell and Caroline Parker for their participation on this project.

References

Gray J (1971) Psychology of fear and stress. Weidenfeld & Nicolson, London
Kaplan S (1987) Aesthetics, affect, and cognition. Environ Behav 19(1):3–32
Schachter S, Singer J (1962) Cognitive, social and physiological determinants of emotional state. Psychol Rev 63:379–399
Wiseman R, Watt C, Stevens P, Greening E, O'Keeffe C (2002) An investigation into alleged hauntings. Br J Psychol (accepted for publication)

Chapter 35
Interfacing the Narrative Experience

Jennica Falk

2003 Chapter

1 Introduction

I have a research interest in interfaces to games that are played, not on computers, but in the physical environment and that ultimately transform the world into a game board for computer games. Of specific interest to this agenda are activities that are narrative and social in their nature, not only in the interaction between people, but also in people's interaction with the physical world. Having spent time role-playing in online environments, in awe of their mechanisms for story generation and interactive game worlds, I still have to argue their failure to provide convincing and truly interactive environments for narrative experiences. Put differently, the unmistakable division between character and player, and between character environment and player environment in online role-playing, characteristically fail to induce a desirable level of suspension of disbelief. In contrast, *live role-playing games* offer particularly relevant examples of games where the physical world is adapted as a mature interface to an engaging and creative immersion in an interactive, social, and narrative context. They support social and collective exercises in emergent narrative creation where every participant, is part of the design effort. These narratives take place in a magical and imaginary domain in the cross-section between physical reality and fantastic fiction offering the kind of immersion that most interactive narratives promise as a technical goal, but have yet to deliver, where there is no physical division between player, character, and narrative. Some might argue that this level of immersion is the holy grail of interactive fiction and

J. Falk (✉)
NNIT, Copenhagen, Denmark
e-mail: jennica.falk@gmail.com

© Springer International Publishing AG, part of Springer Nature 2018
M. Blythe and A. Monk (eds.), *Funology 2*,
Human–Computer Interaction Series,
https://doi.org/10.1007/978-3-319-68213-6_35

indeed entertainment, where the narrative thread, or content if you will, is embedded in physical locations and in objects around us, creating a tangible, ubiquitous, and even context-sensitive interface for the participants or players to unleash at.

In this chapter, I share observations and analysis drawn from participating in live role-playing events, primarily in the *Lorien Trust* game system. By studying the use of artefacts and physical game locations in this process, and observing how stories emerge from the interaction between players I hope to inform design thinking about interfaces to *narrative experiences* and to provide insights relevant to the topic of this book—*design for enjoyment*.

2 Live Role-Playing Games

To define live role-playing (LRP) games is a knotty task, but for the purposes of this text, a definition that puts forward its essence is that LRP is a *dramatic* and *narrative* game form in which players portray fictional characters that come to life in a *web of stories*. The narrative emerges in the interaction between *characters*, *objects*, and *physical locations*. By dramatic, it is implied that roles are assumed in person rather than through virtual or abstract means, and by narrative it is implied that a main product or goal of these games is of a story nature (Fig. 1).

Fig. 1 Live role-players at Lorien Trust's 'The Gathering' in 2002

Almost exclusively fictitious, the purpose of LRP games is primarily the dramatization of a make-believe world. They are fiction adventures, and although governed by a body of rules and background information that frame both individual role-play and the progression of the overall storyline, they are predominantly improvisational. Characters with ambitions and professions, dreams and hopes, come together to interact, react, and impact the narrative outcome in an unrehearsed but still measured fashion. While games are not stories, the narrative element of LRP is nothing less than pervasive, and the stories that are generated in the role-play are interactive stories that are lived and experienced with all senses. LRP spawns a highly engaging and immersive narrative environment in which the story is read and written simultaneously, inviting participation and providing the guidance to allow players to perform and partake by putting their creative imagination to work.

2.1 *Live Versus Online Role-Playing*

An important difference between online role-playing games (RPG) and LRP is clearly noticeable in how LRP players interact directly with the narrative, while online environments do not allow for the same sophisticated sensory engagement. This is primarily due to the fact that our means of interacting with the real world (our perception of and navigation in space, sensory input, how we organize and manipulate artefacts, and so on) are transformed into abstractions in online environments. This highlights the substantial difference in using the world as a *metaphor* for interaction, as in the case of online games, and using it as a *medium* for interaction, as in the case of LRP where the immediacy of the physical world is extremely relevant. The online role-player is a puppeteer while the LRP player is a person going through a transformation into a character. One of the reasons LRP worlds are so engaging is that the experiences a character is subjected to also happen to its player.

3 LRP Case Study

Drawing from the understanding that the physical world is a powerful facilitator of LRP games, a compelling design challenge for digital narratives and games is to extend them into the physical domain. If we create game worlds where points of interaction are not confined to a virtual environment and a personal interface, but rather support distributed and tangible interaction qualities, we can make advances towards truly immersive narrative experiences. Some previous work within this agenda includes the *Tangible MUD* (Falk 2002) project where computer game mechanisms were designed to reside in physical objects—a spell-book and a desk lamp—and the *Pirates!* project (Björk et al. 2001) in which physical locations were mapped to computer game locations. The key motivation in the Tangible MUD project was to unveil what kinds of sensory gratification tangible interaction points can add to computer gameplay. The key motivation with *Pirates!* was to restore the social dimension of play to computer games by bringing the players back to the physical environment. Integrating tangible game objects and locations, future entertainment and interactive narratives will provide a "sensory proof" of its reality that is in stark contrast to the reality the interaction space of graphical games offer. LRP games offer the opportunity to further this research in that they allow us to study how the richness of the physical world supports and enhances engagement and story creation within the game context.

3.1 Players, Costume, and Character Identity

The nature of character interaction in LRP games typically causes them to depend on social structures that cannot be formalized under a rules system. LRP players bring their personal attributes and social skills to their characters, which become highly viable resources in their role-play. Additionally, a character concept typically evolves through the interaction with other characters. This suggests a strong dependence on active and interactive players, which in turn indicates that the challenge of the game is less technical, and more of an interpersonal nature.

Three aspects of how players transform into game personas are of particular relevance to this study. Firstly, attention to detail in costume and accessories and other personal props is typically great, to the point of a player using very different costumes when playing different characters. Costumes are important instruments for supporting a player's transformation into and identification with their character. Secondly, as costume is tailor-made for a character, it serves as an outward statement about the character, thus not only strengthening the individual role, but also the interaction with others. Thirdly, a lot of effort is spent on preparation; most players will have spent a significant amount of time making costumes, planning strategies, synchronising actions with other players, and so forth. The time and attention spent on taking on a role reflects the dedication to the player culture, which ultimately is highly appreciated by the community they are part of. Contrast this transformation with the graphical representation avatars provide players of computer games. Donning a costume and performing a role in person is a representation of character that has yet to be made possible in computer games (Fig. 2).

3.1.1   Objects and Locations

The physical environment is the game world, or the stage on which the narrative is performed. This integration of game space and physical space creates a graspable game environment that players have to literally navigate in order to reveal the narrative content. Physical structures may be erected to enhance the game world,

Fig. 2 The narrative emerges in the interaction between players

serving as specific locations of importance to the narrative, and where location-dependant role-play takes place. The physicality of the game world contributes to creating a highly immersive and tangible experience, in which the narrative induces very real physical sensations such as fear and excitement.

LRP players frequently use physical artefacts as props and tools in their role-play, primarily to back up their character roles. Commonly referred to as *physical representations* (or *physreps*), they represent game objects with tangible presence and functionality in the game. Mechanisms named *lammies* (because they are laminated pieces of paper) formalize physreps' functionality in the game. Figure 3 shows an example, an amber talisman that protects its wearer from certain diseases. The numbers printed on the lammie are codes referring to properties of the artefact, such as its value, origin, if it is magical, and so forth, which players with the appropriate skills can check against so called lore-sheets. In this fashion, a lammie is something of a *plug-and-play* feature of the game world. It offers a way to sanctify and transform arbitrary objects into official game artefacts (Fig. 4).

Fig. 3 Physical artefacts populate the game environment

Fig. 4 Physical artefacts populate the game environment

4 Principles for Design

To computer game designers, the game engine is a piece of software that simulates and renders the game world. It deals with e.g. visual effects, such as the animation of characters and objects, the texture of surfaces and other details in the environment. In LRP games, what I refer to as the game engine, in a relaxed sense of the term, are mechanisms that render textures of a more cultural nature. Understanding these mechanisms may be an important step in understanding what factors make successful interactive or game narratives. Which factors make up more and which make up less enjoyable or engaging experiences are still subject for further research, but there are properties of LRP games that are instrumental to creating the conditions for engaging and creative role-playing, which suggests implications for design. They are primarily motivated by the ways in which LRP environments extend and transform the physical world into environments that nurture and encourage players' engagement.

4.1   A Believable Game World

Computer gaming environments are increasingly realistic in using the physical world as a model for their game worlds. Many games have as a feature next to photographic graphical representations of the game environment in their attention to detail in scenery. LRP environments are founded on a different attention to detail, where the game world is believable and convincing because there is no separation between the game world and the physical world. We can note that physreps, including the environment itself, rarely take token shapes or forms, but are instead carefully crafted to convey purpose through physical manifestation. If an amber pendant is needed to make an amulet—such as the one in Fig. 3—players use an amber pendant, not a feather or a stone. Elaborately populating the environment with theatrical props and game artefacts, as exemplified by the old library filled witch books (Fig. 5), and an alchemy lab (Fig. 6), is one way to make the players believe in and agree with what happens to their characters. Allowing the game to extend into the physical world is key to fostering coherent and meaningful role-playing relationships between characters and the game world.

4.2   Magical Interfaces

While believability is important, at least on the level of physical form, LRP worlds are typically rendered fantastic rather than realistic in regards to functionality. Therefore, what you see may not necessarily be what you get, and if that message is encoded and reflected in the design of an artefact, it often sparks curiosity and beckons the player to interact with the object. As an example of magical objects, take the puzzle in Fig. 4, which when solved not only spells out a message, but also functions as a key that unlocks the vessel that contains a particularly nasty lich creature. In this example, the player will know what to do, or how to interact with the puzzle, but cannot be certain what the result of that interaction is. The fact that it begins to suggest its functionality—the word "Death" is being spelled out when the

Fig. 5 Convincing game environments, populated with game artefacts

Fig. 6 Convincing game environments, populated with game artefacts

pieces are put together—is part of encoding this particular artefact's magical message. When the game world is designed with mystery and concealed facts in mind, it adds to creating an alluring, if not seductive, environment that strengthens players' interest and commitment to engage with the game world. Interestingly, players are habitually sensitive to the fact that game artefacts often have unexpected effects, and their interaction with them is reflected in their typically curious but very careful approach to them.

4.3 Tangible and Aesthetic Interfaces

Aesthetics play an important role in creating engagement and maintaining the appeal of the environment. It deals with the expressive identity of things, their form and shape and how we experience them with our senses, and is of great consequence in rendering the reality of the LRP game world and making it meaningful to the players. While aesthetics traditionally deals with what appeals to the senses in terms of e.g. shapes and colours, what is emphasised here are the aspects that aid the *elegance of make-believe*. They are part of making the unreal real and giving integrity to the game world. Tangible interfaces, props, and costume, play a significant role, not only in that they are physical details that support the extension and manifestation of the game world in the physical world, but also because humans are inherently good at relating to and manipulating such objects.

4.4 Dedicated Versus Token Representation

Most objects and locations are incorporated into game play with context of use in mind, which again is reflected in the design. It is noteworthy that important game artefacts are often highly dedicated, specialised, and articulated tools for role-play, as in the case of the puzzle in Fig. 4, which is a unique item created for one specific purpose. This tends to put emphasis on the design of interfaces that communicate contextual functionality, rather than being generic or universal in their physical appearance. When designed with their context of use in mind, they are powerful tools in transforming the physical world into a game world. The costumes players don and the accessories they choose to illustrate their characters' positions or professions are some examples of this principle, as suggested by Figs. 1 and 2.

5 Conclusion

Our knowledge of the physical world and the skills with which we engage with it are powerful facilitators to LRP games. The artefacts, costumes, game-specific locations and buildings transform the physical world into a magic place where fantastic narratives are spawned. The level of engagement such a game environment creates—with no physical division between player, character, space, and narrative—is the kind of immersion many interactive narratives and computer games seek to achieve but where they also fail. By looking at the appropriation of artefacts and physical game locations in LRP games, and observing how the stories emerge from the interaction between all these components, we can inform the design process for interfaces to interactive narrative applications.

Acknowledgements This research is carried out in the Story Networks group at Media Lab Europe in Dublin, MIT Media Laboratory's European Research Partner. I want to acknowledge the support from the member's of this group, particularly that of Glorianna Davenport who has offered invaluable comments and support. Mark Blythe and Peter Wright were instrumental to the process of writing. Lastly I want to acknowledge the passion of LRP gamers and organizers—wherever you are.

References

Björk S, Falk J, Hansson R, Ljungstrand P (2001) Pirates!: using the physical world as a game board. In: Proceedings of Interact'01. Tokyo, Japan

Costikyan G (1994) I have no words & i must design. http://www.costik.com/nowords.html (2002-11-20)

Costikyan G (1994) Where stories end and games begin. http://www.costik.com/gamnstry.html (2002-11-20)

Crawford C (1994) The art of computer game design. http://www.erasmatazz.com/Library.html (2002-11-20)

Falk J (2002) The world as game board. In: Position statement, workshop on funology at CHI 2002

Chapter 36
Whose Line Is It Anyway? Enabling Creative Appropriation of Television

Erik Blankinship and Pilapa Esara Carroll

Authors' Note, Funology 2
What is presented in this paper is the foretelling of a shorthand form of communication now used by millions of people. It is now common to see video clips from popular television shows interspersed into both public and private digital conversations. Crowdsourced libraries of annotated video, mostly encoded as animated gifs with superimposed subtitles, can be searched for snippets to slip into online exchanges with the ease of auto-complete. These media epistles result in collaborative narratives, and feelings of empowerment, not so different from those described in our paper.
2003 Chapter

1 Introduction

"Live long and prosper" and "beam me up" are popular quotes from the Star Trek television series. Quoting television can be considered a form of media appropriation. If people are given the ability to manipulate video clips of television directly will they construct their own media works? A software tool called talkTV provides this form of access by chunking television into video clips of dialogue that can be re-sequenced. An evaluation of this software tool occurred with Star Trek fans at an annual science fiction convention. The exploratory findings of this study suggest that if given access and the ability to re-purpose television, people enjoy constructing short fan films.

E. Blankinship (✉)
Media Modifications, Cambridge, USA
e-mail: erikb@mediamods.com

P. Esara Carroll
Department of Anthropology, The College at Brockport, New York, USA
e-mail: pesara@brockport.edu

© Springer International Publishing AG, part of Springer Nature 2018
M. Blythe and A. Monk (eds.), *Funology 2*,
Human–Computer Interaction Series,
https://doi.org/10.1007/978-3-319-68213-6_36

1.1   Television Fans as Producers

Television shows come as packaged entertainment: well-assembled stories, professional actors, and soundtracks all fitted together. They are generally designed as passive entertainment. The audience might discuss the show around the water cooler or online, but for the most part the audience is intended to *consume* television.

Not content with just watching their favourite shows, television fans have a history of *producing* their own fictional stories. Fans of the Star Trek television series are the most well known for constructing original fictions derived from the show (Jenkins 1992). Some fans' short films involve appropriating soundtracks, sound effects, costumes, and digital star ship models. Music videos are another format for fan appropriation in which video clips from the show are re-edited to music so as to create new meanings dictated by the chosen song (Jenkins 1992). These activities fall under the banner of "fan fiction", which is replete with its own sub-genres such as slash fiction, erotic stories between two characters separated by a/in the title (such as Kirk/Spock romances) (Jenkins 1992). Fans create meanings separate from those intended by the producers of the television series. In this way, they claim ownership of media by integrating it into their own creations.

Not all fans have the time or skills to easily create their own films. How can people's media appropriation be facilitated? As a research problem, the construction of fan movies was conceptualised in terms of its component parts. Our intuition is that television dialogue is a well-established way of appropriating media—people quote television shows often—so that with access to source material, fans could easily construct films. talkTV is a software tool designed to search television clips via dialogue. Our testing of this application among a group of Star Trek fans reveals useful insights into how fans appropriate media. Our work is a contribution toward a future in which this activity might be commonplace (Davis 1997) and possibly automated (Sack and Davis 1994).

1.2   Engineering

Dialogue is encoded into most television broadcasts as subtitled Closed Captions (CC), providing a ready index into a television show from which to segment video clips. talkTV requires fans to enter a text query into a database of television programs that retrieves video clips in which the queried words are spoken. For example, a search for "warp factor" retrieves clips in which characters speak these words. talkTV's database consists of four seasons of *Star Trek: The Next Generation*, providing about 100 h of searchable dialogue. The tool also allows fans to create title slides by superimposing text over selected background videos. The backgrounds were pre-selected to provide different establishing shots featuring the star ship (Fig. 1).

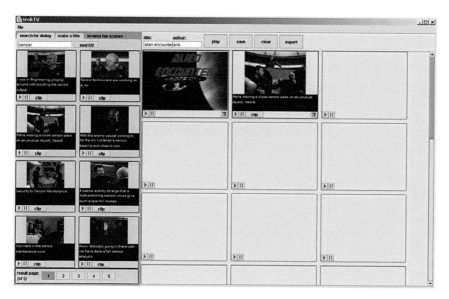

Fig. 1 The talkTV user interface. On the left are the search results in the form of video thumbnails. Clips are dragged to the storyboard panel on the right

Video clips of dialogue and titles can be dragged into and sequenced in talkTV's storyboard panel. Once added to the storyboard, video clips can be previewed in sequence as a short film. For example, a title slide reading, "Alien Encounter!" could be followed by a video clip of a character saying, "Sensor report!" followed by a clip of another character saying "An enemy ship!" or "Weird energy signatures..." depending on the fan's story. Fans can also browse completed fan films and modify them, so that multiple endings can be generated from one initial film.

2 To Boldly Go...

To study if talkTV enables fan to be producers of television, we decided to make it available to science fiction fans at an annual convention. Each participant was given a few words of instruction and was seated at their own computer terminal. Observations were taken of the participant's keystrokes and mouse movements, their body language and their comments. Completed movies were queued to a display monitor which looped prior completed movies. Immediately after participants viewed their movie publicly, they were asked to be interviewed. The informal interview was brief using open-ended questions to elicit information regarding people's initial intents, experiences and opinions.

In total, seventeen men and women were observed and interviewed. None of the participants appeared distracted or annoyed by either our note-taking or the comments of those in the crowd regarding their editing. Participants varied in ages from early twenties to early fifties with the average age being 33. The majority of participants were male (88%). Seven participants (41%) returned to either finish their movies or to create new movies. Although session times varied, on average, participants spent approximately 50 min in total working on their movies. Participants had varying degrees of computer experience. Some people used computers occasionally for e-mail whereas others used computers as part of their jobs. Several participants were experienced with video-editing software whereas others found the concept of pull-down menus and key word searches to be unfamiliar.

2.1 *Findings*

In total, 21 fan movies were created. The majority of participants endeavoured to create "mini-episodes" where continuity was key. These mini-episodes attempted to emulate the stereotypical Star Trek episode by beginning with a Captain's log, followed by a semblance of plot and conclusion. A refinement of the "mini-episode" was structured around characters asking questions and varied responses repeated to humorous effect. Several participants created films centred upon a keyword (i.e. "Klingon") without any attempt to create a plot or to maintain continuity between locations or situations. This kind of film was prevalent among the earliest users implying a development in film construction, as later users were able to view other's movies.

Interviews and observations reveal aspects of the participants' motives, strategies and thoughts regarding their experiences. Initially, participants began their sessions with one of two intents: to explore and see what could be done, or to create a narrative based on a theme or situation. Once they started, they began to form secondary goals. For example, some users found that they were forced to create a serious story rather than a humorous one. Others endeavoured to achieve continuity in their film by establishing a flow between the visual images, the lines of dialogue, the story line, or all three together.

While participants approached their filmmaking in various ways, common strategies emerged. For instance, it became clear to most of the participants that the software was dialogue-driven. Since participants were given minimal instructions and a few were not familiar with key words searches, this characteristic of the software was not readily apparent. Thus the selection of keywords to query became in itself a particular strategy. Another strategy derived from the lack of space to store clips. Some participants used their storyboard area as a kind of "image bank" placing their selected clips in that area and arranging them into a sequence later. Others placed clips on the storyboard in a linear fashion often having to use one keyword or memorizing keywords, in order to bring up intended clips when they were needed.

All the participants stated that they enjoyed their experiences. Commonalities between the participant's responses highlight three aspects of the talkTV experience which participants found to be "fun." The technical possibilities facilitated by the software, the experience of the actual editing process and the end product are all aspects of what made talkTV enjoyable. Many participants stated how fun it was as a technical possibility to actually be able to "make a mini episode". This was particularly true for a couple of users, who said they had never seen anything like this.

In addition to what one could do with the software, people enjoyed editing the television clips:

- "[It's] fun to query clips and then you can cut and paste…my own editing lab at my fingertips."
- "It was funny to select [clips] and make them fit. [And] taking them out of context."

These quotes emphasize the high degree to which participants enjoyed the editing process.

To edit effectively it is important to be able to query for meaningful (i.e.: usable) clips. Those persons with a comprehensive knowledge of the television series found that it was fun to test themselves. As one fan commented, "It's a good way of testing your knowledge of the show. It triggers recall of episodes". Several of the users had memorized whole seasons of the television show. Their ability to remember the scripts of various scenes enabled them to query efficiently for specific characters or locales. Less familiar fans still found the query searches "fun" because they were able to re-watch the television clips.

Although enjoyable, the editing process required thought and was in varying degrees, challenging. As another user described it:

- "I couldn't be pro-active today. I had an idea for [the character] Wes, had to search for people by chance. I was lucky if I found [a clip of] Wes speaking"

This participant's frustration with trying to query for clips with certain characters emphasizes part of the challenge of searching dialogue for visual images. He later described his experience as "fun" once he found a "rhythm." His comments reveal an additional aspect of what makes talkTV "fun"—being able to master the software. Be it "coming up with the idea" or "finding a way to tie it together", the participants were proud of their ability to effectively mine the dialogue to create an actual movie. Another user described it as a "problem-solving" challenge. Once they mastered the "challenge", it became "fun." The participants often described their sense of accomplishment in terms of feeling like an actual director or screenwriter, occupations which affirm their appropriation of the media.

A final aspect of talkTV's fun is the creation of an "end product." When asked, participants generally responded favourably to the public feedback from their movie:

- "It was flattering that [one of the co-authors] laughed...I could sit for hours and tweak [my movie]."
- "Heck, I was a little embarrassed and a little proud. [It was] something I had sat down and I did."
- "[It's] cool getting to hear people laugh. It was fun when finished."

A couple of participants who were dissatisfied with their movies faulted their own inexperience rather than the software. Several users felt they needed to "play with it" more, and many felt that greater familiarity with the software would enable them to make better movies, in their eyes.

Although using talkTV was enjoyable, it could benefit from refinement. Common complaints about the software included the desire for a bigger clip library, the need for the interface to be more intuitive and the inability to search for characters or certain categories of clips. These complaints suggest that users would like to continue working with talkTV and to do more with it. Rather than criticizing the functionality of the application, the users were more concerned with how to improve its pre-existing interface and capabilities.

In summary, the participants of this study appeared to enjoy the use of talkTV. Not only were users able to make movies, but they described their experiences as "obviously fun". In this context, talkTV was fun because (a) it enabled people to make "mini episodes" and to be "a director", (b) the editing process was enjoyable and facilitated creativity, and (c) their completed movie showed mastery of the software and was typically well received by their peers.

3 Discussion

We tested if talkTV utilizes the popular Star Trek series as its source material. In order to find suitable participants for the software's evaluation, we implemented our study at a convention based on the series. Although we cannot posit that talkTV *enabled* fans to make movies, we have gained insights about how it was used to make movies. While participants actively engaged the dialogue-search, there was a desire to be able to search for non-dialogue based items, such as explosions. This suggests that other forms of indexing video for repurposing may be useful complements to a dialogue-based search (Davis 1995; Bove et al. 2000; Mills et al. 2000).

In addition, we found support for the common-sense notion that if television re-editing is fun, television viewers will appropriate the media to make their own films. The various aspects of fun revealed in this study are related to the quality and popularity of the source material and the manner in which their access to this material is facilitated. For example, participants were able to make films out of the actual clips from a popular television series rather than pre-made facsimiles, thereby enabling users to make "authentic" mini-episodes. An insight gained from this study is that users find it enjoyable to have access to original source material.

Connected to this aspect of fun is the shared knowledge about Star Trek between the participant filmmaker and their audience. The filmmaker takes for granted that their audience will meet them half way in making sense of their efforts. A brief introduction of a character or reference can have large significance for the right audience. For example, among the fan films made, a few fans inserted the character "Q" without any apparent introduction or plot connection. This character would appear seemingly out-of-place to viewers unfamiliar with his established role as an unpredictable omni-powerful being. In this way, part of the fun of talkTV derives from creating well-produced pieces, which others can appreciate.

Given the effort people put into establishing a meaningful plot and maintaining continuity within their short films, we argue that talkTV provides "hard fun". That does not mean "it's fun in spite of being hard… [but rather] it's fun because it's hard" (Papert 1998). It is an activity that engages the subjects and compels them to do their best work. This kind of fun motivates television viewers to actively re-appropriate their favourite show. In a broader sense, this study highlights the possibility that television production is not just for fans but also for the typical television viewer. A different television audience not known for media appropriation might enjoy using a tool like talkTV to author their own show of choice. If television is rendered malleable and accessible, a wave of fan films for other genres might begin to supplant what is already playing.

Acknowledgements Special thanks to AlmaMedia and the sponsors of the MIT Media Laboratory.

References

Bove M, Jr Dakss J, Chalom E, Agamanolis S (2000) Hyperlinked video research at the MIT media laboratory. IBM Syst J 39(3–4)

Davis M (1995) Media streams: representing video for retrieval and repurposing. Massachusetts Institute of Technology, Cambridge

Davis M (1997) Garage cinema and the future of media technology. Commun ACM 40(2):42–48

Jenkins H (1992) Textual poachers: television fans & participatory culture. Routledge, New York

Mills TJ, Pye D, Hollinghurst NJ, Wood KR (2000) AT&TV: broadcast television and radio retrieval. In: Paper presented at the RIAO 2000 (Recherche d'Informations Assistée par Ordinateur; Computer Assisted Information Retrieval), Paris

Papert S (1998) Does easy do it? Children, games, and learning. Game Developer 88

Sack W, Davis M (1994) IDIC: Assembling video sequences from story plans and content annotations. In: Paper presented at the proceedings of IEEE international conference on multimedia computing and systems. Boston, Massachusetts

Chapter 37
The Interactive Installation ISH: In Search of Resonant Human Product Interaction

Caroline Hummels, Kees Overbeeke and Aadjan Van Der Helm

2003 Chapter

1 The Human as a Whole

> The history of HCI can, in many ways, be seen as an ongoing attempt to capitalize on the full range of human skills and abilities. (Dourish 2001)

In the beginning, computer science and HCI manifested themselves through encoded patterns (e.g. punch cards) and command lines, thus calling upon the cognitive skills of users. The shift to visual computing with a desktop and a mouse that was tightly coupled to an on-screen cursor, expanded the interaction range towards perceptual-motor skills. If we look at the current developments within HCI, like tangible interaction, affective and social computing, we see a refinement towards the use of perceptual-motor skills and the urge to incorporate emotional skills. It seems that respect for the human as a whole has come into vogue, at least within a part of the HCI research community.

This emphasis on the human as a whole can also be seen in the shift of contextual focus. The computer is leaving the sphere of the workplace, thus widening the spectrum of efficiency, productivity and 'getting things done' with values like curiosity, playfulness, intimacy and creativity (Caenepeel 2002; Overbeeke et al. 1999). The computer has entered our daily and social life. It is no longer just a

C. Hummels (✉) · K. Overbeeke
Eindhoven University of Technology, Eindhoven, The Netherlands
e-mail: c.c.m.hummels@tue.nl

A. Van Der Helm
Delft University of Technology, Delft, The Netherlands

© Springer International Publishing AG, part of Springer Nature 2018
M. Blythe and A. Monk (eds.), *Funology 2*,
Human–Computer Interaction Series,
https://doi.org/10.1007/978-3-319-68213-6_37

means to perform our work; it helps us to pursue our lives (Gaver 2002). In this way, the world of HCI has united with the world of product design.

Although this is an interesting and challenging way to go, it isn't an easy one, especially in our contemporary culture, which has lost its unifying ideology (Branzi 1989). We do not only have to develop the next generation of digital products with which we can pursue our lives, we also have to decide what kind of life and society we want these products to support. Buchanan (1998), Marzano (1996), Borgmann (1987) and Saul (1997) all plead for respect and humanism; for 'real' individualism, in which the individual is part of society and takes responsibility for that society. This implies that we shouldn't design products for a universal audience, or "the consumer". Products should be personal pathways that allow individuals to find and create their own experiences (Hummels 2000).

'Capitalizing on the full range of human skills and abilities', as Dourish (2001) mentioned in light of the history of HCI, is a condition for designing 'contexts for experiences'. However, we would like to expand the focus from human skills to the concept of *resonance*. In the remaining part of this text, we will explain this concept and our reasons for advancing it. Moreover, we discuss on the basis of the interactive installation called *ISH*, how to find salient aspects of resonance.

2   Resonance

Resonance stems from the theory of ecological or direct perception, which also engendered the term affordance; a term that Norman introduced to the HCI community. Gibson (1979) used the term in combination with a radio metaphor to clarify the directness of our perceptual system. A radio station broadcasts information, i.e. waves with a particular radio frequency that is used by that particular station. The detection of radio waves is based on the principle of resonance. Given that many frequencies (stations) reach a receiver from the antenna, proper tuning of the receiver causes a current in it to resonate in response to one of the incoming signals, and not others.

In case of e.g. visual perception, the radio waves in this metaphor stand for light that is reflected (broadcasted) by our environment (the radio station). Our eyes (the antenna) let the signals pass through, and we (the radio) must tune into the information. For example, if we want to write a message, we are tuned into information in our environment that affords us to write. Thus when a pencil comes into view, our perceptual system resonates to that information (Michaels and Carello 1981).

However, resonance does not only relate to our perceptual-motor skills. It relates to our cognitive and emotional skills too. Moreover, it is not only a temporal response, e.g. we want to write, so we resonate with a pen. We also resonate with products because we are people with certain needs, desires and intentions, a social and cultural history and position etc. Consequently, we do not resonate to the same products. To elaborate on the writing example, one person might resonate with a cheap disposable pen, another person with that fountain pen he got from his

Fig. 1 The first author's
Sunbeam toaster

grandpa and another person might resonate with the I-Mac that he bought with his savings. What's the *real* difference? (Fig. 1).

Let us explore the concept of resonance a bit further with an example. The first author resonates with a Sunbeam toaster that she bought approximately ten years ago at a jumble sale. She resonates with it, because it functions better than well. It has a small catch which causes the slices to be automatically transported downwards at a calm pace and upwards again when they have a nice tan. This calm pace enhances the feeling of luxury and Sunday morning relaxation, which she associates with toast. She considers the toaster to be visually and tactilely pleasant, simple and easy to clean. It expresses for her respect and friendliness, which triggers her vivid imagination. The slow transportation of slices gives the impression that the toaster is almost saying: *"Come, hand me your bread. I will take good care of it and produce the most delicious toast, specially for you."* Due to this invitation, she places the slices of bread with a gentle and elegant gesture into the toaster. Finally, she bought it relatively cheap after some haggling, which she experienced as an additional advantage, especially in the beginning.

This example shows, that it is not just about tuning a product to one's skills, which makes a person resonate with a product. A resonant interaction is the result of a mixture of different ingredients like usability, human skills (cognitive, perceptual-motor and emotional), richness of the senses, individual and social needs, desires and interests, personal history, ways of acquiring the product, context of use (situation, timing, environment, social setting), aesthetics of interaction, intimacy, engagement and openness to find and create one's own meaning, story and ritual.

This implies three things. Firstly, resonance can be a concept that provides respectful and humanistic HCI/products, which allow individuals to find and create their own experiences.

Secondly, resonance can only be found in the ensemble of ingredients, thus requiring a holistic design approach. For example, Norman (2002) argues that pleasant and attractive products actually work better, providing that they are not

Fig. 2 The interactive installation called ISH (Image and Sound Handling)

used in emergency situations. Thus the whole is greater than the sum of its parts. This implies that people involved in the development of HCI and product design, e.g. designers, computer scientists, engineers, psychologists, marketers, should work together, combine their knowledge and develop integral solutions, in order to attain resonant products and interaction.

Thirdly, because of the personal character of resonant interaction, HCI and product developers should involve people for whom they are developing products right from the start, for inspiration, information, discussion, evaluation, testing and validation of resonant interaction.

Resonance is rather an unexplored area due to its complexity. Why do some people resonate with a certain product, while others do not? Can one formulate guidelines for designing products that evoke resonance? Within our own research we study resonance by building and testing interactive installations and a variety of products with the same function. Let us discuss one of these installations called *ISH* (Fig. 2).

3   A Design Example: The Interactive Installation Called *ISH*

ISH (Image and Sound Handling) is an interactive multi-media installation that allows a group of people to create together an atmosphere through visuals and music. It is a dynamic research environment which allows us to evaluate resonant interaction through loops of (re)designing, building and testing. At the moment *ISH* consists of eight tangible products and a projection screen. Every product has its own character with respect to feed forward, feed back, time-delay, temptation, clarity etc., which allows us to evaluate different aspects of resonance. For example, is it necessary that a product shows what the user can expect after he carries out an action? Should a product seduce a person to explore it? How devastating is time-delay for resonance? Should one pursue subtle interaction? What makes interaction engaging and beautiful?

Hitherto, *ISH* focuses on tangible interaction. This doesn't mean that this is the only way to interact with products. We will explore other forms of multi-modal interaction in the future through *ISH*. However, we started with tangible interaction, because you inescapably handle objects in an expressive way, thus linking two important aspects of resonance.

We describe shortly the different components, before evaluating the installation with respect to resonance.

A person selects audio samples and images by moving his hands above and through the sand of *Gatherish*. The position of the hands and the character of the movements determine the expression of the sounds and images.

Smallish alters the volume of the audio part and the size of the image in the active layer (which is selected through *Compositish*). The volume and the size increase by pushing the square plane.

Four sequentially placed images containing holes create the visual environment. A person determines with four tangible transparent cards the order of the four layers. Moreover, one of the layers can be made active by placing a banner next to it. One can manipulate the image in the active layer using *Gatherish*, *Smallish*, *Stirish* and *Jitterish*.

The difference in force exerted to the pillows of *Stirish* alters the position and orientation of the selected image.

ISH has its own character, which means that the images have their own movements, depending on the mood of *ISH*. These movements are influenced by the kind and number of actions the users make on the eight products. Moreover, one can set the mood (tensed–relaxed) by increasing or decreasing the tension (curve) of *Jitterish*.

The audio part consists of sounds and rhythms, which can be interactively manipulated. The expression of the sounds and the rhythm can be altered with *Acoustish* by making bridges between four infrared senders and receivers, using several reflectors.

The expression of the visual pattern of tokens that is created with *Rhythmish* fits the expression of the rhythm section. For example, a low number of tokens placed in an orderly way, creates a simple and relaxed rhythm.

Mixish determines the balance between *Acoustish* and *Rhythmish*. Shifting the cylinder towards *Acoustish* puts an emphasis on *Acoustish*, and shifting it towards *Rhythmish* fades *Acoustish* away.

4   Evaluation *ISH* and Conclusions with Respect to Resonance

ISH was primarily evaluated by observing people interacting with *ISH* and each other. During open days, demonstrations and conferences, we observed the behaviour and remarks of an audience that was unacquainted with our research.

Their behavior and remarks showed that the overall installation, enabling people to generate visuals and music, resonate with most spectators during interaction. However, the social aspect of it, i.e. creating this atmosphere together, is still underexposed. This is partly the result of the set-up. All products faced towards a vertical screen, thus complicating natural interaction between people. The successor of the present set-up will arrange all products in a circle around a horizontal projection of the visuals.

The individual products show different reactions with respect to resonance. They can be divided into three groups.

For the vast majority *Smallish* (Fig. 3) and *Rhythmish* (Fig. 4) seem dead on target with respect to resonance. People considered them to be extremely clear and pleasurable. The mapping was considered natural in both cases: for *Smallish* between distance/force and size/volume and for *Rhythmish* between the visual

Fig. 3 Smallish

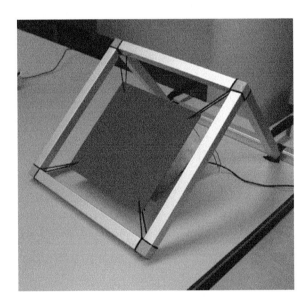

Fig. 4 Rhythmish

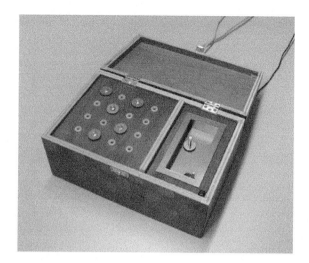

Fig. 5 Gatherish

pattern and the resulting rhythm. Moreover, the subtle flexibility when pushing the square of *Smallish*, caused by the elastic suspension was experienced as very pleasurable and resonant. The intimate and expectant moment of closing the box and hearing the created rhythm, seem to enhance the resonance with *Rhythmish*.

The feeling of sand through one's fingers, made *Gatherish* (Fig. 5) at least for a short period of time very attractive, similar to *Stirish* (Fig. 6) and *Acoustish* (Fig. 7). However, only a minority of the users experienced these three products as

Fig. 6 Stirish

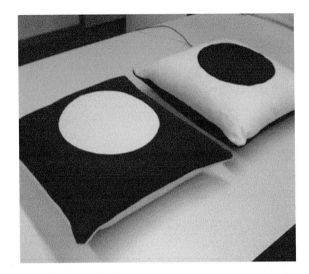

Fig. 7 Acoustish

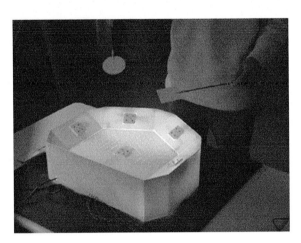

resonant over a longer period of time, because they require exploration. Cause and effect are not immediately clear. It appeared that exploration seekers—less goal-oriented and more imaginative people—find themselves attracted to this kind of interaction.

Most people experienced *Compositish* (Fig. 8), *Jitterish* (Fig. 9) and *Mixish* (Fig. 10) clear and simple to operate, with appropriate tactile and kinaesthetic feedback. They were considered pleasurable with respect to interaction, functionality and appearance. Nevertheless, they scarcely challenged extensive interaction, due to their simple and functional character. This made them less resonant for the exploration seekers in contrast with the goal seekers.

Fig. 8 Compositish

Fig. 9 Jitterish

Fig. 10 Mixish

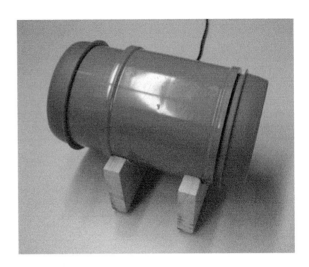

ISH shows the importance of pursuing diversity within product interaction; not all people resonate to the same things. In future work we will expand *ISH* with respect to diversity, by incorporating others forms of modality, like speech (intonation, volume) and gestures, and add/alter products.

ISH also revealed a major drawback: it is predominantly suited for short-term interaction. The example of the Sunbeam toaster showed that time has an important impact on resonance. Therefore, we are pursuing several other projects next to *ISH* to gather knowledge about resonance. For example, *Coppia Espressiva* offers digital musical instruments to individuals to create their own music (Hummels et al. submitted). Another project expands the idea of creating an atmosphere, but this time at home. How can a person create an atmosphere at home (music, lighting etc.) that fits his mood, through resonant interaction with a product? Our goal is to design and build several versions of an atmosphere controller, which people can experience at home over a period of time.

Over the years, we have explored the concept of resonance further (Hummels 2007). Since people are all different, one could wonder why so many digital products are alike, especially at an interaction level? We believe that resonance could be a principle for designers to address the diversity, since "resonant interaction is the perfect and unique interplay between a person and a product (or object, another person, ...,). It evokes strong positive emotions, unique accompanying behaviour, heightened awareness and cognitive processing and it is presumed that these feelings, behaviour and thoughts last after the actual moment of interaction." (Hummels 2007).

Through ISH we did the first explorations, and also through other studies we see some salient aspects arise that can evoke a prolongation of pleasure and offer a potential fit for specific people, which can be the first step towards designing for resonant interaction. As stated before, resonance is not an easy goal to achieve, but it is certainly worth pursuing. The HCI and design community can help each other designing, building and testing experiential prototypes. After all, the proof of the pudding is in the eating.

Acknowledgements We thank Rob Luxen, Philip Ross, Frans Levering, Rudolf Wormgoor and Joep Frens for their effort to prototype *ISH*.

References

Borgmann A (1987) Technology and the character of contemporary life. University of Chicago Press, Chicago

Branzi A (1989) We are the primitives. In: Margolin V (ed) Design discourse: history theory criticism. University of Chicago Press, Chicago

Buchanan (1998) Branzi's dilemma: design in contemporary culture. Des Issues 14(1):3–20

Caenepeel M (2002) Summer issue: technology on the right side of the brain. I^3 magazine, No. 12, June 2002, p 1

Dourish P (2001) Where the action is: the foundations of embodied interaction. MIT Press, Boston, MA

Gaver B (2002) Designing for homo ludens. I^3 magazine, No. 12, June 2002, pp 2–6

Gibson JJ (1979) An ecological approach to visual perception. Lawrence Erlbaum Associates, London (reprinted in 1986)

Hummels CCM (2000) Gestural design tools: prototypes, experiments and scenarios. Doctoral dissertation, Delft University of Technology, Delft, The Netherlands

Hummels C, Ross P, Overbeeke CJ (2003) In search of resonant human computer interaction: Building and testing aesthetic installations. In: Rauterberg M, Menozzi M, Wesson J (eds) Human Computer Interaction - Interact '03. Amsterdam: IOS Press, pp 399–406

Hummels C (2007) Searching for salient aspects of resonant interaction. Knowl Technol & Policy 20(1):19–29

Marzano (1996) Introduction. In: Philips. Vision of the Future. V + K Publishing, Bussum

Michaels CF, Carello C (1981) Direct perception. Prentice-Hall, Englewood Cliffs

Norman DA (2002) Emotion and design: attractive things work better. Interactions IX(4):36–42

Overbeeke CJ, Djajadiningrat JP, Wensveen SAG, Hummels CCM (1999) Experiential and respectful. In: Proceedings of the international conference useful and critical: the position of research and design, 9–11 Sept 1999, U.I.A.H.

Saul JR (1997) The unconscious civilization. The Free Press, New York

Chapter 38
Fun with Your Alarm Clock: Designing for Engaging Experiences Through Emotionally Rich Interaction

Stephan Wensveen and Kees Overbeeke

2003 Chapter

To strive for the incorporation of fun in product use is to design for engaging experiences (Fig. 1). In order to do this we should respect all human skills in human-product interaction. So while most current electronic products appeal to cognitive skills, we believe that a person's perceptual-motor and emotional skills should be taken into account as well. One way of opening up such an experience is to allow people to use their natural expressive powers by permitting them to use their perceptual-motor skills. Most current products do not tap into these skills because their functionality is accessible in just one way, and often a very poor way indeed. For example, to set Sophie's alarm clock, you have to push a tiny button several times, while holding another tiny button. Why not go one step further and try to design products that can adapt to a person's emotions and feelings to enrich the experience? If Sophie were able to express her feelings to the product, it could read these feelings and consider them when reacting to her (Fig. 2).

For many people waking up and getting out of bed is not the most pleasant experience. The accompanying product, the alarm clock doesn't really help. Yes, it does wake you up, when you have set it properly, but it doesn't adapt at all to different situations. It is a perfect example of a product that should adapt to the diversity of emotional experiences. It is also a product with a simple functionality yet it has all the features of the current interface malaise, like a lack of politeness and nonsensical buttons (Cooper 1999). That is why we chose the alarm clock as a vehicle for our research through design approach. In our research towards designing emotionally intelligent products we advocate designs that both allow for the recognition and expression of emotion while avoiding anthropomorphic design (talking heads) and physiological sensors. This product design-driven approach

S. Wensveen (✉) · K. Overbeeke
Eindhoven University of Technology, Eindhoven, The Netherlands
e-mail: c.c.m.hummels@tue.nl

© Springer International Publishing AG, part of Springer Nature 2018
M. Blythe and A. Monk (eds.), *Funology 2*,
Human–Computer Interaction Series,
https://doi.org/10.1007/978-3-319-68213-6_38

Fig. 1 From the edge of her bed Sophie throws her high heels in the corner of the room. It has been her first week at her first real job and she's not used to wearing them every day. She's also not used to working such long hours. Getting up at six and not being home before eight, is not her thing. Like today, that annoying Phil guy had her work till nine, on a Friday! And then that terrible train ride. Tonight she can get some real sleep and lie in. She still sets her alarm clock, because tomorrow she arranged to meet with her Mum. They'll go shopping in the spring sale. While one finger is pressing down the button that reads 'alarm', another is pushing the 'hour' button three times and then the 'minute' button another 30 times. The display reads 9:31; she shrugs, releases the 'alarm' button and flicks the tiny switch to 'on'. "Who designs these dumb products?" she mutters…

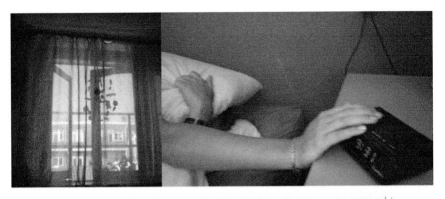

Fig. 2 The sunny Saturday morning is rudely disturbed by an annoying sound. Tuut…tuut… tuut…tuut…tuut…tu.. Sophie smacks the alarm. With her stretched out hand still resting on the alarm clock she tries to regain her consciousness. Through the slits in her eyes she sees sunlight, startled, she sits up straight, but then remembers it's only Saturday. She smiles, shakes her head thinking: "I hate these stupid products!"

Fig. 3 Sophie slams the door behind her and stumbles into the bedroom. While fidgeting with the strap of her high heels she falls on her bed. At the cocktail party celebrating Phil's farewell she had behaved like a fool. At first she had ignored him, giving him the cold shoulder. But after a few drinks she confronted him and tried to tell him what she really thought of him. Phil just gave her a blank look and that made her angrier. Luckily, a colleague saw her behaving badly and took her away from the party just in time. She gave Sophie a lift home. But tomorrow she has to take that terrible train again. With one hand she reaches for her alarm. It looks like a purple disc with slider knobs. Her dad gave it to her a few weeks ago after she had wrecked her old one hitting it too hard out of frustration

takes the interaction with the product as the starting point for the detection of emotion (Fig. 3).

The mood or emotional state you are in colours the way you interact with the world. For human-human communication this expression of emotion is essential. People express emotion through behaviour. In human-product communication people express their emotion as well, e.g., by slamming a door, shoving a chair away, or encouraging the printer with 'come on you can do it!'. Yet this behaviour does not enhance communication between user and product at all. On the contrary, if we fully express our negative emotion we might break the product.

In our product design-driven approach we take the interaction with the product as the starting point for the detection of emotion. While you interact with the product to communicate 'factual' information like the alarm time, the product senses your emotions from the way you handle it.

To make this possible we designed an alarm clock that meets the following three conditions:

1. It elicits rich emotional behaviour while the user communicates 'factual' information.
2. It has the ability to recognise this emotional behaviour.
3. It reflects and understands the expressed emotion.

Fig. 4 She feels the round knobs and slides a couple of them towards the middle. At the party did she and Phil…? No?! She sits up and randomly starts to slide one or two sliders with each action. Using one hand she keeps sliding until she notices that the display shows 7:56. She pauses for a moment and thinks about tomorrow's 11 o'clock meeting with her new boss. She then slides the bottom two sliders all the way back to the outside. The display now shows 6:30. She waits and then firmly pushes the central button

Expressing emotions presupposes freedom of expression, and we therefore designed the alarm clock to allow for freedom in interaction (Fig. 4). It offers a myriad of ways of setting the "factual" information i.e., the wake up time. People can choose to set it by displacing as many sliders as they can grasp or by sliding one slider at a time. This behavioural freedom affords emotions to influence and colour behaviour. The freedom of interaction is further enhanced by the fact that sliders can go back and forth. It stimulates playful interaction, as sliding actions are easily reversed and don't have serious consequences.

We demonstrated in an experiment the alarm clock's ability to recognise this behaviour and identify a person's mood from the interaction. We refer the interested reader to Wensveen et al. (2002) as this goes beyond the scope of this chapter.

When expressing emotions it is important that the receiver gives some sort of feedback that the communication has succeeded. When Sophie expressed her feelings to Phil, his blank look only made her more angry. Just like people, products too should give some sort of reflection of the emotion, a sign of empathy, a sign of understanding. We therefore believe that different emotions should leave different behavioural traces on the product. A slap or a caress leaves a different trace on a face. Likewise, setting the time in a different mood leaves a different trace on the alarm clock. In our design the central display offers augmented feedback about the wake up time (factual information). But it is the successive patterns of the sliders that reflect the influence the emotion had on the setting behaviour leading to this wake up time. It is because of the richness of the inherent feedback that these traces can be perceived. Inherent feedback can be defined as "information provided as a

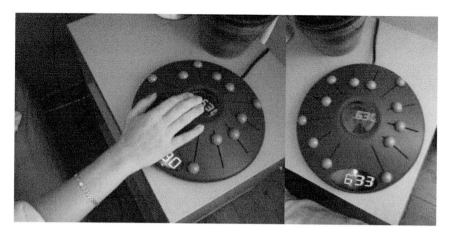

Fig. 5 Rooooo… roo… Sophie hears the sound but it stopped before she realises what it is. She pulls the blanket a bit higher over her head. Roooo… rooo… roo… It seems to be more urgent this time. Rooo… roo… roo… By now she realises it's her alarm clock. She stretches her hand and touches the snooze button. The sound stops. When she looks at the alarm she sees the display showing 6:30. The pattern of sliders looks chaotic. She smiles, thinking that it looks a bit how it feels inside her head

natural consequence of making an action. It arises from the movement itself" (Laurillard 1993).

Based on the interaction of setting the alarm in the late evening and the time-related aspects (alarm time, hours of sleep) the alarm clock makes a decision about what it believes is an appropriate sound. The next morning the alarm wakes you with this sound. The choice of this sound makes it clear that the alarm understood you. It shows its ability to adapt to the situation in an appropriate way (Fig. 5).

It is important that the alarm clock knows the essential information, at what time you need to wake up. It is of less importance that the product exactly knows your emotions as long as you can teach the product how you function. In order for the product to learn about the decisions it took, whether they are appropriate or not, it needs feedback about these decisions. Again through a person's behaviour the product can receive this feedback by the way the snooze button is pressed by the user to turn off the alarm sound. The combination of the delay time (the time between starting and turning off the sound) and a person's behaviour of pressing or hitting the snooze button provides valuable information for the decision making system.

Next morning the inherent feedback in the form of the final slider pattern proves its importance again. Because the end pattern is still present and it is a reflection of last night's behaviour it provides feedback about the decision. It offers people the possibility of linking the alarm sound and the expression of the end pattern together.

This doesn't imply a one to one relationship between the expression of the sound and the expression of the end pattern. After all setting the alarm after a stressful

Fig. 6 Sophie draws the curtains from her balcony door. After she came home from work she drank some Italian rosé, watched the sunset and just enjoyed a warm autumn night. Setting her alarm she uses both hands and with gentle even actions makes a smooth and symmetrical pattern to set the alarm time to the usual 7:15

night leading to a disorderly pattern should not result in a chaotic sounding alarm the next morning.

The slider pattern changes the appearance of the alarm clock and provides feedback for the user and insight into how decisions are made. Since it is the only perceivable change in the alarm clock it provides a reason for the user to believe there is causality between the slider pattern and the alarm sound (Fig. 6).

The design of the alarm clock illustrates the importance of a tight coupling between action and appearance in interaction design. It distinguishes itself from current electronic products through traces and inherent feedback. Because of the inherent feedback the traces become visible, are made explicit for the user and guide her behaviour. For example, when using both hands on the sliders in an even and balanced way the resulting pattern is symmetrical and smooth. The way this pattern looks will push the user to either heighten the symmetry and smoothness or disrupt them depending on how she feels. Traces and inherent feedback thus work in synergy. Without inherent feedback using traces is meaningless, as the product cannot guide the user's behaviour: the trace is invisible and cannot invite the user to act in an emotionally rich manner. Next morning the inherent feedback also offers valuable information for the user and gives insight into the decision-making system of the alarm clock.

From our product design perspective, the appearance of interactive products can no longer be considered as arbitrary. Appearance and interaction need to be designed concurrently (Fig. 7).

When you combine freedom of interaction, rich inherent feedback (slider patterns) loosely coupled with generated feedback (changing alarm sounds) and a system that tries to adapt to and can learn from specific situations you have a good recipe for an engaging experience. It invokes curiosity and playful interaction and maybe, maybe it offers us a recipe for having fun with our alarm clock too.

Twiiiiingwiiiiing... priit... twiiiiiiingwiiiiing ... priit... the soft sound reaches Sophie. When she hears the sound appear for a second time she gently strokes the

Fig. 7 Being a bit tipsy and feeling naughty she dents the smooth pattern with one swift move. The display shoots from 7:15 to 7:43. She adjusts one slider to set it to 7:25. "I wonder what that will do?"

snooze button. She pulls the blankets away, sits up straight and reaches for her alarm. She looks at the pattern while replaying the sound in her head. "Funny…" she thinks and pushes all the sliders to the outside to avoid the sound from playing again. "…at least we understand each other."

She steps over the empty bottle of rosé, giggles and walks to the shower.

References

Cooper A (1999) The inmates are running the asylum. SAMS McMillan, Indianapolis

Laurillard D (1993) Rethinking university teaching: a framework for the effective use of educational technology. Routledge, London

Wensveen SAG, Overbeeke CJ, Djajadiningrat JP (2002) Push me, shove me and I show you how you feel. Recognising mood from emotionally rich interaction. In: Macdonald N (ed) Proceedings of DIS2002, London, 25–28 June 2002, pp 335–340